THE ARCHEAN

THE ARCHEAN
Search for the Beginning

Edited by

G. J. H. McCALL

Senior Consultant Geologist
Paragon-Contech, Teheran

Dowden, Hutchinson & Ross, Inc.

STROUDSBURG, PENNSYLVANIA

Library of Congress Catalog Card Number: 76-11015
ISBN: 0-87933-128-3

79 78 77 1 2 3 4 5
Manufactured in the United States of America.

Library of Congress Cataloging in Publication Data

Main entry under title:
The Archean
Includes bibliographical references and indexes.
1. Geology, Stratigraphic—Archaean—Addresses, essays, lectures. I. McCall, Gerald Joseph Home.
QE653.A72 551.7'12 76-11015
ISBN 0-87933-128-3

Exclusive Distributor: **Halsted Press**
A Division of John Wiley & Sons, Inc.
ISBN: 0-470-99053-8

PREFACE

The topic "Archean rocks," originally chosen for this volume, is one of exceptional breadth, and it soon became apparent that the subject could not be covered by one volume. Although literature in the form of textbooks is remarkably scanty, journal contributions and the like dealing with the Archean are prolific, and so it was found preferable to restrict this volume to the topic *The Search for the Beginning,* by which subtitle we mean *that beginning of which the pioneer geologist James Hutton could find no vestige in the geological record.*

The papers selected are designed to cover the gap in time between the first major thermal event that is known to have closely followed accretion, about 4600 million years ago, and the appearance of the earliest well-preserved, Proterozoic-type platform sequences in the geological record, about 2550 million years ago. The papers are geological rather than geophysical: some have a geochemical emphasis, but these form only a subordinate part of the collection. Those papers were selected that are believed to make fundamental contributions to the store of evidence and arguments concerning events and processes of the Archean time span, which comprises nearly half of the geologic history of our planet. Unavoidably, many important papers have had to be omitted; this applies in particular to papers of specialist-topic emphasis, such as these on sedimentology, petrology, structural geology, paleontology, and economic geology.

G. J. H. McCall

CONTENTS

THE ARCHEAN

INTRODUCTION

Two renowned geologic dicta may be cited to introduce our topic:

- The result therefore of our present enquiry is that we find no vestige of a beginning (James Hutton, 1788).
- We look across the epi-Archaean unconformity into the ruins of an Archaic world (A. C. Lawson, 1913).

The first dictum has inspired our subtitle, for this volume is no more or less than a form of enquiry effected by citing and commenting upon outstanding published works in order to reveal what evidence, if any, of the beginning processes of the planet is recognizable in the geological record of the earliest rocks. In this enquiry we can call on the results of 200 years of progress in the earth sciences since Hutton's pronouncement.

Hutton was impressed, and geologists of our time continue to be impressed, by the remarkable duplication within very ancient Precambrian rocks of familiar geological structures seen in Recent formations: current and graded bedding, ripple marks, soft rock folding, pillows, and so on. Lawson was impressed by the lack of system evident in the Archean, particularly its lack of continuity in time and space. To some extent this problem can now be overcome. With the aid of our expanded, radiometrically derived time scale, we can now probe backward by direct geologic investigation into an era that extends for five times the span that has succeeded the appearance of the first trilobites in the Lower Cambrian. That event was the effective lower limit of accurate stratigraphical geology in Hutton's day, and, indeed, also in Lawson's time, only a few decades ago.

Lawson worked at a time when the base of the fossiliferous Cambrian (now fixed at less than 600 million years) was seen as the boundary of an era that was completely different from the Phanerozoic (Cambrian to Recent). The essentially nonfossiliferous Upper Precambrian—the *Proterozoic*—contained structures that are familiar in more modern rocks, but this was thought to be separated from the early Precambrian—the *Archean*—by an even more intense revolution. The Archean was seen as a primordial melange or basement complex, hopelessly jumbled up by that revolution, and, thus, not a possible subject for geological unraveling, except in isolated detail in time and space. An ordered sense could not be made of the Archean record, lacking both fossils and continuity. It will be evident from the papers included in this selection that this is now far from the case thanks to the aid of the magic tool of radiometric geochronology, which shares with air photography the distinction of being the greatest methodological advance of twentieth-century geology.

In this volume, emphasis has been placed on publications of the last decade. Although these are many important pioneer works of earlier date, such works tend to have become dated by virtue of the revolution in thinking on the Archean that has accompanied the intense geologic activity of the last decade. Of the earlier literature, only J. Tuzo Wilson's 1949 paper, because of its prophetic quality, is included.

There is a wealth of up-to-date information on the Archean contained in this volume; but if the reader believes that research into the Archean has really progressed very far, let him consider the warning by Cloud (1971). The following are some of the areas of investigation that call for much more detailed study in the immediate future:

1. New data in geochronology, leading to more exact geochronological definition of each stage in the development of each separate greenstone belt, and its enclosing granitic terrain.

2. New data in geochronology, leading to more accurate definition both of the backward extension of platforms and platform-type sediments into the arbitrarily defined Archean era and of the forward extension of greenstone development into the Proterozoic era.*

3. Geochronologic and petrologic studies of all Archean and early Proterozoic tensional effects superimposed on stable platforms after stabilization: in particular, regional dike and flood basalt ("trappean") suites. The mechanism of emplacement of subparallel, regional dike swarms needs investigating.

4. New data bearing upon the nature of the primordial crust, including documentation of very ancient granite occurrences and the nature of those granites.

5. Geotectonic studies, leading to a statement of a rationale compatible with the apparent lack of the mass overturn effects, such as appear to be responsible for plate tectonics in the Phanerozoic, and the apparent emphasis of vertical transfer and magmatogenic (permobile) effects in the Archean.

6. Investigation of megacyclicity, in particular the relationship of the Gastil cycles and chelogenic cycles to possible shift in global convection patterns, and changes in style and rate of convection, or to cosmic influences.

7. More study of possible rift zone analogues in the Archean.

8. Further comparison between greenstone volcanic assemblages and those of modern ocean floors, and of calc-alkaline suites and modern island arc suites, delineating their occurrence on a global scale.

9. More data concerning igneous bodies within the Archean: many of them are little modified by metamorphism, and even those that are extensively modified can be studied profitably using modern instrumental techniques. Igneous manifestations of a "unique" character—anorthosites, high magnesia basalts, ultrabasic flows, and layered intrusions of almost complete refractory composition—all these call for closer investigation.

10. Detailed studies of Archean volcanic and intrusive processes and metamorphic effects of emplacement, with special reference to enhanced heat flow and thermal gradient.

*The use of the term "era" within the Precambrian has not yet been formally defined and is employed here in its nontechnical sense—*R. W. Fairbridge, Consulting Editor.*

11. Geochemical and petrologic studies pertaining to the nature of the Archean mantle and the derivation of the eruptives from it. Search for kimberlites and the like carrying mantle-derived nodules might produce evidence extending backward in time the story already derived from the kimberlites of North America, Yakutia, and South Africa, which provides a sectional view of the deep mantle through the last 2000 million years.

12. Search for more evidence of unconformities within the Archean.

13. Investigations of the depth, configuration, chemistry, and temperatures of the Archean oceans, to which sedimentology and volcanic studies could contribute.

14. Paleocurrent and other analyses to determine how closely the Archean greenstone synclines correspond in their boundaries to the basins of deposition.

15. Search for evidence pertaining to changed global physical controls; paleomagnetic studies extended to the Archean; changed tidal and gravitational effects; varve effects indicating astronomical cycles and astronomical controls acting in the Archean, in particular the position of the moon relative to the earth; evidence, if any, of global expansion.

16. Study of microbiotas and biogeochemistry, especially in carbonaceous sediments.

17. Detailed studies of the nature of the sedimentary, volcanic, and extrusive rocks that form the greenstone belts, both by means of specialist regional studies and highly detailed studies of small areas, such as have proved so valuable in Canada, Australia, and South Africa.

18. Thorough studies of the "granite seas" that enclose the greenstone belts and separation of these granite terrains into their various elements, as has been successfully done in South Africa and, more recently, Western Australia.

19. New crust (mantle-derived), regenerated older sialic crust, and metasomatized greenstone terrain need differentiating. Comparisons among Archean granite zones and Grenville-, Peruvian-, and Himalayan-type granite zones, and with rift zones.

20. Specialist studies of banded ironstones, detrital pyrite and uraninite occurrences, red beds and evaporites, all of which provide clues to the nature of the Archean atmosphere. Reevaluation of the role of chemical sedimentation, and, in particular, derivation of ancient chemical sediments from weathered crusts; consideration of the significance of alumina-rich metasedimentary suites in the very old Archean systems.

21. Extension of mapping to little-known Archean regions (West Africa, South America, India, Central and North Africa and, Antarctica) and little-known areas of well-studied regions (Western Australia, Canada, Russia, and Greenland).

22. Regional structural syntheses in poorly exposed and glacial Archean terrains, especially Canada, to link up isolated, well-documented exposures.

23. Overall global syntheses in the manner of Dearnley (1966); at present we have data on many pieces of the puzzle, but more syntheses are needed to supply a holistic view of the Archean world.

24. Search for evidence of meteorite impact in Archean terrains.

25. Specialist study of the nuclear terrains, gneiss in folded ovals, and gneissic domes, such as recognized by Choubert and Salop, with possible analogy to magmatogenic structures of the lunar terrae kept in mind.

26. More geochronologic data pertaining to the overlap in the lunar and Archean records.

References

Cloud, P. E. (1971). The Precambrian. *Science, 173,* 851–853.

Dearnley, R. (1966). Orogenic fold belts and a hypothesis of earth evolution. *Phys. Chem. Earth, 7,* 1–114.

Hutton, J. (1788). Theory of the earth. *Trans. Roy. Soc. Edinburgh.*

Lawson, A. C. (1913). A standard scale for the Precambrian rocks of North America. *Compt. Rend. 12th Int. Geol. Congr.,* 23 p.

I: GENERAL PAPERS CONCERNING THE BEGINNINGS OF THE GEOLOGICAL RECORD

Step by step, geochronologic advances have pushed the age of the oldest known rocks back toward the 4600-million-year "age of the earth," the age of the "starting of the radiometric clocks" at the end of the first postaccretional major heating event, or at some point in the cooling off after its maximum peak. Discoveries by V. R. McGregor (1973) in Greenland have disclosed a granite–gneiss terrain displaying characteristic protocontinental character and yielding a metamorphic Rb/Sr age of 3787 (±85) years (Oxford Isotope Geology Laboratory and McGregor, 1971; Windley, Paper 38). This indicates that the first major heating event had its maximum peak well before this date, at which time the first protocontinental terrain was probably already forming. In South Africa, a single granite boulder has yielded an age of 4100 million years, which indicates that, prior to the deposition of the Moodies Group of the Swaziland System (which included the boulder), there was indeed a granite source in existence—whether a primordial sialic crustal layer, a protocontinental nucleus even older than that found by McGregor, or a differentiate of a basic magma (Condie et al., Paper 17A). Confirmation of this age dating would signify that such a granite source surface was available at or near the surface of the earth only 500 million years after the date of the starting of the "radiometric time clocks" about 4600 million years ago.

In Paper 1, Wilson, writing in 1949, put the age of the earth between 4000 and 2500 million years ago, and favored 3300 million years. The "age of the earth" is now confidently fixed at a point in time about 600 million years prior to his upper limit. Later writers (Cloud, Paper 2, in 1968, and Wetherill, Paper 3, in 1972) tentatively assigned the end of the last major, thermal event that was global in effect to 3500 million years ago, at which point in time Cloud infers catastrophic heating due to lunar capture; on the other hand, Wetherill believed there was radiogenic and gravitational heating that survived at a high level up to this date. However, even these models may already be superseded.

Stable continents appear to have been forming, even if in protocontinental microcosm, as far back as 3750 million years ago, and possibly even 4100 million years ago. The processes of geology familiar in Phanerozoic examples seem to have been initiated and the preceding major thermal event to have been waning in intensity well before the date of lunar capture suggested by Cloud. Lunar capture (if it occurred at all) must certainly have intensely affected the surface of the earth thermally and caused melting, leaving a strong global imprint (Singer, 1971) in the form of fusion. *There seems to be no real reason to suppose that there was any catastrophic thermal event of major magnitude during the 4000–3000 million year time span,* although the effects of the first major thermal event could also have extended into this time span in a residual manner. There is every reason now to believe that the earth–moon system was already established 4000 million years ago (Singer, 1971). It is all the more surprising, therefore, that certain expected imprints are not found in the Archean terrains of this age; they should be recognizable if present widely accepted interpretations of the lunar surface are correct. Equivalents of the supposedly impact produced lunar maria do not appear to be recognizable in Archean terrains.

The dark or "Hadean" period of Cloud (Paper 2) has dwindled as geochronologists have extended the Katarchean backward, and we now know that we cannot explain this anomaly by simply supposing that no terrain of the right age is preserved. The Katarchean covers the period of development of the maria (supposedly impact excavated basins, later infilled with basalts, having an active history that extended from 3900–2900 million years ago; Green, 1971). If such mega-impacts occurred on the moon during this period, surely they must have occurred also on an earth already in station nearby. It is not acceptable to cite the absence of lunar atmosphere to explain this anomaly (such immense bodies as are indicated by theory would not have been seriously impeded by the earth's atmosphere). Thus, a major anomaly has been revealed by the combination of planetologic and geologic studies.

Green (Paper 4) attempts to resolve the problem by supposing that the depositories of the volcanic–sedimentary sequences that now comprise the Archean greenstone belts were impact-excavated basins analogous with the lunar maria. It is both an original and ingenious solution, but it seems to be at variance with observed facts concerning the development of these depositional troughs. We should, perhaps, look elsewhere to resolve this anomaly, even to the point of reexamining the evidence for and against impact origin for the lunar maria and highland craters (McCall, 1972).

Cloud seeks to explain the "Hadean obscurity" by invoking catastrophic heating due to tidal effects at the time of lunar capture. If such effects have any validity, this heating must have occurred well before the 3500 million year date suggested, for the maximum intensity of these effects would have been experienced more than 4000 million years ago, the revised lower time limit for lunar capture. This would make capture coincident with the first major heating of the planet, but it is not acceptable to simply refer this heating solely to lunar capture, for initial heating at the 4600 million year date is evidenced for other members of the solar system, the meteorite parent bodies of the asteroid belt. Whether there were two closely integrated events or only a single major heating event prior to 4000 million years ago, there is no need to invoke a subsequent catastrophic event to explain the "Hadean obscurity." The gap is in reality only partial and may be explained by any one of the following four causes or a combination of them:

1. The gap is simply a product of the mean rate at which geologic processes erase the record of older events; by the very nature of things, there is likely to be

less and less preservation of older and older geological surfaces, as Wasserburg (quoted by Wetherill, Paper 3) has suggested. There was only a limited area of Archean continental surface, and more and more of it will be obliterated with the succession of more and more chelogenic cycles.

2. The earth may well have expanded. The picture commonly presented of the Archean Protopangea as a dwarf island within a vast surrounding sea of ocean appears highly unlikely. Both oceanic and continental crust have probably increased vastly in area since Archean times, and the extent of the Katarchean stable platform surface may have been very small; hence the small amount of preservation.

3. Modifying the suggestion by Wetherill (Paper 3), the crust may not have cooled from the first thermal event sufficiently to preserve more than vestiges of the stable crust up to 3500 million years old, the radiometric clocks being reset metamorphically over most of it. This is tantamount to saying that prolongation "in a plateau" of effects of tidal, gravitational, or radioactive heating produced a greater heat flow and steeper thermal gradient in this period than later. Diminution of the effects of long-term radioactive origin seem most likely to be the correct explanation for this heat contrast, but this diminution may well have occurred over a long span of the Archean and not abruptly at 3500 million years ago.

4. The nature of each successive "granitization" cycle in the Archean was to obliterate earlier cycles, moving outward, step by step.

There is no need to invoke any *catastrophic* thermal event 3500 million years ago or even just before 4000 million years ago, although lunar capture *is* now favored*; it certainly would have produced a thermal event of great magnitude. There is therefore likely to have been more than one *very early* major thermal event *if the moon was in fact captured.* There is no need to invoke extraterrestrially induced heating, due either to capture or to impact, to explain the probable residual effect of increased heat flow during the 4000- to 2500-million-year time span (the period of the komatiites or high-magnesia basalts; see Paper 4). The "Hadean obscurity" is complete only over a span of 500–850 million years, and we can entertain a quite simple picture of the development of the earth through the Archean:

1. Initial, quite brief postaccretion heating: short-lived radionucleids, gravitation (4600 million years). Possibly, at somewhere around 4000 million years, additional intense heating due to lunar capture.

2. Initiation of the familiar processes and cycles of geology, involving geosynclines, crustal thickening, and stabilization, albeit in rudimentary fashion (4100+ to 3750 million years).

3. General destruction (i.e., very limited preservation) of these initial stabilization products because of either the statistical effect, the expanding earth effect, residual thermal effects of metamorphism of the near-surface Archean rocks, successive obliteration by outward progressing granitization cycles, or a combination of these.

On the moon, the formation of cratered highlands occurred more than 4100 million years ago (ages up to 4300 million years have been obtained) with the immense extrusion of anorthosite basalts (possibly as breccias). There is a popular

*This was the case at the time of writing, but opinion has since swung back toward the other view—*Editor.*

belief that the cratered highlands of the moon reflect a period of megaimpact that was orders of magnitude greater than during the events that succeeded it in the mare period—a bombardment that was due to a sweeping up of an early prolific generation of planetesimal meteoroids residual to the accretion process. It may be questioned, however, whether the lunar highlands represent anything of the sort. It is strange that the few large maria postdate the highlands cratering and indicate a sequence from large to immense impacting bodies, and by no means a smooth progression. This appears astronomically improbable. The lack of any analogue to the craters of the lunar highlands on the surface of the earth may be explained by their great age, but they still seem more likely to be products of the first major heating and degassing of the moon. Fielder et al. (1971) have suggested that statistical analysis indicates an internal origin for many of the maria craters, a suggestion made earlier on less detailed evidence for all lunar craters by McCall (1965). It may well be that similar rigorous analysis of highland craters will indicate a similar origin.

As a corollory to this suggestion, we should reexamine carefully the large circular structures recognized by Choubert and Salop in Katarchean terrains (see Paper 22: Commentary). Could not these be terrestrial analogues of the cirques of the lunar highlands, for they are not of vastly different scale? This idea has been hinted at by Khain (Paper 30B), independently from the editor; both types of cirques could be responses to early degassing.

References

Fielder, G., Fryer, R. J., Titulaer, C., Herring, A. K., and Wise, B. (1972). Lunar crater origin in the maria from analysis of Orbiter photographs. *Phil. Trans. Roy. Soc. London, A 271,* 361–409.

Green, D. H. (1971). The magmatic activity as the major process in the chemical evolution of the earth's crust and mantle. *Tectonophysics, 13,* 47–71.

McCall, G. J. H. (1965). The caldera analogy in selenology. *Ann. N.Y. Acad. Sci., 123,* Art. 2, 843–875.

_____. (1972). A new look at the origin of lunar surface breccias. *Astron. Space, 2,* 5–17.

McGregor, V. R. (1973). The early Precambrian gneisses of the Godthaab district, West Greenland. *Phil. Trans. Roy. Soc. London, A 273,* 343–358.

Oxford Isotope Geology Laboratory and McGregor, V. R. (1971). Isotope dating of very early amphibolite facies gneisses from the Godthaab district, West Greenland. *Earth Planetary Sci. Letters, 12,* 245–259.

Singer, S. F. (1970). Origin of the moon by capture and its consequences. *EOS Trans. Am. Geophys. Union, 51,* 637–641.

PAPERS 1 THROUGH 4: COMMENTARY

As early as 1949, Wilson emphasized the importance of studying the little-known Precambrian rocks to obtain information concerning the origins of the continents, information that cannot be readily derived from studies of the Phanerozoic. Problems of growth or diminution of continents could be resolved in this way. He believed such studies must be included in and, indeed, form a major part of fundamental studies of the origin of the earth.

His preface and the section dealing with the origin of the continents are reproduced here and provide a good starting point for Part I. Like Gill (Paper 36), Wilson does not favor the traditional concept of a primordial sialic crustal layer. He foreshadows subsequent, more detailed observations of other Canadian geologists concerning the possible role of volcanic centers amid oceanic outpourings of noncentral volcanics in providing the protocontinental nucleii. He favors progressive growth of continents, by virtue of the addition of successive fringing geosynclines, but notes that "primary" and "secondary" fold systems are of a different kind; the latter are shallow rooted and founded on continental crust. In this, Wilson foreshadows the greenstone-belt and stable-platform trough distinction now widely accepted, for example, by Russian authorities (Papers 22, 27, 28, 29, and 30), by Anhaeusser (Paper 35), and by Sutton (Papers 31A and 31B). He doubts whether the term Proterozoic has any real meaning in the sense of age, and, although rocks of Proterozoic character do in general tend to succeed Archean greenstones and their enveloping granites, it does now seem that, in the sense of an absolute age division, the significance we give to the Proterozoic is purely arbitrary; there is no single global break between the Archean and Proterozoic type of sequence marking a worldwide revolution. It should also be noted that the sense of the term Proterozoic that implied earliest life has long been invalid; there are life forms even in Katarchean systems.

Wilson's summary of the then existing theories concerning the origin of crust and continents is quite useful. On page 177 he foreshadows ideas concerning primitive chemical sedimentation, which are covered in more detail by later Russian authorities (Papers 30 and 32). He also hints at the theories of granitic underplating of the continents discussed in Papers 5 and 35.

The depletion of noble gases relative to their cosmic abundances seems to require the earth's atmosphere to be of secondary origin. Concepts of gradual, episodic, or abrupt degassing from internal sources have stemmed from various authorities, and dispute has focused not on the fact of outgassing, but rather on the original proportions and subsequent changes in the proportions of oxygen, nitrogen, carbon dioxide, and hydrogen in the earth's atmosphere. The oldest mineral dates were about 3500 million years (at the time Cloud wrote), and there is no geological evidence concerning the atmosphere prior to that date. The first Archean sediments of that age provide evidence for the existence of an atmosphere and hydrosphere (including seas).

Cloud postulates a cataclysmic thermal event 3500 million years ago in addition to and quite distinct from the first major thermal event 4600 million years ago. However, the evidence for this event, also considered by Wetherill (Paper 3), is questionable, whether it be from the Kola Peninsula, the Ukraine, or Rhodesia. Cloud considers that the atmosphere and hydrosphere, if they existed prior to this postulated event, amounted to but a fraction of their present extent, and that they were lost as a consequence of this catastrophic event.

The evidence of detrital uraninite and pyrite allows for little oxygen in the early Precambrian atmosphere (see K. A. Kvenvolden, *Geochemistry and the Origin of Life,* and H. Lepp, *Geochemistry of Iron,* both in this Benchmark series). The absence of limestones and dolomites and the abundance of bedded chemical silicates points to little ammonia content. Evidence is also given for a paucity of methane. An early atmosphere consisting of gases typically occluded in igneous rocks or present in juvenile gases is indicated: hydrogen, nitrogen, carbon monoxide and dioxide, sulfur dioxide, and hydrochloric acid.

Accretion is believed to have occurred at no more than an outer temperature of a few hundred degrees Kelvin. Raised temperature would be required to initiate atmosphere–hydrosphere segregation by sufficient melting to initiate the release of volatiles to the surface.

Possible sources of heat are (1) radiogenic energy plus gravitational heat, and (2) tidal friction on lunar capture. One or both may have operated. Cloud calls for a study of stromatolites to detect tidal amplitudes and to find out when the moon appeared in orbit; sedimentological studies of very ancient estuarine sediments might provide the same information. Evidence of this sort might tell us whether the second heat source theory is valid. Cloud favored 2000 million years ago as the minimum date for lunar capture and believed 3000 million years to be more likely. Density and inclination of orbit suggest that the Moon *was* captured at some time, but the theory of capture presents problems in the likely vaporization of the earth and disintegration of the moon; large-scale volcanism, plutonism, and metamorphism and resetting of nuclear clocks are concomitants of lunar capture. Cloud tentatively suggests that magma welled up to fill impact-excavated maria as related to this event; he mentions difficulties in equating such a hypothesis of capture at that date with the visible sequence of major modifications of the lunar surface. Singer (1971) has suggested that the moon possibly appeared on the scene earlier than Cloud supposes, perhaps 4000–4500 million years ago.

As to the beginnings of life in the Archean, Cloud envisages oxygen produced by organisms as having been first accepted by ferrous iron to form the extensive hematitic ironstones of the 3000–1800 million year old formations (see Lepp's volume); oxygen then invaded the hydrosphere and, finally, the atmosphere, producing ozone and dissociated oxygen. Lacking an ozone screen to ultraviolet rays, this would have led to extensive "red bed" development. That such beds make their appearance just after the banded iron formations seems compatible with his expectations.

From the sedimentary siderites, sulfide-rich sediments, and bedded cherts, evidence is supplied that the pH and dissolved salt mix of the Archean hydrosphere were not significantly different from that of today, although pH may have been slightly lower and sulfite iron almost absent. Iron formation cannot occur in modern seas under an oxygenic atmosphere, but it possibly could under the reducing atmosphere of the Archean; thus, it does *not* provide evidence of fresh-water origin. The configuration of the Archean hydrosphere is seen as, initially, one of isolated basins of intermediate depth; however, much more evidence is needed concerning the limits and details of these early basins. If truly isolated, it would appear that there would be latitudinal variations in their chemistry.

A hypothesis is called for to explain the lack of rocks older than 3500 million years, which is taken by Wetherill to be the maximum metamorphic age established by radiometric dating. The following alternatives have been suggested:

1. The age of the oldest known rocks represents a kind of steady-state age indicative of the mean rate at which normal geologic processes erase the record of older events (as suggested by G. J. Wasserburg).

2. The earth formed as a cold planet and heated slowly by the agency of long-lived radioactivity. Melting began about 3500 million years ago, and the sinking of iron into the core released sufficient additional energy to initiate fractionation of crust, mantle, and core. The oldest crustal rocks postdate this event (suggested by Fairbridge).

3. Lunar capture took place 3500 million years ago and released tidal energy that destroyed the prior record.

4. The earth was formed at a high temperature and did not cool sufficiently to preserve stable crust until 3500 million years ago.

The Archean rocks, those more than 2400 million years old, form 5 percent of the earth's land surface. Areas of Archean rocks, once they became established as continents, were by no means entirely protected from later disturbance, which might have led to possible disruption and obliteration. Cratons were continuously subject to marginal deformation, the development of internal fractures and geosynclines (mobile belts), and consequent breakup. Some cratonic blocks, as is the case in Montana–Wyoming, have been caught up in much later overridings of plates. Nevertheless, cratons do tend to surround themselves with fold belts of younger age, and so to be largely protected from rejuvenation; at least some of the Archean cratonic nucleii have been preserved in this way. Although Wetherill appears to be uncertain as to the sum total result of protection as against fragmentation and obliteration, it does seem that the effect of superimposed chelogenic cycles (Papers 31A and 31B) reflecting new convection patterns would have obliterated much of the original surface of stabilized Archean cratonic crust, and the first alternative may account for much of the scarcity of areas of early Archean rocks.

The major fractionation of Rb and Sr on the moon occurred 4600 million years ago, apparently just after its formation. Heating was too great to be solely attributable to long-lived radioactivity sources. Short-lived sources such as ^{26}Al and electrical induction have also been considered; but, although an intensely active pre-mare sequence did occur on the sun at just the right time for planetary formation, there is evidence against ^{26}Al as an additional heat source in the lack of excess ^{26}Mg, the daughter isotope. Also, both these sources require the moon to be hotter at its center, but Wetherill accepts an interpretation of the controversial lunar electrical conductivity data which indicates that this is not the case.

Wetherill regards gravitational energy involved in lunar accretion as a much more acceptable additional heat source; if accretion was very rapid (taking only 10^3 years), such a heat source could be reconciled with a hotter outer zone and cool center to the moon (one must, however, bear in mind that this is by no means the only possible thermal interpretation of the lunar evidence, but only one favored model).

Mare Imbrium is supposed to have been formed 3.9×10^9 years ago, and its formation must have been preceded by a much more intense energy accession to the lunar surface, if the prolific cratering of the terrae or cratered highlands is of impact origin, as Wetherill supposes. Although he discusses the subject of meteorite impacts and possible energy involvements at length, Wetherill concludes that the main cause of the deficiency of rocks more than 3500 million years old is the survival of the original heating of radiogenic and gravitational origin for about 10^9 years in a steady state after accretion. He believes that, whereas at the present time there is incipient melting at 100-km depth, 3500 million years ago this took place at 30-km depth; thus, the Archean lithospheric plates were thin and subject to rifting and transport down to the region of melting by subduction. After 10^9 years had passed, the effect of the decay of the radiogenic sources ^{40}K and ^{235}U diminished, and thus the total heat source diminished sufficiently to allow thick sialic blocks to form and develop into stable continental crust. Adjacent regions of a character analogous to present-day oceanic crust continued to participate actively in these ongoing tectonic processes.

Wetherill regards the greenstone belts not as the oldest crust of the earth, but rather as the oldest crust that has been able to survive remelting. He does not attribute these belts to impact agencies (see Paper 4), but does regard the period prior to 3500 million years ago as likely to have been one of intense planetesimal bombardment on earth and moon. Such bombardment would have resulted in additional accession of heat, augmenting that derived from internal causes, and, presumably, would have aided in the obliteration of the crust formed more than 3500 million years ago.

The preservation for thousands of millions of years of Archean greenstone belts in cuspate synclinal troughs is quite at variance with plate tectonics theory, which requires rapid reprocessing of ocean-floor eruptives in subduction zones to form calc-alkaline island-arc suites. Green notes the likelihood that the Archean regime was one of thin sialic plates, shallow subduction zones, and high geothermal gradients. He considers the uniqueness of high-magnesian mafic to ultramafic eruptives of the Archean greenstone suites, and the difficulty in explaining access of 60–80 percent melting products to the surface at 1600–1650°C from the mantle. Rapid upward movement of the liquid–crystal diapir is necessary to explain this, he entertains both extremely rapid internal convection and mantle melting by impact triggering.

The answer to the first apparent anomaly may be that the Archean was a regime of vertical lithospheric generation, as Ringwood has suggested (see Part VII); the mass overturn process simply did not operate. The answer to the second anomaly may lie in endogenetic rather than exogenetic processes, in very rapid convection, or in different physical environments.

The suggestion that the greenstone depositories are lunar mare analogues is original and must be considered seriously; yet objections may be raised on the following grounds:

1. With very few exceptions, the evidence suggests that granite immediately enveloping greenstone belts is a *later* emplacement, greenstone belts having the essential nature of paleosomes; this does not, of course preclude the granite terrains from including older reworked granite material.

2. The nuclear Katarchean terrain (Papers 27, 28, 29, 30, and 32) is commonly attributed to uprise of granite domes; yet the greenstone basins must have subsided to accommodate the vast accumulations. The presence of calc-alkaline suites in Canada suggests that the first volcanics were depressed and reworked in the manner of the later subduction zones, although possibly in a vertical tectonic regime. No evidence at all of collapse inward of the basins is found in the tectonic style of the greenstones. Nuclear terrains are by no means restricted to the Katarchean, nor are high-magnesia basalts, which appear in the Canadian and West Australian late Archean and Proterozoic Jimberlana Dike (2420 million years ago) (Paper 15 and 16C).

3. Evidence from Western Australia indicates considerable thicknesses of greenstone assemblage *underlying* the high-magnesia basalt zone and *devoid of* such basalts (at Edjudina and Norseman). The basal connotation of these basalts is neither supported by this evidence nor by the reappearance of such basalts at no less than four levels in the Yilgarn succession.

4. The high-magnesia basalts are intimately associated with tholeiites of abyssal type at every level in Western Australia and in the Jimberlana Dike. They are clearly congenetic with the tholeiites and cannot be divorced from them.

5. Arguments as to whether the Archean greenstones are geosynclinal or not are largely semantic; they reproduce so closely, in both sedimentology and in eruptive patterns, the features of later geosynclinal systems that there can be little doubt that they are the Archean analogues of the later geosynclinal sequences—the *protogeosynclines.* Pillow lavas, turbidites, littoral conglomerates, thick contemporaneous differentiated sills, serpentinites, cherts, dark shales, and island-arc-type volcanics are all represented. It is difficult to see the need for as special an explanation as Green offers, although it is clear that the Archean protogeosynclines may have differed in not being derived under a regime of lateral transfer and subduction of oceanic lithospheric plates.

The first critical fact brought forward by Green is the inexplicable character of the accession of high-temperature ultramafic flows to the surface in terms of present-day conditions at the earth's surface. The answer to this must surely lie in the quite different nature of Archean eruptive processes and in the quite different physical environment; it must surely be an endogenetic answer. The second critical fact is the lack of terrestrial maria analogues. This anomaly remains unexplained, but its resolution may well call for a much more radical reappraisal of *lunar* theory, not a reappraisal of the terrestrial greenstone belts.

Reference

Singer, S. F. (1970). Origin of the moon by capture and its consequences. *EOS Trans. Am. Geophys. Union, 51,* 637–641.

1: THE ORIGIN OF CONTINENTS AND PRECAMBRIAN HISTORY

J. Tuzo Wilson

TWO years ago in his presidential address to the American Geophysical Union Dr. Leason H. Adams (1947) discussed six unsolved problems of geophysics. Of the two first which he chose, he said: "The problem of the origin of geosynclines is closely related to the problem of the origin of mountain-building. Indeed, geologists, noting the general association of mountains with vast sedimentary deposits that must have been laid down in geosynclines, are likely to say that we have here two aspects of one and the same problem; and going back to the dawn of geologic history, we may think of the formation of continents and ocean basins as the really fundamental problem."

In no aspect of geology are these matters of more importance than in Precambrian geology.

There is now a fairly general agreement that the earth is likely to be between 2.5 and 4×10^9 years old. If we accept the age of 3.3×10^9 years which was recently determined by Holmes (1949), then the time since the beginning of the Cambrian constitutes only about 15 per cent of that. This period, for which the geological history is well established, provides a rather short base from which to extrapolate backwards and generalize upon the whole history of the earth. In particular, it has proved difficult from a study only of Palaeozoic and later rocks to tell very much about the origin of continents and to be certain whether they have grown, diminished, or remained nearly constant in size.

Precambrian rocks represent a much greater period for the oldest ages determined among them are about 2×10^9 years (Ellsworth, 1932 and Ahrens, 1949). Thus rocks have been preserved from about two-thirds of the earth's history, but our understanding of the history and structure of the Precambrian era is very imperfect compared with that of the fossiliferous part of the record. Ideas concerning the origin of the earth and continents could influence and assist our interpretation of early Precambrian history. Moreover, when the structure and relationships of the older rocks are worked out they can be expected to show us much about the origin of continents and about their possible migrations and supposed growings and shrinkings. Although our knowledge of both the origin of the earth and of Precambrian history is fragmentary, we will increase our chance of arriving at an acceptable explanation if we combine the two studies together. That is the object of this paper.

By good fortune the methods of attack on the two problems are quite dissimilar. The knowledge which can be applied to the joint solution is therefore greater and the conditions which any theory must meet are therefore more stringent than when either problem is studied alone.

In the next section of this paper, some of the theories of the origin of the solar system and of the probable state of the crust of the earth soon after its formation are discussed. Reference is particularly made to evidence from astronomy, from seismology, and from meteorites.

In the third section some of the possible manners of evolution of continents are reviewed and examined in the light of what is known of the geology of the Canadian Shield.

It may be said at once that the conclusion reached follows Lawson (1932). It is that the earth cooled and became solid without continents or sialic layer, that the earth's outer crust has repeatedly failed, due probably to continued cooling. Each failure has allowed lighter phases of igneous material in the form of hot solutions, intrusives and extrusives to rise towards the surface.

The first failures formed volcanic islands. Erosion formed marginal dumps or shelves of sediments about them and later about continents. Continued contraction sought out these shelves as zones of weakness and formed fresh failures in them. Failure, which allowed hot solutions and magmas to rise, transformed these thick accumulations of sediments into marginal mountain ranges.

Water was the agent that dissolved some of the basic oxides from igneous rocks, gathered the residues and emplaced them as sedimentary formations. Without water the earth's cooling would have proceeded in different fashion and continents as we know them would never have formed.

Hot acidic solutions and magmas rising from the depths both consolidated the existing sediments into gneisses and provided the additional material necessary for enlarging the continents.

The erosion of each marginal range provided on the outer side of the range material for the next shelf and on the inner side a trough of sediments destined to become a secondary range (Holmes, 1946, pp. 401-2). It is considered that the older Appalachians are an example of a primary marginal range and that the Atlantic coastal shelf and the younger Appalachians are respectively a shelf and a secondary range derived from it.

The secondary ranges differ from the primary in that their structures are simpler and they were laid down on continents not on sea-floors. This probably is the reason why secondary ranges do not become granitized as do primary ranges. It suggests that the arguments about whether granite has been formed from sedimentary rocks or from basic igneous rocks may be resolved by showing that both are necessary and that most granite has only been formed where great thickness of the one rests upon the other.

The roots of primary mountains are deep and may extend to the Mohorovicic discontinuity but secondary deposits are shallow although widespread. The proportion of contorted metamorphic rocks exposed in the younger eras is therefore small but it increases in the older periods until in the Precambrian only limited areas of Proterozoic or secondary rocks lie upon the Archaean, as the roots of former primary ranges are called.

If the continents have grown during geological time by this process of marginal accretion about some of the first formed nuclei and by occasional coalescence of two expanding continents, then it should be possible to divide the metamorphic welts and the Archaean parts of the shield into orogenic belts progressively younger towards the margins.

This theory of continental growth is essentially that proposed by Lawson (1932). In so far as the later eras are concerned it has been accepted by many geologists for a century. As Bucher (1933, p. 379) has stated: "In 1849 Dana wrote, 'all continents have had their laws of growth'. . . . American geologists have been accustomed to think of continental growth, especially that of North America, as due to marginal accretion."

One mechanism by which the failures necessary to cause mountain building might have occurred was suggested by Lake (1931). His hypothesis has recently been extended (Wilson, 1949a), but Scheidegger, a mathematician at the University of Toronto, who has been examining Lake's hypothesis and other possible methods of failure of spherical shells has advanced an alternate and improved theory of the failure of the earth (Scheidegger and Wilson, in preparation).

The objectives of the present paper are limited. It is one of several written to examine, extend, and remove objections to a theory of continental growth built upon Lawson's proposals. The details of mechanisms involved and of geological interpretation are reserved for treatment elsewhere. This paper is only intended to show that the concept of continental growth already widely accepted for later eras may also have occurred in the Precambrian and indeed that the creation and entire growth of continents during geological time is compatible with the evidence from all sources.

[*Editor's Note:* We have omitted pages 160–172 and 174, which deal at some length with the origin of the earth and solar system—subjects about which a wealth of new data has accrued since publication of this article.]

The Origin of Continents

Umbgrove (1947, pp. 241-50) has discussed the problems of the origin of continents so recently and with such numerous references to previous work that a short statement will suffice here. He divided the theories, which have been put forward, into three groups.

In the order in which they will be discussed these are: theories involving an original sialic crust of which part was removed by the birth of the moon; theories involving a thin, original sialic crust which gradually coalesced into continents, and theories involving an original basic crust from which or through which the sialic materials of the continents were derived and accumulated.

Theories depending upon removal of part of a sialic crust. The first group of theories were sufficiently discussed in the previous section to indicate that, even if the separation of the moon from the earth during its early history is physically reasonable, which is now open to doubt (Jeffreys, 1948), difficulties are involved in using that disruption to form the continents. These difficulties would be sufficiently great if the continental fragments had taken up their present positions immediately after the time of the moon's birth. They are made in many ways more difficult in Wegener's theory of continental drift. The physical forces that could cause a single continent to break up and separate in relatively recent geological time are quite unknown. Why such an event should happen only once in geological history is unexplained. Even if the theory provided a satisfactory explanation of the world's recent mountain ranges, some other explanation would have to be advanced to explain the older ranges including those whose roots are now considered to make up the shields (Jeffreys, 1929, p. 304).

Theories depending upon coalescence of a sialic crust. The second group of theories gives the earth a thin, sialic crust after its origin. This is considered to have become thicker in some places to form continents and thinner elsewhere to form ocean basins. This group includes ideas favoured by Umbgrove.

Such a thin uniform crust may have been physically possible at the time of the earth's origin, but it has been noted that, although it may have once existed, there is no certain proof that it did. No portion of it has ever been identified. There is no clear evidence that the planet supposed to have been parent to the meteorites had such a granitic crust. Although both the moon and Mars show surface markings, they are probably not like the earth's continents. Any analogies existing between planetary crusts and the tops of lopoliths can be regarded as favouring the next group of theories, for lopoliths are large sills formed under cover. The granite layer is not a surface phenomena in them. Granite layers do not form at the tops of thick, basaltic lava flows.

Umbgrove suggests that the convection forces which he supposes to have operated to cause the sialic crust to be thinned or thickened in different places had ceased to affect the growth of continents before Cambrian time and that "from that time onwards the terrestrial forces were imprisoned and became operative as *subcrustal* processes" (Umbgrove, 1947, p. 247). He bases this upon the observation that later geosynclines originate upon a basement which had already been folded. He thus visualizes two processes to have been at work, one earlier and the other later. But many late Precambrian rocks were also laid down upon a crust which had already been folded. Therefore as the history of the Precambrian era is worked out one might expect to find evidence of such a change. Radioactive age determinations suggest that the Keewatin rocks are much the oldest in the shield for all determinations in them give an age of about 2 billion years (Ellsworth, 1932, Ahrens, 1949 and Hurley, 1949). They differ in some respects from younger Precambrian rocks like the Grenville, but few Canadian geologists would admit that the Keewatin rocks represent the folded, original sialic crust. The origin of continents is thus pushed back into at least the first third of postulated geological time according to these theories.

Concerning the origin of the shields by convection currents Hills has recently expressed the view that "the formation of the continents by felspars as phenocrysts in the magma appears to offer an explanation of the structure of the original gneiss. Pressure in the entangled magma, while still liquid, may cause the felspars and mica to take up a linear arrangement and thus to assume the well-known gneissic structure. The subsequent crumpling of gneiss may be caused by the "set" of the currents, as is the case with ice floes at the pole in an arctic winter freeze-up." (Hills, 1947, p. 11). It is doubtful if many geologists familiar with the Canadian Precambrian would accept this.

Umbgrove and Hills like many other recent writers favour deep-seated convection currents as the mechanism causing the formation of continents and mountain systems. The suggestions put forward have many attractive features, but they do not explain the structures of the Canadian shield. They also leave at least two major geological difficulties unexplained.

In the first place the distribution of these currents has not been worked out in plan. Griggs (1939), and Vening Meinesz (1948) have discussed the formation of cross-sections of mountain ranges formed in this manner in a most convincing fashion, but it is very difficult to visualize a system of world-wide currents which could explain in plan the arcuate mountain belts of the world. Where arcs like the Aleutians and Kuriles join the currents have to intersect one another. Until an explanation is offered the theory is seriously incomplete.

Another difficulty with currents is to understand why some parts of the world should be so active and yet leave other parts formerly active now so stable. For example, the greater part of the Canadian shield is a relatively low plain rising less than 2,000 feet. At many places near the borders and around Hudson Bay there are outliers of Ordovician and Silurian age. Several of these are large enough to be shown on the geological map of Canada and there is a small but important one near Lake Dubawnt, in the heart of the Northwest Territories (Tyrrell, 1898). These infaulted patches are sufficiently numerous and widespread to suggest that the Shield was extensively covered during parts of Ordovician, Silurian and perhaps Devonian time, but the absence of inliers of all other ages (except lacustrine Cretaceous south of Hudson Bay) suggests that the Shield was not covered by the sea at other periods or some record would have been left (Cooke, 1931, pp. 177-9). This suggests, if it does not prove, that the Shield has been very stable, neither depressed enough to allow the sea to enter often, nor raised sufficiently for erosion to have removed all trace of those few principal inundations.

It is considered that there is neither any complete theory nor any convincing evidence of any mechanism such as would be necessary to form continents in the manner required by this group of theories.

Theories based upon growth of the continents during geological time. Turning now to the last group of theories, they are those which involve the growth of continents during geological time from or upon an originally more basic crust. It has already been shown that the formation of the earth with no sialic crust is compatible with the evidence. If there was no original acid crust, it is easy to understand why no evidence of one has ever been found.

To form continents from an initially basic crust requires the accumulation of acid material. Four processes have been suggested to bring this about. It has been supposed that sea-water has dissolved some of the basic oxides from basaltic rocks to leave a siliceous residue. It has been suggested that acid intrusives broke through into the crust from below. The rise of acid or at least intermediate volcanics is a third possible source, which may have been connected with this. It has been proposed that gases rose and reacted to increase the acid content of overlying rocks.

Considering the first of these proposals, it was Lawson (1932) who emphasized that sea water would attack islands of basic rock formed by volcanism and raised above the waters of the primitive ocean. Water would dissolve bases and leave a more siliceous residue. Of the bases much sodium was left in solution. Calcium, magnesium, and perhaps iron were deposited separately as less dense rocks like carbonates and perhaps deposited also on the floors of the ocean. Lawson considered that in course of time the sialic waste products accumulated as geosynclines on the continental borders. These failed from time to time, became fused together to form new marginal mountain ranges with granitic cores. He visualized a rude differentiation occurring in the sial as it formed which separated it into two vaguely defined layers above the Mohorovicic discontinuity. He held that the continents had grown by the addition of successive marginal ranges which were gradually eroded until they were added to the shields.

A second possible source of more acidic materials has already been mentioned. It was Buddington's suggestion that a layer of granitic composition, which had formed by differentiation beneath an earlier gabbroic crust, rose in places through that crust. Jeffreys has favoured a somewhat similar mechanism as an explanation of the formation of continents. It is well known that acid intrusions frequently follow mountain building. The formation of granodiorite cores in mountains on the continents is well established. Granites and granodiorites are found in oceanic islands of moderate dimensions like Japan and Iceland, but not in small oceanic islands.

Evidence of the third source of acid rock is the andesitic lava which has been abundantly produced all around the Pacific. Volcanism producing such lavas has been associated with many former mountain ranges. Eardley (1947) and Kay (1948) have recently produced evidence of the importance of these additions to the continents. They have pointed out that a large part of the geosynclinal sediments in the Coast ranges and the Appalachians were not derived from the continents but from off-shore volcanoes. Eardley draws a comparison between the Cordilleran geosyncline and the present situation off the east coast of Asia. In each case detritus has been added to shallow seas from the continent and also new material of volcanic origin has been added farther off shore. Here then is a definite source of new materials of intermediate to acid composition causing continental growth. Is there any reason to believe that there could not have been such sources in the Precambrian era?

The fourth source of more acid materials was suggested by Rittmann (1939) who is quoted by Umbgrove (1947, pp. 241-50) as suggesting that it was not granite, but gases emanating from the substratum that rose and that assisted in the granitization of sediments. The source of gases is in keeping with the view that a layer of hyperfusibles was trapped in the crust at the time of its formation. The existence of such a layer of hyperfusibles rich in radioactive elements has been mentioned as a source of heat for volcanism under the oceans (Jeffreys, 1929, p. 159). If the escape of these hyperfusibles during granitization brought the radioactive elements to the surface, the heat generated by these elements would thereafter escape. Thus the lack of volcanism in shields compared with the frequency of volcanoes in ocean basins might be thus explained.

The relative importance of these four processes in producing new sialic material cannot be assessed, but there seems to be good evidence that they have produced large quantities of new sialic material. Some growth of continents can thus be explained and is indeed already widely believed in for the period since the Precambrian. As Holmes (1946, p. 402) has said: "The continents have developed by successive orogenic additions to their shields." There is no reason to believe that the behaviour of the earth suddenly changed at the close of Precambrian time. The evidence of continental growth should therefore be preserved in shields if it has occurred.

Some aspects of the structure of the Canadian Shield have already been discussed elsewhere (Wilson, 1949b) and it will only be possible to review the salient points here. In the first place, it has been pointed out that the Proterozoic rocks may be considered to be secondary deposits in the sense used by Holmes (1946, p. 402). Although folded they have not been subjected to the same intense deformation as have rocks of Archaean type. In considering the growth of the Shield they confuse the picture for it is only the Archaean or primary belts that are successively younger from nuclei outwards towards the margins. Such an arrangement is an essential part of this theory which it should be easy to prove or disprove. It is considered that the arrangement and ages determined for the Superior, Grenville, older Appalachian and Atlantic provinces already support the theory. In Fig. 3 it is considered that the Superior province (Gill, 1948) and the Yellowknife province (Jolliffe, 1948) represent the earliest parts of the continent, although not necessarily exactly equal in age. They are marked by an unusually high proportion of basic lavas and by peculiar greywacke and conglomeratic sedimentary rocks. Limestones and pure quartzites are lacking in them. It is considered that these rocks may have been formed about volcanic islands, that the large proportions of basic lavas may indicate that the crust was thinner and the earth hotter and

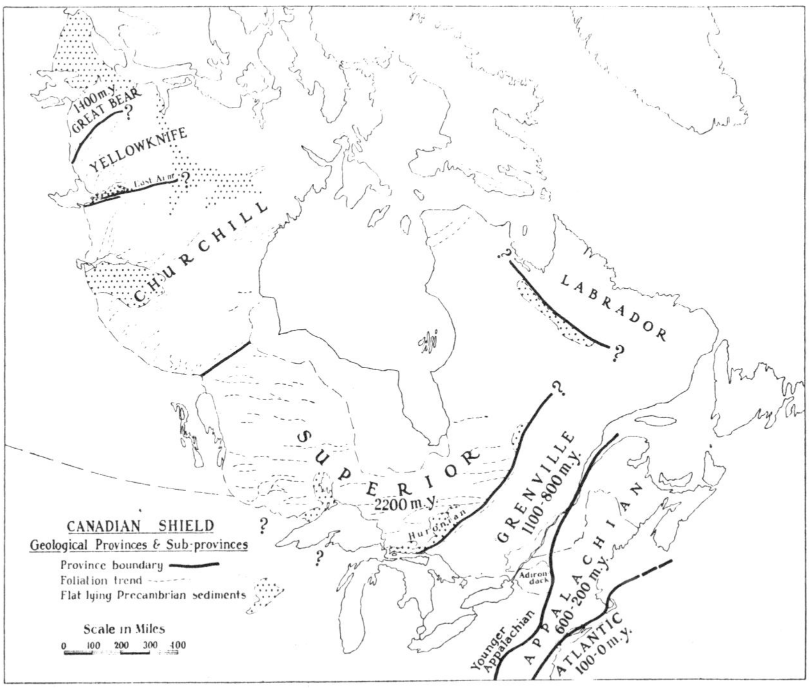

FIGURE 3.—Proposed orogenic belts or provinces for part of the Canadian Shield and Appalachian region with age determinations in millions of years (m.y.) where known.

more readily able to discharge basic lavas. The source of the granite pebbles in these rocks does not necessarily demand a continent nearby nor an earlier orogeny because granite stocks occur in the unfolded Cenozoic basalts of Iceland. Thus granite pebbles can be found there which imply neither an orogeny nor a continental origin. Presumably similar sources of granite pebbles could have existed in Keewatin rocks. It is difficult to explain the method of emplacement of granitic bodies in the basic lavas of Iceland. The relation between granites and Keewatin and Timiskaming rocks in Canada is also obscure, but a knowledge of the origin and mode of occurrence of granite is not essential to this discussion.

It is suggested that these continental nuclei of the Canadian Shield grew for a time under slightly different conditions from those prevailing later, but that, after these nuclei had grown to considerable size in the course of hundreds of millions of years, marginal shelves of large size accumulated beside and between them. These in due course failed, became mountains, and were eroded to form the Great Bear (Jolliffe, 1948), Churchill (M. E. Wilson, 1941), and Grenville (Gill, 1948) provinces. They probably did not all fail at the same time and the process of change from marginal geosynclinal shelves to mountain ranges also took hundreds of millions of years. Little of the Great Bear province is exposed. The Churchill province is not for the most part well examined or known and may be a complex of more than one orogenic belt. The Grenville province has long been considered to be the roots of a mountain range. The age of pegmatites in it suggests that the process of failure occurred during the interval from 1100 to 800 million years ago (Ellsworth, 1932).

In the far north and also covered under the interior plains other unidentified and unstudied nuclei and orogenic belts presumably occur, but by the close of Precambrian time the shield had coalesced to form a single continent (except probably for Greenland) and fresh marginal shelves had started to form in the sites of the future older Appalachian and Coast Range mountains. Radioactive age determinations on pegmatites and palaeontological evidence (based indirectly on the same time scale) agree in assigning an age of latest Precambrian to Carboniferous for the period of failure and intrusion in the first of these orogenies and a Mesozoic to Tertiary age for the second. Since the close of the Appalachian orogeny another marginal shelf, the Atlantic coastal plain, has started to form of Mesozoic and Tertiary sedimentary rocks. It has not yet failed at all, but it is suggested that it may be expected to do so. By analogy with the east coast of Asia, which is visualized as having progressed farther in the same cycle, it is suggested that failure will be marked, first by the formation of off-shore andesitic volcanic arcs and later by folding, invasion of granites and uplift to form another marginal range of mountains.

It is considered that the Front Range Rocky Mountains through Colorado and Wyoming are not a fundamental part of the Coast Range cycle. The Palaeozoic rocks there are thin, not geosynclinal. Some age determinations in their cores are middle Precambrian. They are considered to be blocks of the older foreland or shield fractured and uplifted by the Coast Range orogeny. That orogeny has been so recent that no marginal geosyncline has yet had time to start to form beyond it in the Pacific.

Leith (1935) has pointed out that Precambrian time is not necessarily divided into a younger Proterozoic era and an older Archaean era, but that these names should be applied to different types of rock assemblages. This view is accepted and the Archaean held to be the roots of primary mountains, while the Proterozoic rocks are secondary deposits. It may be that those names should be abandoned and others substituted, for it is felt that this great division is not unique to Precambrian rocks, but that all exposed cores of primary mountain belts formed out of continental shelf geosynclines are of Archaean type regardless of age from early Precambrian to Tertiary and that all secondary deposits derived from those mountains and deposited inland upon the old continents are of another type which includes the Proterozoic rocks.

As examples, it is suggested that the Huronian rocks present a secondary deposit derived from the rising primary Grenville mountains and deposited upon an older foreland, while the younger Appalachians are a similar deposit derived from the older Appalachian welt.

It is not possible to extend this brief sketch interpreting the structure of the Canadian Shield in terms of continental growth, but it is only fair to attempt to answer Umbgrove's principal objection to it (Umbgrove, 1947, pp. 241-50). Apparently, he considered that it was necessary to have primary basaltic continents from which to form more acid ones by the processes described. He rightly pointed out that isostasy would have prevented such continents from emerging above the sea. In the theory presented here this difficulty has been avoided by considering that the sialic continents did not form from more basic ones but grew in the oceans around groups or arcs of volcanic islands too small initially to have been in isostatic balance. It is therefore considered that Umbgrove's objection is not necessarily valid.

The theory of continental growth needs development, but this paper will have served its purpose if it encourages consideration of that theory by geologists. What has been emphasized here is not exposition or proof of the theory but rather a removal of some of the difficulties previously preventing its consideration. Attention has been devoted to showing that the evidence from astronomy, from meteorites, and from seismology does not make such a theory untenable. On the contrary, it is considered to be the most reasonable theory for the formation of continents and the only one which can explain Precambrian structures as well as the sialic composition of continental rocks. An effort has been made to indicate that the production, during geological time, of the required sialic material from which to form continents is reasonable. It has been suggested that the granite-gneiss of which continental platforms are chiefly built is a product of orogeny and that both great thicknesses of sediments and an underlying crust of basic rocks are involved. Heat and the products of differentiation released from below first contribute volcanic material for the formation of the sediments and then serve during failure to consolidate them into granite-gneiss.

It is considered that the theory of continental growth offers interesting possibilities which can be checked in the Canadian Shield. The idea was first suggested by Lawson, whose early training was there. No other place seems better suited for the study of this fundamental problem in geology.

It is not claimed that this theory is yet proved to be correct and it has certainly not been worked out in detail. Changes in it and in the deductions from it will certainly be required, but further attempts are being made to see whether it is not possible to establish a hypothesis of continental growth as a uniform and general theory. If it is entirely wrong it should be easy to disprove, but in that case it is hoped that the discussion will aid a little in establishing a better theory.

It is believed that any attempts to do this are justified, because it is apparent that what the science of geology greatly lacks is a general idea or theory of the behaviour of the earth. Without it splendid pieces of detailed work lack cohesion, while general works upon geology remain essentially descriptive. Until more of the fundamental problems discussed by Adams (1947) can begin to be answered, the science of geology will remain immature and its pursuit will be lacking in direction.

But this present perplexity need not long remain unresolved. The great accumulation of sound geological observations made during the past century, which is now being extended by new techniques into hitherto unexplored parts of the earth, will make possible unparalleled progress in interpreting the history of the globe as soon as the fundamental laws are discovered, by which the development of the earth proceeds. When we consider the brilliant flowering of results achieved in other sciences by the discovery or invention of sound theory, we may hope for the same clarification of our own science and a correspondingly rich harvest from the labours of our predecessors.

REFERENCES

ADAMS, L. H. (1947). Some unsolved problems in geophysics. Trans. Amer. Geoph. Union, **28:** 673-9.

AHRENS, L. (1949). Measuring geologic time by the strontium method. Bull. Geol. Soc. Amer., **60:** 217-66.

BAUER, C. (1949). Lectures at University of Toronto.

BROWN, H. and PATTERSON, C. (1947 and 1948). The composition of meteoritic matter. Geol., **55:** 405-11, 508-10; **56:** 85-111.

BUCHER, W. H. (1933). The deformation of the earth's crust. Princeton.

BUDDINGTON, A. F. (1943). Some petrological concepts and the interior of the earth. Amer. Mineral., **28:** 119-40.

BULLEN, K. E. (1940). The problem of the earth's density variation. Seis. Soc. Amer. Bull., **30:** 235-50.

COOKE, H. C. (1931). Studies of the physiography of the Canadian Shield. Trans. Roy. Soc. Can., **25**, sec. IV: 127-81.

EARDLEY, A. E. (1947). Paleozoic Cordilleran geosynclines and related orogeny. J. Geol., **55:** 309-42.

ELLSWORTH, H. V. (1932). Rare-element minerals of Canada. Econ. Geol. Series No. 11, Geol. Surv. of Canada.

ESCHER, B. G. (1949). Origin of the asymmetrical shape of the earth's surface and its consequences upon volcanism of the earth and moon. Bull. Geol. Soc. Amer., **60:** 353-62.

GILL, J. E. (1948). Structural geology of Canadian ore deposits, p. 29. 50th Anniversary vol., Can. Inst. of Min. and Met. Montreal.

GRIGGS, DAVID (1939). A theory of mountain building. Amer. J. Sci., **237:** 609-60.

HILLS, G. F. S. (1947). The formation of the continents by convection. London.

HODGSON, J. H. (1947). Analysis of travel times from rock bursts at Kirkland Lake, Ontario. Bull. Seis. Soc. Amer., **37:** 5-17.

HOLMES, A. (1949). Lead isotopes and the age of the earth. Nat., **163:** 453-6.

——— (1946). Principles of physical geology. New York.

HURLEY, P. M. (1949). Age of Canada's principal gold-bearing belt. Science, **110:** no. 2845, 49-50.

JEFFREYS, H. (1948). The origin of the solar system. Monthly Notices Roy. Astron. Soc., **108:** 94-103.

——— (1929). The earth. (2nd ed.) Cambridge.

JOLLIFFE, A. W. (1948). The northwestern part of the Canadian Shield. Vol. of titles and abstracts, p. 93. Inter. Geol. Cong., London.

KAY, M. (1948). Paleozoic North American geosynclines and island arcs. Vol. of titles and abstracts, p. 93. Inter. Geol. Cong., London.

LAKE, PHILIP (1931). Island arcs and mountain building. Geog. J., **78:** 149-60.

LAWSON, A. C. (1932). Insular arcs, foredeeps and geosynclinal seas of the Asiatic coast. Bull. Geol. Soc. Amer., **43:** 353-81.

LEITH, C. K., LUND, R. J. and LEITH, A. (1935). Precambrian rocks of the Lake Superior region. U.S. Geol. Surv. Prof. Paper 184, p. 10.

MEINESZ, F. A. VENING (1948). Major tectonic phenomena and the hypothesis of convection currents in the earth. Quart. J. Geol. Soc. London, **103:** 191-207.

MINTROP, L. (1949). On the stratification of the earth's crust according to seismic studies of a large explosion and of earthquakes. Geophysics, **14,** no. 3: 321-35.

RITTMAN, A. (1939). Über die Herkunft der Vulkanischen Energie und die Entstehung des Sials. Geol. Rundschau 30.

SHAND, S. J. (1947). Eruptive rocks. (3rd ed.) London.

SHAPLEY, H. (1945). On the astronomical dating of the earth's crust. Amer. J. Sci., **243-A** (Daly volume): 508-22.

SMITH, F. G. and PEACH, P. A. (1949). Recent advances in the laboratory study of ore. Can. Min. Metal. Bull., **42:** 351-3.

STONELEY, ROBERT (1948). The continental layers of Europe. Bull. Seis. Soc. Amer., **38:** 263-74.

TUVE, M. A. (1948). Annual report of the director of the Department of Terrestrial Magnetism, Carnegie Institution of Washington, Year Book No. 47, pp. 60-5.

———, GORANSON, R. W., GREIG, J. W., ROONEY, W. J., DOAK, J. D. and ENGLAND, J. L. (1948). Studies of deep crustal layers by explosive shots. Trans. Amer. Geoph. Union, **29:** 772.

TYRRELL, J. B. (1898). Report of the Doobaunt, Kazan and Ferguson Rivers. Geol. Surv. of Canada, Ann. Rept. 1896, pp. 53-5.

UMBGROVE, J. H. F. (1947). The pulse of the earth, 2nd ed. The Hague.

URRY, W. D. (1949). Significance of radioactivity in geophysics: thermal history of the earth. Trans. Amer. Geoph. Union, **30:** 171-9.

WHIPPLE, F. L. (1941). Earth, moon and planets, p. 250. Philadelphia.

WILLMORE, P. L. (1949). Seismic experiments on the north German explosions, 1946 to 1947. Phil. Trans. Royal Soc. London, ser. A, **242:** 123 51.

WILSON, JAMES T. and BAYKAL, O. (1948). Crustal structure of the North Atlantic basin as determined from Rayleigh-wave dispersion. Bull. Seis. Soc. Amer., **38:** 41-53.

WILSON, J. TUZO (1949a). An extension of Lake's hypothesis concerning Mountain and island arcs. Nature. **164:** 147-8.

——— (1949b). Some major structures of the Canadian shield. Can. Min. Metal. Bull., **42,** no. 451: 547-54.

WILSON, MORLEY E. (1941). Precambrian. 50th Anniv. vol. Geol. Soc. Amer., pp. 269-306.

2: ATMOSPHERIC AND HYDROSPHERIC EVOLUTION ON THE PRIMITIVE EARTH

Preston E. Cloud, Jr.

This discussion focuses on the interactions that necessarily took place between biospheric, atmospheric, lithospheric, and hydrospheric evolution on the primitive earth and perhaps the moon. How can evidence and conjecture about each of these different kinds of evolution limit or illuminate hypotheses about the others, and how can all such lines of thought be integrated to bring us closer to a consistent and plausible model of early terrestrial events?

A salient and long-appreciated aspect of the terrestrial atmosphere is its great depletion in the noble gases, relative to their cosmic abundances (*1*). This seems to require the conclusion that the atmosphere as we know it is of secondary origin. Either the earth originated without a primary atmosphere, or such an atmosphere was mainly lost in a subsequent thermal episode.

Actually, an internal source for our atmosphere as a result of gradual, episodic, or rapid volcanic outgassing and weathering was proposed long before the depletion in noble gases was recognized. The Swedish geologist Högbom (*2*) suggested this in 1894, and his countryman Arrhenius (*3*), in 1896. Later Chamberlin (*4*) recognized a post-accumulational generation of the atmosphere from occluded gases as a necessary consequence of his and Moulton's planetesimal hypothesis for the origin of the earth. Rubey's critical and comprehensive assessment (*5*) of possible sources for the volatiles that comprise the atmosphere and hydrosphere has led to wide acceptance of the concept of accumulation of both from juvenile sources.

As far as the atmosphere is concerned, dispute focuses on the composition and time of origin of the primitive atmosphere and the original proportions and changes in the proportions of O_2, N_2, CO_2, and H in it.

The atmosphere cannot be older than the earth. Hence an outside limit on its time of origin is the time of origin of the earth. Much evidence has been adduced in recent years, based on the ages of meteorites (*6*, *7*) and on isotopic composition of terrestrial leads (*7*, *8*), to suggest that both meteorites and leads were involved in some kind of homogenization event about 4.6 × 10^9 years ago (10^9 years is hereafter referred to as an aeon). Small excesses of xenon-129 in some meteorites (*9*) support the inference that this event closely approximates the time of origin of the solar system and hence the approximate time of formation of the earth. Such an interpretation, however attractive and highly probable as it seems, is not the

The author is professor of geology, University of California, Los Angeles. This article is adapted from a paper presented 27 December 1967 at the New York meeting of the AAAS, as a part of the Lloyd V. Berkner Memorial Symposium of the Section on Astronomy. The last half is also a greatly condensed and evolved version of a 1967 Sigma Xi National Lecture and of a longer paper, "Pre-metazoan evolution and the origins of the Metazoa," in *Evolution and Environment* (Yale University Press, New Haven, in press).

only possible one. On the one hand, available evidence does not completely eliminate the possibility that the earth as a solid body in the general form we know it might be as much as an aeon younger than the parts from which it accumulated (*10*). On the other hand, Baranov (*11*) suggests ways of looking at the lead isotope data which *might* require as much as 7.5 aeons for the radiogenic lead isotopes to reach current crustal ratios.

A minimal age for the earth is, of course, given by the most ancient minerals dated. This age appears to be around 3.5 to 3.6 aeons. Zircons from gneisses in Minnesota show uranium-lead isotope ratios which, on a concordia curve, indicate an age of 3.5 aeons (*12*). Amphibolites from the central Ukraine are said to give potassium-argon ages exceeding 3.5 aeons (*13*). Granites which are apparently not the oldest in South Africa (*14*) yield rubidium-strontium ages of 3.2 aeons or more, and some pegmatites give whole-rock ages as great as 3.4 aeons (*15*), although the larger figures are very uncertain. The basement complex of the Congo is said to yield lead-isotope ages approaching 3.3 to 3.4 aeons (*16*). And in Western Australia, rubidium-strontium ages indicate that granites and greenstones were metamorphosed 2.7 aeons ago (*17*) and probably originated 3 aeons or more ago (*17, 18*). Thus the earliest dates so far obtained in the rather intensively studied older parts of North America and Europe are remarkably close to each other and are approached by the ages of the oldest dated rocks in the Southern Hemisphere. Although the record is distressingly sparse, it underscores the question: What was happening for the first aeon or so of the earth's history, if accumulation of the earth was completed not long after the homogenization event that took place approximately 4.6 aeons ago? We will return to that question, and to the fact that a later marked uranium-lead differentiation, presumably denoting a significant episode in crustal evolution, is dated at about 3.5 aeons (*7, 19*)—coincidentally, about the same age as that suggested by the regrettably sparse record of dated minerals in the most ancient rocks.

The immediate point is that, inasmuch as we do not have rocks known to be older than about 3.5 aeons, geology can as yet do little to place limitations on conjecture about the nature and time of origin of an atmosphere or hydrosphere that might have existed before that time. Indeed, these oldest dated rocks have only an indirect bearing on the questions before us, and the ages obtained for them may well date only an early episode of metamorphism and not their time of primary origin.

Table 1. Stromatolites of large amplitude.

Approximate amplitude (m)	Approximate age (10^9 years)	Geologic unit	Locality
5	> 1.1	Belt Series, Helena Dolomite	Helena, Montana
6	(Proterozoic)	Otavi Series, Abenab Formation	Kombat Mine, South-West Africa
2.5–3	> 2	Dolomite Series	Boetsap, South Africa
2.5–3	1.5–1.7	Paradise Creek Formation	Northwest Queensland

Some Limitations on Conjecture

What are the oldest rocks that tell us something reasonably unequivocal about the existence and nature of the contemporaneous hydrosphere and atmosphere? They are sedimentary rocks, considered from several lines of converging evidence to have a minimum age of about 3 aeons in South Africa and Swaziland (*15*), to be more than 2.7 aeons old in Minnesota (*20*), and mainly younger than 2.5 to 2.7 aeons at other places. These rocks consist of water-laid detrital and chemical sediments that clearly could not have originated without the prior existence of atmospheric weathering and a substantial hydrosphere (see *21* for questionable older records).

As Holland (*22*), Abelson (*23*), Cloud (*24*), and others cited by them have pointed out, these rocks and associated minerals also place limitations on conjecture about the composition of the atmosphere beneath which they were deposited. Apparently detrital uraninite and pyrite at various localities combine with more esoteric geochemical evidence to indicate that the atmosphere before 1.8 to 2 aeons ago could have contained little or no free oxygen. The abundance of bedded chemical silicates, representing the chert component of the banded iron formation, and the relative rarity of limestone and dolomite among the oldest sedimentary rocks also imply that there could have been little ammonia in the atmosphere of those times. For ammonia would raise the *p*H of the hydrosphere and favor an abundance of carbonate rocks and a rarity of bedded chert. Abelson (*23*), moreover, has shown that, in an atmosphere devoid or nearly devoid of free oxygen, the dissociation of methane, had it been abundant, should have given rise to a rain of carbon of nonvital origin; yet carbon, although not uncommon, is not a conspicuous component of the oldest sediments. Thus it may be concluded that the atmosphere, from about 3 aeons onward at least, was not, as some have suggested, rich in methane or ammonia. The sedimentary record, rather, supports Rubey's conjecture (*5*) that the early atmospheric gases would be those which are occluded in igneous rocks, or which on other grounds are recognized as juvenile components of volcanic and hot-spring gases. These are H_2O, CO_2, CO, N_2, SO_2, HCl, and a few other trace gases.

Possible Timing and Significance of Postaccumulational Melting

How far back might such an atmosphere go? How would its accretion have become initiated? And what can we postulate about its evolution from the known records of biospheric and lithospheric evolution? Here we will become involved with some paleontologic and geologic evidence bearing on the appearance of the moon in orbit around the earth, not because the evidence for a postaccumulational melting episode depends on this, but because there could be a relation between these phenomena and it is interesting to examine that possibility.

In order for an atmosphere and hydrosphere to begin to accrete on an earth which, during accumulation, probably had an outer temperature of no more than a few hundred degrees Kelvin, some significant part of the outer shell or mantle of that earth would have had to undergo melting sufficient to start release of its volatiles to the surface. Such melting might have resulted from either or both of two postulated thermal episodes. One is the combined effect, early in the earth's history, of the normal evolution of radiogenic energy and the conversion of gravitational energy to heat. The other is the generation of heat from tidal friction incident to lunar capture (*25–28*), abetted by the two other factors.

Direct geologic evidence, we have seen, as yet tells us little about the nature of the earth and its enveloping hydrosphere and atmosphere before about 3 aeons ago. However, if we could find among the older rocks or fossils some that provided clues as to tidal amplitudes, this information would have a bearing on whether or not the moon might have been in orbit at any particular time. Such data, then, would constitute a test as to whether it would be consistent with the geologic record to hypothesize that lunar capture might have stimulated an early pulse of melting and outgassing, as Singer (*27*) has suggested.

Such fossils and rocks do, in fact, exist. They include the variously shaped but often domal sedimentary structures of algal origin known as stromatolites, plus various cross-bedded and desiccation-cracked sediments indicative of extensive tidal-flat deposition. The stromatolites are the most impressive evidence bearing on tidal amplitude for rocks up to somewhat more than 2 aeons old.

Stromatolites in process of formation are known from recent intertidal and shallow subtidal environments of Western Australia (*29, 30*), the Bahama Islands (*30, 31*), and Bermuda (*32*). From the few records known in modern seas, it seems that subtidal stromatolites characteristically take the form of undulating crusts and small nodules. Stromatolites that rise conspicuously above the surface on which they grow have so far been found only in the intertidal environment, where they reach a maximum known relief among recent forms of 0.7 meter in Western Australia. Logan (*29*) found that the height at maturity was "determined by the tidal range and the position of the structure in the intertidal zone." If this is true generally it would give us important clues to tidal ranges in the geologic past.

We are not entirely limited to analogy with present structures. Stromatolites in the geologic past are commonly associated with interstitial breccias and oölites suggesting turbulent waters, and with contraction-cracked sediments and truncated flat-topped ripple marks implying exposure to the atmosphere. Indeed some stromatolitic domes from 2-aeon-old rocks in South Africa (Schmidts Drift, northeastern Cape Province) show polygonal patterns of desiccation cracking on their own planed-off upper surfaces. The geologic record is thus consistent with their characteristic habitat being an intertidal one. It is also a feature of the geologic record that stromatolites of pre-Paleozoic and even early Paleozoic age often have far greater heights than do younger ones, and over areas far too extensive to represent abnormal local tidal ranges. If stromatolites, however, are to give us useful clues to ancient tidal amplitudes, we must be careful about what we measure. Some stromatolitic domes build up through the rock section. We should measure only the demonstrable height of a particular unit structure above a related bedding surface, or, better yet, the amplitude of a particular upwardly convex and complete stromatolitic lamina above the surrounding contemporary sea floor. If the characteristic habitat and manner of growth are correctly understood, this then implies, for such stromatolites, that tidal amplitude at a given time and place was at least as great as the amplitude of individual stromatolitic laminae.

What are some of these stromatolite amplitudes? I had observed that stromatolites of large amplitude were widely distributed in the pre-Paleozoic long before it occurred to me that it might be significant to keep a careful record of such measurements. A great many pre-Paleozoic stromatolites, however, have amplitudes, above the growing surface, of a meter or more. Four occurrences that show amplitudes from 2.5 to 6 meters or more and have stratigraphic associations that convey little likelihood of locally exceptional tidal ranges are listed in Table 1. These stromatolites are found in dolomites and limestones ranging in age from about 1 to more than 2 aeons, and it seems likely, from my own field reconnaissance and scanning of published accounts, that concerted effort would reveal other, similar records in this age range on all continents where such rocks occur. In addition, extensive deposits of muddy siltstone and shale of the same general age show polygonal patterns of desiccation cracking and truncated ripple marks indicative of equally extensive tidal-flat deposits (the Belt and Grand Canyon Series, for example). The giant individual stromatolite domes in the Belt Series near Helena, Montana, are illustrated by Knopf (*33*, p. 85), who describes them as "up to 15 feet thick [that is, in amplitude] and traceable for thousands of feet along the strike . . . from the bottom to the top of the Helena dolomite," a unit more than 1200 meters thick.

The implication I draw from this and other records is that tidal amplitudes too great to be accounted for by the solar component alone existed at least 2 aeons ago, that tides generally were probably greater then than they are now, and that the moon was, therefore, at that time already in orbit and presumably somewhat closer to the earth than it is now. This timing, of course, is contrary to some theoretical expectations and to recent reports, such as that of Olson (*34*), suggesting that sedimentary textures in younger rocks call for lunar origin in more recent times.

Stromatolites of large amplitude (about 2 meters) in ankeritic dolomite at Steeprock Lake, Ontario, may give a record of large tides of even greater age. Although not radiometrically established, geologic relations suggest that the age of these rocks may approach 2.5 aeons. The evidence of the approximately 3-aeon-old (*15*) rocks of the Swaziland System is even more tenuous, but also consistent with the moon's then being already in orbit around the earth. This succession of rocks, beginning with a typical, and the oldest known, marine ophiolitic and eugeosynclinal succession, is terminated by the Moodies Series of characteristic molasse facies (*35*). This includes an abundance of markedly cross-bedded and ripple-marked sandstone and conglomerate sheets, and desiccation-cracked shales. The association implies extensive intertidal deposits and again points to tidal amplitudes too great to be explained by the solar component alone (although, like most of these older pre-Paleozoic records, it needs more study from this aspect).

If the moon was in orbit 3 aeons or more ago, and if it did not originate concurrently with the accumulation of the earth, perhaps its origin had something to do with the thermal event of 3.5 to 3.6 aeons ago suggested by the oldest known metamorphic and granitic rocks and the lead isotope composition of various minerals (Fig. 1).

If our moon was captured, as is suggested by its density (*36*, p. 143) and the inclination of its orbit (*36*, p. 124), tidal friction sufficient to induce subcrustal melting would have been likely (see *26*, pp. 772–780, for a review of the theories; *36–38* for general background). Indeed, the theoretical problem is how to avoid vaporization of the earth and concomitant disintegration of the moon. Singer (*27, 28*) has suggested how such events might be circumvented, by postulating capture in a prograde rather than a retrograde orbit. This would extend the initial spin period of the earth to

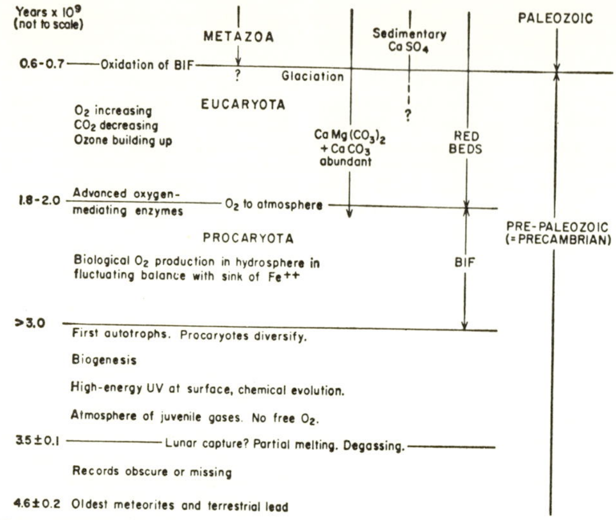

Fig. 1. Postulated main features of interacting biospheric, lithospheric, and atmospheric evolution on the primitive earth.

about 5 hours (rather than less than 1 hour) but still produce enough melting to generate a considerable concurrent accretion of atmosphere and hydrosphere. Likely related consequences for lithospheric evolution would include large-scale volcanism, plutonism, and metamorphism of preexisting cosmic debris (and of any preexisting volcanic and sedimentary rocks) as well as the general resetting of nuclear decay series.

Thus the limited chronology of the oldest rocks, and the hints from lead isotope data of the concomitant beginning or peaking of internal differentiation of the earth, can be seen as consistent with a near approach of the moon to the earth about 3.5 to 3.6 aeons ago. What little is known about lunar evolution also seems to be consistent with such an interpretation (although by no means proof of it). Just as the oldest terrestrial rocks, and conceivably the initiation or climaxing of core and mantle formation, may be related to processes incident to lunar capture, so also may the surfaces of the lunar maria, indicating as they do, from observations by Shoemaker (*38*), a welling-up from beneath to a "nearly common level." What would account for such a widespread welling-up to fill the presumably collisional maria and some smaller preexisting impact craters (for example, Archimedes) to "an approximately hydrostatic level" (*38*)? This would seem to require both a major episode, or a sequence of related episodes, of partial internal melting and a widely interconnected series of conduits along which molten or fluidized particulate matter could travel to such a nearly general level. In the absence of a core or other evidence of major internal differentiation of the moon, tidal friction provides a source of heat for fluidization otherwise difficult to visualize. In addition, stresses incident to capture might well account for an extensive interconnected lunar fracture system, along which lava, ash, or both, could ascend [as well as the impacting masses that produced the maria (*27*, *28*)]. Difficulties arise in trying to explain the preservation of large-scale relief in the lunar highlands, the absence of surface offsets due to faulting, and the lesser relief within the maria themselves. If a major melting episode did occur as postulated, it could not have been a general liquefaction of the interior of the moon nor could it have been narrowly synchronous at all places. Something more like a pervasive partial melting is called for. Interconnecting fractures along which effusive matter would have traveled outward cannot have had large offsets. And results of statistical treatments suggesting high relief within the maria (*39*) would require some other explanation. These difficulties circumscribe the model envisaged. Like it, they must undergo further critical examination, hopefully with more abundant and better evidence.

The rocks we now know cannot of themselves tell us *how* the moon originated. Among hypotheses available to choose from, however, the testimony that can be wrung from the rocks is consistent with Singer's suggested prograde lunar capture and not consistent with an origin of the moon in relatively late pre-Paleozoic time. This testimony can also be interpreted as suggesting that, however the moon originated, a near approach of moon to earth may have taken place closer to 3.5 to 3.6 aeons ago than to the 4 to 4.5 aeons suggested by Singer from tenuous geochronological inferences. And, after all, a considerable pre-mare (or pre-Imbrian) lunar history would be in order to account for the extensive lunar highlands whose infrapositional relations to debris seemingly splashed from the maria imply a significantly greater age for these highlands (*38*).

A special problem that arises in connection with a date as late as 3.5 to 3.6 aeons ago for lunar capture is that of where the moon could have been stored for a whole aeon if the earth is about 4.6 aeons old. Getting things out of the asteroid belt is no longer as fashionable as it once was, but at least one piece of evidence points to such a source for the moon. This unique planetary body, with density intermediate between that of the inner and outer planets, is the most inconsistent item in the distribution of planetary densities (*36*, p. 143) unless it did originate in the asteroid belt. If it did, and if it could be transferred from the asteroid belt to the vicinity of the earth, the storage problem would vanish.

If these conjectures should happen to be correct (and they certainly have their shaky aspects), then, when samples of the mare fillings are obtained from beneath any superficial covering debris of cosmic or local origin, such samples should give radiometric ages of about 3.5 to 3.6 aeons. Another likely corollary is that any preexisting terrestrial atmosphere and hydrosphere (*21*) would have been largely lost at that time and a new or first atmosphere and hydrosphere started, with a sizable input of volatiles. At the beginning this would probably have been no more than about 10 percent of the present mass of the atmosphere and hydrosphere, however, on the basis of Rubey's calculations showing that no more than about 10 percent of the present hydrosphere can be retained as gaseous H_2O in equilibrium with even a generally molten earth. Nevertheless, a prolonged episode of accelerated degassing could have led to fairly rapid growth of terrestrial atmosphere and hydrosphere in the following few tens or hundreds of millions of years.

It is also conceivable that the postulated near approach of moon to earth and accompanying thermal episode might have given rise to a temporary lunar atmosphere and hydrosphere. Whether this might have lasted long enough for life to evolve on the moon is a question that cannot be settled until lunar sedimentary rocks are found and examined for microfossils. If such microfossils ever are found, they might well be similar to the early pre-Paleozoic microfossils now being found on earth. Thus the development of terrestrial paleomicrobiology may offer the best foundation for the search for life in lunar materials.

Inferences from Biospheric and Lithospheric Evolution

Such reflections suggest that the present terrestrial atmosphere and hydrosphere may have evolved, with additions, from one that began or underwent great increase as a result of a major thermal episode about 3.5 to 3.6 aeons ago (Fig. 1). Life as we know it, then, would have had to originate within the next few hundred million years, if we accept as evidence of once-living things the structures described by Pflug (*40*) and by Barghoorn and Schopf (*41*) from the Swaziland System. (In Fig. 1, *biogenesis* denotes simply the origin of life in the strict etymological sense of the word and not any particular theory of origin. Pirie has used the term *biopoesis* for this, to avoid the connotation that *biogenesis* once held—of life only from preexisting life. It no longer seems helpful, however, to give up a good term like *biogenesis* merely because of historical precedent, and there is now no other reason to do so.)

In any case, first life would presumably have been anaerobic in the absence of oxygen and heterotrophic (that is, dependent on external food sources) short of a highly improbable coincidence. But it could not have continued, to give rise to the observed evolutionary record, without the emergence of an organism that could manufacture its own substance—an autotroph. However many unsuccessful starts there may have been, it is clear that one start eventually did give rise to an autotroph. But what kind? Since the biochemical complexity of the chemoautotrophs suggests a late origin for them, it was probably a photoautotroph—an organism that manufactures its food by photosynthesis. And photosynthesis is the likely process whereby oxygen might eventually appear and be generated fast enough to oxidize most of the reduced substances of the hydrosphere, atmosphere, and surface of the earth and begin to accumulate as a free gas.

Only one of the three known types of photosynthesis, however, releases free oxygen, and it is unlikely that *it* was primitive. Nevertheless, oxygen-releasing photosynthesizers did arise, and when they did they would have faced the problem of disposing of oxygen in such a way as not to burn themselves up. Unless advanced oxygen- and peroxide-mediating enzymes, therefore, arose simultaneously with, or preceded, the origin of oxygen-releasing green-plant photosynthesizers, such organisms would probably have been dependent on an associated oxygen acceptor in the physical environment.

The abundance of hematitic banded iron formation (BIF in Fig. 1) among sediments deposited between about 3 and 1.8 to 2 aeons ago suggests that the available oxygen acceptor may have been the ferrous ion. Banded iron formation is a rhythmically banded chemical sediment of large, open water bodies that takes different aspects but most characteristically consists of alternating layers of iron-rich and iron-poor silica. Nothing like it of any thickness or regional extent is found in younger rocks; and younger (but not time-restricted) types of sedimentary iron deposits appear to be mainly replacement bodies of a quite different sort. The geochemical problem of the banded iron formation has been that of explaining transport of the iron in solution under oxidizing conditions, or precipitation of the iron under anoxidizing conditions. This problem, as well as that of an oxygen acceptor for the primitive green-plant photosynthesizers, is largely resolved by the concept of a balanced relationship between organisms and banded iron formation. The iron could be transported in solution in the ferrous state and precipitated as ferric or ferro-ferric iron upon combining with biological oxygen. The rhythmic banding could result from a fluctuating balance between oxygen-producing biotas and supply of ferrous ion. The facies of iron formation can be visualized as incidental products of this regime. The oxygen presumably became locked in chemical sediments precipitated from the hydrosphere and did not appear in the atmosphere except for small quantities from photolytic dissociation of H_2O and CO_2, rapidly scavenged by reduced substances then abundant in the atmosphere and at the surface of the lithosphere.

Eventually, however, efficient oxygen- and peroxide-mediating enzymes did arise, and when that happened the postulated balance would have collapsed. Green-plant photosynthesizers, equipped with such enzymes, could then spread as widely through the hydrosphere as light penetration and high-energy ultraviolet (UV in Fig. 1) shielding mechanisms would permit. The hydrosphere would be swept free of ferrous ion in a last great episode of banded iron formation. And O_2 would accumulate in excess in the hydrosphere and begin to evade to the atmosphere. The last great episode of banded iron formation, about 1.8 to 2 aeons ago, may mark such an event, and blue-green algae are known to have been associated with it.

What would happen when O_2 began to evade to the atmosphere? At that time, in the absence of an ozone screen, ultraviolet light in the range of 2000 to 2900 angstroms would impinge on the surface of the earth, some of the molecular oxygen (O_2) would be converted to atomic oxygen (O) and ozone (O_3), and iron would be retained in the weathering profile of the earth in the ferric state. Because of the high chemical activity of O and O_3, even a low rate of evasion of O_2 to the atmosphere should give rise to a high rate of oxidation of surface materials. Red beds—detrital continental or marginal marine sediments in which the individual grains are coated with ferric oxides—should appear in abundance in the geological column at that time.

Available records indicate that the oldest thick and extensive red beds are about 1.8 to 2 aeons old—a little younger than, or overlapping slightly with, the youngest banded iron formation. The evidence of lithospheric evolution suggests that this may mark the time in atmospheric evolution when free O_2 began to accumulate, perhaps following the appearance of advanced oxygen- and peroxide-mediating enzymes in biospheric evolution.

This would also set the stage for the appearance of a new type of cell and organism. Paleontological evidence (*24*) implies that until then all organisms were procaryotes, having no nuclear wall or clearly structured chromosomes, and being incapable, therefore, of mitotic cell division and sexual reproduction in the usual sense. The presence of free oxygen, even in small quantities, was presumably followed by the evolution of the eucaryotic cell, with nuclear wall, well-defined chromosomes, mitotic cell division, and the capacity for sexual reproduction. (The oxygen requirements of a single eucaryotic cell, to be sure, are much more easily satisfied than those of a differentiated multicellular animal.)

How fast did O_2 accumulate in the atmosphere once it began? Probably slowly at first. The green-plant photosynthesizers would still be restricted by high-energy ultraviolet radiation to protected sites in stromatolite-forming sedimentary mats, or at locations where they would not be circulated into surface waters, until such time as ozone built up to a sufficient density to exclude DNA-inactivating radiation in the neighborhood of 2600 angstroms. Berkner and Marshall (*42*) found that this happens when the atmospheric level of oxygen reaches about 1 percent of the present level. In addition, both they and, earlier, the Canadian biologist Nursall (*43*) suggested that the appearance of the Metazoa (differentiated multicelled animal life) was a consequence of the reaching of atmospheric oxygen concentrations sufficient to support a metazoan level of oxidative metabolism.

Certainly the appearance of the Metazoa in the geologic record does signify the previous fulfillment of two necessary if not sufficient preconditions. One is the origin of the eucaryotic cell, of which all metazoans are constituted. The other is a sufficient level of free oxygen —although perhaps closer to 3 percent of the present atmospheric level than to 1 percent (better data on this are needed). Now the oldest rocks in which eucaryotic fossils are known are probably more than 0.7 aeon old (*44*), although the minimal conditions necessary for primitive eucaryotes may have appeared much earlier (*24*). At least the precondition of the eucaryotic cell, therefore, was satisfied well before the dawn of the Paleozoic; and I have elsewhere (*24*) provided documentation for the conclusion that there are as yet no records of unequivocal Metazoa in rocks of undoubted pre-Paleozoic age.

This, then, can be interpreted as suggesting that the precondition of sufficient free oxygen may have been the triggering event. The necessary compression of early metazoan evolution may be partially explained by postulating a polyphyletic origin. This can be visualized as a wave of multicellularization affecting different pre-metazoan ancestors almost simultaneously in the geological sense. Since, moreover, all ecologic niches that could ever be occupied by Metazoa were unoccupied when the Metazoa first arose, adaptive radiation probably contributed to a geologically rapid diversification of the metazoan root stocks. This may have taken place over an interval of, say, 100 million years, more or less—perhaps somewhat less than the time required for chemical evolution leading to the origin of life itself, and somewhat more than that required for the Cenozoic diversification of the mammals following extinction of the dinosaurs.

Is there evidence other than the geologically rapid evolution of the earliest Metazoa at this time to suggest that the dawn of the Paleozoic might indeed have closely followed the appearance of a level of free O_2 adequate for metazoan metabolism? First consider what might happen when ozone reached a level sufficient to exclude the DNA-inactivating ultraviolet radiation at about 1 percent of the present atmospheric level of O_2. As Berkner and Marshall recognized (*42*), this would open up the surface waters of the entire hydrosphere to occupation by photosynthesizing phytoplankton, and such occupation, in turn, could lead to rapid and large increase in the amount of O_2 in the atmosphere. Such a step-increase in atmospheric O_2 might be related to three features in lithospheric evolution — late pre-Paleozoic glaciation, oxidative enrichment of the banded iron formation to produce high-grade ores, and the appearance of thick and extensive deposits of sedimentary calcium sulfate in the geologic record.

Inasmuch as the increase in O_2 would presumably have been paralleled by a decrease in long-radiation-reflecting CO_2, this could have led to a temperature decrease that might have triggered the widespread late pre-Paleozoic glaciation often suggested to account for certain rocks of that age having resemblance to deposits of modern and Pleistocene glaciers (another matter needing more study).

Enrichment of the emerged banded iron formation by surface oxidation during late pre-Paleozoic or earliest Paleozoic time converted the lean primary deposits into the bonanza ores that were first mined in the Lake Superior region. This event also would be consistent with a substantial contemporaneous increase in the amount of oxygen in the atmosphere, and it may have coincided with similar oxidative enrichments in other regions to denote a broadly contemporaneous episode of surface oxidation and laterization in late pre-Paleozoic or earliest Paleozoic time.

The record of sedimentary calcium sulfate may also be consistent with the postulated late pre-Paleozoic increase in O_2. Thick and extensive deposits of gypsum and anhydrite require large amounts of sulfate ion, and the only likely adequate sources are from the oxidation of sulfides and sulfurous volcanic gases (SO_2, SO_3, H_2S). Sedimentary calcium sulfate is abundant from the base of the Paleozoic onward, yet occurrences of appreciable volume appear to be rare in, or absent from, all but perhaps the uppermost part of the pre-Paleozoic (*24*). Anhydrite in the Grenville, for instance, could be a later replacement of a tabular marble body (*45*). The most likely candidates yet proposed for thick and extensive pre-Paleozoic sedimentary sulfates are deposits in the Shaler Group of arctic Canada (*46*), although they appear to be very late pre-Paleozoic indeed (*47*, p. 55). On balance it seems that the appearance of abundant sedimentary sulfates in the geologic record may precede by only a little the origin of the Metazoa, and that both, together with oxidative enrichment of the iron protores and a reputed episode of glaciation, could be partial consequences of a relatively large-scale increase of free O_2.

The relative abundance of limestone and dolomite among the younger pre-Paleozoic rocks, as contrasted to their near absence from sequences older than about 2 aeons, may, in turn, reflect a gradual diminution of CO_2 and increased *p*H of the hydrosphere, accompanying the relatively slow buildup of O_2 in the atmosphere after it began to accumulate but before ozone reached effective ultraviolet-screening levels.

These are the kinds of things we can

postulate about the composition and evolution of the primitive atmosphere by calling on the combined evidence of biospheric and lithospheric evolution.

The Hydrosphere

As for the hydrosphere, there is little to add to what Rubey (*5*) has already so brilliantly developed. On grounds of the occasional occurrence of glauconite in rocks as old as 1.2 to 1.5 aeons (*48*)—glauconite being a mineral characteristic of the diagenesis of various siliceous and claylike parent materials in contact with potassium-rich solutions—the hydrosphere is presumed to have had saline components at least that long ago. The prevalence of dolomite among rocks as old as 2 aeons also implies normal marine to hypersaline environments at least that far back. The evidence of the older rocks is less clear; but the occurrence of sedimentary siderite, sulfide-rich sediments, bedded chert, and so on, is consistent with the presence of a hydrosphere not drastically different in its pH and mix of dissolved salts from the present one. The Eh, at least before about 2 aeons ago, was probably negative or neutral, however, and the pH may have been slightly lower than now. Sulfate ion was apparently present in solution only in quantities too small to account for much $CaSO_4$ until the latest pre-Paleozoic.

Evidence that might suggest nonsaline seas includes the interesting fact that, among the commoner elements of the microbiotas now being found in rocks of various ages in the pre-Paleozoic, many are morphologically close to living freshwater forms. Such organisms, however, are characterized by a combination of morphological conservatism and ecological plasticity. Similar living forms tolerate a wide range of salinity. Their pre-Paleozoic ancestors probably lived in salt water. The iron formation itself, however, has been taken as evidence of a freshwater environment. That is because iron in modern marine waters occurs only in vanishingly small quantities. Under anoxygenous conditions, though, dissolved ferrous ion could have been, and evidently was, relatively common in older pre-Paleozoic seas of approximately normal salinity, and probably of a range of pH and temperature not greatly different from the range for younger seas.

There remains the question of the configuration of the pre-Paleozoic hydrosphere. If, as suggested, it accumulated gradually, while the continents and ocean basins were still differentiating, the main part of it may in early times have occupied isolated basins of intermediate depth rather than deep interconnected ones, such as comprise the present world ocean. If this was the case, the sequence and timing of events postulated might have varied from one basin to another. The record as now understood does not support such a variation, but differences in the range of tens of millions of years are not yet resolvable in pre-Paleozoic time. Among the many unfinished jobs ahead for geologists, one of the most important is to map and date the shorelines and to estimate the physical structure and environments of the pre-Paleozoic sedimentary basins. When such information is available over very large regions we should be able to make better informed guesses as to the configurations, volumes, tidal range, and history of the evolving early hydrosphere and its relation to other parts of the primitive earth.

References and Notes

1. N. Russell and H. Menzel, *Proc. Nat. Acad. Sci. U.S.* **19**, 997 (1933); H. Brown, in *The Atmospheres of the Earth and Planets*, G. P. Kuiper, Ed. (Univ. of Chicago Press, Chicago, ed. 2, 1952), pp. 258–266.
2. A. G. Högbom, *Svensk. Kem Tidskr.* **6**, 169 (1894) [see Arrhenius (*3*), pp. 269–273, for summary].
3. S. Arrhenius, *Phil. Mag.* **41**, 237 (1896).
4. T. C. Chamberlin, *J. Geol.* **5**, 653 (1897).
5. W. W. Rubey, *Bull. Geol. Soc. Amer.* **62**, 111 (1951).
6. E. Anders, in *The Moon, Meteorites, and Comets*, vol. **4**, *The Solar System*, V. M. Middlehurst and G. P. Kuiper, Eds. (Univ. of Chicago Press, Chicago, 1963), pp. 402–495.
7. C. Patterson, *Geochim. Cosmochim. Acta* **10**, 230 (1956). C. Patterson, in *Isotopic and Cosmic Chemistry*, H. Craig, S. L. Miller, G. J. Wasserburg, Eds. (North-Holland, Amsterdam, 1963), pp. 244–268.
8. G. R. Tilton and R. H. Steiger, *Science* **150**, 1805 (1965).
9. J. H. Reynolds, *Sci. Amer.* **203**, 171 (1960).
10. L. T. Silver, oral communication, December 1967.
11. V. J. Baranov, *Soviet Astron.—AJ* (*English Trans.*) **10**, No. 5, 860 (1967).
12. E. J. Catanzaro, *J. Geophys. Res.* **68**, 2045 (1963); S. S. Goldich, E. G. Lidiak, C. E. Hedge, F. G. Walthall, *ibid.* **71**, 5389 (1966).
13. N. P. Semenenko, *UNESCO Bull. Intern. Hydrol. Decade* **3**, No. 3, 23 (1967); A. J. Tugarinov, oral communication, 9 April 1965.
14. A. E. J. Engel, "The Barberton Mountain Land—Clues to the Differentiation of the Earth," *Univ. Witwatersrand Econ. Geol. Res. Unit Inform. Circ. No. 27* (1966).
15. L. O. Nicolaysen, in *Petrologic Studies—A Volume to Honor A. F. Buddington*, A. E. J. Engel, H. L. James, B. F. Leonard, Eds. (Geological Society of America, Denver, 1962), pp. 569–598; H. L. Allsopp, T. J. Ulrych, L. O. Nicolaysen, paper presented at the University of Alberta in 1967 at a conference of the International Union of Geological Sciences, Geochronology Commission.
16. W. L. Donn, B. D. Donn, W. G. Valentine, *Bull. Geol. Soc. Amer.* **76**, 287 (1965); R. Furon, *Geology of Africa* (Oliver and Boyd, Edinburgh, 1963), p. 5.
17. A. F. Wilson, W. H. Compston, P. M. Jeffery, G. H. Riley, *J. Geol. Soc. Australia* **6**, 179 (1960). Some refinements of these numbers are reported by W. Compston and P. A. Arriens in a paper presented at the University of Alberta in 1967 at a conference of the International Union of Geological Sciences, Geochronology Commission.
18. R. P. Dunn, K. A. Plumb, H. G. Roberts, *J. Geol. Soc. Australia* **13**, 593 (1966).
19. M. Tatsumoto and C. Patterson, *J. Geol.* **72**, 232 (1964).
20. P. E. Cloud, Jr., J. W. Gruner, H. Hagen, *Science* **148**, 1713 (1965).
21. A. O. Woodford [*Historical Geology* pp. 204–205 (Freeman, San Francisco, 1965)] and Donn, Donn, and Valentine (see *16*, pp. 292–293) have summarized available data on reportedly 3.4 to 3.5×10^9-year-old associated "sedimentary" and plutonic rocks in the Kola Penninsula, U.S.S.R. The validity of the older dates from the Kola Penninsula, however, is much disputed, both in the U.S.S.R. and the United States, and the nature of the "sedimentary" rocks and their relations to the dated rocks is not clear from descriptions available. Similarly, Semenenko (*13*) refers to the most ancient Ukranian (amphibolite) schists as "normal geosynclinal complexes of sedimentary volcanogenic formations." And Donn, Donn, and Valentine (*16*, pp. 291–92) report that the undoubtedly sedimentary Sebakwian Series of Rhodesia is younger than an unconfirmed date of 3.44 aeons obtained by the potassium-argon technique on micas from an associated pegmatite. Should it turn out that one or more of these interpretations (and the associated ages) is correct, and that there exist sedimentary or truly metasedimentary rocks that predate, or were metamorphosed by, the postulated thermal event of 3.5 to 3.6 aeons ago, the prior existence of an atmosphere or hydrosphere will be demonstrated. Such an atmosphere and hydrosphere, however, could probably not have been more than a small fraction of the present one and might well have been lost as a result of the thermal event. There is much need for further study and supplementation of the relevant data.
22. H. D. Holland, in *Petrologic Studies—A Volume to Honor A. F. Buddington*, A. E. J. Engel, H. L. James, B. F. Leonard, Eds. (Geological Society of America, Denver, 1962), pp. 447–477.
23. P. H. Abelson, *Proc. Nat. Acad. Sci. U.S.* **55**, 1365 (1966).
24. P. E. Cloud, Jr., in *Evolution and Environment*, E. T. Drake, Ed. (Yale Univ. Press, New Haven, in press).
25. G. J. F. MacDonald, *Science* **145**, 881 (1964).
26. ———. *Ann. N.Y. Acad. Sci.* **118**, 739 (1965).
27. S. Singer, *The Origin and Dynamical Evolution of the Moon* (American Astronautical Society, Tarzana, Calif., 1966), pp. 66–192.
28. ———, *Sci. Res.* **1967**, 69 (Nov. 1967).
29. B. W. Logan, *J. Geol.* **69**, 517 (1961).
30. ———, R. Rezak, R. N. Ginsburg, *J. Geol.* **72**, 68 (1964).
31. C. Monty, *Ann. Soc. Geol. Belg. Bull.* **88**, 269 (1965).
32. C. D. Gebelein, paper presented at the annual meeting of the Geological Society of America, 1967.
33. A. Knopf, *Amer. J. Sci.* **255**, 81 (1957).
34. W. S. Olson, *Amer. Scientist* **54**, 454 (1966).
35. C. R. Anhaeusser, C. Roering, M. J. Viljoen, R. P. Viljoen, "The Barberton Mountain Land—A Model of the Elements and Evolution of an Archean Fold Belt," *Univ. Witwatersrand Econ. Geol. Res. Unit Inform. Circ. No. 38* (1967).
36. H. D. Urey, in *Space Science*, D. P. Le Galley, Ed. (Wiley, New York, 1963), pp. 123–168.
37. G. P. Kuiper, in *Advances in Earth Science*, P. M. Hurley, Ed. (M.I.T. Press, Cambridge, Mass., 1966), pp. 21–70.
38. E. Shoemaker, *Sci. Amer.* **211**, 38 (Dec. 1964).

39. E. R. DeFresne, *Astrophys. J.* **124**, 638 (1956).
40. H. D. Pflug, "Structured Organic Remains from the Fig Tree Series of the Barberton Mountain Land," *Univ. Witwatersrand Econ. Geol. Res. Unit Inform. Circ. No. 28* (1966).
41. E. S. Barghoorn and J. W. Schopf, *Science* **152**, 758 (1966).
42. L. V. Berkner and L. C. Marshall, *Discussions Farady Soc.* **37**, 122 (1964); *Proc. Nat. Acad. Sci. U.S.* **53**, 1215 (1965).
43. J. R. Nursall, *Nature* **183**, 1170 (1959).
44. E. S. Barghoorn and J. W. Schopf, *Science* **150**, 337 (1965); P. E. Cloud, Jr., G. R. Licari, L. Wright, L. T. Silver, in preparation.
45. A. E. J. Engel, letter of 5 October 1966.
46. R. Thorsteinsson and E. L. Tozer, "Banks, Victoria, and Stefansson Islands, Arctic Archipelago," *Geol. Surv. Can. Mem. 330* (1962).
47. G. B. Leech, J. A. Lowdon, C. H. Stockwell, R. K. Wanless, "Age Determinations and Geologic Studies," *Geol. Surv. Can. Paper 63-17* (1963).
48. R. A. Gulbrandsen, S. S. Goldich, H. H. Thomas, *Science* **140**, 390 (1963).

49. This article had the advantage of critical review by Leon T. Silver, S. F. Singer, Louis Slichter, J. H. Mackin, and W. W. Rubey. I am particularly obliged to Silver for critical discussion of the radiometric ages and their limitations, as well as the implications of the oldest rocks and meteorites. I wish also to acknowledge a special indebtedness to W. W. Rubey, G. E. Hutchinson, J. B. S. Haldane, Harold Urey, and Lloyd Berkner, whose pioneering researches have provided so much of the foundation and inspiration for my own efforts in this field. Preparation of the article was supported by grant No. NGR-05-007-169 from the National Aeronautics and Space Administration.

3: THE BEGINNING OF CONTINENTAL EVOLUTION

George W. Wetherill

Department of Planetary and Space Science, University of California, Los Angeles, California

ABSTRACT

Wetherill, G.W., 1972. The beginning of continental evolution. In: A.R. Ritsema (Editor), *The Upper Mantle. Tectonophysics*, 13(1–4): 31–45.

Lunar studies carried out in connection with the Apollo program permit considerable understanding of the early history of this planetary body. The resulting thermal history involves primary fractionation of the outer half of the lunar mass by accretional energy followed by subsequent formation of the mare basalts by radioactive heating at the boundary between the fractionated and unfractionated regions.

Application of this same model to the earth implies the entire earth was initially melted and geochemically fractionated. Review of the geological and geochronological evidence for the most ancient rocks indicates that not until 3400 m.y. ago did the earth cool sufficiently to permit the formation of extensive areas of stable crust, and that this is the reason more ancient rocks have not been found more abundantly.

This same line of reasoning implies that there is a class of planetary bodies intermediate in mass between the earth and the moon, possibly including Mars, which are less magmatically active than either the earth or moon, whereas the tectonic activity of others, possibly Venus, resembles that of the early earth.

INTRODUCTION

Since the work of Patterson (1956) it has been known that the age of the earth (4600 ± 100 million years) was considerably greater than that of the oldest (~ 2800 m.y.) dated terrestrial rocks (e.g. see compilation by Ahrens (1955)). Although at the time it seemed quite possible that many older rocks would be found, this appears to be increasingly unlikely.

During the last 15 years the development and widespread application of modern geochronological methods have opened the Precambrian to detailed investigation: the distinction can now be made between the relatively young (e.g. 1000 m.y. old) granitic basement complexes and the truly ancient 2800 m.y. old rocks which had previously been lumped together as "Archean". With a single exception these much more extensive geochronological surveys have succeeded in pushing back the age of the oldest rocks only to about 3400 m.y. and the number of well-documented occurrences of that age is small. It therefore appears likely that the scarcity of older rocks is a fact which requires explanation.

Some possible hypotheses to explain this phenomenon are:

(1) The age of the oldest rocks represent a "steady-state age", as discussed by Wasserburg (1961), and is merely indicative of the mean rate at which normal geological processes erase the record of older events.

(2) The earth was formed as a cold planet and was heated slowly by long-lived radioactivity. About 3500 m.y. ago, melting began. The sinking of iron to the core released sufficient additional energy to initiate the major chemical fractionation of the earth into crust, mantle, and core. The oldest rocks of the crust necessarily postdate this event.

(3) The capture of the moon took place about 3500 m.y. ago. The close approach of the earth and moon at that time resulted in the release of sufficient tidal energy in the earth's interior to destroy the record of previous earth history.

(4) The earth was formed at a high temperature and did not cool sufficiently to preserve extensive stable crust until 3500 m.y. ago.

During the last 2 years, measurements made on samples of rock returned from the moon, together with geophysical measurements performed on the moon, have permitted us to understand for the first time the early history of an observed planetary body. While this understanding is not complete, it is sufficiently advanced to permit new insights into the early history of the earth, and other planets of the solar system.

In this paper, I shall review the evidence obtained from terrestrial rocks, discuss the early history of the moon, as revealed by lunar studies, and interpret the terrestrial data in the light of this new knowledge. This discussion will also lead to some inferences regarding what one may reasonably expect to find on other planets.

THE OLDEST ROCKS OF THE EARTH'S CRUST

On every continent there are large contiguous areas ($\sim 10^6$ km^2) underlain by igneous and metamorphic rocks greater than about 2400 m.y. in age. Typically, the age of these rocks is 2600–2800 m.y. Altogether, these areas cover about 5% of the earth's land surface (Hurley and Rand, 1969). Within many of these areas, there are smaller regions in which rocks as old as 3900 m.y. have been reported.

Although the term has sometimes been used in different ways, and consequently could lead to misunderstanding, it is convenient to refer to these > 2400 m.y. areas as *cratons.*

There is a general belief that these cratons have become stabilized and are relatively invulnerable to rejuvenation by younger tectonic events. This is certainly not always the case: the 2700 m.y. rocks in Wyoming and Montana are in a region of the North American continent which has been profoundly affected by recent tectonic processes, presumably resulting from the overriding of an East Pacific plate by the North American–West Atlantic plate. Furthermore, the presence of cratons on the margins of present-day continents should expose them to the tectonic processes associated with such margins: development of trenches and subduction zones and collision of continental lithospheric masses. Whether or not this present exposure has characterized their past history is related to the question of whether or not Laurasia and Gondwanaland existed as super-continents until their recent breakup during the Mesozoic. The fact that cratons are frequently surrounded by not too much younger (~ 2000 m.y.) mobile belts, rather than by rocks of arbitrary age, suggests that at least in some cases they have been protected in some way from rejuvenation, although the exact nature of this protection is at present an open and important question.

In a few cases the existence and age of > 3000 m.y. rocks within the cratons is fairly well established; in other cases such certainty is not possible. At least in some cases the reported occurrence of > 2800 m.y. old rocks has been shown to be incorrect. On the other hand, evidence is abundant that these older rocks have been affected by the later events which occurred about 2700 m.y. ago, and it is plausible that sometimes these later events have completely obscured the evidence for the older age of the rock.

Probably the best documented case for the existence and nature of rocks greater than 3000 m.y. in age is in the Barberton Mountain Land on the boundary between the Transvaal and Swaziland in southern Africa. The oldest rocks comprise a greenstone terrane consisting of low grade metavolcanic and metasedimentary rocks, which have many lithologic and chemical similarities to greenstone belts found within cratons on other continents (Anhaeusser et al., 1969). The stratified rocks are divided into a lower Swaziland System, overlain by the upper Moodies System. The Swaziland System is in turn divided into the principally volcanic Onverwacht Series, overlain by the predominantly sedimentary Figtree Series. These stratified rocks are surrounded and intruded by great masses of granitic rocks, some of which have been named and correlated (Anhaeusser et al., 1969).

The 2920 m.y. age found for the post-Swaziland Kubuta pegmatite (Nicolaysen, 1954, Aldrich et al., 1958)* supported the geological inference that these were indeed very ancient rocks. The first clear evidence that most of these rocks were more than 3000 m.y. old was given by Allsopp et al. (1962). He obtained a Rb–Sr whole rock isochron age of 3070 m.y. for granites designated by Hunter (1957) as "G4", i.e., a younger granite series intrusive into post-Swaziland gneisses and migmatites. Although only four samples were measured, they were highly enriched in radiogenic strontium, and fell very accurately on the isochron.

More recently, Oosthuyzen (1970) has made an extensive geochronological study of granitic rocks from the Transvaal which post-date the Swaziland and Moodies Systems. His most interesting results are those obtained by isotopic U–Pb measurements on zircon, sphene, and apatite. His data for two of these granites, Salisbury Kop and Dalmein, are shown in Fig.1.

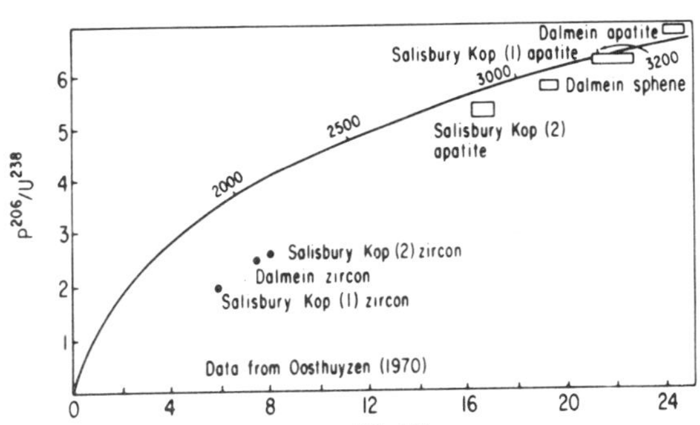

Fig.1. Concordia plot of U–Pb measurements on minerals separated from ~ 3250 m.y. old granites of southern Africa.

* These, as well as all other Rb-Sr ages discussed in this paper, are calculated using a decay constant of $1.39 \cdot 10^{-11}$/year.

As usual, these minerals yield discordant U–Pb ages. However, when the analyses are plotted on a concordia diagram (Wetherill, 1956) they are most plausibly interpreted as ~ 3250 m.y. minerals which have lost lead at some subsequent time in their history. This implies that the granites and the greenstone sequence intruded by the granites are at least this old, and possibly significantly older. Oosthuyzen has obtained similar U–Pb ages for other granites in this region. Although the Rb–Sr data on these granites yield a somewhat younger age (2990 m.y.), presumably because of the effect of later events, Allsopp (1961) has obtained a high quality Rb–Sr whole rock isochron of 3200 m.y. on granite near Johannesburg, about 300 km to the east, thereby providing further evidence for ~ 3200 m.y. granites in this area.

The exact age of the older greenstone sequence is less clear. Van Niekerk and Burger (1969) have published isotopic data for a single zircon separate from the quartz porphyry phase of the Onverwacht Series, the lower series of the Swaziland System. The U–Pb ages are discordant; the $^{207}Pb/^{206}Pb$ age of 3360 m.y. is interpreted as the minimum age for the formation of these volcanic rocks. Some further evidence for the existence of 3400–3500 m.y. rocks in this area has been based on the isotopic composition of lead separated from sulfide minerals in the Onverwacht lavas (Van Nieke k and Burger, 1969), whole rock lead from these lavas (Oosthuyzen, 1970) and lead from galenas from gold mines in the Barberton region (Ulrych et al., 1967). These latter ages are strongly model-dependent, and consequently it should be concluded that although it is quite likely that the greenstone sequence is 3400–3500 m.y. old, this age remains to be definitely established.

Although the evidence is less clear, it is rather probable that portions of other cratons are equally old. Data suggesting ages > 3000 m.y. have been reported for the Aldan craton of Siberia (Rudnik et al., 1969), the Fennoscandian shield of northern Europe (Sobotovich et al., 1963), western Australia (Leggo et al., 1965), India (Crawford, 1969) and Antarctica (Halpern, 1970). In some cases these ages are based on measurements on lead minerals, and are thus model-dependent; in other cases the experimental and geological uncertainties in the determinations preclude definite assignment of these ages to the > 3000 m.y. range. In none of these cases does the evidence approach the completeness of the African measurements discussed previously.

The evidence for rocks of this age in North America is somewhat puzzling.

In the Canadian Shield there are many occurrences of greenstone belts resembling those found on other continents and there has been a considerable quantity of geochronological work done in these areas, particularly on the granitic rocks intruding these metasedimentary and metavolcanic rocks. Despite considerable effort and careful study, none of these rocks have been found to be older than about 2800 m.y. (Gast et al., 1958; Aldrich and Wetherill, 1960; Moore et al., 1960; Osborne, 1964; Van Schmus, 1965; Wanless et al., 1966; Wanless et al., 1967; Green et al., 1968; Purdy and York, 1968; Stockwell, 1968; Wanless et al., 1968.)

On the other hand Goldich et al. (1970) have reported ages of 3550 m.y. for granitic gneisses in the Minnesota River Valley, not obviously in a greenstone terrane, although the exposure of the rocks is not good and their limited exposure might prevent observing the entire geological picture. The best evidence for this great age is obtained by isotopic measurements on zircons, which on a concordia diagram lie along a straight line between 1850 m.y. and 3550 m.y. This high value for the lower intercept is itself not usual, on the other hand it offers no firm basis for rejecting the data. Indeed it is what one might expect as a consequence of the known events in this region of the continent about 1850 m.y. ago.

Although Rb–Sr whole rock ages up to 3800 m.y. are reported, these data are not conclusive, as they are in disagreement with one another, and fail to define an isochron.

Recently Black et al. (1971) have reported apparently convincing ages of 3980 ± 170 m.y. in West Greenland.

Further work is needed in these areas to confirm or disprove this evidence for what could prove to be the oldest surviving terrestrial rocks.

[*Editor's Note:* Material has been omitted at this point.]

IMPLICATIONS WITH REGARD TO THE EARLY HISTORY OF THE EARTH

The proposed thermal history of the moon, invoking in an essential way gravitational heating accompanying rapid accretion, has obvious implications with regard to the thermal history of the earth.

If it is assumed that the earth and moon accreted in the same region of the solar system, then it is possible to estimate the initial temperature distribution of the earth. Under this assumption the time scale for the earth and the moon will be of the same order of magnitude. If the rate of accretion is proportional to the surface area, then the radius of the accreting body will increase linearly, and the accretion of the earth will take four times as long as that of the moon. However, the radius which is significant in accretion is the gravitational radius r_g:

$$r_g = R\sqrt{1 + U_e^2/U^2} \qquad (2)$$

where R is the physical radius; U_e is the escape velocity; U is the geocentric velocity of the accreting material. Because the escape velocity also increases with R, the accretion of the earth will proceed very rapidly during its terminal stages, and it is therefore a satisfactory assumption that the earth and moon accreted "competitively" in essentially the same period of time.

Therefore melting of the earth would also take place when its radius was about 1500 km, and eq.1 would then imply that material accreted at its final radius of 6400 km would be at a temperature of about 20,000° K. Actually, it would probably not be this hot, because once the melting point was reached mixing would be much more rapid, and the entire body would tend to have a more uniform temperature of about 6000° K. These extremely high temperatures would suffice to melt the earth throughout, and would not leave a central undifferentiated region as was the case for the moon. Chemical differentiation would proceed very rapidly, producting a surface region about 50 km thick highly enriched in radioactive elements. As may be seen from Fig.3, this surface layer will be sufficiently thick to maintain a steady-state temperature of $> 1500^\circ$ at its base, assuming an average radioactive element content equivalent to that of alkalic basalts. Furthermore, only about 10^8 years would be required to achieve this steady state. The actual temperature will be significantly higher because of heat flow from the melted interior. As the surficial regions of the interior cools to the melting point, the temperature in the radioactive layer will decrease, and the depth at which melted rock occurs will become progressively deeper. After $\sim 5 \cdot 10^8$ years the loss of heat sources due to radioactive decay will also become a significant factor in lowering temperatures near the surface.

At the present time, the depth of incipient melting in the earth is about 100 km, and this presumably is the depth at which the rocks are sufficiently weak to permit large scale motions of lithospheric plates. Again, assuming the radioactive heat production of an alkalic basalt to represent a lithospheric average, the rate of heat production would be three times as great as $3.5 \cdot 10^9$ years ago, and it seems likely that the depth of incipient melting would have been about 30 km at that time.

Geological studies of the greenstone belts (Anhaeusser et al., 1969; Martin, 1969) have provided additional reasons for believing that these represent the remnants of small, thin, and unstable lithospheric plates. Such thin plates would be especially vulnerable to rifting and being destroyed as a consequence of being carried down into the region of melting by subduction as well as by bombardment by planetesimals similar to these which produced the lunar maria. The formation and destruction of such thin plates probably characterized the surface of the earth during the first $\sim 10^9$ years of earth history, after which time the diminished heat production resulting from the decay of ^{40}K and ^{235}U permitted the growth of sufficiently thick sialic blocks to more often survive these processes and remain "afloat" while adjacent regions, analogous to the present ocean basins, participated more actively in these tectonic processes.

Further evidence that the primary geochemical fractionation of the earth took place early in its history is provided by the work of Hart and Brooks (1969) on the evolution of the isotopic composition of Sr in the earth's mantle. These measurements yield a $^{87}Sr/^{86}Sr$ ratio of 0.7017 ± 0.0002 for 2700 m.y. old archean metavolcanics and 0.7026 ± 0.0002 for modern basalts from oceanic ridges. Assuming, as is reasonable from meteoritic and lunar studies, that the initial $^{87}Sr/^{86}Sr$ ratio of the earth was essentially equal to that of the basaltic achondrite meteorites, i.e. 0.6990, this implies that the $^{87}Sr/^{86}Sr$ ratio of the source of the basalts increased by 0.0027 during the first 1900 m.y. of earth and only 1/3 as much during the remaining 2700 m.y. While these results do not lead to a unique model for the evolution of the mantle Rb/Sr ratio, they are consistent only with models in which mantle depletion in Rb relative to Sr took place prior to the formation of the most abundant ancient rocks, i.e. those of about 2700 m.y. in age.

In summary, the evidence is accumulating for the earth being very hot during the first 10^9 years of its history. This conclusion is not entirely tied to a particular model for the accretion of the earth and moon. For example, the alternative "precipitation hypothesis" of lunar origin (Ringwood, 1970) also requires an earth which was initially melted throughout. Therefore, although these ancient greenstone belts do not actually represent "original crust", they are the oldest crustal material which has been able to survive the active tectonism of early earth history. From this point of view they come closest to the "vestige of a beginning" that we are likely to find. More extensive and complete geological, geochemical, and geochronological studies of these regions should permit us to go beyond uniformitarianism and gain understanding of an earth quite different in many ways than the one we see today. It is an interesting question as to whether or not all of these original crustal blocks should be of nearly the same age. The present, admittedly fragmentary, evidence suggests they are not, but may range from 3900 m.y. to 2800 m.y. in age. Whether or not this represents a problem for the ideas discussed here is worthy of considerable thought.

EARLY HISTORY OF OTHER PLANETS

This same line of reasoning leads to conclusions regarding the tectonic development of other planets.

Consider first planets formed at approximately the same distance from the sun as the earth and moon, e.g. Mars and Venus. Planetary bodies significantly smaller than the moon (e.g. 1300 km in radius) will undergo insufficient gravitational heating to reach the melting point, and will heat slowly as a consequence of their approximately uniform concentration of long-lived radioactivity. For lunar concentrations of radioactive elements this source will fail to provide sufficient heat to differentiate the body; however, after ~ 1000 m.y. the temperature could rise sufficiently to permit redistribution of strontium isotopes and loss of inert gases. The absence of meteorites exhibiting this thermal history argues either that bodies that have undergone this thermal history do not exist in our solar system, or that the somewhat uncertain processes which place meteorites into earth-intersecting orbits have failed to provide them to us. Heating by electrical induction should be most effective for these smaller planets (Sonett and Colburn, 1968) and may have resulted in their complete differentiation $4.6 \cdot 10^9$ years ago, in spite of their lower gravitational energy of accretion.

In planets larger than the moon, gravitational melting will involve an ever larger fraction of the planetary mass, resulting in complete melting at a radius of about 2500 km. For such planets, the heat carried to the central region by sinking iron and high temperature silicates will be sufficient to melt and chemically fractionate the previously undifferentiated core. Planets of this size will have their radioactivity concentrated in a surface layer too thin to maintain steady-state melting beneath the surficial radioactive layer, and will also be greatly depleted in internal sources of heat. Although the deep interiors of such planets will still remain at or slightly below the melting point, the absence of internal heat sources will preclude significant tectonic or magmatic activity following their initial differentiation. Mars may be such a planet. Larger planets, such as the earth and Venus will have a surficial layer of radioactivity sufficiently thick to maintain high temperatures at its base, which together with the increase in melting point due to pressure and the additional gravitational energy released by differentiation, will continue to be tectonically active throughout their history. The high surface temperatures of Venus of about 500° C may be expected to bring the depth of incipient melting closer to the surface of that planet. It is quite likely that the tectonic activity of Venus is still characterized by low relief and "thin plate tectonics", as was the case for primitive earth.

For planets accreting at greater distances from the sun it may be expected that accretion times will be longer. According to the collision formula of Öpik (1951), either in its original form or modified for collisions of bodies in eccentric orbits (Wetherill, 1967), the collision probability varies essentially as the inverse cube of the distance from the sun, and directly proportional to the mass of material in orbit at that distance.

These circumstances will lead to very low initial temperatures for bodies forming in the asteroid belt, unless some alternative heating mechanism was present. In the case of Jupiter, the greater distance from the sun will be offset by the greater mass in orbit at that distance, and accretion times for that planet should be similar to the earth, resulting in extremely high initial temperatures for that planet. If the Galilean satellites of Jupiter accreted "competitively" with Jupiter, thermal and tectonic histories similar to either the moon or Mars may be anticipated.

The smaller mass and greater distance of the other major planets will result in longer accretion times and lower initial temperatures for these planets and their satellites. Smaller planets, such as possibly Pluto and other undiscovered bodies in the outer solar system could well require 10^8 years for their formation, and gravitational heating during accretion will be of negligible importance. At these great distances, electrical heating by the solar wind will also be of minor importance, and radioactive heating will probably play the major role in the differentiation of these bodies.

Many of the statements of this paper are obviously speculative. However, it is important to recognize the relevance of lunar and planetary studies to our understanding of the earth, and it may be expected that future discoveries will permit discussions of this kind to be made on a much firmer basis.

ACKNOWLEDGEMENT

This work has been supported by NASA Grant NGL 05-007-005.

REFERENCES

Ahrens, L.H., 1955. Oldest Rocks Exposed. *Geol. Soc. Am., Spec. Pap.*, 62: 155–168.

Aldrich, L.T. and Wetherill, G.W., 1960. Rb–Sr and K–A ages of rocks in Ontario and northern Minnesota. *J. Geophys. Res.*, 65: 337–340.

Aldrich, L.T., Wetherill, G.W., Davis, G.L. and Tilton, G.R., 1958. Radioactive ages of micas from granitic rocks by Rb–Sr and K–A methods. *Trans. Am. Geophys. Union*, 39: 1124–1134.

Allsopp, H.L., 1961. Rb–Sr age measurements on total rock and separated-mineral fractions from the Old Granite of the Central Transvaal. *J. Geophys. Res.*, 66: 1499–1508.

Allsopp, H.L., Roberts, H.R., Schreiner, G.D.L. and Hunter, D.R., 1962. Rb–Sr age measurements on various Swaziland granites. *J. Geophys. Res.*, 67: 5307–5313.

Anhaeusser, C.R., Mason, R., Viljoen, M.J. and Viljoen, R.P., 1969. A reappraisal of some aspects of Precambrian shield geology. *Bull. Geol. Soc. Am.*, 80: 2175–2200.

Arnold, J.R., 1965. The origin of meteorites as small bodies, 3. General considerations. *Astrophys. J.*, 141: 1548–1556.

Asylum, L., 1970. Mineralogic and isotopic investigations on lunar rock 12013. *Earth Planet. Sci. Lett.*, 9: 137–163.

Black, L.P., Gale, N.H., Moorbath, S., Pankhurst, R.J. and McGregor, V.R., 1971. Isotopic dating of very early Precambrian amphibolite facies gneisses from the Godthaab District, West Greenland. *Earth Planet. Sci. Lett.*, 12: 245–259.

Cliff, R.A., Lee-Hu, C. and Wetherill, G.W., 1971. Rb–Sr and U, Th–Pb measurements on Apollo 12 material. *Proc. Apollo 12 Lunar Sci. Conf., Suppl. 2, Geochim. Cosmochim. Acta*, in press.

Crawford, A.R., 1969. Reconnaissance Rb–Sr dating of the Precambrian rocks of southern peninsular India. *J. Geol. Soc. India*, 10: 117–166.

Dohnanyi, J.S., 1969. Collisional model of asteroids and their debris. *J. Geophys. Res.*, 74: 2531–2554.

Gast, P.W., Kulp, J.L. and Long, L.E., 1958. Absolute age of early Precambrian rocks in the Bighorn Basin of Wyoming and Montana, and Southeastern Manitoba. *Trans. Am. Geoph. Union*, 39: 322–334.

Gault, D.E., 1970. Saturation and equilibrium conditions for impact craters in the Lunar Surface. *Radio Sci.*, 5: 273–291.

Gault, D.E. and Wedekind, J.A., 1969. The destruction of tektites by micrometeoroid impacts. *J. Geophys. Res.*, 74: 6780–6794.

Goldich, S.S., Hedge, C.E. and Stern, T.W., 1970. Age of the Morton and Montevideo gneisses and related rocks, Southwestern Minnesota. *Bull. Geol. Soc. Am.*, 81: 3671–3695.

Gopalan, K., Kaushal, S., Lee-Hu, C. and Wetherill, G.W., 1970. Rb–Sr and U, Th–Pb ages of lunar materials. *Proc. Apollo 11 Lunar Sci. Conf., Suppl. 1, Geochim. Cosmochim. Acta*, 2: 1195–1205.

Green, D.C., Baadsgaard, H. and Cumming, G.L., 1968. Geochronology of the Yellowknife area, Northwest Territories, Canada. *Can. J. Earth Sci.*, 5: 725–735.

Halpern, M., 1970. Rb–Sr date of possibly 3 billion years for a granitic rock from Antarctica. *Science*, 169: 977–978.

Hart, S.R. and Brooks, C., 1969. Rb–Sr mantle evolutionary models. *Carnegie Inst. Wash. Ann. Rep. Dep. Terr. Mag.*, 1968–1969: 426–429.

Hays, J.F., 1971. Radioactive heat sources in the lunar interior. *Preprint.*

Huneke, J.C., Papanastassiou, D.A., Podosek, F.A., Tera, F., Turner, G. and Wasserburg, G.J., 1971. Age determinations on samples from the Apollo 14 landing site. *Proc. Seattle COSPAR Meeting, Space Research*, XII, in press.

Hunter, D.R., 1957. The geology, petrology, and classification of the Swaziland granites and gneisses. *Trans. Geol. Soc. S. Afr.*, 60: 85–125.

Hurley, P.M. and Rand, J.R., 1969. Pre-drift continental nuclei. *Science*, 164: 1229–1242.

Leggo, P.S., Compston, W. and Trendall, A.F., 1965. Radiometric ages of some Precambrian rocks from the Northwest Division of Western Australia. *J. Geol. Soc. Aust.*, 12: 53–65.

Martin, H., 1969. Problems of age relations and structure in some metamorphic belts of southern Africa. *Geol. Assoc. Can. Spec. Pap.*, 5: 17–26.

Moore, J.M., Hart, S.R., Barnett, C.C. and Hurley, P.M., 1960. Potassium–argon ages in northern Manitoba. *Bull. Geol. Soc. Am.*, 71: 225–230.

Nicolaysen, L.O., 1954. *Age Determinations on African Pre-Cambrian Minerals.* Thesis, Princeton Univ., Princeton, N.J., 101 pp., not published.

Oosthuyzen, E.J., 1970. *The Geochronology of a Suite of Rocks from the granitic terrain surrounding the Barberton Mountain Land.* Thesis, Univ. Witwatersrand, Johannesburg, 110 pp., not published.

Öpik, E.J., 1951. Collision probabilities with the planets and the distribution of interplanetary matter. *Proc. R. Irish Acad.*, 54A: 164–199.

Osborne, F.F., 1964. *Geochronology in Canada. R. Soc. Can. Spec. Publ. No. 8*, Univ. Toronto Press, Toronto, Ont., 156 pp.

Papanastassiou, D.A. and Wasserburg, G.J., 1970. Rb–Sr ages from the Ocean of Storms. *Earth Planet. Sci. Lett.*, 8: 269–278.

Papanastassiou, D.A. and Wasserburg, G.J., 1971. Lunar chronology and evolution from Rb–Sr studies of Apollo 11 and 12 samples. *Earth Planet. Sci. Lett.*, 11: 37–62.

Papanastassiou, D.A., Wasserburg, G.J. and Burnett, D.S., 1970. Rb–Sr ages of lunar rocks from the Sea of Tranquillity. *Earth Planet. Sci. Lett.*, 8: 1–19.

Patterson, C.C., 1956. Age of meteorites and the earth. *Geochim. Cosmochim. Acta*, 10: 230–237.

Purdy, J.W. and York, D., 1968. Rb–Sr whole-rock and K–Ar mineral ages of rocks from the Superior Province near Kirkland Lake, northeastern Ontario, Canada. *Can. J. Earth Sci.*, 5: 699–705.

Ringwood, A.E., 1970. Origin of the moon: The precipitation hypothesis. *Earth Planet. Sci. Lett.*, 8: 131–140.

Rudnik, V.A., Sobotovich, E.B. and Terent'yev, V.M., 1969. Archean age of the oldest rocks of the Aldan complex. *Dokl. Akad. Nauk S.S.S.R.*, 188: 897–900.

Schramm, D.N., Tera, F. and Wasserburg, G.J., 1970. The isotopic composition of ^{26}Mg and limits on ^{26}Al in the early solar system. *Earth Planet. Sci. Lett.*, 10: 44–59.

Silver, L.T., 1970. Uranium-thorium-lead isotopes in some Tranquillity Base samples and their implications for lunar history. *Proc. Apollo 11 Lunar Sci. Conf., Suppl. 1., Cheochim. Cosmochim. Acta*, 2: 1533–1574.

Sobotovich, E.V., Graschenkc, S.M., Alexandruk, V.M. and Shats, M.M., 1963. Age determination of ancient rocks by lead isochron and spectral strontium methods. *Izv. Akad. Nauk S.S.S.R., Ser. Geol.*, 10: 3–14.

Sonett, C.P. and Colburn, D.S., 1968. The principle of solar wind induced planetary dynamos. *Phys. Earth Planet. Inter.*, 1: 326–346.

Sonett, C.P., Smith, B.F., Colburn, D.S., Dyal, P., Parkin, C.W., Schubert, G. and Schwartz, K., 1971. Lunar electrical conductivity profile. *Nature*, 230: 359–362.

Stockwell, C.H., 1968. Geochronology of stratified rocks of the Canadian Shield. *Can. J. Earth Sci.*, 5: 693–698.

Tatsumoto, M., 1970. Age of the moon: An isotopic study of U – Th – Pb systematics of Apollo 11 lunar - samples - II. *Proc. Apollo 11 Lunar Sci. Conf. Suppl. 1, Geochim. Cosmochim. Acta*, 2: 1595–1612.

Ulrych, T.J., Burger, A. and Nicolaysen, L.O., 1967. Least radiogenic terrestrial leads. *Earth Planet. Sci. Lett.*, 2: 179–184.

Van Niekerk, C.B. and Burger, A.J., 1969. A note on the minimum age of the acid lava of the Onverwacht Series of the Swaziland System. *Trans. Geol. Soc. S. Afr.*, 72: 9–21.

Van Schmus, R., 1965. The geochronology of the Blind River – Bruce Mines Area, Ontario, Canada. *J. Geol.*, 75: 755–780.

Wanless, R.K., Stevens, R.D., Lachance, G.R. and Rimsaite, J.Y.H., 1966. Age determinations and geological studies – K–Ar isotope ages, Report 6. *Geol. Surv. Can., Pap.*, 65–17, 101 pp.

Wanless, R.K., Stevens, R.D., Lachance, G.R. and Edwards, C.M., 1967. Age determinations and geological studies: K–Ar isotope ages, Report 7. *Geol. Surv. Can., Pap.*, 66–17, 120 pp.

Wanless, R.K., Stevens, R.D., Lachance, G.R. and Edwards, C.M., 1968. Age determinations and geological studies, K–Ar isotopic ages, Report 8. *Geol. Surv. Can., Pap.*, 67–2, *Part A*, 141 pp.

Wasserburg, G.J., 1961. Crustal history and Pre-Cambrian Time Scale. *Ann. N.Y. Acad. Sci.*, 91: 583–594.

Wetherill, G.W., 1956. Discordant uranium–lead ages I. *Trans. Am. Geophys. Union*, 37: 320–326.

Wetherill, G.W., 1967. Collisions in the asteroid belt. *J. Geophys. Res.*, 72: 2429–2444.

Wetherill, G.W., 1969. Relationships between orbits and sources of chondritic meteorites. In P. Millman (Editor), *Meteorite Research*. Reidel, Dordrecht, pp. 573–589.

Wood, John A., unpublished. Thermal history and early magmatism in the moon.

4: ARCHEAN GREENSTONE BELTS MAY INCLUDE TERRESTRIAL EQUIVALENTS OF LUNAR MARIA

D. H. Green

Department of Geophysics and Geochemistry, Australian National University, Canberra, Australia

The lower portions of the volcanic sequence of some Archaean greenstone belts include members with crystallized from ultramafic liquids extruded at the earth's surface at 1600–1650°C. These liquids are interpreted as products of 60–80% melting of their mantle source composition which implies more catastrophic conditions of mantle melting than obtained in Palaeozoic, Mesozoic or Recent crust-mantle dynamics. Such conditions may be a consequence of major impacts on the surface of the primitive earth. It is suggested that the production of the lunar maria basins was accompanied by similar impacts on the earth and that such terrestrial maria played an important role in early stages of chemical differentiation of the crust and upper mantle. An hypothesis is presented in which some Archaean greenstone belts are interpreted as very large impact scars, initially filled with impact-triggered melts of ultramafic to mafic composition and thereafter evolving with further magmatism, deformation and metamorphism to the present Archaean greenstone belts.

1. Archaean greenstone belts

An important characteristic of the earth's continental crust is the presence of regions commonly referred to as Archaean granite–greenstone terrains [1–3]. These include irregularly arcuate, synclinal regions dominated by ultramafic, basic and andesitic volcanics ('greenstone belts') which occur within inhomogeneous granodioritic gneiss. Age relationships between the volcanics and the gneisses are not clear. The regional relationships suggest that the gneisses form a basement to the greenstones but there is evidence that intrusions and doming by gneissic acid intrusions is the major cause of the synclinal and arcuate form of the greenstone belts [1, 4–7].

These regions of the earth's crust appear to have been remarkably stable for $> 2.5 \times 10^9$ yr with only thin and slightly deformed sediment cover and without great elevation or extensive erosion. The greenstone belts in many areas have suffered only very low grade metamorphism and preservation of primary minerals and igneous textures is common. In addition, the mineral assemblages of the gneisses and granites are not indicative of high pressure conditions of crystallization, and seismic refraction studies show that normal crustal seismic velocities extend to depths of 30–40 km in shield areas. Thus, these appear to be regions which attained relatively thick, stable, crustal character very early in the earth's history and which preserve evidence of these primitive processes of crustal evolution.

Recent studies of geochemistry of the volcanics of Archaean greenstone belts have emphasized their similarity to basaltic rocks of the ocean floor (particularly the low-K tholeiite magma types) or to mixtures of such rock types with the volcanic characteristic of island environments [8–11]. Such comparisons have emphasized the mantle origin for the volcanics, and the similarity to modern oceanic crustal environments. The authors have, in general, advocated the absence of continental crust or the presence of only very thin sialic plates and presented models of early Archaean geodynamics in terms of the modern theory of plate tectonics involving thin lithospheric plates, shallow subduction zones, and high geothermal gradients [8, 9].

However, a characteristic feature of plate tectonics theory is that the oceanic crust and litosphere is unstable and within relatively short geological time is returned into the mantle along subduction zones.

The oceanic crust and lithosphere undergo further differentiation resulting in island arc magmatism and the gradual accretion and modification of such island arc sequences on continental margins.

In attempting to interpret the Mesozoic and Palaeozoic geological record in terms of plate tectonics theory, the identification of rock types of the oceanic crust and of 'fossil' subduction zones usually rests on the presence of melanges of mixed sedimentary, basaltic and ultramafic sequences, commonly in blueschist and partly in eclogite metamorphic facies, of paired or highly asymmetric metamorphic and lithologic belts and of very linear belts of intense deformation and overthrusting. Thus plate tectonics does not appear to produce cuspate, synclinal troughs dominated by almost unmetamorphosed basaltic volcanics, enclosed and marginally intruded by granodioritic gneiss, and with an absence of high pressure metamorphism (of either low temperature or high temperature character). An instructive analysis of the differences between the early Proterozoic Circum-Ungava Geosyncline, with close analogies to Alpine geosynclines, and the more ancient Archaean greenstone belts is given by Dimroth et al. [12].

It may be argued that the volcanic sequences of at least some of the most ancient greenstone belts are developed on or within primitive sialic crust. Geological evidence is contradictory or not definitive [4–7], and geochronological evidence is as yet inadequate. Geochronological evidence for the existence of metamorphosed sialic (gneissic) terrains of 3.9×10^9 yr in western Greenland [12], of $\geqslant 3.5 \times 10^9$ yr in Minnesota [13] and of $\geqslant 3.4 \times 10^9$ yr in Africa (6) argues for the existence of some areas of moderately thick sailic crust at times $> 3 \times 10^9$ yr. Geochronological studies of upper members of greenstone volcanic sequences and of later clearly transgressive granodiorites and granites shows that some of these volcanic sequences are $\geqslant 3.4 \times 10^9$ yr (Swaziland, S. Africa [14, 15, 17]) $\geqslant 3.3 \times 10^9$ yr (Sebakwian and/or Bulawayan, Rhodesia [6] and $\geqslant 3.0 \times 10^9$ yr (Pilbara, W. Australia [16]). These age determinations are the present older limit on the chronological sequence of exposed surface rocks but, rather intriguingly, these overlap onto the younger limit of lunar chronology as defined by ages of returned crystalline lunar rocks. This aspect of terrestrial and lunar chronology is further discussed in a later section.

2. Implications of unique Archaean volcanism

While a number of workers [8–11] have emphasized the geochemical similarity of basaltic rocks of the Archaean greenstone belts to modern ocean floor basalts and/or to island arc volcanic sequences, it has recently become clear that in *some* Archaean greenstone belts there are ultramafic and mafic volcanics of very distinctive character which are apparently restricted to these ancient magmatic sequences and do not occur in Proterozoic, Palaeozoic or younger sequences. The distinctive character of the lowermost parts of some greenstone sequences suggests that there are important differences between the magmatic activity characteristic of the modern earth and that present, at least locally, $> 3 \times 10^9$ yr ago.

Viljoen and Viljoen [18] have presented convincing evidence for the existence of *peridotitic liquids*, extruded as pillowed lava flows, and forming a major rock type of the lowest units of the greenstone sequence in the Barberton area of Swaziland, southern Africa. The name peridotitic komatiite was given to this distinctive volcanic rock. The same rock type, distinguished by characteristic olivine quench textures and ultramafic composition, has been observed as a major component of the greenstone sequences of the Yilgarn shield of Western Australia [19] and as intrusive sills or extrusive flows in Archaean rocks in Canada [20, 21]. It has been shown experimentally [22] that the temperature of extrusion of ultramafic liquid was at least 1600–1650°C, and that the magma contained < 1% water. Compositions of peridotitic komatiite magmas are similar to estimates of the bulk composition of the upper mantle in regions acting as sources for modern basalts [23, 24]. The principal differences in $Mg/(Mg + Fe^{2+})$ and in pyroxene/olivine ratios, are consistent with derivation of the peridotitic komatiite by very high degrees of melting (60–80%) of upper mantle pyrolite leaving only olivine (Fo_{93}) as the residual phase. Less olivine-rich magmas (basaltic komatiites [18] or high-magnesian basalts [19]) which overlie or are interbedded with the peridotite flows, may be interpreted as resulting from lower degrees of melting (< 40% melting) of similar upper mantle pyrolite or possibly as derivative liquids after olivine or olivine + clinopyroxene fractionation from parental peridotitic komatiite magmas. Thus the characteristics of the distinctive ultramafic and basic magmas of some Archaean greenstone belts are consistent with melting of a mantle of composition close to that of pyrolite [23, 24] but require conditions of melting differing greatly from those operating to yield modern basalts, developed by < 40% melting under various P, T, P_{H_2O} conditions [24]. If these conclusions are correct, then we are faced with the problem of understanding how at times > 2.5–3 × 10^9 yr ago, the earth's mantle was able to generate liquids reaching the surface at 1600–1650°C and derived by 60–80% melting of their source rock. In the partial melting of a pyrolite source rock, if the degree of melting exceeds ~30%, then residual olivine and orthopyroxene crystals would be separated and 'floating' in the basaltic or picritic magma. The densities of crystals (ρ = 3.3–3.4) and liquid ($\rho \simeq 2.7$) are such that, in the earth's gravitational field, liquid and crystals will readily segregate. This is considered to act as a major constraint in modern magma genesis so that parental magmas contain < 30% olivine and are generated at temperatures of 1200–1400°C by < 40% melting of the source rock [9, 10]. It appears that to overcome this constraint, extremely rapid upward movement of a crystal + liquid diapir is necessary, so that the liquid is able to entrain residual crystals. The degree of melting may then increase to a very high percentage melting due to pressure decrease along an approximately adiabatic cooling path. If the characteristics of Archaean volcanism as exemplified by peridotitic komatiite and komatiite magmas, are due entirely to processes of internal origin in the Archaean upper mantle, then it may be inferred that convective movements within the upper mantle were very much more rapid and derived from much higher temperature regions (?' greater depths) than those characteristic of modern crust–mantle dynamics. This hypothesis encounters severe difficulties in the constraints imposed by the solidus of the upper mantle and the mechanical constraints of lateral movement of primitive lithosphere complementary to magmatic upwelling in the upper mantle.

However, recent lunar studies, arising directly from the Apollo 11–15 landings, have placed new constraints on the origin and early history of the earth–moon system. Geochronological studies on lunar samples and knowledge of the relative cratering history of the lunar surface, have led to the deduction [26] that very high rates of impact cratering prior to 3.8–4.0 × 10^9 yr ago would allow only the survival of extensive lava surfaces (such as the lunar maria) formed after 3.85 × 10^9 yr ago. The early high cratering rate and its exponential decay are attributed to a family of planetesimals in eccentric and inclined solar orbit which was swept up primarily by collision with the earth [26]. If this interpretation of the early history of the earth–moon system is correct then the evolution of the earth's surface at times > 3 × 10^9 yr may be strongly influenced by major impact events. The geochronology of lunar samples [27–30] has shown that major impacts (of size producing maria basins > 300 km in diameter) occurred as late as 3.8–3.9 × 10^9 yr ago and that filling of maria basins by volcanism occurred up to at least 3.2 × 10^9 yr ago. The production of very large impact-produced depressions on the lunar surface on this time scale would be expected to be accompanied by similar impacting of the earth.

In the following discussion a hypothesis is presented suggesting that major impacts on the primitive earth, at times > 3.0 × 10^9 yr ago, played a fundamental role in triggering a distinctive type of volcanism and producing large impact-scars filled with ultramafic and mafic volcanics ('terrestrial maria'). Whereas the lunar maria have been preserved for > 3 × 10^9 yr on the moon, the mobility of the earth's crust caused deformation, metamorphism and further evolution to yield sequences now included amongst Archaean greenstones.

3. Terrestrial maria?

It is envisaged that the primitive earth, following segregation of an iron core and some degree of chemical differentiation of the mantle, possessed a primitive crust with mineralogy and composition determined by low pressure partial melting and crystal fractionation processes (sequence peridotite → quartz tholeiite → tholeiitic andesite → rhyodacite or rhyolite), and by deformation and metamorphism of such rocks [12–14]. While the Na -depleted character of the lunar composition is consistent with anorthitic plagioclase as a major phase of the primitive lunar crust, the primitive terrestrial crust was probably dominated by intermediate and sodic plagioclase, quartz, K-feldspar and ferromagnesian minerals. Acceptance of Shoemaker's [26] model of lunar impact history would imply that it is most improbable that surface rocks of crystallization age > 3.8–4 × 10^9 yr would occur on the earth (though metamorphic or whole rock Rb/Sr 'ages' greater than this could occur, cf. the 4.5 × 10^9 yr 'age' for lunar regolith samples [27]). Also it is *not* implied that a primitive sialic crust would be uniformly distributed over the earth nor that convective motions of internal origin and consequent geochemical differentiation of the crust and upper mantle were absent during the pre- 3.8 × 10^9 yr history of the earth–moon system. It is suggested that if any areas of sialic crust were hit by a very large impact, then a distinctive sequence of events would occur. The location of such an impact scar with a 'stable' sialic region might allow preservation of its distinctive character in spite of the erosive processes and mobility of the earth's crust. In a region of sialic crust, a very large impact structure of 300–500 km diameter (cf. Mare Humorum, Nectaris, Moscoviense) or larger (cf. Mare Orientale, Serenitatis, Imbrium) would result in instantaneous unloading of the impacted region and throwout loading of the surrounding regions. The effect on pressure–temperature covariance in the excavated region relative to the surrounding region is illustrated schematically in fig. 1. The geotherm illustrated in this figure is hypothetical but is much steeper than the estimated modern oceanic geotherm, consistent with a higher radioactive heat production early in the earth's history. A large negative gravity anomaly must result over the impacted area and intensive fracturing would occur below the impacted area, possibly to depths of 60–80 km.

It is suggested that two main processes act to eliminate the instability established by a very large impact. Firstly, the load pressure drop beneath the impact site means that melting of the mantle may occur (figs. 1, 2) at quite deep levels and the degree of melting may be quite high. Uplift of the central floor of the impact requires vertical movement at deeper levels and the presence of a partially molten region underlying a highly fractured zone probably creates favourable conditions for very rapid ascent of large peridotitic diapirs. Such diapirs may reach shallow levels and there separate into liquid and crystal fractions with very high degrees of melting. This process

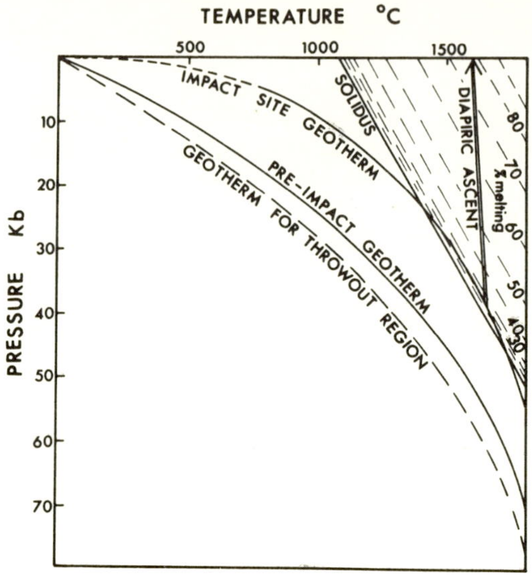

Fig. 1. Hypothetical diagram illustrating the instantaneous effect on a terrestrial geothermal gradient following major impact and unloading (by throwout and brecciation) of a large impact basin. The geotherm is illustrative but is drawn as a steeper geothermal gradient than estimates of the modern oceanic geotherm [34, 35]. Thermal effects of impact are neglected but would produce a transient high temperature at and very near the surface (P < 5kb), including impact melting in the ejecta blanket. A region at ~ 100–200 km depth is taken above its solidus by the pressure drop and the diagram shows adiabatic ascent of diapirs from this region to yield magmas reaching the earth's surface at temperatures above 1500°C.

would produce highly fluid peridotitic komatiite magmas leaving minor residual dunite (fig. 2). It is emphasized that the magmas discussed are not direct impact melts. In this model the impact-produced *unloading* produces partial malting at depth, the uplift of the fractured floor initiates diapiric ascent of partially molten peridotite and the process accelerates. The cycles of peridotitic komatiite to basaltic komatiite reported by Viljoen and Viljoen [18] possibly imply the ascent of individual diapirs (fig. 2) rather than a single regional upwelling. Slowing down of the process as the impact basin became filled would lead to power degrees of melting in ascending diapirs or to increased opportunity for crystal fractionation at low pressure with magma types changing to olivine-poor and low-Mg basalts, and possibly to tholeiitic andesite and dacitic derivatives.

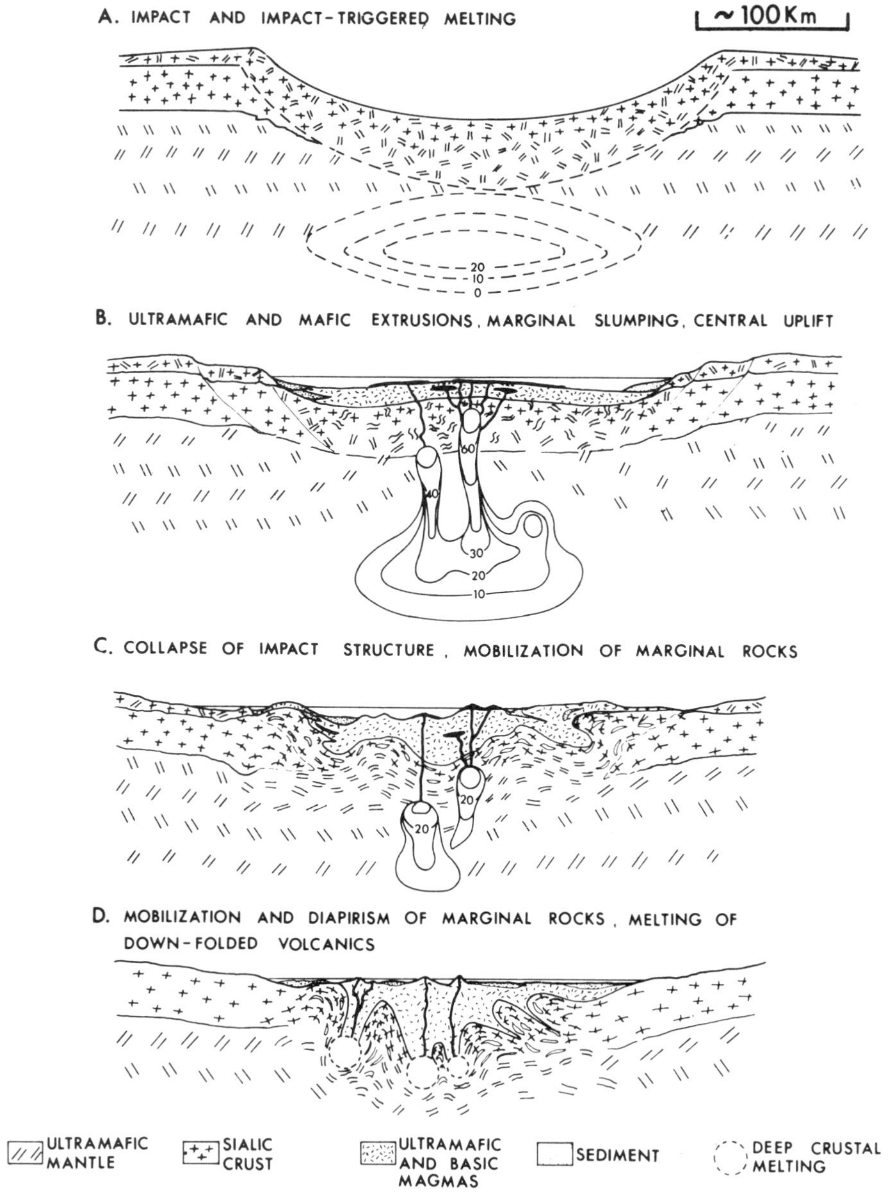

Fig. 2. Diagram illustrating the hypothesis in which some Archaean greenstone belts are suggested as terrestrial equivalents of lunar maria. (A) Primitive sialic crust impacted by very large body excavating large structure of ~ 30–50 km depth, floored with an ejecta blanket overlying brecciated upper mantle rocks. Instantaneous unloading (pressure drop) beneath the impact produces melting and dashed curves illustrate the degree of melting (at ~ 150 km depth). An extremely large negative gravity anomaly would be formed over the impact. (B) Partially melted peridotite rises rapidly into the brecciated impact floor with very great increase in degree of melting and extrusion of very high temperature ultramafic ($\gtrsim$ 1600°C) and basaltic magmas (komatiites). Some marginal slumping and sedimentation. (C) Accelerated inward collapse of impact structure, possibly influenced by regional crustal stresses, but mainly a consequence of increasing temperature and mobility of underlying and marginal mixed ejecta blanket. It is considered that the original crustal component (? granititc) of this would recrystallize, begin to partially melt and intrude or dome the overlying dense ultramafic and mafic lavas. Volcanism changes to more normal basaltic volcanism. (D) Dominant process is the infolding of the ultramafic and mafic volcanics accentuated by upward movement and intrusion of gnèiss domes – the basement to the original ultramafic and mafic lava sequence becomes intrusive into that sequence. In the deeper regions, temperatures are sufficiently high to produce remelting (under $P_{H_2O} < P_{Total}$ conditions) of mixed basaltic and 'basement' materials, yielding a second stage of volcanism characterized by andesite and dacitic magmas.

While magmatic activity is considered to be the most rapid and major mechanism responsible for filling the postulated terrestrial 'maria', there are processes which would operate on the earth to eliminate the maria but which have not operated on the moon (at least for the last 3.0 by approx.). With the excavation of a large impact scar through sialic crust and creation of a large gravity anomaly, there is a strong tendency to eliminate the impact structure by lateral flow, or slumping, of the marginal 'highlands' into the maria basin. It is considered that the high gravitational field of the earth, the higher terrestrial geothermal gradient and the lower melting temperature of the postulated primitive terrestrial crust (andesitic) relative to the lunar crust (gabbroic anorthosite) all create conditions favourable to elimination of the terrestrial maria by largely lateral displacement or overthrusting by the primitive sialic crust. The process is illustrated schematically in fig. 2 and is invoked as an explanation for the observation of conformable gneissic granites, intrusive into the base of the greenstone sequences yet regionally underlying the synclinal greenstone sequences and appearing to structurally form the basement to these sequences [4–7]. The effects of erosion and sedimentation (consequent on the existence of a primitive hydrosphere and atmosphere) may also aid in the encroachment of the margins of the impact structure on the central area. In their description of the Onverwacht Group, the lowest group of the Swaziland Sequence, Viljoen and Viljoen [18, 31] note that the upper formation of the Onverwacht Group (Komati Formation) is 3.5 km thick in the type area and consists entirely of altered ultramafic and basic lavas. Interlayered sediments or acid rocks are absent although later felsic intrusives do occur. It is suggested that this formation represents the central or deeper portion of the impact structure whereas the Theespruit Formation, which is in fault contact with the Komati Formation but is interpreted to underlie the Komati Formation, may be in part a facies equivalent of the Komati Formation, formed closer to the margins of the impact structure. The Theespruit Formation consists of 1.9 km of metamorphosed mafic lavas with significant though minor quantities of felsic tuffs and associated poorly bedded chert. There appear to be no rocks described from the Swaziland System which might correspond to the veneer of impact melt, impact tuffs and impact breccia which would initially form the floor to the impact structures (fig. 2). In this context, the thermal effects of frequent and voluminous flows of very high temperature ultramafic and basic magmas may cause metamorphism, partial melting and mobilization of marginal sedimentary intercalations or adjacent older 'sial'.

The result of the process outlined above would be to transform an initially circular impact scar through a 'terrestrial maria' filled with ultramafic and basic lavas and a minimum amount of sediment, to a synclinal but irregularly cuspate or lobate greenstone belt. As the lava sequence of the maria first thickened rapidly and then became infolded with encroaching sial, the possibility of partial melting of the deeper parts of the greenstone sequences would increase. Thus the evolution of the 'terrestrial maria' might continue with the ultramafic and basaltic sequences being overlain by diacitic or rhyodacitic, andesitic and tholeiitic basaltic volcanics and related intrusives, developed by partial melting of peridotitic and basaltic rocks under variable P_{H_2O} conditions at depths of about 30–60 km. If horizontal movements or stresses were as important $> 2.5 \times 10^9$ yr ago as in the Palaeozoic or Mesozoic, then the end result of deformation of large terrestrial maria might be elongate folded greenstone belts such as those of the western Australian Yilgarn shield, rather than the irregular form of the Swaziland Barberton area.

4. Summary

The suggestion that the major impact scars are preserved on Archaean shield regions is not new. An impact origin, with the impact triggering magmatic intrusion and extrusion, has been suggested for Sudbury Lopolith ($\sim 1.7 \times 10^9$ yr old), Vredefort Dome and the Bushveld Complex ($\sim 2 \times 10^9$ yr old) [32, 33]. In this paper, I have used the distinctive characters of the ultramafic and mafic intrusives of some Archaean greenstone belts to argue for a particularly catastrophic type of magmatism adding the hypothesis that this can best be interpreted in terms of major impact. I have attempted to support this hypothesis with the argument that events recorded in some Archaean greenstone belts are nearly synchronous with very large impacts forming the maria basins on the lunar surface. If this hypothesis (relating a distinctive Archaean magma type (peridotitic komatiite) to an origin by impact-triggered melting) is valid, then it implies that the real age of Archaean greenstone belts in which these distinctive magmas are present would be greater than approximately 3.5×10^9 yr (from present understanding of lunar impact chronology [26]). This aspect of the model is open to testing. Failure of the hypothesis carries the implication of very different upper mantle dynamics and temperature distribution within the Archaean crust and upper mantle. This paper advocates a rather fundamental role for large exogenous impacting bodies in activating chemical differentiation of the crust–mantle system through the volcanism and evolutionary sequences of Archaean greenstone belts. It is emphasized that endogenous processes of crust–mantle differentiation, possibly the precursors of the petrogenetic processes occurring as an integral part of plate tectonics, were probably also in evidence $\sim 3.0 \times 10^9$ yr ago. It is postulated that prior to 3 or 3.5×10^9 yr ago, the *major* geochemical differentiation of the earth was due to initial accretionary processes ($\sim 4.5 \times 10^9$ yr ago) followed by impact-working of the surface and impact-triggered processes in the upper mantle (to $3–3.5 \times 10^9$ yr ago). Since $\sim 3 \times 10^9$ yr, the magnitude and frequency of large impacts on the earth decreased very greatly [26] so that crustal evolution has been dominated by geochemical processes of internal origin (e.g. plate tectonics).

Acknowledgements

The author thanks Drs. M.J. and R.P. Viljoen for their initial stimulation of his interest in the Swaziland rocks, for supply of samples for experimental work and for continuing discussions. The author also acknowledges helpful discussions and manuscript criticism by Professor A.E. Ringwood, Drs. K. Frederiksson, A. Glikson, I.A. Nicholls, V.M. Oversby and W. Compston.

References

[1] C.R. Anhaesser, R. Mason, M.J. Viljoen and R.P. Viljoen, A reappraisal of some aspects of Precambrian Shield geology, Geol. Soc. Am. Bull. 80 (1969) 2175.

[2] A.Y. Glikson, Geosynclinal evolution and the geochemical evolution of early Precambrian systems, Tectonophys. 9 (1970) 397.

[3] A.M. Goodwin, Archaean protocontinental growth and early crustal history of the Canadian Shield, Rept. 23rd Internat. Geol. Congress, Prague 1 (1968) 69.

[4] M.J. Viljoen and R.P. Viljoen, An introduction to the geology of the Barberton granite–greenstone terrain, Geol. Soc. S. Africa Spec. Publ. 2: Upper Mantle Project (1969) 9.

[5] M.J. Viljoen and R.P. Viljoen, A reappraisal of the granite–greenstone terrains of shield areas based on the Barberton model, Geol. Soc. S. Africa Spec. Publ. 2: Upper Mantle Project (1969) 245.

[6] N.W. Bliss and P.A. Stidolph, A review of the Rhodesian Basement Complex, Geol. Soc. S. Africa Spec. Publ. 2: Upper Mantle Project (1969) 305.

[7] A.M. Goodwin and R.H. Ridler, The Abitibi orogenic belt, in: Symposium on Basins and Geosynclines of the Canadian Shield, Geol. Surv. Canada paper 70–40 (1970) 1.

[8] A.Y. Glikson, Primitive Archaean element distribution patterns: Chemical evidence and geotectonic significance, Earth Planet. Sci. Letters 12 (1971) 309.

[9] S.R. Hart, C. Brooks, T.E. Krogh, G.L. Davis and D. Nova, Ancient and modern volcanic rocks: A trace element model, Earth Planet. Sci. Letters 10 (1970) 17.

[10] W.R.A. Baragar, Major element geochemistry of the Noranda volcanic belt, Quebec–Ontario, Can. J. Earth Sci. 5 (1968) 773.

[11] D.C. Green and H. Baadsgaard, Temporal evolution and petrogenesis of an Archaean crustal segment at Yellowknife N.W.T., Canada, J. Petrol. 12 (1971) 177.

[12] L.P. Black, N.H. Gale, S. Moorbath, R.J. Pankhurst and V.R. McGregor, Isotopic dating of very early Precambrian amphibolite facies gneisses from the Godthaab District, West Greenland, Earth Planet. Sci. Letters 12 (1971) 245.

[13] S.S. Goldich, C.E. Hedge and T.W. Stern, Age of the Morton and Montevideo gneisses and related rocks, southwestern Minnesota, Geol. Soc. Am. Bull. 81 (1970) 3671.

[14] H.L. Allsopp, R.D. Davies, A.A.A. de Gasparis and L.O. Nicolaysen, Review of Rb/Sr age measurements from the early Precambrian terrain in the southeastern Transvaal and Swaziland, Geol. Soc. S. Africa Spec. Publ. 2: Upper Mantle Proeject (1969) 433.

[15] P.M. Hurley, B. Nagy and W.H. Pinson, Age of the Middle Marker Horizon, Onverwacht Group, Swaziland Sequence, South Africa (abstract), Geol. Soc. Am. Abst. with Programs 3 (1971) 609.

[16] W. Compston and P.A. Arriens, The Precambrian geochronology of Australia, Can. J. Earth Sci. 5 (1968) 561.

[17] M.J. Viljoen and R.P. Viljoen, A proposed new classification of the granite rocks of the Barberton region, Geol. Soc. S. Africa Spec. Publ. 2: Upper Mantle Project (1969) 153.

[18] M.J. Viljoen and R.P. Viljoen, Evidence for the existence of a mobile extrusive peridotite magma from the Komati Formation of the Onverwacht Group, Geol. Soc. S. Africa Spec. Publ. 2: Upper Mantle Project (1969) 87.

[19] R.W. Nesbitt, Skeletal crystals forms in the ultramafic rocks of the Yilgarn block, Western Australia; evidence for an Archaean ultramafic liquid, Geol. Soc. Aust. Spec. Publ. 3 (1972) in press.

[20] A.J. Naldrett and G.D. Mason, Contrasting Archaean ultramafic igneous bodies in Dundonald and Clergue Townships, Ontario, Can. J. Earth Sci. 5 (1968) 111.

[21] A.J. Naldrett, Ultramafic and related mafic rocks of the Abitibi origin, in: Symposium on Basins and Geosynclines of the Canadian Shield, ed. A.J. Baer, Geol. Surv. Canada Paper 70-40 (1970) 24.

[22] D.H. Green, I.A. Nicholls, M.J. Viljoen and R.P. Viljoen, Experimental determination of magmatic temperature and crystallization sequence of Archaean peridotitic magma (in preparation).

[23] A.E. Ringwood, Chemical composition and origin of the earth, in: Advances in Earth Science, ed. P.M. Hurley (MIT Press, Cambridge, Mass., 1966) 287.

[24] D.H. Green, The origin of basaltic and nephelinitic magmas (Bennett Lecture), Trans. Leicester Lit. Philos. Soc. 64 (1970) 28.

[25] D.H. Green and A.E. Ringwood, The genesis of basaltic magmas, Contrib. Mineral. Petrol. 15 (1967) 103.

[26] E.M. Shoemaker, Cratering history and early evolution of the moon (abstract), Revised Abstracts of the Third Lunar Science Conference, Lunar Sci. Inst., Houston Publication 88 (1972) 696.

[27] W. Compston, H. Berry, M.J. Vernon, B.W. Chappell and M.J. Kaye, Rubidium-strontium chronology and chemistry of lunar material from the Ocean of Storms, Proc. 2nd Lunar Sci. Conf. 2 (1971) 1471.

[28] D.A. Papanastassiou and G.J. Wasserburg, Rb-Sr ages of igneous rocks from the Apollo 14 mission and the age of the Fra Mauro formation, Earth Planet. Sci. Letters 12 (1971) 36.

[29] G. Turner, J.C. Huneke, F.A. Podosek and G.J. Wasserburg, $^{40}Ar/^{39}Ar$ ages and cosmic-ray exposure ages of Apollo 14 samples, Earth Planet. Sci. Letters 12 (1971) 19.

[30] W. Compston, M.J. Vernon, H. Berry and R. Rudowski, The age of the Fra Mauro formation: A radiometric older limit, Earth Planet. Sci. Letters 12 (1971) 55.

[31] M.J. Viljoen and R.P. Viljoen, The geology and geochemistry of the Lower Ultramafic Unit of the Onverwacht Group and a proposed new class of igneous rock, Geol. Soc. S. Africa Spec. Publ. 2 (1969) 55.

[32] R.S. Dietz, Sudbury structure as an astrobleme, J. Geol. 72 (1964) 412.

[33] W. Hamilton, Bushveld Complex – product of impacts?, Geol. Soc. S. Africa Spec. Publ. 1 (1970) 367.

[34] S.P. Clark and A.E. Ringwood, Density distribution and constitution of the mantle, Rev. Geophys. 2 (1964) 35.

[35] A.E. Ringwood, I.D. MacGregor and F.R. Boyd, Petrological constitution of the Upper Mantle, Ann. Rept. Geophys. Lab. Carnegie Inst. Year Book 63 (1964) 147.

II: THE GREENSTONE ASSOCIATIONS

In Part II are assembled a number of significant papers that supply details on the more important of the world's Archean greenstone belts. The selection begins with three treatments of the best documented of the Katarchean belts, all in southern Africa. Four papers covering the late Archean of Canada, the best-documented region of its kind, follow. In Western Australia, both ages are represented, and although only one paper of sufficient detail can be offered covering the Katarchean of the Pilbara, the late Archean of the Yilgarn Province has lately been extensively covered by members of the editor's former research team at the University of Western Australia, five of whose papers are included, and by officers of the Geological Survey of Western Australia (one paper). The editor regrets not being able to do more than mention the papers by Pyke et al. (1973), dealing with ultrabasic flows in Canada, and by Gemuts and Theron (1973), which present a realistic analysis of the key area south of Kalgoorlie, Australia, and provide the most up-to-date addition to the Barberton, Abitibi, and Slave models for greenstone belts.

The reader may ask why nothing was selected to represent other important regions such as Russia (Ukraine), India, West Africa, and Guiana. There seems to be much literature of a general nature, including ideas of great importance, available on the Archean from Russia, but detailed descriptions of the greenstone belts remain elusive, at least to Western geologists, in translation or in the Russian language. The Indian geochronology has only recently become rationalized (Sarkar, 1972), and the editor has found it difficult to determine the real age of sequences discussed in detailed descriptions. He has tried throughout this treatment to limit the selection to belts the age of which we can fix within narrow limits. The West African and Guiana cratons contain Archean and even Katarchean remnants, but appreciable portions of the Archean-type belts appear not to be truly Archean in age, so these regions have been reluctantly omitted, despite the availability of excellent treatments on the latter by McConnell and Williams (1970).

Geochronologic advances have now clearly defined at least two age clusters of well-formed greenstone belts at around 3400 and 2700 million years; the Pilbara belts of Australia (Paper 14) may well represent another age cluster, around 3050 million years (Paper 24).

The belts of the older, Katarchean cluster are not significantly different in character from those of the late Archean, yet there is significant difference in the tectonic patterns; the older group of belts show more emphatically developed the so-called nuclear characteristics, the dome and brachysyncline patterns of the permobile stage (Papers 29 and 32). The late Archean belts show subdued patterns of this type, but possess an overriding linearity of grain. Despite the disparity of tectonic style, the older belts (Swaziland, Rhodesia, Pilbara) possess all the features characteristic of the later belts (Superior, Slave, Kalgoorlie, Nyanza), and the abrupt distinction into two quite different types of geological development (permobile–nuclear and protogeosynclinal) (Paper 29) appears unrealistic.

Argument continues regarding the possible primary nature of the older brachysynclines and younger linear synclines. Are the synclines as we now see them (manifestly secondary tectonic features) developments that closely correspond to the actual depositories in which volcanics and sediments were formed, or do they have no geometrical correspondence, being purely secondary entities? The evidence seems to favor the first solution (Papers 7 and 9) although contrary evidence that the brachysynclines represent only an infolded preservation of a remnant of much more extensive depositories is given in Papers 17A and 35. Whatever the answer, it is clear that all Archean greenstone depositories must be regarded as protogeosynclines, although this implies no very strict genetic inference. These are thick accumulations of contemporaneous volcanic, intrusive, and sedimentary rocks laid down in extensive but narrow, fault-controlled, subsiding oceanic troughs. It seems blind to deny that they are forerunners of the classic geosynclines, the features of which and the components of which they, to a large extent, share.

It is popular to equate these greenstone troughs with rift zones (Paper 38), but the true rift zone is primarily a major, geotectonic upwarp, and the renowned graben features are secondary, postmagmatic adjustments. A rift zone occupies a raised welt in the crust and is analogous to the Proterozoic mobile belts. The greenstone associations of the Archean were deposited in subsiding troughs. Upwarping complications may well have been present, internally, but the negative movements that dominated these zones were probably primary, compensating positive movements in intervening protocratonic areas. The greenstone depository seems to have been a primary, negative feature, like a geosyncline, and not strictly analogous to a modern rift zone. That such depositories suffered continuous subsidence is shown by the enormously thick accumulations within them (up to 30 km); otherwise, these could not conceivably have been accommodated.

We may divide Archean greenstone belts into (1) early vestiges, (2) well-formed Katarchean belts, and (3) late Archean belts. The oldest well-formed belts are those of the Ukraine, Swaziland, Rhodesia, and the Pilbara (also, possibly, Bihar–Orissa). These belts are fortuitously preserved from later superimposed metamorphism and are in a relatively fresh, unmetamorphosed state. Despite their great age, they can be unraveled using orthodox geologic methods. They include basic, ultramafic, and intermediate–acid volcanics, layered intrusions of basic–ultramafic composition (but not chromititic anorthosites), banded ironstones, other chemogenic sediments, and thick cappings of clastic graywackes and conglomerates, which reflect progression from chemogenic to clastic dominance. These are the first clastic sequences of any significance in the geologic record (Fig Tree and Moodies groups, Papers 5 and 26, Shamvaian of Rhodesia, Paper 7; Gorge Creek and Mosquito Creek of the

Pilbara, Australia, Paper 14). Cyclicity has its most perfect expression in some of these Katarchean belts (Swaziland and Pilbara).

The late Archean belts may show cyclicity, as in the case of the Abitibi belt (Paper 8), but more commonly display hiatal rhythmicity (Papers 13 and 34). The nature of these late Archean belts is varied, contrasts being seen within the limits of one shield (Superior and Slave). There seems to be no universally applicable model. They are the most important economically; the Abitibi belt contains a cluster of large acid volcanic centers associated with large-scale copper–zinc mineralizations. The Kalgoorlie belt has failed to yield a similar mineralization on any significant scale (perhaps because of the paucity of development of large acid volcanic centers of the calc-alkaline association), but has yielded immense gold and nickel mineralizations. The Yellowknife belt has yielded significant gold mineralization. The variation in mineralization from belt to belt may be a factor of the variation of internal development of the eruptive suite or dependent on mantle inhomogeneity (Sillitoe, 1972).

The late Archean belts show less well defined cyclicity as a general rule, and the chemogenic to clastic progression in sedimentation is repeated, but less concisely defined than in the Katarchean belts.

The early vestiges are discussed elsewhere, in comments on papers dealing with Greenland (Papers 19, 20, and 38). Supracrustal rocks, among which basites are dominant, form belts within very ancient Katarchean gneiss terrains. They have been most extensively studied in Greenland, but analogues may well be present in granite–gneiss terrains of Rhodesia, South Africa (Johannesburg Dome, Ancient Gneiss Complex), India (Bihar–Orissa), Madagascar, East Africa, Angola, Sierra Leone, Montana, Wyoming, Manitoba, Western Australia, and the Kola Peninsula. Yet there is little or no firm evidence from any of these areas affording geochronologic proof that rocks formed prior to the Katarchean belts mentioned above do exist in these tracts of very old rocks. Only in Greenland has this been incontrovertibly established. Very ancient residuals in Katarchean high-grade granite–gneiss–granulite terrain will have rarely escaped modification by later metamorphisms, will tend, even if present, to yield only superimposed geochronologic numbers, and will be difficult to unravel in their primary nature. Such vestiges may be more numerous than realized, and yet be largely modified.

It is doubtful if any of the really old vestiges ($\geqslant$ 3500 million years) are truly analogous with the well-formed Katarchean and late Archean greenstone belts; the evidence is discussed elsewhere (see editor's commentary on Papers 19, 20, and 38).

References

Gemuts, I., and Theron, A. (1973). Stratigraphy and mineralisation of a portion of the Archaean between Norseman and Coolgardie, Western Australia. *Abstract Vol., Sec. 3, ANZAAS, 45th Congr., Perth,* p. 88.

McConnell, R. B., and Williams, E. (1970). Distribution and provisional correlation of the Precambrian of the Guiana shield. *Proc. 8th Guiana Geol. Conf., 1969,* 20p.

Pyke, D. R., Naldrett, A. J., and Eckstrand, O. R. (1973). Archaean ultramafic flows in Munro Township, Ontario. *Geol. Soc. Am. Bull., 84,* 955–978.

Sillitoe, R. H. (1972). A plate tectonic model for the origin of porphyry copper deposits. *Econ. Geol., 67,* 184–197.

PAPERS 5 THROUGH 7: COMMENTARY

Paper 5 is a classic description of a low-grade Katarchean greenstone assemblage about 16,000 m thick. The Barberton Mountain Land covers an area of little more than 250 square miles; it comprises a complex brachysyncline set within granitic rocks and forms a small section of a nuclear terrain (Paper 29). The *Onverwacht* basal volcanic group includes several cycles, the ultramafic–mafic basal section including pillow lavas and pillow breccias; basic schists predominate over ultramafic within it. Salic tuffs (now sericite schists) and cherts cap this section, marking periods of quiescence. The authors recognize 10 major cycles of this sort and up to 20 minor cycles. One to five lava flows, each between 2 and 100 m thick, constitute the extrusive phase of a cycle. Pillow lavas alternate with unpillowed material, as in Canada (Paper 8), and with ultrabasics in the middle Onverwacht Group.

The upper part of the group is 1200 m thick, and consists of at least five cycles, each 150 to 1100 m thick and diminishing in thickness upward. Such cycles are characterized by an initial outpouring of 5 to 20 flows of basic to intermediate lavas, followed by a thin effusion of acid lava, which is capped by a thin zone of banded, black and white, and locally ferruginous chert. Extensive masses of salic, volcanogenic sediment and lava at the top of the Onverwacht Group are overlain by thick sequences of similar cherts.

In this initial paper, analogy with ophiolite sequences of modern geosynclines and with island-arc systems was suggested. Deposition is thought to have occurred in a fault-bounded, deep-sea graben on a crust only a fraction of its present thickness. Widespread outpourings of ultramafic flows (including true ultrabasics) are recognized, and potash-poor oceanic tholeiites are also abundant.

The overlying *Fig Tree Group* (3500 m thick) consists of siliceous chemical precipitates, pelitic sediments, graywackes, and banded cherts, and contains primitive fossils (see Paper 17A and Part V).

The *Moodies Group* above (2000 m thick) consists of arenites with intercalated shales and a thick basal conglomerate.

The Fig Tree Group is attributed to deposition in a deep-water geosynclinal environment; turbidites alternate with the chemical sediments formed in quieter periods. Photosynthesizing organisms in the chemical sediments probably played a part in the deposition of the banded ironstones. Sedimentary structures are widespread and well preserved: slump bedding, current marks, load casts, flute casts, convolute and graded beds, and rhythms characteristic of turbidite sequences.

The Moodies Group shows abundant cross bedding, ripple marks, and mudcracks (in shales). Trough cross bedding preserves braided stream channels, and, with the presence of conglomerates, indicates a shallow-water environment. Denudation of the Onverwacht and Fig Tree Groups and erosion of nearby unroofed, elevated granite terrain is thought to have produced these sediments. The granite in question was probably the basement beneath the Swaziland System. The Moodies Group is regarded as a form of *molasse* deposit. A narrow band of amygdaloidal lava indicates minor, contemporaneous volcanism (later investigation has shown it to be trachytic). The marginal granite may include a gneissic and migmatitic basement (see Paper 26), but the main component is an intrusive suite of tonalites more than 3000 million years old, younger than the greenstones. This suite includes diapiric intrusions, deformed supracrustal paleosomes, and a late potassic, high-level hood granite. As in the Pilbara Katarchean (Paper 14), there are minor occurrences of a late Archean potassic, porphyritic granite.

The authors argue the case for the thick basal ultramafic assemblage of the Onverwacht Group acting as a resistor to granitization, a possibility that seems to be commonly overlooked in Archean studies. The status of the Ancient Gneiss Complex as a possible older basement is discussed in Papers 26 and 35.

This study has been regarded as presenting a model for Archean greenstone belts, but, despite the excellence of the detailed investigation, Barberton must be correctly regarded as *one variant of the variable pattern* of greenstone-belt development. It is the most complete description of any Katarchean belt yet published but, again, Katarchean greenstone belts are likely to show considerable variation in detail. It is noteworthy that this classic paper includes but the merest mention of intermediate volcanics, and doubts must remain as to whether there is really a complete calc-alkaline *edifice* cycle superimposed on the tholeiitic and high-magnesia basalt (*komatiitic*) platform cycle, as is the case in the late Archean Abitibi Belt of Canada (Paper 8).

The authors describe in more detail (see Paper 5) the lower three divisions of the Onverwacht Group, erecting a new igneous rock class embracing the *peridotitic komatiites* and three subtypes of *basaltic komatiites.* These are high-magnesia ultrabasic and basic rocks, distinguished on this and other criteria from the associated potassium-poor *abyssal tholeiites.* Differentiated, layered ultramafic intrusions derived from basaltic komatiite stem magmas are also described.

Of the three types of intrusions described, all are remarkably deficient in felspathic (gabbroic) end members, the intrusions being almost entirely composed of refractory minerals (olivine, pyroxenes). Only subordinate amounts of gabbro are present (see also Papers 14 and 15 for descriptions of similar megalayered, repetitive sill complexes in the Katarchean of Western Australia). Such bodies may be the "hallmark" of the very ancient Katarchean greenstone belts, as so far they have not been recorded from the late Archean.

Discussing the chemistry of the komatiites, the authors emphasize the lime–alumina inversion factor, as well as the unusually high magnesia content. The three types of basaltic komatiite were distinguished on variation diagrams; such diagrams reveal no overlap between komatiites and tholeiites, and, despite their intimate association in occurrence, there seems to be a distinct geochemical hiatus between komatiites and tholeiites.

The suggestion is advanced that these unusual ultramafic and basic volcanic associations represent a primordial oceanic crust, more or less transitional to the upper mantle, and that little, if any, sialic crust existed at the time of these lava outpourings. The later, extensive development of tonalitic granite nearby, 3400–3200 million years ago, is attributed to a single-cell convective overturn process, a major catastrophic event in the Earth's development that has been hypothesized by Vening-Meinesz and Hess as likely to have been responsible for the segregation of the nickel–iron core and the ascent of low-melting silicates to form continents.

Viljoen and Viljoen appear to envisage a primordial oceanic crust prior to the formation by this process of a primitive, sialic Pangea, with the basal Onverwacht Group as a sample of that crust.

In the few years that have elapsed since this statement was published much more evidence has accrued concerning the nature of the komatiites and the existence of very ancient developments of protocontinental granitic terrains. The komatiites have been found to have analogues in both Australia (see Papers 16B and 16C) and Canada (Pyke et al., 1972); in Australia they are widespread, but in Canada they are sparsely developed. The lime–alumina inversion is not usually displayed by otherwise similar rocks in Australia, and the high magnesia content would appear to be the main diagnostic feature of the komatiite equivalents, which are now more commonly referred to as high-magnesia basalts (Papers 16B and 16C), although alumina contents do tend to be lower than those of abyssal tholeiites. These rocks are constant associates of tholeiites and like them are notably potassium-deficient.

It has been found in Australia (McCall and Hudson, 1970) that developments thousands of meters thick of monotonous tholeiite pillow lavas occur within nickel-bearing komatiite equivalents (e.g., at Mt. Morgan, Widgiemooltha). There are two or three major cycles in this greenstone belt, each equivalent to the Onverwacht Group. High-magnesia basalts occur with tholeiites at the base of each such major cycle, overlain by clastic sediments and acid volcanics (Paper 13; Gemuts and Theron, 1973), and the entire greenstone sequence shows an overall trend from ultramafic–basic volcanic dominance to sediment–acid volcanic dominance. Other evidence from the same region suggests that the komatiite equivalents are by no means everywhere situated at the base of the sequence in the greenstone belt. Locally, at least, in outcrops at Norseman and Edjudina, tholeiitic and acid volcanic sequences stratigraphically underlie their main development and are devoid of komatite equivalents.

Another fact that has emerged from Australian and Canadian studies is that the komatiite equlvalents occur in both Katarchean and late Archean greenstone associations. This gives a span of 1000 million years for their appearance in the record, but they have also been recognized in the complex, differentiated Jimberlana Dike (Paper 16C) of Proterozoic age (2420 million years), erupted into a fully stabilized Archean cratonic block during a late poststabilization phase of regional dike development.

Certainly, granite and gneiss of protocontinental nucleii were in existence several hundred million years prior to the Onverwacht eruptivity (Paper 20). The concept of a relationship between these rocks and a single, catastrophic layer overturn event, as originally stated by Viljoen and Viljoen in this paper, now requires radical reappraisal in the light of all this new evidence.

The high-magnesia basalts are certainly not continuous with the oceanic mantle, but were erupted onto downfaulted segments of crust, possibly protocontinental and granitic, possibly oceanic in newly formed gulfs that separated blocks of the earliest sialic cratonic areas. They are invariably interleaved with minor sedimentary intercalations and with these formed an oceanic crust, which may have been in no sense primordial. Their unusual composition may be due to the thinness of the crust, the enhanced geothermal gradient and the rapidity of uprise of partial melting processes in the mantle, or a very deep-seated differentiation of the primary product of such melting.

Stowe, in a highly condensed treatment, describes the Rhodesian Archean craton, which yields radiometric ages from 3400 to 2300 million years. The youngest possible age for the Great Dyke (2530 million years) provides an upper limit for the cessation of deformation within the craton, which is bounded by younger mobile belts, all three of which incorporate Archean residuals: Limpopo (modifying events at 2600 and 2000 million years), Zambezi (modifying events at 2300–2100, 1900–1650, 1300–1100, and 600–400 million years), and Mozambique (last and main event, 600–400 million years).

MacGregor (1951) first drew attention to the characteristic structural pattern of the Rhodesian craton, one of basement schists between separating "gregarious" batholithic domes. This is now recognized as the characteristic *nuclear* pattern of the very ancient Archean (Papers 27, 29, 30, and 32). This cratonic terrain of Rhodesia is now revealed as a most complex Archean terrain, and there are undoubtedly several superimposed Archean systems and deformations therein.

The geochronology may be summarized as follows:

1. Paleosomes (relics in ancient gneisses?).
2. *Sebakwian* metamorphic event, +3400 million years.
3. *Bulawayan* metamorphic event, 3370 million years.
4. Main granite event (dome emplacement), 2900 million years.
5. *Shamvaian* sedimentation.
6. Post-Shamvaian granite invasion, 2700–2300 million years.

Stowe recognizes a central unit of gneiss basement with a north–northeasterly fold trend, a protocraton that was later incorporated in three separate domes. He describes one locality in which migmatites of this ancient nucleus, characterized by much basic content, are overlain unconformably by a jaspilite of Sebakwian I. A 3400-million-year-old igneous event concluded the formation of the oldest protocratonic nucleus.

In the subsequent geosynclinal phase, small Bulawayan geosynclines were formed across and on either side of this nucleus, being controlled by fracture lines. Locally, an older Sebakwian II system is recognized, which is deposited in deep, narrow geosynclines and includes sediments that incorporate clasts derived from the older gneissic terrain. Sebakwian II has been differentiated on account of unconformable relationships and higher metamorphic grade from the Bulawayan.

Essentially, there are two separate Bulawayan geosynclinal belts. The Gneiss Dome phase, which produced the structures noted by MacGregor, was the second tectonic phase, and the principal invasive component was tonalitic granite, dated at about 2900 million years. The domical structures are typical products of second-stage vertical tectonics and, although much modified by later deformations, retain their essential pattern. Shamvaian sediments were probably at least in part deposited in rim depressions produced in this doming. The Shamvaian was succeeded by invasion of late granites (of the early cratonic stage) intruded at the unstable margins of the craton. These include vast floodings of granite and small transgressive bodies.

There appear to be two distinct divisions of the Sebakwian separated by an unconformity, and two distinct divisions of the Bulawayan separated by folding and unconformity; the latter system was folded and deeply eroded before the superimposition of the Shamvaian, which itself includes three distinct formations. The Bulawayan, which attains a thickness of 12,000 m, is a typical Archean greenstone sequence in which pillow lavas are prominent, together with banded ironstones, conglomerates, and acid and intermediate volcanics. The Sebakwian also includes typical greenstone-belt members, volcanic schists, serpentinites, and sericitic grits. The Shamvaian, on the other hand, is a sedimentary sequence of conglomerates, phyllites, and banded ironstones.

Of particular interest is Stowe's recognition of a local development of Alpine nappe fold structures (Stowe, 1968). It is not possible to reproduce this important paper in facsimile because of lack of space, but it documents, for the first time, a nappe complex in the Katarchean. A questioner asked during discussion of this paper, "Whatever happened to gregarious granite batholiths?," but Stowe seems to make clear the fact that the story commenced with the usual greenstone accumulation; the first stage of orogenesis involved this extreme manifestation, and subsequent stages in the eruptive and tectonic development of the craton involved vertical tectonics coupled with granite dome uprise.

Archean shield areas characteristically display folds of open to closed style, commonly isoclinal, with steep limbs and axial planes. Inverted limbs, where present, tend to be steep, and there is a lack of flat-lying or gently inclined inverted limbs, which are a feature of Alpine or Caledonian nappe styles. Complex nappes are well documented from the Proterozoic of Central Australia (Rickard and Shaw, 1972), but were hitherto unknown in the Archean. Their rarity may, perhaps, be attributed to the thin nature of the crust and the lack of extreme altitude differentials throughout the globe in this ancient time; continental areas probably did not stand high above the sea floor. Also, if one introduces considerations of plate tectonics, it seems that this regime allowed little or no large-scale overriding movements of large plates one over the other, and there were probably neither high-standing nor extensive continental plates in existence (see Part VII). The Selukwe evidence of nappes suggests an unusual local condition of intense lateral shoving, which contrasts with the more usual tectonic pattern of the early Archean of limited lateral displacements.

References

Gemuts, I., and Theron, A. (1973). Stratigraphy and mineralisation of a portion of the Archaean between Norseman and Coolgardie, Western Australia. *Abstract Vol., Sec. 3, ANZAAS 45th Congr., Perth, 88* (abstract only).

McCall, G. J. H., and Hudson, D. R. (1970). Excursion guide, Symposium on the Archaean Rocks, Geological Society of Australia, Perth, 1970.

MacGregor, A. M. (1951). Some milestones in the Precambrian of southern Rhodesia. *Proc. Geol. Soc. South Africa, 54,* xxvii–lxxi.

Pyke, B. R , Naldrett, A. J., and Eckstrand, O. R. (1973). Archaean ultramafic flows in Munro Township, Ontario, *Geol. Soc. Am. Bull., 84,* 955–978.

Rickard, M. J., and Shaw, R. (1972). Preliminary interpretation of the basement structures north of the Arltunga nappe complex. *Abstract Vol., Geological Society of Australia Joint Specialist Groups Meeting, Canberra, Ia,* 13–16.

Stowe, C. (1968). Intersecting fold trends in the Rhodesian basement south and east of Selukwe. *Trans. Geol. Soc. South Africa, 71* (annexure), 53–78.

5: THE BARBERTON MOUNTAIN LAND—A MODEL OF THE ELEMENTS AND EVOLUTION OF AN ARCHAEAN FOLD BELT

C. R. Anhaeusser, C. Roering, M. J. Viljoen, and R. P. Viljoen

The Barberton Mountain Land forms one of the best developed and best preserved remnants of Archaean strata in South Africa. The volcanics and sediments are similar to those associated with geosynclines and island arcs. The Onverwacht Series of the Swaziland System, consists mainly of basaltic and ultrabasic rocks together with siliceous sediments and constitutes the initial magmatic or ophiolite assemblage of the Swaziland geosyncline. The Fig Tree Series, composed primarily of greywackes, shales, banded cherts and banded ironstones, displays characteristics of a flysch assemblage of rocks while finally, the overlying Moodies Series, made up of conglomerates, quartzites, shales and jaspilites, displays the characteristics of a molasse assemblage of a geosyncline.

The entire area is surrounded and swamped by granites and gneisses, thereby making it impossible to identify the miogeosynclinal and the cratonic areas associated with the Barberton Mountain Land remnant. Furthermore, the granites are not typical of normal geosynclinal areas and constitute an anomalous event, probably much later than the orogenic episode.

Evidence is presented that the Onverwacht and Fig Tree Series of rocks were deposited in a deep trough (pillow lavas, turbidites). A brief account of the structural history of the area is given, and some notes on the mineralization are included. A model is suggested for the major evolutionary events in the development of the Barberton Mountain Land. The first event was the geosynclinal cycle, while the granites make up the second event. It is suggested that the Barberton Mountain Land represents a fold belt in which the geosynclinal and granite cycles are well-displayed, and, as a result, may assist in a better understanding of the evolution and development of other, less well-developed schist belt remnants in Archaean granite terrains of continental shield areas.

CONTENTS

I INTRODUCTION

With the aim of discovering further sources of mineable ore the Barberton Mountain Land has been subjected to intensive geological investigations over the last six years. This work, undertaken by the Economic Geology Research Unit of the University of the Witwatersrand, in conjunction with various mining companies and the Swaziland Geological Survey, has given rise to a wealth of basic data. In addition, the area is at present being examined as part of South Africa's contribution to the International Upper Mantle Project. This work is primarily concerned with a study of the various ancient lavas and ultrabasic rocks of the Onverwacht Series. In places, these rocks are exceptionally well preserved, and it is hoped that field evidence, coupled with geochemistry, will provide valuable information of the early evolution of the earth's crust and mantle, and the differentiation of continents.

Prior to the renewed interest in the Barberton Mountain Land, there existed a detailed regional geological map and explanation of the area compiled by the South African Geological Survey (Visser *et al.*, 1956). This early work provided a useful foundation upon which the subsequent studies have been built.

The geology of the Barberton Mountain Land is important primarily because

[*Editor's Note:* Plates 1 through 3 have not been reproduced because they are oversized.]

of its great age, the volcanics and sediments having been intruded by granites having ages of 3,000 million years. In spite of this great antiquity, the supracrustal rocks are relatively unmetamorphosed over the greater part of the area, and thus provide leading clues to deciphering some of the mechanisms operating in the ancient crust, as it existed in these times. The supracrustal rocks are almost identical in type to those associated with younger geosynclines and emergent island arcs. They display isoclinal folding and ubiquitous thrusting. However, the source of the latest deforming stresses appears to have been the vast masses of intrusive granite that completely surround and intrude the rocks of the Mountain Land. These ancient granites do not have analogues in comparable younger geologic terrains, and it is suggested that their genesis is intimatedly related to the formation of the stable cratons. A further significance of these ancient rocks lies in the fact that very primitive forms of life have been identified in them.

II REGIONAL SETTING AND GENERAL GEOLOGY

The Barberton Mountain Land forms one of the best developed and best preserved remnants of ancient Precambrian strata in South Africa. It occurs in the Eastern Transvaal Lowveld, immediately east of the younger cover of the Transvaal Drakensberg Escarpment, and occupies a wedge-shaped, triangular tract of country

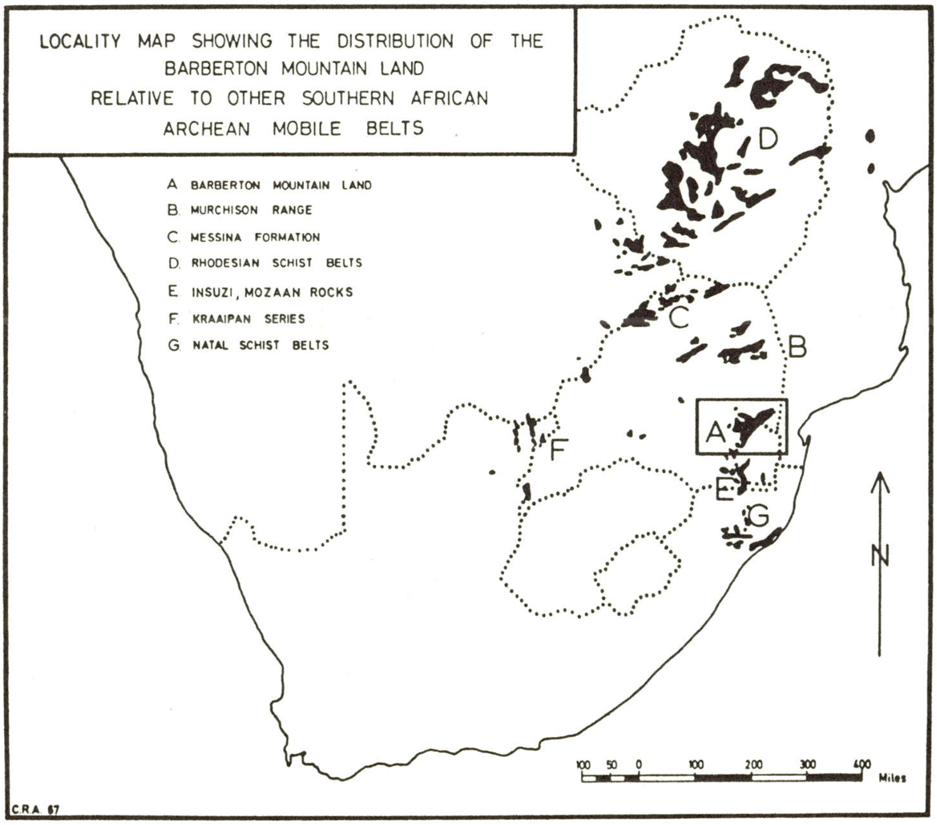

Figure 1
Locality map showing the distribution of the Barberton Mountain Land relative to the other Southern African Archaean mobile belts.

extending northeastwards from its broad base between Badplaas and Mbabane in the southwest, to its point of disappearance beneath the Lebombo Mountains on the Mozambique border near Komatipoort. Most of the area falls within the Transvaal, but the southeastern edge of the wedge straddles the Swaziland border and occupies the northwestern segment of the territory (Figure 1). As the name implies, the best development of the Barberton Mountain Land occurs in the area flanking the town of Barberton.

Geologically the area is underlain by a variety of rock-types collectively grouped as belonging to the Swaziland System (Plate I). The Onverwacht Series forms the oldest phase, and consists mainly of metamorphosed basaltic and ultrabasic rocks, together with siliceous sediments. The assemblage is essentially volcanic in origin, and has suffered considerable alteration in places, producing a variety of basic and siliceous schists. The rocks in the Series constitute the initial magmatic, or ophiolite, assemblage of the Swaziland geosyncline. The Fig Tree Series, which is essentially a sedimentary sequence, overlies the Onverwacht rocks, and consists mainly of fine-grained shales, grits, greywackes, banded cherts, and banded ironstones. Some lava horizons may also be developed in places. The sediments display characteristics of having been deposited in deep water by turbidity currents, and form an assemblage of rocks typical of the flysch sequence of geosynclines. Finally, the Moodies Series overlies the Fig Tree strata unconformably in places, and is made up of conglomerates, quartzites, shales, and a few bands of magnetic jaspilite and lava. These rocks provide evidence of shallow-water deposition, thus contrasting sharply with the essentially deep-water environment of deposition of the Fig Tree succession, and are likened to molasse sediments of geosynclines.

The Mountain Land is surrounded on all sides by a variety of granites and gneisses, some of which have produced profound metamorphic and structural alteration of the rocks of the Swaziland System (Hunter, 1965; Anhaeusser, 1966a; Roering, 1965, 1967; Anhaeusser and Viljoen, 1965; Urie and Jones, 1965; van Eeden and Marshall, 1965).

The Geological Survey (Visser *et al.*, 1956) envisaged a further event — that of intrusion of a vast, differentiated suite of basic rocks, referred to by them as the Jamestown Igneous Complex. Subsequent studies have questioned this interpretation, and the proposal has been made that the Jamestown rocks merely represent the metamorphosed and structurally altered equivalents of Onverwacht rocks (Anhaeusser and Viljoen, 1965; Cooke, 1965; Anhaeusser, Viljoen, and Viljoen, 1966; and Viljoen and Viljoen, 1967).

The rocks of the Barberton Mountain Land have suffered a considerable degree of structural deformation, with the consequent formation of numerous fold- and fault-structures. Many structural data have become available, following the investigations of the Geological Survey (Visser *et al.*, 1956; Ramsay, 1963; Herget, 1963; Anhaeusser, 1964, 1965, 1966b; Poole, 1964; Van Vuuren, 1964; Viljoen, 1964; Roering, 1965; Urie, 1965).

A critical examination of the stratigraphy of the relatively small remnant of Archaean rocks of the Barberton Mountain Land is indicated if we are to attempt construction of a model of the evolutionary events which led up to the complex development of a fold-belt. Evidence, in the form of other small schist belt remnants, far removed from the Barberton area, tends to suggest a once extensive development of Archaean rocks over the Southern African subcontinent. These remnants cannot, at present, be incorporated in the model, as very little is known about them. The model, of necessity, has to be constructed from all the available stratigraphic, sedimentological, and structural data that can be derived from the Barberton Mountain Land itself. The excellent exposure and preservation, coupled with the detailed knowledge that has recently been acquired from the area, have facilitated the construction of a model that appears, from the study of available literature, to be one of the most complete examples of an Archaean geosyncline, or fold-belt, anywhere in the world. All the fundamental elements are believed to be contained in the Barberton fold-belt, and it is contended that other, similar Archaean belts represent more poorly exposed, more intensely metamorphosed, and less complete fragments in which all the elements might not necessarily have been present, and where the stratigraphic succession and structure are not as clear.

The view is held that the majority of Archaean greenstone belts possess fundamentally similar elements (however obscure these may be), and experienced a broadly similar history to that of the Barberton Mountain Land, and that these belts may, in general, bear a close resemblance, or similarity, to the model formulated in this paper.

III DETAILED DESCRIPTIVE STRATIGRAPHY OF THE BARBERTON MOUNTAIN LAND

A The Onverwacht Series

Lying at the base of the Swaziland System is a sequence of rocks comprising basic pillow-lavas, ultramafites, and associated siliceous sediments. These rocks, collectively termed the Onverwacht Series, are best developed and preserved in the southwestern part of the Mountain Land, where they attain a thickness of over 35,000 feet. This essentially volcanic series can be traced right around the Mountain Land, and is everywhere in contact with the various intrusive granites which surround the whole area. Proceeding northwards, from the type-area in the south, the series thins rapidly, and also becomes more strongly thermally and dynamically metamorphosed. This metamorphism has resulted in the obliteration of many of the diagnostic lava structures, such as pillows and amygdales, which are so well developed in the south. It has also led to the previous grouping of these metamorphosed basic rocks into the Jamestown Igneous Complex (Visser *et al.*, 1956), which was considered to represent a post-Swaziland intrusive sequence of basic and ultrabaisc rocks. As pointed out above, however, it has been shown (Anhaeusser and Viljoen, 1965; Anhaeusser *et al.*, 1966; and Viljoen and Viljoen, 1967a and b), that these metamorphosed rocks, consisting essentially of amphibolites, and talc, talc-carbonate, and chorite schists, are the equivalents of the Onverwacht Series, as defined in the south, and together they constitute the initial magmatic, or ophiolite, assemblage of the Swaziland geosyncline.

1. *Stratigraphy*

In the type-area, in the southwestern part of the Mountain Land, the Series has been divided into three stages (Viljoen and Viljoen, 1967a). The Lower Onverwacht, or Theespruit Stage, which attains a thickness of 9,000 feet, consists of a sequence of metamorphosed basic lavas with interlayered siliceous sediments and salic tuffs, and occasional, thin, carbonaceous chert horizons and bands and lenses of serpentinized ultramafites. The main distinguishing feature of the Lower Onverwacht sequence is the occurrence of numerous siliceous, sedimentary or salic, tuffaceous horizons in the form of white, often friable, fine-grained quartz-sericite rocks, interlayered with the amphibolitized pillow lavas. Quartz and sericite are the main minerals in these rocks, but smaller amounts of andalusite, staurolite, biotite, muscovite, and sodic feldspar are also present. Most of these horizons are capped by thin, black, carbonaceous chert bands. The associated lavas, which frequently contain signs of pillow structures and amygdales, are dark green to black in colour, the main minerals being actinolite, hornblende, grunerite and cummingtonite, with smaller amounts of quartz and sodic plagioclase. Chlorite and tremolite-actinolite are more common farther away from the granite contact. Numerous, thin, apparently interlayered, serpentinized ultrabasic bodies are typical of the lower part of the Stage. These are composed mainly of antigorite and tremolite, with relict olivine crystals and magnetite being fairly common.

The Middle Onverwacht, or Komati River Stage, attains a thickness of 11,500 feet in the type-area, and comprises an alternating sequence of amphibolitized pillow basalts and ultrabasic bands, with the exclusion of any siliceous material. The ultrabasic rocks predominate in the lower half of the Stage, and a striking feature is the sympathetic thinning of individual ultrabasic and pillow basalt horizons. The ultrabasic rocks are remarkably fresh looking and are usually dark blue to black in colour. They are, however, generally serpentinized to a greater or lesser degree, although in many cases, only partly serpentinized olivine crystals form a mesh constituting much of the rock. Besides antigorite, which is the main mineral, varying amounts of tremolite, magnetite, clinopyroxene, and talc also occur. These rocks represent altered dunites. The main minerals encountered in the altered basalts are tremolite-actinolite, cummingtonite, chlorite, quartz, and feldspar. Important in the Middle Onverwacht Stage are numerous intrusive bodies of quartz- and feldspar-porphyry, which do not appear to be related to the surrounding intrusive granites, and are thought to represent an integral part of the Onverwacht succession.

The Upper Onverwacht, or Hooggenoeg Stage, reaches a thickness of 15,500 feet in the type-area, and consists of a sequence of pillow basalts with numerous, interlayered zones of andesitic to acid lavas, the latter often associated with well-developed chert horizons. The above assemblage constitutes the lower part of the Hooggenoeg Stage, and is unconformably overlain by a broad, irregular zone of "acid looking material" in the form of acid lavas and pyroclasts, salic tuffaceous material, and probable sediments. The lower part of the Hooggenoeg succession comprises a number of cycles, each commencing with a broad zone of andesitic to basaltic pillow lavas. The latter are followed by thinner, intermediate to acid lavas, and the cycles are frequently terminated by narrow, persistent, well-banded, and often carbonaceous chert horizons. Various calc-silicate and banded carbonate rocks are usually associated with these chert horizons. The predominant mineral in the more basic lavas is chlorite, accompanied by altered sodic plagioclase and carbonate. In the more acidic lavas, soda plagioclase, quartz, chlorite, sericite, and carbonate are found.

The Upper Onverwacht is developed in the southern part of the Mountain Land only, whereas the two lower stages can be recognized right the way around.

2. *Cyclic Evolution of the Onverwacht Series*

One of the most striking features of the Onverwacht Series, is the ordered cyclic nature of the volcanicity and sedimentation. The cyclic nature of deposition is perhaps best displayed in the Onverwacht type-area in the Komati River Valley (Viljoen and Viljoen, 1967a and b).

At least 10 major cycles and up to 20 minor cycles are developed in the Lower Onverwacht Stage. These cycles are characterized dominantly by the occurrence of basic and ultrabasic lavas, the former being in ascendancy over the latter. The lavas give way to rather thin salic tuffs (now quartz-sericite schists), the whole cycle being terminated by the deposition of narrow chert bands in quiescent conditions before the onset of the subsequent cycle. One to five lava flows, varying in thickness from 5 to

300 feet, constitute the extrusive volcanic phase of a particular cycle. In the Middle Onverwacht Stage, the cyclic nature of the volcanicity is manifested by alternating pillow lavas and ultrabasic rocks. In the Upper Onverwacht sequence the cyclic nature of the volcanicity and sedimentation is remarkably well demonstrated, and at least five cycles, varying in thickness from 500 to 5,000 feet, and diminishing in size upwards, have been mapped. The latter cycles are characterized by the initial outpouring of basic to intermediate lavas (5 to 20 flows or more), followed by a thin zone of acid lava (one to three flows) which, in turn, is followed by the deposition of a substantial zone of black and white, sometimes slightly ferruginous, banded chert. Judging by the thickness and extent of these cherts, they were deposited over fairly long periods of quiescence before the initiation of the following cycle. The upper division of the Upper Onverwacht consists of a huge mass of salic volcanic sediment and lava overlain by a very substantial banded black and white chert (also sometimes slightly ferruginous). This cycle, characterized by a marked period of quiescence during which the chert was deposited, marks the end of the Onverwacht Series, as defined in the type-area in the southwestern part of the Mountain Land.

3. *The Environment, Conditions of Deposition, and Geological Significance*

Most geosynclines are characterized by an initial volcanic episode during their early evolution. Abundant outpourings of lava, known as ophiolites, are characteristic. These rocks are liable to metamorphism and structural disturbance during the subsequent orogenic phase of the geosyncline, as well as during the intrusion of granite (in the case of the Barberton Mountain Land, at a much later date). These processes give rise, as a result, to a complex association of basic rocks — the universal "greenstones" of English geologists, and the "grünes gestein" and "pietri verdi" of European geologists. Despite the severe structural disturbances which most ophiolite assemblages have undergone, three main rock-types — basic lavas, ultrabasic bodies, and cherts — emerge as fundamentally important units. The recognition of the association of these rocks by Steinmann (1905) was of classic importance, and subsequently became known as the Steinmann Trinity. Most workers have employed the presence of pillow-structures in the lavas as evidence of submarine extrusion. The cherts, on the other hand, are thought to represent chemical precipitates, deposited during periods of quiescence between the outpourings of lava. The origin of the ultrabasic rock (the so-called Alpine peridotites) remains, however, one of the most enigmatic of geological problems.

From the description of the stratigraphy of the Onverwacht Series, presented above and described fully by Viljoen and Viljoen (1967a), it is abundantly clear that the rocks of the Series have their more recent analogues in the ophiolite assemblages of geosynclines and present-day emergent island arcs. This being the case, it is conceivable, and likely, because of the huge thickness of the sequence and the occurrence of chemically precipitated sediments, that the environment of deposition of the Onverwacht Series was that of a deep-sea trough, or graben, in a crust which was only a fraction of its present thickness.

Despite the age of the rocks, a number of factors, including the relatively unaltered nature, the low degree of thermal and dynamic metamorphism, and the excellent exposures, makes the southern part of the Mountain Land, in the Komati River Valley, a remarkable area for the study of a complete ophiolite sequence. Many of the problems associated with ophiolites of more recent geosynclinal belts, and the problems of the greenstones of the shield areas may find a ready explanation in the model presented by the Onverwacht Series. Initial results (Viljoen and Viljoen, 1967b) indicate, for example, the close association of cherts with acid lavas or acid tuffs, and a relationship between the silica (volcanic in origin) content of the water and the deposition of cherts. Furthermore, mounting evidence from the most undisturbed portions of the Mountain Land indicates the possibility of widespread outpourings of ultrabasic magma in the form of flows or extrusive pods which, in some places, show distinct signs of differentiation. Many apparently intrusive ultrabasic bodies are thought to represent tectonically transported portions of the Onverwacht sequence.

In addition, the tholeiitic nature of many of the lavas is becoming apparent. These low-potassium tholeiites appear to have their modern counterparts in the present-day ocean basins (Engel *et al.*, 1965; Engel, 1966), and the "primitive" (uncontaminated) nature of the volcanics is attributed to the passage of magma through a relatively thin sialic crust. Chemical data on these lavas should, therefore, give valuable information as regards the composition of the upper mantle of the earth some 3.4×10^9 years ago.

B The Fig Tree Series

Immediately and conformably overlying the predominantly volcanic successions of the Onverwacht Series is a group of rocks composed essentially of siliceous chemical precipitates, together with a great accumulation of pelitic sediments. Where best developed in the Ulundi Syncline, north of Barberton, this Fig Tree succession has been subdivided into a lower division, which includes rocks that are at times difficult to distinguish from the metamorphic products of the underlying Onverwacht Series; a middle division which includes the typical Fig Tree rocks, notably turbidites (greywackes) and banded cherts; and an upper division, characterized by a reworked, unconformable grit horizon, more greywackes, and a volcanic horizon. The Fig Tree rocks have recently yielded evidence of primitive life dating back more than 3,000 m.y.

1. *Distribution of the Fig Tree Series*

Rocks belonging to the Fig Tree Series are best developed towards the interior regions of the Mountain Land, where they are sandwiched between the underlying Onverwacht sequence and the overlying group of Moodies strata. The Fig Tree sequence attains its fullest development northeast of Barberton, in the Ulundi Syncline, where Ramsay (1963) estimated a total thickness of at least 10,000 feet for the tectonically flattened strata. Fig Tree rocks are also well developed in the area south of Malelane, in the area southwest of Barberton, and in the northwestern segment of Swaziland. The resistant cherts of the Series build lofty mountain chains while the softer shales occupy deeply incised V-shaped valleys.

2. *Rock Types and Stratigraphic Subdivisions*

The Fig Tree Series is composed of a highly varied group of sediments composed essentially of banded cherts, banded ironstones, banded jaspers, shales, greywackes, quartzites, grits, and conglomerates. The Zwartkoppie Stage, or lower division of the Fig Tree Series, consists of three main members, viz., talc-carbonate rocks, quartz-sericite rocks, and banded chert. At present, it is not certain whether the Zwartkoppie Stage forms the true base of the Series. In the southwestern parts of the Mountain Land, shales and cherts of the Fig Tree Series overlie either acid or basic volcanic rock-types, and at no place can the typical assemblage of the Zwartkoppie Stage be identified in contact with the Onverwacht Series. In the Ulundi Syncline, carbonate-chlorite-talc rocks occupy the lowest stratigraphic position, and occur in the cores of anticlinal structures (Van Vuuren, 1964). Herget (1963) noted serpentinites occurring as cores in the talc-carbonate schists of the Zwartkoppie Stage in the Montrose area, southwest of Barberton. Van Vuuren (1964) showed that the Ni and Cr content of these carbonate-bearing rocks is suggestive of their being of volcanic origin, and suggested that they had moved into the anticlinal cores by means of tectonic transport, i.e. the upfolding of the volcanic rocks of the underlying Onverwacht Series. The "Zwartkoppie Horizon" occurs as a marker in the Zwartkoppie Stage. In places the horizon is comprised solely of massive and banded chert, but well-bedded cherts and black shales are also developed. In addition, ferruginous shales and banded cherts, referred to as "banded ironstones", occur here, as well as higher in the succession. Koen (1947) concluded that the greenschists in the Zwartkoppie Stage are markedly similar mineralogically to greywackes, and that they owed their origin to intense mylonitization of greywackes and shales, sandwiched between the competent banded cherts and the underlying carbonate-rocks. Van Vuuren (1964) recorded a thin but persistent, sulphide band at the top of the Zwartkoppie Stage, the areal extent of which suggested a syngenetic origin with the banded chert. Overlying the "Zwartkoppie Horizon" is a carbonate-bearing sandy phyllite, followed by an oolite layer.

The middle Fig Tree sequence consists of a very thick succession of alternating greywackes, shales, and grits, with thin chert bands. Higher in the sequence more greywackes, shales, banded ironstones, and gritty greywacke layers are developed.

The upper Fig Tree sequence consists of still more greywackes and shales, banded chert, banded ironstone, and grits. At the top of the succession is a felspathic, tuffaceous greywacke. A greywacke conglomerate is also locally developed.

The mineralogical composition of the greywackes throughout the Series indicates material clearly derived from basic and acid volcanics, as well as from crystalline bodies such as granites, gneisses, or migmatites.

3. *Evidence of Primitive Life Forms*

The cherts and shales of the Fig Tree Series have yielded evidence of very primitive life-forms dating back more than 3,000 million years. Ramsay (1963) reported oolitic structures, resembling primitive algae (Calcisphaera) from the Sheba Mine area. Hoering (in: Pretorius, 1965) reported the presence of straight-chained hydrocarbons with 16 to 24 carbon atoms in Fig Tree shales. These, he suggested, were diagnostic of biologically produced petroleums. More recent studies by Pflug (1966) and by Barghoorn and Schopf (1966) have revealed microfossils preserved mainly in the dark banded cherts. Pflug (1966), furthermore, suggested that the algal bodies were able to precipitate metal salts, with cations such as copper, iron, and calcium, from water, by the action of their life processes. Biological processes of such a type, he believed, could have played an important factor in the formation of sedimentary iron formations.

Cloud (1965), in discussing the Gunflint microflora from Ontario, presented similar ideas concerning the origin of Precambrian banded ironstone formations. The Gunflint thallophytes at 1,900 m.y. were, up to the time of the discoveries of similar fossil organisms in the Barberton cherts by Pflug (1966) and by Barghoorn and Schopf (1966), the oldest recorded structures closely resembling specific living organisms. Cloud (1965) suggested that the thallophytes may have been photosynthesizers which attached any free oxygen that was generated to some convenient oxygen acceptor in the hydrosphere. Ferrous iron, transported to the basins of deposition in reducing surface-waters, would have been such a convenient acceptor for any available oxygen. The oxidized iron may then have formed the typical banded ironstone formations. This would explain the apparently anomalous situation of ferric (banded ironstone) formations being deposited in an early atmosphere that has widely come to be regarded as of a reducing type (Rubey, 1955; also: Symposium on the Evolution of the Earth's Atmosphere, 1965).

4. *Sedimentation Pulses*

The Fig Tree succession can broadly be divided into three phases or pulses of sedimentation, which cannot be defined as clearly as those of the underlying Onverwacht rocks. At the base of the Series the rocks are composed essentially of a considerable thickness of precipitated cherts, banded cherts, shales and other fine argillites of the Zwartkoppie Stage. This period of quiescence seems to have been carried over from late Onverwacht times as many of the rock types in the lower portion of the Zwartkoppie Stage are similar in character to the underlying, essentially volcanic assemblages.

The next pulse resulted in the rapid accumulation of the "greywacke suite" and it began with the deposition of coarse greywackes, grits, and shales, with some minor chert interlayers. Later in the sequence, the greywackes become much finer than lower down, and there is more shale and numerous banded chert and banded ironstone layers towards the top.

The final, broadly recognizable sedimentation pulse of the Fig Tree Series again began with the accumulation of grits and coarse greywackes and, at the end of Fig Tree times, the deposition took place of a tuffaceous, feldspathic greywacke and a greywacke conglomerate, the latter merging in places with the basal conglomerate of the Moodies Series.

5. *Environment and Conditions of Deposition*

The greywacke suite, or flysch, is a well-established facies of a geosynclina environment, and is generally closely associated with spilitic and pillowed greenstones In the Barberton Mountain Land, the Fig Tree assemblage of rocks appears to conform with this pattern, being closely associated in time with the Onverwacht volcanic successions. Typically, the Fig Tree rocks consist of a considerable thickness of argillites. These are essentially clastic, although there were apparently quieter periods when bedded cherts and banded ironstones were also laid down, particularly in the early stages of the sequence. As pointed out previously the presence of primitive photosynthesizing organisms in the early depository probably played an important part in the formation of the banded ironstones.

Deposition of the sediment was nearly continuous and uninterrupted, with the bedding being rhythmic and well marked. Graded bedding is frequently seen in the greywackes, microbreccias and grits, and poorly sorted greywackes reflect incursions of slump-generated turbidity-flows. A great variety of sedimentary structures is of use in establishing the environment and style of deposition of these rocks. Slump-bedding, current-marks, load-casts, flute-casts, convolute bedding, and graded bedding are characteristic of the sequence. The rhythmically bedded formations of alternating shales and well-graded greywackes led Kuenen (1963) to classifying the Fig Tree Series as a turbidite formation. He recognized numerous sedimentary structures indicating that the greywackes were formed by turbidity currents. Both Van Vuuren (1964), who identified flute-casts, load-casts, and possible slide-marks in the greywackes of the Ulundi Syncline, and Herget (1963), who observed flute-casts and convolute-bedding in the Montrose area, confirmed Kuenen's earlier findings, thus making the Barberton turbidity formations the most ancient yet recorded anywhere in the world.

A gradual increase in the coarseness of the sediments is evident towards the top of the Fig Tree sequence, with grits, coarse greywackes, and minor conglomerates making an appearance. The conclusions that can be drawn, therefore, suggest that an extremely thick formation of sediments accumulated rapidly in a deep-water marine environment. Texturally and mineralogically, the sediments are very immature, and must have been derived from erosion of a high land mass believed, in part, to consist of a crystalline terrain, as well as of strata deposited earlier in the same geosyncline, or trough.

C The Moodies Series

1. *Distribution*

The Moodies Series follows, in places conformably, in others unconformably, on the underlying Fig Tree succession. It consists mainly of arenaceous rocks, with minor sandy-shale horizons and a characteristic and persistent basal conglomerate. The rocks of this Series cover large areas, especially in the interior of the Mountain Land, where the formations build lofty ranges responsible for much of the scenic beauty of the area. Although the Series has a wide distribution, particularly in the central and eastern portions of the area, the Moodies rocks are generally located in synclines that are separated from one another by large, longitudinal thrust-faults that trend in a northeast-southwest direction and which divide the Mountain Land into a number of fault-bounded "cells". The Lily, Eureka, Saddleback, and Stolzburg Synclines are but a few examples of major structures in which Moodies rocks are preserved.

2. *Lithology*

The Moodies Series is composed predominantly of psammitic and psephitic sediments. The dominant rock types are boulder-beds, conglomerates, quartzo-feldspathic sandstones, subgreywackes, siltstones, and shales, with local developments of magnetic banded ironstones and jaspilites. Also included in the stratigraphy of the Moodies rocks of the Eureka and Saddleback Synclines is a thin amygdaloidal lava horizon, now mainly amphibolitized.

The conglomerates consist of well-rounded pebbles and boulders. The conglomerate is polymictic, containing pebbles of black and white chert, banded ironstone, jasper, grit, granite, quartzite, micropegmatite, quartz porphyry, shale, and quartz-sericite schist. In addition, pebbles of conglomerate material have also been found in the basal conglomerate of the Eureka Syncline. The pebbles occur in an arenaceous matrix of feldspathic quartzite, greywacke, and sandy shales. The conglomerate is generally followed by a carbonate-rich quartzite.

The most complete succession of Moodies strata is preserved in the Eureka Syncline, where it has been broadly subdivided by Visser *et al.* (1956), as follows:

Upper Shale
Upper Quartzite (Baviaanskop)
Middle Shale
Middle Quartzite (Joe's Luck)
Lower Shale
Lower Quartzite
Basal Conglomerate

The thickness of sediments along the southern limb of this syncline is about 10,300 feet, of which argillaceous rocks represent 46 per cent. Along the northern limb the thickness is approximately 6,500 feet, of which 75 per cent consists of argillaceous material. Magnetic shales and jaspillites occur near the base of the Lower Shale horizon. The

Middle Shale also has a magnetic shale and jasper horizon, as well as a thin band of amygdaloidal lava. The Upper Shale is capped by a thin, small-pebble conglomerate, with a matrix of coarse, gritty sandstone. The three quartzite horizons have locally developed conglomeratic bands, but these never approach the dimensions and character of the basal conglomerate. The Moodies rocks of the central portion of the Mountain Land appear to consist of vast thicknesses of quartzitic material, and it is not possible to correlate these masses with the Moodies stratigraphy as defined above. In Swaziland, only the basal conglomerate and the Lower Quartzite are represented.

3. *Sedimentation Pulses*

There appear to have been several stages of sedimentation of the Moodies succession. In the type-section of the Eureka Syncline, three main pulses can be recognized. The first of these began with the deposition of the basal conglomerate, followed by calcareous orthoquartzites and feldspathic quartzites, and ended with shales and banded magnetic jaspilites. The presence of calcareous quartzites alternating with impure limestone and marble bands (orthoquartzite-carbonate facies) in the lower portion of the Moodies Series is suggestive of a shallow-water depositional environment, marginal to a low-lying stable land surface.

The next pulse began with quartzite development (only locally conglomeratic), followed by a minor period of volcanism and a second, banded magnetic jasper horizon, together with shales. The third and last pulse began with minor (local) conglomerate and quartzite development, which was followed by an alternating sequence of shales, subgreywackes, grits, and, finally, a small-pebble conglomerate.

Each pulse appears to have started with a great influx of coarse detritus. Gradually, the depositional pattern trended towards an environment that allowed for the settling out of shales, fine-grained argillites, and chemically precipitated jaspilites and banded ironstones.

4. *Conditions of Deposition*

All the quartzites show abundant cross-bedding and ripple-marks. Mud-cracks frequently occur in the shales. In the Eureka Syncline, trough cross-bedding is particularly prominent in some areas, and appears to be indicative of transportation of sediments in braided stream channels. Graded bedding and cross-bedding can frequently be used to determine the direction of younging of the sediments. Bedding in all rock-types is clearly evident, although in some of the shales in the Eureka Syncline it is, at times, obscured by a strong cleavage developed obliquely to the stratification. The sedimentary features, together with the development of conglomerates, which locally display poor sorting, are indicative of deposition in shallow water. This contrasts sharply with the deep-water sedimentational environment of the Fig Tree shales, banded cherts, and greywackes.

Much of the Moodies sedimentation can be ascribed to the denuding of previously deformed Fig Tree and Onverwacht Series rocks. Evidence that the Fig Tree succession had been subjected to folding prior to the deposition of the Moodies Series has been extended by Visser *et al.* (1956), who reported the presence of contorted banded chert pebbles in the Moodies conglomerates. These contorted banded cherts are, however, not entirely diagnostic of an earlier deformation, for they could equally well have been formed by slumping within the depository itself. The polymictic nature of the basal conglomerates is ascribed to the influx of pebbles of black chert, quartz porphyry and quartz-sericite schists from denuded Onverwacht successions, while the banded ironstone, jasper, grit, and banded chert pebbles could have been derived from the Fig

Tree Series. Furthermore, the presence of granitic, micropegmatitic and quartz porphyry pebbles are ascribed to erosion of an elevated granitic terrain close to the Mountain Land. This terrain is also regarded as the basement upon which the Swaziland System sediments and volcanics were deposited. Microscopic examination of the sediments in the Moodies Series of the Eureka Syncline has disclosed the presence of abundant sodic and potassic feldspar in the form of oligoclase, albite, microcline, and perthite, prevalent particularly in the impure quartzites and sandstones. These minerals are all abundantly encountered in the granite gneiss terrain of the Eastern Transvaal Lowveld.

Although much of the material in the Moodies succession appears to be essentially the product of a high-energy turbulent environment, there must have been short periods of quiescence in which the banded magnetic shales and jaspilites and other finely laminated shales were deposited. The depositional cycle was also interrupted by a period of mild volcanism, during which time the narrow band of amygdaloidal lava was extruded.

The clastic sediments of the Moodies Series, consisting mainly of conglomerates, quartzites, sandstones, and shales, can be defined as a molasse association. Furthermore, the over-all character of the Moodies rocks conforms with the molasse requirements outlined by Pettijohn (1957). These requirements include the presence of calcareous sands and shales, marked cross-bedding, and sediments formed in varying environments, that are differentiated texturally (and hence chemically and mineralogically) according to the energy controls of the depositional agencies. In addition, the molasse of the Moodies frequently transgresses the flysch of the Fig Tree Series and the underlying ophiolite sequence of the Onverwacht volcanic assemblage of the Barberton geosyncline.

IV DYKES

Hypabyssal rocks, mainly in the form of dykes, are prolifically distributed over the entire area of the Barberton Mountain Land, but attain their greatest development in the areas underlain by granitic rocks. Practically all the dykes are vertical, and many give rise to striking topographical features. The dykes are of several varieties, each of which tends to prefer a certain well-defined strike direction. The dykes vary widely in age and in mineralogical and chemical composition. Sheet- or sill-like intrusions occur, but are less frequent than the dykes.

The dykes generally conform with definite strike directions. A prominent direction is from 30° to 35° west of north to approximately northwest (Visser *et al.*, 1956). A second direction is roughly at right angles to the former, i.e. from 30° east of north to northeast. Some dykes trend north-south, others parallel the strike of the formations, and yet others deviate completely from these directions. Zones of faulting and jointing appear to have had the major controlling influence on the occurrence and trend of the dykes.

Highly deformed early dykes occur in the Nelspruit Granite, the Badplaas Granite, and in the Ancient Gneisses of Swaziland. These dykes are often folded, the axial planar structure of the migmatites being identical to the axial plane of the dykes (Roering, 1967). The dykes are locally cross-cut and injected by the granitic matrix into which they at one time intruded. The dykes are now amphibolitized.

There are several ages of early dykes; some have been folded, others have not, but both types have been involved in the palingenesis of the surrounding migmatitic and gneissic material. The ancient gneisses and associated dykes were possibly reactivated during the G.4, or synorogenic, granitic cycle which is dated at 3,070±60 m.y. by Allsopp *et al.* (1962). The inference to be gained from the involvement of the early dykes in the migmatite gneiss palingenesis is evidence in support of this terrain being representative of an ancient, pre-Swaziland System basement.

Dykes later than the Swaziland System have been divided into several categories (Visser *et al.*, 1956) which reflect the age relationships between the dykes themselves and the various stratigraphic successions of the Mountain Land and the surrounding areas.

The dykes have been grouped as follows:

(1) Intrusions older than the Moodies Series;
(2) Intrusions older than the Godwan Formation;
(3) Intrusions younger than the Transvaal System;
(4) Late dykes of Karroo dolerite.

The majority of dykes in the area are diabasic in composition. Some of the larger diabase bodies have differentiated into a wide variety of rock-types. These include peridotite, olivine gabbro, quartz gabbro, quartz diorite, and granophyric end-members with micrographic intergrowths of quartz and feldspar. Some diabase dykes are porphyritic and contain large feldspar phenocrysts. Pyroxene-amphibole dykes and ultrabasic dykes occur less frequently. Karroo dolerite dykes mark the last phase of igneous activity in the area.

V GRANITES

The supracrustal rocks of the Mountain Land are surrounded by granites which are all clearly intrusive, and which are also responsible for contact-metamorphic effects. The contact rocks are most frequently members of the Onverwacht Series (Jamestown of the Geological Survey). These rocks, generally varying from basic to ultrabasic types, were considered to have acted as barriers or basic resisters against granitization, thus preserving the sediments higher up in the stratigraphy from metamorphism (Visser *et al.*, 1956). However, if the basement on which the Onverwacht rocks were deposited were granitic and made rheomorphic, to produce diapiric intrusive domes, the latter would be draped by Onverwacht rocks in a manner similar to the mantled gneiss domes of Eskola (1948).

Hunter (1965) placed the granitic rocks of Swaziland in a coherent sequence, based on the concept of cycles. He recognized an Ancient Gneiss Complex, comprising an assemblage of gneisses and granulites, with recognizable relics of metasediments. These rocks are intimately associated with the Granodioritic Gneisses, which are thought to be derived from the former group by processes of granitization. Both groups are well foliated, and represent members of the earliest cycle.

The next cycle began with geosynclinal conditions and the deposition of the Swaziland System on rocks belonging to the first cycle. The first deformation of the orogenic phase of this cycle led to the formation of the Synorogenic Granodiorites and Granites. These rocks vary from quartz diorite to leucocratic granodiorite. Locally, there is evidence of complete mobility, with the rocks displaying intrusive contacts and randomly oriented amphibolite fragments. In addition, rocks of this group reveal a definite foliation which is also present in the rocks belonging to the first cycle. Hunter (1965) concluded that the earlier gneisses also reveal an earlier deformation, and that the synorogenic granites crystallized from a magma.

The Late Orogenic Granites are distinguished from the former types in that they are generally devoid of planar or linear fabrics, except close to the margin, where they intrude the supracrustal rocks. They are typically massive, dark grey, medium-grained granites and granodiorites, the latter being quantitatively less abundant than the former. The association of pegmatites with this late granitic variety is frequent, and most of the mineralization of Swaziland (Au, Sn, Nb, Ta, Sb, Wo, and Mo) is related to this granite event.

A fundamental feature, noted by Hunter (1965), is that the Late Orogenic Granites occupy a higher position in space than do the previously described granites. They are thus sheet-like in shape.

Several large plutons of coarse-grained porphyritic granite occur in the territory, and reveal sharply transgressive contacts. These Post-orogenic Granites are not all of the same age, and were emplaced when the Swaziland area ceased to be part of a mobile belt. They were thus intruded after the close of the second cycle, which ended with the emplacement of the Late Orogenic Granite.

A less rigorous breakdown of granites in South Africa has led to the recognition of a number of distinct types, viz. the Kaap Valley, Nelspruit, Nelshoogte, M'pageni, Salisbury Kop, and Dalmein granites.

The Kaap Valley Granite forms a huge pluton surrounded by Swaziland rocks (Plate I). It is essentially a hornblende granodiorite, often containing biotite. A distinct, foliated fabric is developed close to the contact area with the supracrustal rocks. The granite is clearly intrusive into the Swaziland rocks. The Nelspruit Granite is the name given to vast areas of granitic rock that are distinct from the other varieties. It is made up of highly variegated rock-types, ranging from homogenized granites without any visible fabric, through gneisses, to complex migmatites. On the northern side of the belt of supracrustal rocks, this granite reveals a homogenised phase, several thousand feet wide, which is confined to the contact zone. Pegmatites are clearly related to this homogenized phase. The Nelshoogte Granite, like the Kaap Valley Granite, forms a pluton north of Badplaas, that is surrounded by Swaziland System rocks in the east, but is covered by younger formations in the west. It is essentially a granodiorite, containing biotite, and has myrmekitic intergrowths of quartz and feldspar. The granite is intrusive into the Swaziland System rocks and possess a well-defined foliated fabric near the contacts.

The M'pageni Granite occurs as a discordant pluton in Nelspruit Granite. It is characteristically a coarse-grained granite somewhat similar to the post-orogenic granites described by Hunter (1965). The Salisbury Kop Granite is characterized by having no visible foliation, and is somewhat similar in appearance to the M'pageni Granite, in places. It is generally a medium- to coarse-grained granite which intrudes Swaziland rocks. Its exact relationship to the Nelspruit Granite is not established, as yet. Van Eeden and Marshal (1965) consider this granite to be a rheomorphosed equivalent of the basement on which the supracrustal rocks were deposited. The writers, however, are of the opinion that this body is a younger granite pluton, similar to the M'pageni and post-orogenic granites (formerly known as the G.5 granites), described by Hunter (1961, 1965). The granite is typically homogeneous, and has a narrow chill-phase of aplitic and alaskitic granite around the edges, in contact with the Swaziland System rocks (Viljoen and Viljoen, 1967b). Another, essentially similar, post-orogenic granite pluton, called the Dalmein Granite, invades the Onverwacht successions in the Komati River Valley (Plate I), and truncates the sequence abruptly. The granite is a coarse-textured rock, and possess no contact foliation (Viljoen and Viljoen, 1967b).

Table I compares the three major groups of granitic rocks in Swaziland and South Africa, and includes the interpretation currently favoured by the authors.

TABLE I

A COMPARISON OF THE THREE MAIN GROUPS OF GRANITIC ROCKS—PREVIOUS VIEWS AND PRESENT INTERPRETATION

SWAZILAND Hunter (1965)	SOUTH AFRICA (pre-1967) Ramsay (1963); Anhaeusser (1964, 1966a); Anhaeusser and Viljoen (1965); Roering (1965); (1965); Viljoen (1964).	PRESENT PAPER
3. Post-orogenic Granites	M'pageni and Salisbury Kop Granites	Younger high-level porphyritic granites, such as the M'pageni, Salisbury Kop, Dalmein, and G.5 (or Post-orogenic) of Swaziland; (granites non-foliated; cut cleanly across the trends of the supracrustal rocks).
2. Late-orogenic Granite Synorogenic Granodiorites and Granites	Mobilized border-phase of Nelspruit-type Migmatites and Gneisses Kaap Valley Granite and others—representing rheomorphism of the basement gneisses	3000 m.y. granitic episode superimposed on an already orogenically deformed Swaziland System—invasion of diapiric granite bodies such as the Kaap Valley Granite, Nelshoogte Granite, etc.; (granites foliated; responsible for structural disturbance of the supracrustal Swaziland System rocks)
1. Granodioritic Gneisses Ancient Gneiss Complex	Basement represented by Nelspruit-type Migmatites and Gneisses	Basement Migmatites and Gneisses.

VI STRUCTURE OF THE BARBERTON MOUNTAIN LAND

In broad terms, the Barberton Mountain Land represents a strongly deformed synclinal, boat-shaped remnant of layered, supracrustal rocks, intruded and completely enveloped by a vast and complicated granitic terrain. This synclinorial keel has a length of 80 miles, and a strong east-northeasterly trend. It forms a narrow, elongated triangle, tapering gradually from a maximum base width of 32 miles in the south (between Badplaas and Forbes Reef), to five miles in the extreme northeast, near Hectorspruit, where it disappears under a younger cover of Karroo sediments. The average width of the belt (between Barberton and Pigg's Peak) is 17 miles.

A The Arcuate Schist Belt Remnants

Strikingly apparent from the structural map (Plate II) are the generally arcuate, narrow, tapering protruberances of layered rocks which trend in all directions away from the main, east-northeast-trending grain of the triangular keel. These protruberances, or schist belts, attain their best development on the northwestern, western and southwestern sides of the Mountain Land, but are also present as much smaller and generally more conformable features along the other contacts of the Mountain Land. The most spectacular of these is the Jamestown Schist Belt which trends in a northwesterly direction from Noordkaap to Kaapsehoop. It is slightly arcuate, being concave towards the south, i.e. towards the Kaap Valley Granite. The belt has a strike-length of about 22 miles, from Noordkaap to the Transvaal Drakensberg Escarpment, where it disappears under the younger, flat-lying quartzites of the Black Reef Series and the lavas of the Godwan Series. Another conspicuous, tapering, triangular belt is the Nelshoogte Schist Belt on the southern side of the Kaap Valley Granite. It also trends in a northwesterly direction, and is slightly arcuate. Some of the most spectacular arcuate schist remnants occur on the south side of the Mountain Land, between Badplaas and Oshoek. Although much smaller than the two above-mentioned belts, numerous coalescing tongues form interferring arcuate structures, as can be well seen north of Lochiel (Plate II).

A significant feature of these relatively small, arcuate belts is that almost all of them consist of metamorphosed basic volcanics and ultrabasic rocks of the Onverwacht Series (particularly the two lower stages). This also applies to the outlying, small remnants within the granites near Alkmaar, in the area north of Badplaas, in the area of the Kruger National Park, and in the region extending from Kaapmuiden towards the Consort Mine.

Of importance in connection with the origin of the arcuate belts is the nature and disposition of the surrounding granites. In the southwest, where the arcuate belts are best developed, the granitic terrain consists of a series of generally elliptical or circular, discrete, homogeneous granite plutons or domes, varying in diameter from 25 miles (long axis of the Kaap Valley Granite) to one mile. These granites are intrusive into the Onverwacht lavas and, although not always strikingly apparent, are clearly transgressive in many cases. The plutons are partly separated from one another by the arcuate tongues of basic material, and have clearly been responsible for the strong thermal and dynamic metamorphism of the latter, as well as for their present structural disposition. Evidence of this is afforded by the development of a strong metamorphic fabric in the metavolcanics, which closely parallels a similar foliation in the adjacent granites. The latter foliation is caused by the alignment of platy minerals, and is accentuated by the alignment of metamorphosed basic xenoliths.

Clearly, the metamorphosed arcuate belts have originated as a result of the updoming, and/or intrusion, of more or less discrete, and partly interfering, granite domes of different sizes, which have moulded the metavolcanics to fit their outlines. It follows that the amount of arcuation and, to a lesser extent, the size of every basic schist belt, is a direct function of the size (radius) of the individual, adjacent granite plutons. Thus, the largest and least arcuate of the belts, viz. the Jamestown Schist Belt and the Nelshoogte Schist Belt, surround the largest granite pluton (the Kaap Valley Granite). The smallest and most arcuate belts, viz. those just north of Lochiel, partly envelope the smallest plutons.

Although the Kaap Valley Granite is the largest of the obvious granite plutons, and the Jamestown Schist Belt is the largest of the arcuate belts, it is the opinion of the authors that the entire eastern contact of the Mountain Land has been gently arcuated by the emplacement of an even bigger granite body. In its general characteristics and relationship to the Mountain Land, this granite, the so-called synorogenic granite of Swaziland, is closely analogous to the smaller granite plutons mentioned above. It represents a very large equivalent of the latter, and an idea of its dimensions can be gained from the gentle arc of curvature of the whole eastern contact of the Mountain Land.

The northern granite contact is of interest in that it represents a zone of strong shearing — a phenomenon which continued after the consolidation of the narrow belt of intrusive granite along this contact. This is evidenced by the cold mechanical grinding which these granites suffered following crystallization.

B The Major Fold Structures

Conspicuous within the main northeast-trending keel, and constituting the main structural features of the Mountain Land, are a series of well-developed, predominantly tight, synclinal folds. These trend in a general northeasterly direction, and have steeply dipping limbs, often overfolded to the northwest. The axial planes may be locally folded or warped, as in the Eureka and Ulundi synclines. Other prominent, northeast-trending synclines include the Saddleback, Stolzburg, and Kromberg synclines. One major exception is the Jamestown Syncline which trends northwest (parallel to the superimposed axial plane of the inflected Eureka and Ulundi synclines). This syncline occurs in the largest arcuate schist belt in the Mountain Land.

Anticlinal structures between major synclines are generally absent, having been faulted out in many cases. If anticlines are present, they are very poorly developed in comparison with the adjacent synclines. Two conspicuous anticlines are the Steynsdorp Anticline, in the Komati River Valley, and the Stentor Anticline, to the southeast of Kaapmuiden. As will be shown later, these anticlines have a different origin from those between the major northeast-trending synclines. A large, open anticline (the Onverwacht Anticline) forms a major structural feature in the southern part of the Mountain Land. Unlike the tight, east-northeast-trending, isoclinal fold structures to the north, and those flanking this fold structure on the east and west (Stolzburg and Kromberg synclines), the Onverwacht Anticline is a broad, open fold, and, as such, poses certain structural problems.

C The Major Faults

The Barberton Mountain Land has a striking array of major east-northeast-trending faults. These extensive longitudinal strike-faults divide the Archaean fold belt into a series of long, narrow blocks, trending east-northeastwards, parallel to the regional grain. They are generally of a high-angled thrust type, and often occupy positions between adjacent synclines. Good examples are the Barbrook and Sheba faults, which are of great economic importance, being closely associated with major gold deposits. Another type of fault with which mineralization is also often associated occupies the contact between zones of incompetent and relatively more competent rock types. These are more in the nature of zones of strong shearing, rather than major thrust dislocations. Examples of this type are the Lily Fault and the Consort "Contact" in the north, and the Violet Fault in the southern part of the Mountain Land. Of special note are the Kaap River and Albion faults in the Jamestown Schist Belt. Unlike the majority of other faults which follow the east-northeast grain of the Mountain Land, these trend northwestwards, parallel to the trend of the Jamestown Schist Belt itself. The Kaap River and Albion faults occupy positions in the north and south limbs, respectively, of the Jamestown Syncline. Both converge towards the west, and it is tentatively suggested that they join each other, and represent an early strike-fault that has been folded together with the strata of the Jamestown Belt. A somewhat similar situation exists in the southwestern part of the Mountain Land, where the Moodies and Sheba faults, from the northeast, and the Theespruit and Komati River faults, from the east, converge towards the extreme southwestern tip of the Mountain Land. Although not proved as yet, because of lack of data, it is tentatively put forward that these two sets of faults join with each other. They would thus represent very early faults which have been folded. As mentioned previously, a zone of very strong shearing and mylonitization occurs within a narrow belt of homogeneous intrusive granite along the immediate contact zone in the northern part of the Mountain Land. This contact clearly represents a region of strong movement and dislocation, although the amount and direction of movement are not known.

D Structural History

1. *The Northwestern Part*

The gross structural features of the Mountain Land, mentioned above, have to be satisfactorily explained in any theory of the structural history of the area. The small-scale structural features in the northwestern part of the Mountain Land, summarized by Roering (1965), also have to be accommodated in the synthesis.

On the basis of superimposition of structures, a number of distinct periods of deformation have been recognized within this region. The earliest recognizable structures, deformed by all other structures in the area, are the major east-northeast-trending folds and faults which constitute the so-called Main Phase of folding. These structures are overfolded to the northwest, and there is evidence that they were superimposed on earlier folds about which very little is known. Paralleling this trend along the northern flank of the Mountain Land, is a very strong foliation. The latter is clearly seen in the granite close to the contact, and is manifested in the alignment of metamorphic minerals in the metamorphic aureole of the granite, as well as in a strong cleavage. The latter is best developed in the pelitic sediments of the Fig Tree Series, and in deformed chert pebbles of the Moodies Series. The Main Phase structures, although locally deformed, especially in the vicinity of the granites, are by far the best developed, and give the Mountain Land its grain.

A marked, but local, superimposed phase of deformation occurs in the vicinity of Noordkaap, where the northwest Consort Fold Trend has resulted in the refolding of the Main Phase foliation and cleavage. This is manifest in the arcuation of the Eureka and Ulundi synclines, and the development of the Consort folds and the Jamestown Syncline. The Montrose Fold Trend, south of Barberton, also deforms the Main Phase folds. None of these younger trends can be traced across the Mountain Land, and they are clearly of local origin, being not nearly as pervasive as the Main Phase trend.

A few, very weak phases of deformation post-date all of the above-mentioned deformations. A late-phase, east-northeast fold trend is superimposed on the rotated Main Phase cleavage of the Ulundi Syncline, and is also manifest in the disharmonic fold to the southwest of Noordkaap. A final imprint of deformation is represented by small-scale accordion-folds, with flat-lying axial planes.

2. *The Mountain Land*

Deformation of the Swaziland System, as determined from minor structural analysis, is post-Moodies in age. If the Moodies Series represents a true, late-tectonic molasse deposit, then the original geosyncline must have been subjected to earlier orogeny. The effects of such an orogeny cannot be detected beneath the imprint of the post-Moodies deformation. Any structural analysis, therefore, can refer only to the second of at least two major periods of folding and faulting. It is consequently not possible to present the complete tectonic story.

All the major structures referred to earlier have been caused by the intrusion of post-Moodies granites. Forces acting from the south-southeast, in a north-north-westerly direction, produced the regional fold trend (east-northeast), the parallelism of the regional thrust-faults, and the overturning of the folds to the north. The gneissic terrain of the Nelspruit area probably acted as a resistant, stable buttress against which the layered rocks were compressed. A good indication of the amount of deformation that these rocks have suffered can be obtained from deformed and flattened pebbles of the Moodies Series. Along the northern contact of the Mountain Land, the flattened and elongated ellipsoidal chert pebbles are parallel to a well-developed cleavage. It is possible, using these deformed pebbles, to calculate the approximate amount of shortening suffered by the conglomerate horizon at 61 per cent, in (Anhaeusser, 1966b) the Eureka Syncline. Assuming an average width of 17 miles for the Mountain Land, a shortening of between 40 and 80 per cent, and uniform compression over the whole area, and not taking into account the effects of the thrust-faults, an approximation of the original extent of the unfolded, layered rocks can be set at between 50 and 150 miles.

Soon after the initial deformation, about which little is known, numerous discrete and homogeneous granite bodies intruded the periphery of the Mountain Land. These probably represented high-level, homogenized, mobile fractions of a lower-lying gneissic terrain. Good evidence of this has been obtained in Swaziland (Hunter, 1965), where the homogenized, synorogenic granite of that territory passes gradationally downward into the Ancient Gneiss Complex, exposed in the deeper river valleys. The granites moved up slowly, all at approximately the same time, in the form of rounded or elliptical bodies, varying in size from 25 miles to one mile in diameter, excluding the huge body of synorogenic granite in Swaziland. These diapiric granites forced their way into the basal part of the Onverwacht Series, and produced local thermal, and strong dynamic, metamorphism of the volcanic series. The volcanic rocks became partly or completely draped around the various plutons, to form the tapering arcuate schist belt remnants described earlier. Most of these bodies caused only very local deformation of the Onverwacht rocks, and had hardly any effect on the east-northeast grain of the Mountain Land. Certain of the larger ones, however, generated a more widespread superimposed deformation. The best example is the Kaap Valley Granite, which was responsible for the Consort folds and the arcuations of the Eureka and Ulundi synclines. In addition, it produced the large-scale structure of the Jamestown Syncline, with its northwest trend at right-angles to the general grain of the Mountain Land. Some, well-developed, anticlinal structures were also produced by certain of these rising, homogeneous domes. The best examples of these include the Steynsdorp Anticline and the Stentor Anticline (Viljoen and Viljoen, in preparation).

Sustained pressure from the southeast resulted in the imprint of a younger, distinct, but weakly developed east-northeast trend parallel to the early Main Phase trend. This same force was probably also responsible for the thrusting of the folded rocks on to the stable buttress of the Nelspruit gneiss and migmatite to the north. This resulted in the strong post-crystallization mechanical deformation observed within the granites along the immediate contact zone in the north.

Finally, a sequence of rather brittle accordion-folds and conjugate folds were developed over wide areas in the northwestern part of the Mountain Land. All of these folds indicate a near-vertical stress field. It is possible that the very much younger, porphyritic, homogeneous plutons of the M'pageni-, G.5-, and Dalmein-types were responsible for these folds. These late granites had no part in the major deformations of the Mountain Land, and cut cleanly through all earlier trends and foliations. These granites were, however, responsible for the considerable local faulting of the supracrustal rocks into which they intruded (Viljoen and Viljoen, in preparation).

VII MINERALIZATION IN THE BARBERTON MOUNTAIN LAND

The ancient schist belt of the Barberton Mountain Land has not been merely of scientific interest, for as early as 1883 gold was discovered, and the De Kaap Gold-field came into existence. In the ensuing rush, the Jamestown, Komati, Steynsdorp, and Swaziland fields were opened up, but their lives were short. Several mines in the De Kaap field, however, proved to have major ore-bodies, and have continued production up to the present day. Apart from the gold mineralization, there are several important base metal deposits located within the ancient fold belt. Notable are the occurrences of chrysotile asbestos and iron-ore. In addition, lesser amounts of magnesite, talc, barytes, tin, and ornamental stone (buddstone and verdite) have been mined from the area. Minor occurrences of the nickel mineral, trevorite, have also been found, but have proved to be uneconomic.

Investigations relating to the economic geology of the Barberton Mountain Land have shown, almost without exception, that mineral deposits owe their origin to some form of structural control. Possible exceptions are some of the talc and magnesite occurrences formed as a result of the local intrusion of quartz and feldspar porphyry bodies, and the barytes and iron deposits, both of which owe their origin primarily to stratigraphic control.

A Gold Mineralization

The importance of faults in the development and distribution of gold mineralization has long been recognized. In the northern portion of the Mountain Land, most of the gold mineralization has been located close to the Lily, Sheba, Scotsman, and Barbrook faults. On the Swaziland side, the gold mineralization of the Forbes Reef area is located along, or adjacent to, areas of faulting. The major faults are not mineralized continuously along their strike, and indications are that second-order phenomena, associated with the parent structures, were the main controlling factors for the localization and concentration of ore-forming fluids. Evidence of horizontal movements having taken place along the major faults is available, and it has been demonstrated that much of the gold mineralization is intimately associated with the effects of wrench-faulting (Anhaeusser, 1965).

The gold-producing area around Barberton suffered at least five separate phases of deformation (Roering, 1965). Poole (1964) found that the main control in the localization of the ore-bodies in the Agnes Gold Mine was exerted by structures developed during the second phase of regional deformation. Anhaeusser (1965) suggested that the mineralization was introduced over a wider range of time, beginning during the second phase of deformation, and continuing through into the third phase. It was also suggested that the mineralization is of two different types — the earlier gold being intimately associated with sulphides, and the later gold being virtually devoid of sulphides, and occurring essentially as gold-quartz veins.

The New Consort Gold Mine affords an example of mineralization occurring at a "schist-shale" contact. In the mine the "Consort Contact" is a highly sheared contact between competent Fig Tree shales and incompetent talc and amphibole schists of the underlying Onverwacht Series. Viljoen (1964) considered that the localization of the Consort mineralization was due, partly, to the fact that the contact was also a zone of extensive differential movement, and that the younger Consort folds, produced by the intrusion of the Kaap Valley Granite, attain their best development in the mine area. The Fairview and Sheba mines, two of the largest gold producers in the Mountain Land, are further examples of ore-bodies controlled by structure. The mineralization also occurs in, or adjacent to, the Zwartkoppie Stage, at the base of the Fig Tree Series, in contact with basic schists of the Onverwacht Series. Both these mines are situated close to the Sheba Fault, separating the Moodies rocks of the Eureka Syncline from the Fig Tree rocks of the Ulundi Syncline. The fracture patterns in these mines were considered by Anhaeusser (1965) to have been produced by wrench-movements formed by the reactivation of pre-existing longitudinal thrust-fault planes. The regenerative stresses apparently created suitable second-order structures into which ore migrated. It is noteworthy that, in only a few instances, have the major fault planes themselves provided a favourable environment for the accumulation of payable concentrations of gold mineralization. The wrench movements were also responsible, in a number of mines, for the generation of cymoid structures, i.e. fractures splitting from, and rejoining, major shears. It is possible to demonstrate that these structures are frequently responsible for, or act as hosts to, increased gold and sulphide mineralization (Anhaeusser, 1965).

The gold mines of the Barberton Mountain Land are, as has been previously stated, mainly situated in close proximity with major regional faults, the subsequent movements along which have resulted in the production of second-order faults and fractures. The majority of the ore-bodies are thus located in an environment dictated by the varying influences and effects of fault deformation. Away from the sphere of influence of the major planes of weakness, the incidence of gold mineralization clearly declines, and is either poorly developed or totally absent.

Coupled with the over-all fault control is the effect, on a regional scale, of the various phases of intrusive granite, and the associated contact-metamorphic effects on the rocks of the Swaziland System. Here again, it can be demonstrated that the more significant gold mineralization lies in a broad zone adjacent to the periphery of the Mountain Land and that this mineralization decreases progressively away from the influence of the granite contacts, until it becomes virtually non-existent in the heart of the fold belt proper.

De Villiers (1957) made a general study of the mineralogy of the gold ores of the Barberton area, and recognized four main types of ore, viz.:-

(1) Ore containing arsenopyrite and pyrrhotite;
(2) Pyritic ore;
(3) Lead-bearing ore;
(4) Antimonial ore.

To this list Gribnitz *et al.* (1961) added a fifth type — gold/quartz ore. De Villiers (1957) concluded that all the deposits seemed to have originated at moderate to great depths, but that various temperatures of formation were represented. Hypothermal, mesothermal, and epithermal occurrences are to be found. The ores with arsenopyrite were formed at great depth and high temperature. The pyritic ores also appear to be hypothermal in origin, while the lead-bearing ores belong to the mesothermal class. The antimonial ores seem to have been deposited during the last stages of hypothermal activity, and exhibit characteristics of both mesothermal and hypothermal zones.

Gay (1964) completed a minor- and trace-element study of visible, native gold, obtained from 40 localities over the whole of the Barberton Mountain Land. A regional plot of the fineness values revealed the outlines of a possible pattern. The highest values occur in the vicinities of the Agnes, Sheba, and Consort mines. Gay suggested from this that the possible source of the gold mineralization lay beneath these areas. The Kaap Valley granitic pluton, and not the Nelspruit gneisses and granites, is favoured by him as the possible source of the auriferous solutions. Because most of the deposits investigated seem to be of a hypothermal nature, Gay could not detect a readily apparent zoning of the mineralization.

Anhaeusser (1966c) after studying the Lily gold deposit near Louw's Creek, considered that supergene processes of enrichment of the ore had taken place in the oxidized levels of the mine. He furthermore suggested that the process of supergene enrichment of auriferous deposits may have been a major contributing factor to the numerous surface-enriched gold deposits exploited in the early mining days of the Barberton Gold-field.

B Iron-ore Deposits

The most important iron-ore deposit in the Mountain Land is situated in north-western Swaziland, two miles from the Transvaal border-post at Oshoek. The iron deposit constitutes the spine of Bomvu Ridge, at the southern end of the Ngwenya Range in the Mbabane District. The ore-body is situated in the Fig Tree Series, and is locally composed of sediments comprising alternating bands of siliceous shales and banded cherts. These are followed by a zone of banded ironstones, with subordinate shales, the upper part of which give rise to the ore-body. This is overlain by cherty, feldspathic shales. The deposit of hematite occurs in banded ironstones, the folded nature of which appears to have influenced the varying degrees of secondary enrichment of iron and the emplacement of different types of ore (Bursill *et al.*, 1964). The ore-body has been proved to a depth of over 900 feet below surface, and the average export grade of the ore has thus far been of the order of 64-65 per cent Fe (Swaziland Iron Ore, 1965).

Further iron-ore deposits occur in the Fig Tree successions near Malelane on the Transvaal side of the Mountain Land, but mining of these bodies has ceased. These deposits were also formed as a result of secondary enrichment of banded ironstones.

C Magnesite and Talc Deposits

Magnesite deposits occur in the metamorphosed, ultrabasic assemblages of the Onverwacht Series, in the vicinity of Kaapmuiden and Malelane. The magnesite is confined to a specific horizon, in what appears to be a differentiated assemblage of ultrabasic rocks with a lateral development of many miles (Viljoen and Viljoen, 1967b). The magnesite is considered to have formed in the lowermost, more magnesian-rich, ultrabasic layer at the base of a differentiated body, with the best deposits once again being in areas of structural disturbance. In some of the slightly higher, and more iron-rich, phases of the differentiated succession, there is a preferential development of chrysotile asbestos, which also attains its best development in structurally disturbed areas (Viljoen and Viljoen, 1967b).

Talc has been mined from a number of different deposits, mainly in the northern half of the Mountain Land. The talc deposits all occur in the Onverwacht Series, and were formed mainly as a result of metamorphism and steatitization of ultrabasic rocks by the intrusive granites and quartz and feldspar porphyries.

The ornamental, semi-precious, greenstones, verdite and buddstone, are also found in the serpentinites of the Onverwacht Series, in the same general locality as the talc and magnesite deposits.

D Chrysotile Asbestos Deposits

Several important chrysotile asbestos deposits occur in the serpentinites of the Onverwacht Series. Notable among these are the Havelock occurrence in Swaziland, and the Msauli deposit in the Steynsdorp Valley. In the Jamestown Schist Belt are the workings of the New Amianthus, Munnik Myburgh, Sunnyside (Star), and Marbestos mines. These deposits are thought to be associated with rocks that have undergone some differentiation to form alternating bands of serpentinitic and pyroxenitic material, and differ from the Havelock and Msauli ore-bodies, which occur in serpentinites considered to have been emplaced along fault-zones (Viljoen and Viljoen, 1967b). The Sterkspruit, Stolzburg, and Doyershoek mines, near Badplaas, occur in a fault-bounded mass of ultrabasic rocks, also displaying signs of differentiation into serpentinitic and pyroxenitic bands. Near Malelane, the Barberton Chrysotile Mine affords further evidence of a differentiated mass, with asbestos fibre being developed in the eastward-plunging Koedoe Syncline.

In all cases, structural deformation, combined with suitable host-rocks, have been the contributing factors towards the localization of the asbestos ore-deposits. Faulting, in particular, is believed to have been the main controlling factor for the development of the chrysotile fibre. Van Biljon (1964), after studying numerous asbestos deposits in the Barberton Mountain Land, came to the conclusion that a tensional environment was of prime importance for fibre formation, and that hydrothermal solutions merely assisted the process. The ore-deposits associated with folds, viz. New Amianthus, Barberton Chrysotile, and Marbestos mines, were formed in serpentinites that had been subjected to tensional effects in the hinges of synclinal structures.

VIII TOWARDS AN EVOLUTIONARY MODEL OF THE BARBERTON MOUNTAIN LAND

The supracrustal rocks of the Barberton Mountain Land are identical in sequence and type to those found in younger geosynclines and island arcs. The Onverwacht Series would be equivalent to an ophiolite assemblage, the Fig Tree Series to a flysch assemblage (or greywacke suite), and the Moodies Series to a molasse-type deposit. The question is whether or not the Barberton rocks were deposited in a geosyncline. The latter is defined as an elongate basin, filled with sediments and volcanics, up to several kilometers in thickness, which have been strongly folded.

Geosynclines generally show the following characteristics. They are made up of two major troughs, separated by an intervening ridge. Frequently the geosyncline is situated at the continental margin and comprises a series of ridges and furrows. Passing from the continent, or craton, the miogeosynclinal furrow, the miogeosynclinal ridge, the eugeosynclinal furrow, and the eugeosynclinal ridge, are successively encountered (Aubouin, 1965; De Sitter, 1964). The two furrow-like depositories are generally filled by material derived from the eugeocynclinal ridge area. The eugeosyncline is characterized by the presence of ophiolites, while the miogeosyncline does not act as host to this rock assemblage. A diagnostic characteristic of a geosyncline, that follows the extrusion of the ophiolites, is an enormous thickness of greywackes and shales — often referred to as the "greywacke suite", or flysch, and is generally formed by turbidity currents. According to Aubouin (1965) the flysch fills firstly the eugeosyncline and then the miogeosyncline. The flysch deposition also has a migratory tendency which is in prelude to orogeny and which has its beginnings in the eugeosynclinal ridge and moves gradually outwards in time, towards the stable craton. The orogeny in the eugeosynclinal ridge and furrow area is also associated with metamorphism and granitization. Large granitic batholiths are the result of these processes, and they tend to be confined to the eugeosynclinal area (King, 1959; Eardly, 1962). Molasse deposition, which contrasts strongly with the deep-water turbidite flysch, marks the end of the metamorphic cycle (Sutton, 1965). These deposits comprise shallow-water sediments in the form of conglomerates, quartzites, and shales. Separate molasse basins can occur in various positions within the geosyncline, depending upon which areas were uplifted the most to provide the detritus. The orogeny also gives rise to various fold styles, or types of folds, depending upon the position in the geosyncline.

At least two criteria diagnostic of geosynclines are furnished by the Barberton rocks, viz. the appreciable thickness of the sediments and the volcanics (which are, furthermore, geosynclinal in type), and the ubiquitous, tight, isoclinal folding. What are lacking are the elongate nature of the depository and the characteristic major units of a geosyncline, i.e. eugeosyncline, intervening ridge, miogeosyncline, and continent or craton. This is due to granites entirely surrounding and intruding the supracrustal rocks, obliterating any evidence in support of the depository having been considerably larger. Without the necessary geographic distribution of the required geosynclinal elements in the Mountain Land, the rocks themselves have to be employed to provide clues as to the environment of deposition. Another complicating factor, in deciphering the characteristics of ancient geosynclines, hinges heavily on the significance of ophiolites. Presumably, in ancient times the crust was considerably thinner and, therefore, the distinction between miogeosyncline and eugeosyncline, based on the presence of ophiolites in the latter, may not be valid. With a thinner crust, ophiolites may have occurred in both the miogeosyncline and the eugeosyncline.

It is quite clear that the rocks must have been deposited in a deep trough of some sort. The reason for this is the enormous thickness (seven miles) of Onverwacht rocks found in the southwestern part of the area. This is a minimum value, as the lower contact has been removed by the intrusion of granite. These volcanic rocks are followed by a substantial thickness (at least two miles) of Fig Tree material, and a similar thickness of Moodies sediments. There is very little unconformity between the Fig Tree Series and the Onverwacht Series which then would indicate a down-sagging, of the order of ten miles, or approximately one-third of the thickness of the present continental crust. It is impossible to obtain an exact thickness of conformable sediments in this area which would indicate the true amount of down-sagging. The only equivalent environment, with respect to down-sagging and infilling of volcanics and sediments of the type outlined above, is the geosyncline, or island arc.

At least one side of the trough may have been fault-bounded. The faults must also have been sufficiently deep to allow the flow, from the earth's mantle, of basaltic, untramafic, and more acid magmas into the trough to form the Onverwacht succession. It is proposed, therefore, that the Onverwacht and Fig Tree rocks must have been deposited in a deep, water-covered (pillow-lavas, turbidites), possibly fault-bounded (on at least one side) trough. The faults may also have contributed to the genesis of the flysch-like Fig Tree sediments. Detrital material may have been removed from the fault-zones by means of gravity slides.

As outlined so far, the rock succession is very similar to that of other shield areas (greenstone-greywacke assemblages), which are so characteristic of island arcs (Engel, 1966). However, the next stage in the evolution of the Barberton fold belt contrasts with the earlier events. Earth movements must have been active to account for the folding of strata that took place before the deposition of the Moodies sediments. This is manifested by the unconformable nature of the Moodies rocks at certain localities in the Mountain Land. Moreover, the Moodies sediments are of a continental type, in contrast with the turbidites of the Fig Tree. Similar evolutionary patterns are found in younger geosynclines, where molasse deposits follow closely after major events of orogeny, metamorphism, granitization, and regional uplift. Whether or not these events occurred adjacent to the supracrustal rocks of the Mountain Land will never be known, because the intrusive granites have obliterated most of the other portions of the depository. This hypothetical granitic and orogenic event should not be confused with a similar event which actually took place in post-Moodies times. If such a geosynclinal cycle did not take place, then basement granite must have been exposed adjacent to the trough, in order to account for the granitic detritus found in the Moodies rocks. This would also, probably, indicate earth movements on a significant scale. As already outlined in this paper, three major groups of granitic rocks are present in the area. The oldest are considered to be basement on which the supracrustal rocks were deposited. These comprise numerous migmatites and granitic gneisses found over vast areas.

After the deposition of the Moodies sediments the next important event was the intrusion of the 3,000 m.y. granites. This second group includes the homogenized portions of the Nelspruit and Badplaas Granites, the Kaap Valley Granite, and the Synorogenic and Late-orogenic Granites of Swaziland. At about 3,000 million years ago (Allsopp *et al.*, 1962), much of the earlier basement granite was probably rendered palingenic and rheomorphic. The granites of the second group are probably anorogenic (Roering, 1967), and have the following characteristics:

(1) they were intruded more or less simultaneously (Roering, 1965);
(2) they are diapiric in shape, being capped by a zone of homogeneous granite, which gives way downwards to nebulites, and, ultimately to gneisses and migmatites; the latter probably represent the original granitic basement (Hunter, 1965);
(3) the granites are distinctly intrusive, and give rise to a contact-metamorphic aureole against the supracrustal rocks; this contrasts with anatectic granite generation, where extensive zones of gneisses and metamorphites surround the more homogenized granite masses (Raguin, 1965);
(4) the intrusion of granite is responsible for much of the deformation of the supracrustal rocks of the Barberton Mountain Land.

Granites of this type are thought to have been derived from differentiative processes in the mantle (Roering, 1967; Engel, 1966). Such a process could possibly take place on a continental scale, and, possibly, at these very early times, was directly responsible for the formation, and thickening, of the granitic crust.

The third granitic group are represented by the high-level, medium- to coarse-grained porphyritic post-orogenic granites of Swaziland, and of the M'pageni, Salisbury Kop, and Dalmein plutons. These may have had a similar origin to the 3,000 m.y. granites, but are preserved at a very much higher structural level than the latter.

The dykes are the youngest manifestation of the schist belt. They were intruded after the granitic events, and were formed on a craton. Epeirogenic movements possibly accounted for macro-joints and faults which were subsequently filled with essentially basaltic magma.

From the discussion above it is clear that there were two important events in the evolution of the Barberton Mountain Land. These events may well be applicable to other ancient fold belts. There is firstly, the geosynclinal event, with a period of trough development, the extrusion of an ophiolite suite with its associated sediments, the deposition of a flysch or greywacke suite of sediments, a period of orogenesis associated with the metamorphic and granitic phenomena and, lastly, the deposition of a molasse facies, derived from areas which were exposed during orogenesis and mountain building. The second is the granite event which may be repeated at different intervals of geologic time.

The relationship between the geosynclinal cycle and a particular granite event should be considered first. It has been shown in this paper, and elsewhere, that the 3,000 m.y. granites are post-Moodies and, therefore, post-molasse, in age. This group of granites cannot, thus, be compared with granitic and metamorphic phenomena associated with the classic events in a geosyncline. As Read (1957) has pointed out "the post-orogenic sediments (i.e. the molasse) are clearly the wrong setting for granitization, they are too distant in time and place from the source of granite". Further more, it is generally accepted that the onset of molasse deposition marks the end of the orogenesis, metamorphism, and associated granitic phenomena (Sutton, 1965). It would, therefore, appear that the 3,000 m.y. granite event is independent of the sequence of events in a geosyncline.

The two fundamental elements or episodes which thus emerge from the Barberton area are, the geosynclinal event, and the granite event. The area can be considered as comprising two distinct and apparently unrelated models. These two models are diagrammatically depicted in Figure 4, where the evolution of both is plotted against the relative intensity of earth movements. It can be assumed that some very early earth movements, probably in the form of warping and/or faulting, gave rise to the embryonic down-sag or trough close to, or on the site of, the present Mountain Land. The ophiolites of the Onverwacht Series poured out into a slowly sagging trough, but no signs of any significant earth movements are found. The style of sedimentation and the type of sediment in the Fig Tree Series clearly indicates fairly strong and increasing earth movements, as depicted by turbidite slumps. If we assume that the Barberton area has followed the course of evolution of any typical geosyncline, then the peak of the earth movements would have been reached by orogeny and associated phenomena. However, no evidence of this geosynclinal orogenic event has yet been found in the Barberton area because of the obliteration of all surrounding areas by the younger 3,000 m.y. old granite event. The Moodies Series was then deposited in a declining regime of earth movements as a molasse-type deposit, derived from the postulated folded and uplifted orogen.

A time break of unknown duration then followed, until, with strong earth movements, the 3,000 m.y. old granites were forcefully emplaced. These effectively obliterated almost all of the early history of the geosynclinal model, and were responsible for much of the presently observable, and decipherable, contact-metamorphism, strong folding, and faulting. After crystallization of these granites the Mountain Land and environs, attained very nearly their present-day disposition. Into this stable craton were then intruded the younger granites, which had no significant effect on the structure of the Mountain Land, and, finally at a later date, the various dykes were injected.

Figure 2 depicts a very broad framework of the evolution of the geosynclinal cycle. As was outlined previously, and as can be seen in Plate III, these major divisions can be further broken down into smaller units. These units tend to be cyclic in nature. In the Onverwacht Series this cyclicity is very well displayed in the lower and upper

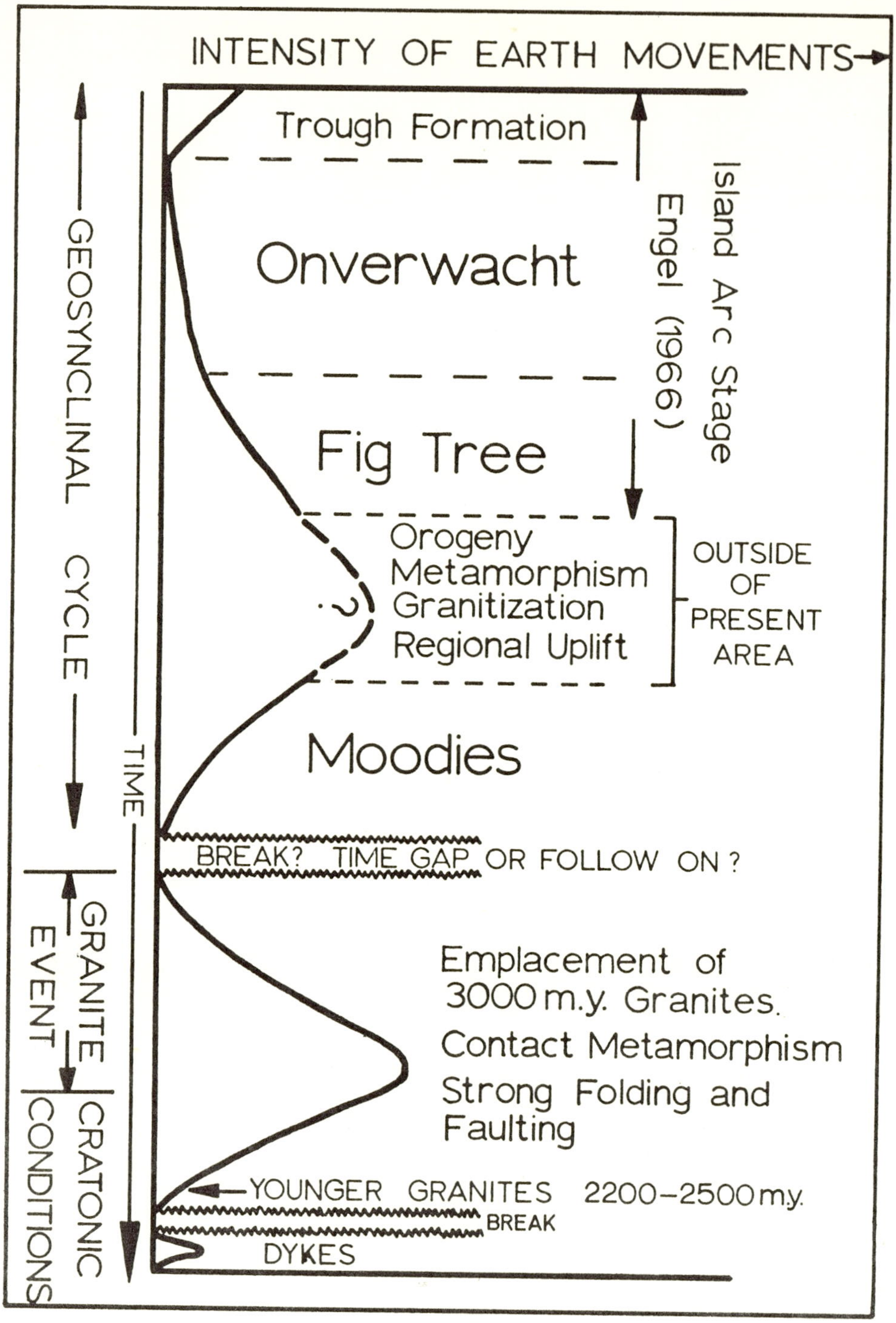

Figure 2
Evolution of the geosynclinal cycle.

stages. In both, basic lavas are followed upwards by salic tuffs and agglomerates, and individual cycles are often capped by cherts. In the Middle Onverwacht no acid material, or siliceous sediments, are present, but here there appears to be a cyclic sequence of ultrabasic material and basic lavas. The origin and significance of the igneous cycles is not fully understood but it seems probable that they may be a result of processes within the upper mantle. There is very frequently a close association between the outflowing of acid lava and tuff, and the formation of cherts. It seems as though the latter have formed by some exhalative process, and that the silica of the cherts has been, in some way, derived from, or closely associated with, this acid material.

In both the Fig Tree and the Moodies Series there are sedimentary cycles, each of which shows a size variation from coarse material at the base, to finer material at the top. The cycles are numerous, and often very small in the Fig Tree Series, and grade generally from greywacke to shale and then chert. In the Moodies Series three larger cycles are present. These grade from coarse conglomerate, through quartzite, shale, to finally jaspilite. The cyclic nature of the Fig Tree Series could be due to turbidity currents, followed by periods of quiescence and chemical precipitation. In the Moodies Series, the cycles are probably to be ascribed to uplift of source-areas initially providing coarse detritus to form the base of the cycle. With further erosion of the uplifted area, the sediments become finer until finally, chemical precipitates accumulate in the form of jaspilites.

If, as in the Barberton area, the 3,000 m.y. old granites were not related to the earlier geosynclinal cycle, then there could be far reaching implications. For example, the granite might appear anywhere in the sequence of events of an Archaean geosyncline. If it were to occur say, after the deposition of the ophiolites, then presumably, the cycle could not continue because the crust had been thickened by the addition of granite from the mantle, and a new geosynclinal cycle would have to start elsewhere. Ancient geosynclinal remnants may, therefore, vary in the sequence of rock types present, e.g. they may consist of greenstone belts, greenstone-greywacke belts, or greenstone-greywacke-molasse belts, as at Barberton. It is suggested that the Barberton Mountain Land is unique by virtue of its preservation quality of exposure and, evidence of the complete geosynclinal cycle.

IX REFERENCES

Allsopp, H. L., Roberts, H. R., Schreiner, G. D. L., and Hunter, D. R., 1962. Rb-Sr Age measurements on various Swaziland Granites. *J. geophys. Res.*, Vol. 67, pp. 5307-13.

Anhaeusser, C. R., 1964. The geology of the Lily Syncline and portion of the Eureka Syncline between Sheba Siding and Louw's Creek Station, Barberton Mountain Land. M.Sc. Thesis, Univ. of the Witwatersrand, Johannesburg.

Anhaevsser, C. R. 1965. Wrench faulting and its relationship to gold mineralization in the Barberton Mountain Land. *Inform. Circ.* No. 24, *Econ. Geol. Res. Unit.*, Univ. of the Witwatersrand, Johannesburg.

Anhaeusser, C. R. and Viljoen, M. J., 1965. The base of the Swaziland System in the Barberton-Noordkaap-Louw's Creek Area, Barberton Mountain Land, *Inform. Circ.* No. 25, *Econ. Geol. Res. Unit*, Univ. of the Witwatersrand, Johannesburg.

Anhaeusser, C. R., 1966a. Facets of the granitic assemblage on the northwest flank of the Barberton Mountain Land. *Inform Circ.* No. 32, *Econ. Geol. Res. Unit*, Univ. of the Witwatersrand, Johannesburg.

———, 1966b. A comparison of pebble and fold deformations in the Nelspruit Granite contact aureole. *Inform. Circ.* No. 30, *Econ. Geol. Res. Unit.*, Univ. of the Witwatersrand, Johannesburg.

———, 1966c. Supergene gold enrichment in the Barberton Mountain Land with particular reference to the Lily Mine. *Inform. Circ.* No. 29, *Econ. Geol. Res. Unit.*, Univ. of the Witwatersrand, Johannesburg.

Anhaeusser, C. R., Viljoen, M. J., and Viljoen, R. P., 1966. A correlation of pre-Fig Tree rocks in the northern and southern part of the Barberton Mountain Land. *Inform. Circ.* No. 31, *Econ. Geol. Res. Unit.*, Univ. of the Witwatersrand, Johannesburg.

AUBOUIN, J., 1965. *Geosynclines*. Elsevier Publishing Company, Amsterdam.

BARGHOORN, E. S. and SCHOPF, J. W., 1966. Microorganisms three billion years old from the Precambrian of South Africa. *Science*, Vol. 152, No. 3723, pp. 758-63.

BURSILL, C., LUYT, J. F. M., and URIE, J. G., 1966. The Bomvu Ridge Iron Ore Deposit. *in* Haughton S. H. (ed.): *The Geology of Some Ore Deposits in Southern Africa*. Geol. Soc. S. Afr., Vol. 2 pp. 405-14.

CLOUD, P. E., 1965. Significance of the Gunflint (Precambrian) micro flora. *Science*, Vol. 148, No. 3666, pp. 27-35.

COOKE, R., 1965. The pre-Fig Tree rocks in and around the Moodies Hills, Barberton Mountain Land. Unpublished report prepared for submission to the Geological Society of South Africa. (8th Ann. Congr.)

DE SITTER, L. U., 1964. *Structural Geology* (2nd ed.). McGraw-Hill Book Company, New York.

DE VILLIERS, J. E., 1957. The minerology of the Barberton Gold Deposits. *Bull. 24, geol. Surv. S. Afr.*

EARDLY, A. J., 1962. *Structural geology of North America* (2nd ed.). Harper and Row, New York.

ENGEL, A. E. J., ENGEL, C. G., and HAVENS, R. G., 1965. Chemical characteristics of oceanic basalts and the Upper Mantle. *Bull. geol. Soc. Amer.*, Vol. 76, pp. 719-34.

ENGEL, A. E. J., 1966. The Barberton Mountain Land: Clues to the differentiation of the Earth. *Inform. Circ.* No. 27, *Econ. Geol. Res. Unit*, Univ. of the Witwatersrand, Johannesburg.

ESKOLA, P. E., 1948. The problem of mantled gneiss domes. *Quart. J. geol. Soc. Lond.*, Vol. 104, pp. 461-76.

GAY, N. C., 1964. The composition of gold from the Barberton Mountain Land. *Inform. Circ.* No. 19, *Econ. Geol. Res. Unit*, Univ. of the Witwatersrand, Johannesburg.

GRIBNITZ, K. H., POOLE, E. J. and VOGES, S., 1961. Geological notes on the Swaziland System. Unpublished Report Geol. Dept., E.T.C. Mines Ltd., Barberton.

HERGET, G., 1963. Beitrage Zur Stratigraphie, Tektonik und Petrographie des Barberton-Berglandes (Transvaal—Südafrika). Unpublished Inaugural-Dissertation zur Erlangung der Doktorwürde, Ludwig-Maximilians-Universität, München.

HOERING, T. C., 1965 *in:* Pretorius D. A., (1965.) Sixth Annual Report for 1964, *Econ. Geol. Res. Unit*, Univ. of the Witwatersrand, Johannesburg.

HUNTER, D. R., 1961. The geology of Swaziland. *Geol. Surv., Swaziland.* Unpublished report prepared for submission to the Geological Society of South Africa. (8th Ann. Congr.)

KING, P. B., 1959. *The Evolution of North America*. Princeton University Press, Princeton.

KOEN, G. M., 1947. Die Geologie in die Omgwing van Sheba, Barbertonse Distrik. M.Sc. Thesis, Univ. Pretoria.

KUENEN, Ph. H., 1963. Turbidites in South Africa. *Trans. geol. Soc. S. Afr.*, Vol. 66, pp. 191-5.

PETTIJOHN, F. J., 1957. *Sedimentary Rocks* (2nd ed.). Harper and Brothers, New York.

PFLUG, H. D., 1966. Structured organic remains from the Fig Tree Series of the Barberton District, Eastern Transvaal. *Inform. Circ.* No. 28, *Econ. Res. Unit.* Univ. of the Witwatersrand, Johannesburg.

POOLE, E. J., 1964. Structural control of mineralization in the Agnes Gold Mine, Barberton Mountain Land. *Inform. Circ.* No. 22, *Econ. Geol. Res. Unit*, Univ. of the Witwatersrand, Johannesburg.

RAGUIN, E., 1965. *Geology of Granite*. Interscience Publishers, John Wiley and Sons, Ltd., London.

RAMSAY, J. G., 1963. Structural investigations in the Barberton Mountain Land, Eastern Transvaal. *Trans. geol. Soc. S. Afr.*, Vol. 66, pp. 353-401.

READ, H. H., 1957. *The Granite Controversy*. Thomas Murby and Company, London.

ROERING, C., 1965. The tectonics of the main gold producing area of the Barberton Mountain Land. *Inform. Circ.* No. 23, *Econ. Geol. Res. Unit*, Univ. of the Witwatersrand, Johannesburg.

ROERING, G. 1967. Non-orogenic granites in the Archaean geosyncline of the Barberton Mountain Land. *Inform. Circ.* No. 35, *Econ. Geol. Res. Unit*, Univ. of the Witwatersrand, Johannesburg.

RUBEY, W. W., 1955. Development of the hydrosphere and atmosphere, with special reference to probable composition of the early atmosphere *in:* Poldervaart, A. (ed.): Crust of the Earth. Special Paper 62, *Geol. Soc. Amer.* Part IV, pp. 631-50.

STEINMANN, G., 1905. Geologische Beobachtungen in den Alpen. II: Die Schardt'sche Uberfaltungstheorie und die Geologische Bedeutung der Tiefseeabsatze unde der ophiolithischen Massenensteine. *Ber. Nat. Ges. Freiburg* i. Bd. 16, pp. 44-65.

SUTTON, J., 1965. Some recent advances in our understanding of the controls of metamorphism. *in: Controls of Metamorphism*. Oliver and Boyd, Edinburgh.

SWAZILAND IRON ORE, 1965. A description of the Swaziland Iron Ore Development Co. Ltd. Mine, Mbabane District, Swaziland. Guide Book to the mine tour of the 8th Ann. Congr. *geol. Soc. S. Afr.*

SYMPOSIUM, 1965. Symposium on the evolution of the Earth's atmosphere. *Proc. Nat. Acad. Sci.*, Vol. 53, pp. 1169-1226.

URIE, J. G., 1965. A reconnaissance structural investigation in the Forbes Reef Area, Swaziland. Unpublished report prepared for submission to the Geological Society of South Africa. (8th Ann. Congr.)

URIE, J. G. and JONES, D. H., 1965. Metamorphic zones of the Archaean fold belt in Northwestern Swaziland. Unpublished report prepared for submission to the Geological Society of South Africa. (8th Ann. Congr.)

VAN BILJOEN, W. J., 1964. The chrysotile deposits of the Eastern Transvaal and Swaziland *in* Haughton, S. H. (ed.): *The Geology of Some Ore Deposits in Southern Africa.* geol. Soc. S. Afr., Vol. 2, pp. 625-69.

VAN EEDEN, O. R. and MARSHALL, C. G. A., 1965. The granitic rocks of the Barberton Mountain Land in the Transvaal. Unpublished report prepared for submission to the Geological Society of South Africa. (8th Ann. Congr.)

VAN VUUREN, C. J. J., 1964. The geology of portion of the Ulundi Syncline between Hislop's Creek and Fig Tree Creek, Barberton Mountain Land. Unpublished report. *Econ. Geol. Res. Unit,* Univ. of the Witwatersrand, Johannesburg.

VILJOEN, M. J., 1964. The geology of the Lily Syncline and portion of the Eureka Syncline between the Consort Mine and Joe's Luck Siding, Barberton Mountain Land. M,Sc. Thesis, Univ. of the Witwatersrand, Johannesburg.

VILJOEN, M. J. and VILJOEN, R. P., 1967a. A reassessment of the Onverwacht Series in the Komati River Valley. *Inform. Circ.* No. 36, *Econ. Geol. Res. Unit,* Univ. of the Witwatersrand, Johannesburg.

———, 1967b. The geological significance of the pre-Fig Tree rocks of the Barberton Mountain Land. (In Press.)

VISSER, D. J. L. *et al.*, 1956. The geology of the Barberton Area. Spec. Pub. No. 15, *geol. Surv. S. Afr.*

ECONOMIC GEOLOGY RESEARCH UNIT,
UNIVERSITY OF THE WITWATERSRAND, JOHANNESBURG.
SOUTH AFRICA

READ ON 27TH OCTOBER, 1967

6: THE GEOLOGY AND GEOCHEMISTRY OF THE LOWER ULTRAMAFIC UNIT OF THE ONVERWACHT GROUP AND A PROPOSED NEW CLASS OF IGNEOUS ROCKS

M. J. Viljoen and R. P. Viljoen

I. INTRODUCTION

As noted in an accompanying paper, tholeiitic successions of the calc-alkaline type constitute the main rock types of the lowermost volcanic assemblages of most early Precambrian greenstone belts. In the Barberton belt, however, a substantial succession of highly distinctive and as yet undescribed or undefined rock types has been encountered below a normal tholeiitic succession of the calc-alkaline type. The tholeiitic succession constitutes the upper three formations of the Onver-

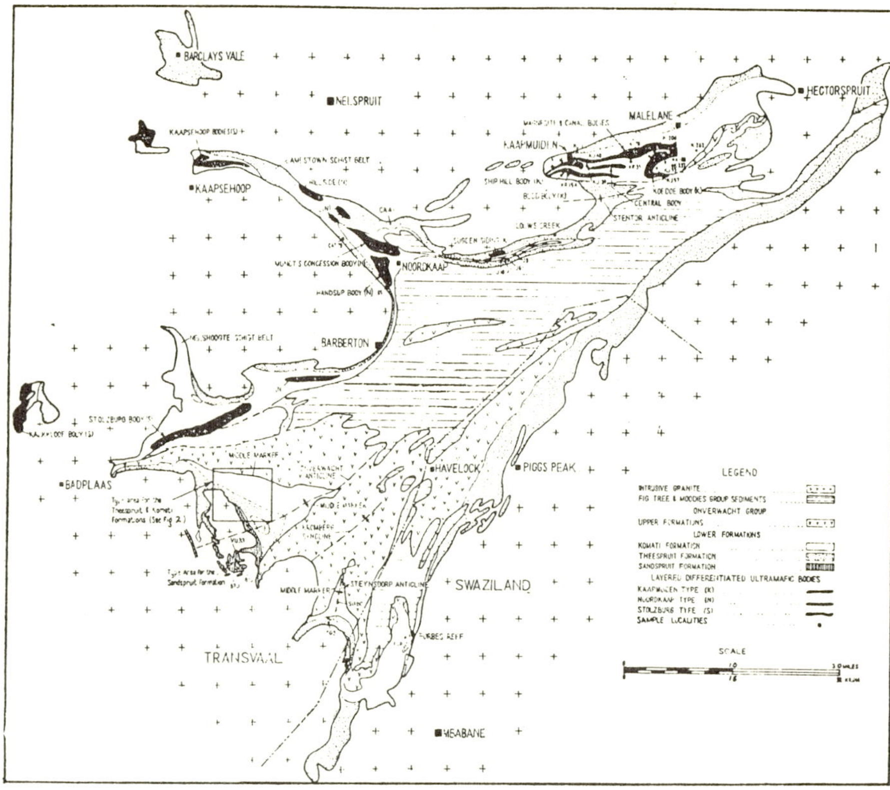

Fig. 1. Distribution of the three lower formations of the Onverwacht Group around the Barberton Mountain Land showing the type area and localities of the chemical analyses quoted.

[*Editor's Note:* The geological map of the type area has not been reproduced because it is oversized.]

wacht Group and is described in a following paper (No. 6). The lower formations of the Onverwacht Group together constitute a distinctive succession of what has been termed the Lower Ultramafic Unit, and are described in detail in this paper.

These formations, known from the base to the top as the Sandspruit, Theespruit and Komati formations, attain their maximum development and are best preserved in the southern part of the Barberton Belt which has consequently been chosen as the type locality. In general the rocks are steeply dipping in this area and occupy the cores of two major anticlinal structures, viz. the Onverwacht and Steynsdorp anticlines. Intrusion of granitic rocks has detached these two anticlines and eliminated the median syncline (now occupied only by rocks of the upper formations) which must have existed previously (Fig. 1).

The Onverwacht anticline is an asymmetrical ENE-trending structure with a somewhat reduced and flattened eastern limb but with a well-developed and well-exposed east-west trending western limb which has consequently been established as the type column for the upper two formations of the Lower Ultramafic Unit (Fig. 1). In this area only a narrow sliver of rocks belonging to the lowermost Sandspruit formation is present so that the type section for the latter has been established some several kilometres to the south-east in a large, well-preserved but detached xenolithic mass within the intrusive tonalitic gneisses (Fig. 1).

II. THE SANDSPRUIT FORMATION

The Sandspruit formation constitutes the lowermost recognisable unit of the Onverwacht Group and has an approximate thickness of 2,134 metres (7,000 ft.), an unknown amount of the stratigraphy, together with possible lower formations, having apparently been eliminated by intrusive tonalitic granites. Most of the rocks constituting the formation occur as xenoliths and stringers largely detached from the main contact of the Mountain Land and almost completely enveloped by granite (Fig. 1). The best developed and preserved of these xenoliths, which is partly traversed by the Sandspruit stream, has been taken as the type area of the formation (Fig. 1).

The rocks consist mainly of ultramafites in the form of bands, lenses and pods, interlayered and closely associated with minor mafic bands and lenses (Figs. 2 and 3). The ultramafics generally have a micaceous and dark green to khaki-green appearance and range petrologically from almost pure serpentinites consisting largely of antigorite with minor amounts of chlorite and magnetite, through antigorite-chlorite-tremolite varieties, to tremolite-chlorite rocks. Well crystallized chlorite, often as large green flakes, is ubiquitous in these rocks and accounts for their distinctive micaceous appearance. It is possible, as suggested by Hess (verbal communication), that much of this chlorite formed from original magnetite and antigorite as a result of the fairly high-grade contact metamorphism to which these rocks have been subjected. Such a reaction could explain the generally minor amounts of magnetite compared to similar rocks in the Komati formation, as well as the distinctive greenish colour of the rocks under consideration. The ultramafics constitute about 60% to 70% of the succession and the metabasalts, in the form of dark green to black, hornblende or actinolitic-hornblende amphibolites with minor amounts of soda plagioclase, the remainder. Occasional narrow, and not very extensive, sedimentary interlayers are present and are composed of quartz and diopside with minor amounts of spinel (pleonaste) and sphene.

Although present, pillow and other volcanic structures are rare in the generally rather massive amphibolites. The ultramafics are usually also massive and structureless, but from their close similarity and analogous setting to the extrusive ultramafics of the Komati formation it is inferred that they represent either lavas or very early contemporaneous near-surface sills.

Selected chemical analyses of ultramafic rocks from the formation are presented in columns 1 to 3 of Table I, and their localities shown in Figs. 1 and 3. These will be discussed later.

The Sandspruit formation has only a limited distribution away from the type area, occurring as isolated xenoliths in the ancient tonalitic gneisses and as a small sliver flanking the eastern side of the Steynsdorp granite in the south-eastern portion of the Mountain Land (Fig. 1). As far as can be ascertained from a reconnaissance investigation the formation does not appear to be developed in other parts of the Mountain Land.

III. THE THEESPRUIT FORMATION

The Theespruit formation is well developed in both the Onverwacht and Steynsdorp anticlines and, as will be shown later, is widespread right around the Mountain Land (Fig. 1). In the type area the formation attains a thickness of 1,890 metres (6,200 ft.) and comprises a succession of metamorphosed mafic lavas characterized by persistent interlayered felsic and minor mafic tuffs. Serpentinized ultramafics, often in the form of conformable and persistent bands, but also occurring as pods and lenses, are fairly well represented, together with talc, chlorite-talc and carbonate schists (Figs. 2 and 3).

The main distinguishing feature of the formation is the occurrence of the felsic tuffs which grade from white fine-grained siliceous often friable rocks consisting almost exclusively of a quartz mosaic with minor amounts of sericite and pyrophyllite to bedded re-worked fine- and coarse-grained felsic tuffs and agglomerates (Plates Ia and Ib). The latter are often aluminous, containing, in addition to quartz and sericite, conspicuous amounts of pyrophyllite together in places with andalusite, chloritoid and staurolite.

In certain instances very siliceous varieties have been completely recrystallized and resemble quartzites. In such cases the sericite has been reconstituted into fairly large muscovite crystals, often with small amounts of staurolite together at times with apatite. Occasional garnets were also noted. The felsic tuffs are frequently associated with and in places apparently gradational into chlorite and talc-chlorite schists.

Analyses of two varieties of felsic tuffs are presented in columns 5 and 6 of Table IV and their localities shown in Fig 2. The former analysis is of the widespread variety of fine-grained friable siliceous quartz-sericite rock of the type shown in Plate IIa, and the latter of the aluminous well-bedded pyrophyllite tuff depicted in Plate Ia. A feature of many of these tuffaceous rocks and particularly the more siliceous varieties is their strongly sheared and lineated nature. This generally increases in intensity towards the granite contact with the lineation taking the form of a steeply plunging "stretch" or "a" variety (Plate IIa). Minute "crinkle" or crenulation folds are often associated with this type of lineation and are widespread in similar rocks around the whole Mountain Land.

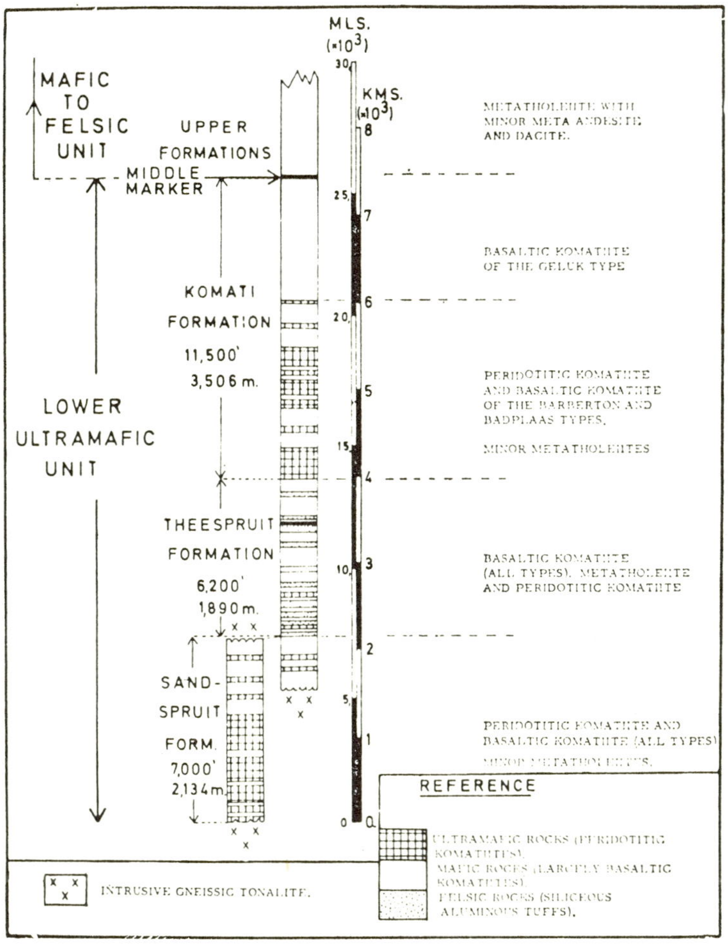

Fig. 3. Stratigraphic column of the three lower formations of the Onverwacht Group.

The felsic horizons are often immediately overlain by, or closely associated with, narrow generally impersistent bands and lenses of black siliceous cherty sediments (Plate Ib). These do not have the fine-grained porcellanous nature of typical chert but are often somewhat coarser grained and duller. They consist dominantly of a fine-grained quartz mosaic with abundant black carbonaceous matter generally scattered throughout, but in places becoming somewhat concentrated into layers. Cherty bands and lenses of this nature often sporadically overlie cycles of tuffaceous material both on a small and a large scale (Plate 1b). This suggests that their origin was closely related to the end phases of felsic pyroclastic activity and is probably to be ascribed to precipitation from silica-charged waters developed as a result of the explosive acid magmatism. A closely analogous situation exists within the upper formations of the Onverwacht Group where acid volcanics are frequently overlain by substantial chert horizons. Although most of the cherts appear to have formed by silica precipitation in the manner outlined above, certain cherty rocks with sedimentary structure such as cross-bedding have been found (Plate IIb). It is suggested that these could have represented fine-grained siliceous tuffs which were deposited by normal sedimentary means and then perhaps further silicified by the silica-charged environment. Similar phenomena have been described by van den Berg (1969) and the authors in some of the sediments from upper formations (Paper No. 6).

Primitive fossil forms have recently been reported in certain carbonaceous cherty rocks from the base of the Theespruit formation, making these the oldest known life forms yet found on earth (Engel *et al.*, 1968). Very recent biogeochemical studies at the National Aeronautics and Space Administration biological laboratories at Moffet Field, Paulo Alto, have revealed evidence for the existence of porphyrins in these cherts (Kvenvolden, 1969), thereby supporting the presence of the primitive fossil morphologies referred to above.

The mafic rocks of the Theespruit formation consist of a varied assemblage of lavas and tuffs. Pillow structures, spherulites and variolites, although not plentiful, are encountered in places, but the lavas are more generally massive. Near the granite contacts these rocks have suffered upper greenschist and more rarely amphibolite facies metamorphism and are generally dark green to almost black in colour. Away from the granite contact the metamorphic grade drops off rapidly to a greenschist assemblage manifested generally by lighter coloured rocks.

At the granite contacts the amphibolitized basalts consist essentially of hornblende or actinolitic hornblende with variable amounts of sodic plagioclase and some quartz. An analysis of a rock of this type is given in Table IIA, column 9, and its locality shown in Fig. 2. Locally the metabasalts are composed almost exclusively of amphibole. Other minerals encountered include pyroxene of a diopsidic composition, as well as epidote occupying fine fractures or as irregular patches resulting from the saussuritization of plagioclase. Accessory minerals include magnetite, sphene, ilmenite and leucoxene. Away from the granite aureole the metabasalts are generally green or grey-green, this colouration often being an indication of the presence of substantial amounts of carbonate and chlorite. The main minerals encountered include actinolite, tremolite, chlorite, epidote, plagioclase and carbonate, all of which vary considerably in their proportions. Some of the carbonated and somewhat schistose rocks have a distinctly pitted appearance due to the weathering-out of carbonate blebs. Petrologically these varieties are composed of a fine-grained felted mass of tremolite-actinolite and chlorite crystals broken every now and then by roughly spherical masses of almost pure carbonate with invariable iron staining. Strongly carbonated layers and bands with commonly associated chlorite and quartz, together with almost pure carbonate lenses, are frequently present in the lava. Similar material is characteristic of the interstices between pillow structures and could in part represent an original impure carbonate-rich oceanic ooze, into which the lavas were extruded. It is also possible that some of the carbonate was "sweated" out of the lavas during metamorphism to form bands and lenses by a process of lateral secretion. The question of carbonitization in general, and particularly its effect on the original chemistry of the Onverwacht volcanic rocks is discussed elsewhere (Paper No. 3).

A variety of layered to finely layered and often sheared mafic schists occur interlayered with the more massive metabasalts described above. On the east limb of the Onverwacht anticline, in the vicinity of the Doornhoek pluton, these schists are closely associated with the felsic schists or tuffs and are considered by the authors to represent sheared mafic tuffs. They consist largely of chlorite and tremolite-actinolite, often with fine bands or layers of carbonate. As noted previously the felsic tuffs are also closely associated with sheared talc-chlorite and talc-chlorite-carbonate schists, also probably representing mafic or ultramafic tuffs.

Ultramafic rocks although fairly well represented in the Theespruit formation are not nearly as widely developed as in the underlying Sandspruit or overlying Komati formations (Figs. 2 and 3). They occur generally as resistant bands, lenses and pods which are largely conformable to the stratigraphy of the formation. Where best preserved they are usually fine-grained dense rocks of a dark blue-green colour and resemble more the ultramafics of the Komati formation. They are, like similar varieties in the latter formation, comprised largely of antigorite and contain variable but generally lesser amounts of tremolite, chlorite and magnetite. They often grade into talcose and talc-chlorite-carbonate rocks.

IV. THE KOMATI FORMATION

The Komati formation attains a thickness of 3,506 metres (11,500 ft.) in the type area in the southern portion of the Mountain Land (Figs. 2 and 3) and consists of an alternating sequence of amphibolitized, pillowed and massive basalt and ultramafic bands. An important feature is the absence of interlayered sediments or acid rocks, although intrusive bodies of felspar and quartz porphyry are plentiful and rather typical. The ultramafics predominate in the lower half of the formation where a striking feature is the sympathetic thinning of individual ultramafic and pillow basalt horizons (Figs. 2 and 3). This, together with other features of the ultramafics described in an accompanying paper, have been employed to prove the extrusive origin of most of these rocks.

A. Ultramafic rocks

The ultramafic rocks will be described in detail in Paper No. 5. To summarize, they are in places remarkably fresh-looking, fine-grained dense rocks with a dark blue-black colour. Microscopic examination, however, reveals extensive alteration and serpentinization with only rarely preserved kernels of olivine in a groundmass of antigorite, tremolite, chlorite and magnetite.

All of these peridotitic rocks have a distinctive and unusual chemical composition. They are unlike any class of peridotitic or picritic rock known and a new name, peridotitic komatiite, after the Komati formation where they are best developed and preserved, has been proposed. An average analysis of peridotitic komatiite from the Komati formation is given in column 5 of Table I.

B. Mafic rocks

Detailed geochemical investigations of the mafic rocks of the lower three formations have been carried out mainly on the Komati formation. Closely similar mafic varieties are present in the lower two formations as is indicated by field and petrological investigations, but only limited geochemical studies were carried out on these because of their generally more highly metamorphosed state. The mafic rocks of the Komati formation (derived almost exclusively from basalts) are generally green to grey-green amphibolites composed of variable amounts of tremolite-actinolite, cummingtonite, anthophyllite, zoisite, chlorite, talc, oligoclase, quartz and carbonate. Detailed field, petrological and geochemical investigations have led to the identification of four general types although it is probable that gradations exist between these. One class, which is of limited extent, is comparable to a known class of basaltic rock, viz. tholeiites. The three remaining and most widespread types, however, are chemically distinctive and unlike any known class of basaltic rocks yet described. Together they constitute a distinctive "primitive" group for which the term Basaltic Komatiite is proposed. The composition of these rocks has generally rendered them very susceptible to shearing and many outcrops take the form of generally low, jagged ridges of the type illustrated in Plate IIIa.

1. Basaltic Komatiite

This class constitutes the bulk of the mafic rocks of the Komati formation and has been divided into three varieties, viz:

(i) The Barberton type,
(ii) The Badplaas type,
(iii) The Geluk type.

(i) *Barberton type*

In the type area basaltic komatiites of the Barberton type are always intimately associated with the ultramafics (peridotitic komatiites) of the lower portion of the Komati formation (Fig. 3). Several pillowed flows associated with massive phases or early, near contemporaneous sills constitute individual mafic horizons which alternate with the ultramafics (peridotitic komatiites).

The pillow structures are of variable shape and size, usually well formed, and often have conspicuous darker and finer, somewhat "crackled"-looking chilled selvedges, generally up to several centimetres wide. A typical group of pillows is illustrated in Plate IIIb. Variolitic and/or spherulitic textures are diagnostic features, being particularly plentiful in the pillowed zones, where they take the form of light-coloured orbicular patches (Plate IVa) often with a clear zonal arrangement within the pillows. The average diameter of these variolites is of the order of a few centimetres, but they may be as large as 10 cm. or more in certain instances. In many cases adjoining variolites merge to form very large irregular light-coloured variolitic patches (e.g. the upper and lower parts of Plate IVa). The etching of the variolites often produces a surface of small craters, rimmed by more resistant ridges, or else the variolites are uniformly resistant to weathering, producing small light-coloured domical features. Both of these modes of weathering are illustrated in Plate IVb. Amygdales are rare, but white vein-quartz frequently occurs in the voids between pillows and is often found occupying large horizontal gas cavities within individual pillows (Plate Va). These invariably occur towards the upper part of the pillows and are therefore useful indicators of the direction of younging.

Microscopically, basaltic komatiites of the Barberton type are composed mainly of amphibole in the tremolite-actinolite range, occurring as stubby laths in the massive varieties and as long slender randomly orientated needles in the pillowed varieties. This mineral represents the alteration of diopsidic pyroxene, remnants of which are occasionally encountered. Varying amounts of soda-plagioclase (generally oligoclase), chlorite, zoisite, epidote, carbonate and quartz are also frequently observed, generally as finer-grained crystals or fibrous aggregates between the amphiboles (Plate Vb). In many places these rocks take the form of massive, tough and resistant, coarsely crystalline amphibolites characterized by huge actinolite needles attaining lengths of up to 10 cm. and more (Plate VIa). Such zones which vary in width from a few metres to about 20 metres always appear to lie parallel to the general layering of the Komati formation. In some instances the large amphibole needles are randomly orientated but frequently they display a very strong parallel alignment, generally giving a marked, near vertical, lineation (Plate VIa). A horizontal or near horizontal growth of parallel actinolite needles was also noted in some places. The material interstitial to the dark green actinolite needles has a light grey-green colour and is comprised largely of finely intergrown chlorite, epidote, clinozoisite and quartz derived mainly from felspathic material. The origin of these coarsely crystalline amphibolites is problematical and two main possibilities exist. In the first instance they could have formed during the period of regional greenschist metamorphism which affected the whole area, in zones of shearing and greater volatile activity under conditions which allowed for large crystal growth. Another possibility is that they formed at the time of subaqueous extrusion of the lava as a result of access of water and rapid cooling or quenching. They could perhaps therefore be closely analogous to the "crystalline quench" structures developed in the associated meta-peridotites (peridotitic komatiites) described in an accompanying paper (No. 5).

Selected chemical analyses of both the pillowed and the massive varieties of basaltic komatiites of the Barberton type from the type area are given in Table IIA columns 1, 2, 3, 7 and 10, and their localities shown in Fig. 2. These analyses will be discussed later.

The lighter-coloured spherules or variolites consist largely of a quartzo-felspathic mosaic often including numerous scattered, partly altered pyroxene crystals and frequently containing scattered carbonate replacement bodies. An analysis of a large spherule of the type shown in Plate IVa is given in Table IIA, column 11.

In some areas the lavas are extensively carbonated and assume a rusty brown appearance. In such instances carbonate, often with minor quartz and chlorite, generally occurs to the near exclusion of all other minerals. The question of carbonization, its cause and the drastic effect it has on the chemistry of the mafic rocks of the Onverwacht Group is discussed in an accompanying paper (No. 2).

(ii) *Badplaas type*

Basaltic komatiites of the Badplaas type are, like the variety discussed above, closely associated with the peridotitic komatiites of the lower part of the Komati formation and can usually be recognised in the field by their generally more coarsely crystalline but even-granular texture, massive pyroxenitic appearance and lighter, somewhat olive green, colour. Although pillow structures have been found, most of the material appears to take the form of massive possible near contemporaneous sills. Rocks of this type are perhaps better termed metapyroxenites as they are composed almost entirely of tremolite-actinolite derived from a diopsidic pyroxene, remnants of which are sometimes encountered. Typical analyses of the Badplaas type are given in columns 1 and 2, Table IIB. The genetic significance and probable origin of these rocks is discussed elsewhere (Paper No. 11).

(iii) *Geluk type*

In the southern portion of the Barberton Mountain Land basaltic rocks of this type appear to be confined to the upper portion of Komati formation in the region which lacks the abundant interlayered ultramafic material encountered lower down (Fig. 3).

Pillow structures are well developed in places but lack the conspicuous spherulitic textures and gas cavities encountered in the two other varieties of basaltic komatiite (Plate VIb). Varying proportions of chlorite, magnesium and calcium amphiboles, talc, epidote, carbonate and very minor plagioclase constitute the main minerals, this assemblage having given rise to incompetent rocks which are strongly sheared in numerous instances. An analysis of the Geluk type from the type area is given in Table IIC, column 9.

In many instances, long amphibole needles of a similar nature to those described from the Barberton type basaltic rocks are also present and have undoubtedly developed in the same manner as the amphiboles in the latter. In the upper part of the Komati formation in both the Onverwacht and Steynsdorp anticlines lenses and blebs of carbonate become conspicuous in the Geluk-type metabasalts, becoming more numerous towards the top and eventually grading into well-banded tuffaceous shales or palagonitic tuffs. These comprise narrow bands of dark green chloritic and carbonaceous material alternating with bands of almost pure carbonate and constitute part of the banded chert and carbonate sediment known as the Middle Marker. This important marker separates the distinctive lower formations of the Onverwacht Group from the upper formations and has been described in detail elsewhere (Viljoen *et al.*, 1969, and accompanying Paper No. 6).

2. Meta-Tholeiite

This group is not widespread in the Komati formation and only appears to be sporadically developed at a few localities, mainly in the vicinity of the basaltic komatiites of the Barberton type. Massive and pillowed varieties are present.

The meta-tholeiites are recognised by a relatively high content of soda plagioclase together with conspicuous green chlorite and minor actinolite. Sulphides and opaque oxides are generally conspicuous and quartz is sometimes encountered. An analysis of a meta-tholeiite from the Komati formation type area is given in column 1 of Table III.

C. Porphyries

Besides the distinctive mafic and ultramafic rocks described above, the Komati formation is characterized by the abundant development of scattered felspar and quartz porphyry bodies. They vary in length from a few metres to about 2,000 metres and generally occur as elongate but somewhat irregular masses largely conformable to the overall layering of the Komati formation. They are readily recognised by their slightly greater resistance to erosion than the surrounding mafic and ultramafic rocks, as well as their distinctive light yellow or yellowish green to yellowish brown colour on weathered surfaces.

Hall (1918) was of the opinion that these rocks formed an integral part of the Onverwacht stratigraphy possibly representing contemporaneous felsic volcanics. In this respect it is of interest to note that the elongated but detached porphyry lenses often tend to remain in one stratigraphic position for long distances. Definite intrusive relationships have however been found so that although it is probable that they are virtually contemporaneous with the mafic and ultramafic vulcanicity they do not represent extrusives. The porphyry bodies appear to have no strong preference for the rock type into which they intrude although to the east of the type region where they are of a much larger size and more extensively developed they appear to be largely confined to the ultramafic rocks.

Felspar-bearing porphyries are the most common variety although quartz phenocrysts are often conspicuous as well. The phenocrysts which average several millimetres in diameter are clearly seen on both fresh and weathered surfaces, the felspars generally taking the form of euhedral to subhedral white patches and the quartz occurring as angular to rounded transparent to translucent blebs. Most of the felspar phenocrysts are albitic in composition, often zoned and generally strongly sericitized, whereas many of the quartz phenocrysts show signs of corrosion. The phenocrysts occur in a matrix comprised largely of fine-grained quartz and sericite. Carbonate is widely present in most instances occurring as disseminated patches which pervade the whole rocks. An analysis of a typical porphyry from the Komati formation is given in Table IV, column 7, and the average of three additional partial analyses for Na_2O and K_2O in column 8. An analysis of a well-preserved, very soda-rich porphyry is given in Table IV, column 9.

V. CORRELATION OF THE LOWER ULTRAMAFIC UNIT

A. Introduction

The description of the stratigraphy of the lower part of the Onverwacht Group as presented above applies specifically to the southern portion of the Barberton belt and more particularly to a well-developed and well-preserved section on the west limb of the Onverwacht anticline where the type section has been established. Previously, as noted in an accompanying paper (No. 2), the well-preserved area of Onverwacht volcanics in the southern portion of the belt was considered to be the only region where this important group was extensively developed (Visser, compiler, 1956). The variety of basic, ultrabasic and associated schistose rocks surrounding the remainder of the belt, and lying conformably below the Fig Tree Group, were all considered to represent part of the basic assemblage of the supposed post-Moodies, called Jamestown Intrusive Igneous Complex (Visse *et al.*, 1956). As has been shown elsewhere (Viljoen and Viljoen, 1967, and accompanying Paper No. 2) the large majority of these rocks actually represent the more highly altered equivalents of the lower formations of the Onverwacht Group and the Jamestown Igneous Complex as previously defined no longer exists (Ferguson, 1967).

Based on a preliminary stratigraphic column for the Onverwacht Group (Viljoen and Viljoen, 1967) a correlation of the pre-Fig Tree rocks in the northern and southern parts of the Mountain Land was suggested (Anhaeusser *et al.*, 1966). With further detailed work by the authors in the southern part of the Mountain Land, the earlier threefold division of the Onverwacht Group has been amended so that the present stratigraphic column contains six formations (see Paper No. 2). Recent investigations, both detailed and reconnaissance by both the authors and Anhaeusser (1969) in other parts of the Barberton Belt, have resulted in a considerable amount of new geological, petrological and particularly geochemical data. In the present section all available data are synthesized and a new correlation of the Onverwacht Group around the entire Mountain Land, based largely on lithological and geochemical grounds, is proposed.

A striking feature is the fact that, almost without exception, the rocks of the Onverwacht Group away from the type area in the south belong almost exclusively to the Lower Ultramafic Unit, and more particularly to the Theespruit and Komati formations of the latter. In the discussion on the lithostratigraphic correlation, given below, in which each formation is discussed in turn, it should be pointed out that, as would be expected, local as well as lateral regional variations and facies changes occur within each of the formations. These are never great enough, however, to alter the identity of each distinctive formation.

B. Sandspruit formation

The only area where the Sandspruit formation appears to be well preserved is in the type area in the southern part of the Mountain Land. Here, as noted previously, it occurs largely in the form of a series of xenoliths almost encircling the Theespruit pluton in the core of the Onverwacht anticline (Fig. 1). A small occurrence is also situated along the Swaziland border on the eastern side of the Steynsdorp tonalitic gneiss pluton in the core of the Steynsdorp anticline (Fig. 1). These are the only two areas where the formation has been positively identified, although it is the authors' contention that many of the mafic and ultramafic bodies within the ancient tonalitic gneiss terrain are probably remnants of the Sandspruit formation.

C. Theespruit formation

The Theespruit formation is one of the most widely developed and most easily recognisable of the entire Onverwacht Group due to the presence of the distinctive siliceous and often aluminous felsic schists. In the southern portion of the belt this formation attains its best development in the core of the Onverwacht anticline where the type area has been established (see previously).

The formation is also well developed, although strongly sheared, in the core of the Steynsdorp anticline where it occurs in the form of a broad arc draped around the Steynsdorp tonalitic gneiss body (Fig. 1).

In most other areas around the entire periphery of the Mountain Land, rock types and assemblages correlated with the Theespruit formation occur in contact with the various granite types which envelop the area. Assemblages which bear a striking similarity to the Theespruit formation in the type area occur along the north-western flank of the Mountain Land. Here the formation constitutes much of the tightly folded Jamestown synclinal belt, is well developed on the northern limbs of the Lily and Eureka synclines in the area between the Consort Mine and Louw's Creek, and flanks the Kaap Valley Granite as a narrow sliver on the eastern margin of the latter (Anhaeusser *et al.*, 1966; Annhaeusser, 1969). In all these areas a narrow zone comprising a variety of dark hornblende or actinolitic hornblende amphibolites, mineralogically identical to similar rocks in the type area of the Onverwacht Group, occur along the immediate granite contacts. Away from the contacts these rocks become greenish in colour, reflecting an increase in the amount of actinolite or chlorite concomitant with a decrease in the grade of metamorphism of the amphibolites. Analyses (quoted in columns 2, 8 and 10 of Table III) clearly indicate the derivation of these amphibolites from original basaltic rocks, confirmed also by the occasional occurrence of well-developed pillow structures as well as variolites (N. Harte, verbal communication, 1967; Anhaeusser, 1969). Forming an integral part of the amphibolite assemblage described above are numerous resistant serpentinite pods and lenses generally associated with, and gradational into, less resistant zones of talcose chlorite, tremolite and carbonate schists (Anhaeusser, 1969). Mineralogically these serpentinitic bodies are identical to similar rocks (komatiites) in the Theespruit and Komati formations of the Onverwacht type area. Associated with the above-mentioned rock types are a variety of highly distinctive quartz-sericite rocks and associated aluminous schists, cherty siliceous sediments and minor shales. These rocks are remarkably similar to the diagnostic siliceous and aluminous felsic tuffs that occur in the Theespruit formation in the Onverwacht type area and it is largely on the basis of this distinctive rock type that a correlation has been effected. Apart from the almost pure quartz-sericite varieties which are the most widely developed, fuchsite-bearing varieties and aluminous schists containing andalusite, pyrophyllite, and more rarely sillimanite, as in the granite contact region to the north of the Consort Mine, are also well represented. Other minerals found in these rocks include chloritoid, staurolite, almandine garnet, plagioclase, muscovite, clinozoisite, leucoxene, magnetite and ilmenite (Anhaeusser and Viljoen, 1965; Anhaeusser, 1969). As in the Onverwacht type area many of these rocks are capped by siliceous cherts or cherty quartzites which often terminate particular depositional cycles and are hence useful indicators of the younging direction of the sequence. Apart from the strong foliation or schistosity generally present in these schists, they frequently also contain a conspicuous near-vertical "stretch" lineation. Numerous minor and accordion folds with near horizontal axial planes are widespread in these rocks and often deform the above-mentioned lineation. Two total silicate analyses of typical aluminous siliceous schists from the eastern portion of the Jamestown schist belt, as well as the average of five partial analyses of similar varieties from the same area, are listed in columns 1 to 3 of Table IV and will be discussed later.

A number of sedimentary horizons not altogether typical of the type area also occur in the Jamestown schist belt. These include black carbonaceous cherts with intercalated slaty shales, ferruginous slates, grits and in places conglomerates or more probably volcanic agglomerates. None of these horizons is very thick or well developed however, and they in no way mask the fundamental identity of the Theespruit formation in this area.

East of Louw's Creek in the Stentor anticlinal structure, mafic rocks, with the exception of the local development of chloritic schists, as well as ultramafic rocks, are only poorly represented. A broad succession of white to grey siliceous and aluminous felsic schists probably derived largely from original tuffs is virtually the only component of the Theespruit formation developed. The main minerals in the latter rocks are quartz, forming a mosaic of small grains, pyrophyllite and sericite. Other minerals include chlorite and soda plagioclase. The latter mineral, together with larger quartz masses, often occurs as conspicuous "knots" around which the schistosity is draped. An interesting feature is the occurrence at the contact between the felsic and mafic schists of barite which is described in an accompanying paper (No. 9).

Assemblages which, although having certain similarities with the Theespruit formation, are not as readily correlated with the latter are encountered in the extreme north-eastern part of the Mountain Land and along its eastern flank in north-west Swaziland.

In a broad strip along the granite contact in the area between Kaapmuiden and Hectorspruit occurs a succession consisting largely of massive and pillowed meta-tholeiitic lavas. Siliceous felsic schist horizons, as well as magnesian-rich basalts and schists, characteristic of the Theespruit formation in the type area, although present are not, however, widely developed in this area. In addition, narrow intercalated layers of banded chert and ironstone, not typical of the Theespruit formation in the type area, are also developed in places.

A variety of actinolite-tremolite-chlorite as well as talc-chlorite schists with narrow intercalated quartzitic, sericitic, and associated cherty and shaly interlayers occurs along the eastern margin of the Mountain Land, adjacent to the granite, in Swaziland and adjoining portions of the Transvaal. Some of the quartz-sericite schists contain andalusite (Jones, personal communication) and serpentinite lenses and bands are sometimes also encountered. The assemblage thus appears to contain most of the components of the Theespruit formation and is therefore correlated with the latter.

In the Forbes Reef area of Swaziland and farther south towards the south-eastern termination of the Barberton belt, much of what has been termed the Metasedimentary Group (Urie and Jones, 1965) has been correlated with the Theespruit formation by the authors. This Group, which occurs at the base of the sequence in this area and in contact with the granite, consists essentially of dark greenish-grey amphibolites constituted largely of actinolite or hornblende with smaller amounts of plagioclase, quartz, chlorite, biotite and diopside (Urie and Jones, 1965). Interlayered siliceous horizons including recrystallised banded cherts, quartzites and quartz-sericite schists also occur within the succession. The main mineral is quartz with varying amounts of sericite, andalusite, biotite, and muscovite with amphibole, felspar, chlorite, epidote and sphene occurring less frequently. The association of andalusite-bearing quartz-sericite schist horizons within actinolitic schists is considered strong evidence for correlating the so-called meta-sedimentary groups with the Theespruit formation.

It is the authors' contention that many of the xenoliths occurring within the granitic terrain surrounding the Mountain Land can be correlated with the Theespruit formation. This applies particularly to the xenoliths containing narrow quartzitic, sericitic or shaly zones in association with mafic and ultramafic rocks. Examples include the xenoliths within the tonalitic gneiss terrain to the south-west of the Mountain Land, especially in the vicinity of Bosmanskop, numerous xenoliths within similar terrain in Swaziland, the large xenolithic body at Barclay Vale, as well as part of a xenolithic mass to the north-west of Badplaas in which the Kalkloof body occurs (Fig. 1).

D. Komati formation

The Komati formation attains its best development and is best preserved in the core of the Onverwacht anticline where it is immediately overlain by the Middle Marker which separates the three lower from the three upper formations of the Onverwacht Group. After being eliminated by intrusive granites in the

Kromberg syncline the marker appears again in the Steynsdorp Valley where it defines the Steynsdorp anticline (Fig. 1) (Viljoen *et al.*, 1969). In this area much of the tight anticlinal structure is comprised of basaltic komatiite of the Geluk type as defined previously with only a reduced amount of ultramafic material in the form of serpentinites and talc-carbonate rocks forming the lower ultramafic part of the Komati formation in this area. Away from the southern area, the Middle Marker has not as yet been identified, although assemblages correlated with the Komati formation are widespread around the Mountain Land (Fig. 1).

It is convenient, because of certain strong similarities, to discuss the correlatives of the Komati formation along the whole western flank of the Mountain Land together. In this region assemblages correlated with the Komati formation generally occur between the Theespruit formation in immediate contact with the intrusive granites and the overlying sedimentary succession of the Fig Tree Group. Rocks of the Komati formation comprise the synclinal Nelshoogte schist belt (Fig. 1), as well as the core of the synclinal Jamestown schist belt at the eastern end of the latter. In addition the formation is extensively developed in the area between Kaapmuiden and Malelane along the northern flank of the Mountain Land. Smaller slivers of Komati formation assemblages partly connect the above-mentioned areas of best development. Other assemblages which have been correlated with the Komati formation also occur at the western extremity of the Jamestown schist belt, in the vicinity of Kaapsehoop, as well as farther west as an inlier, exposed by the erosion of the Transvaal System, in the Elands River valley (Fig. 1). In all these occurrences, the main features are the almost complete absence of interlayered sediments or felsic volcanics and an abundance of ultramafic and magnesian-rich basaltic rocks. The ultramafics take the form of extrusives or near-surface sill-like bodies and occur typically as resistant bands and lenses of green serpentinite or partly serpentinized and metamorphosed green to dark blue-black peridotites, generally lying parallel to the overall stratigraphy. In the more strongly deformed areas such as in the Jamestown schist belt, these serpentinites gives way to zones of tremolite, talc, or talc-carbonate-chlorite rocks. The magnesium-rich schists (meta-basalts), with which the serpentinites are interlayered, include tremolite-actinolite schists, chlorite-talc schists, tremolite-chlorite-talc schists as well as a variety of dolomitic and carbonated rocks, the latter often occurring in zones of strong structural disturbance.

Mineralogically and chemically all of these rocks are closely similar to the variety of so-called komatiites, both of the peridotitic and basaltic types, as defined in the type area of the Komati formation. Although, because of their more highly sheared state, diagnostic volcanic structures such as pillows and variolites are rare, these have been encountered in the less sheared basaltic rocks of the western and eastern ends of the Jamestown schist belt (N. Harte, verbal communication, and Anhaeusser, 1969). No unequivocal evidence supporting an extrusive origin for the ultramafic rocks has been obtained in this region, although the presence of two unequivocal peridotitic dykes in the Jamestown schist belt, well exposed in a gorge cut by the Noordkaap River, supports the existence of a peridotitic magma as adduced also from evidence in the type area of the Komati formation (see accompanying Paper No. 5). These dykes are shown in Plates VII (a) and (b). The largest individual, which averages about 15 cm. in width (Plate VII (a)), could be traced for several tens of metres along an exposure on the river bank. Both dykes have sharply defined contacts and display a slightly darker and finer-textured chill margin clearly seen in Plate VII (b). Both dykes are unfortunately serpentinized and are composed of conspicuous antigorite and lesser amounts of tremolite, chlorite and magnetite. They are intrusive into a large peridotite mass of similar composition. A partial chemical analysis of the largest dyke shown in Plate VII (a) is presented in Table I, column 7. The high water content renders the analysis unsuitable for comparison with the better-preserved specimens from the Komati formation type section. Bearing these limitations in mind, however, and comparing the chemistry of the dyke with the more highly serpentinized peridotites from the Komati formation type area (columns 3 and 4, Table I, of Paper No. 3) it is clear that strong similarities exist. These support the authors' contention of the existence of an early widespread peridotitic magma of distinctive chemistry (Papers No. 3 and 11).

Analyses of basaltic komatiites of all three varieties mainly from the area between Kaapmuiden and Malelane (see Fig. 2) are given in Table II, A, B and C. A description of the basaltic and ultramafic rocks of the latter area has been given elsewhere (Viljoen and Viljoen, 1969).

Except for a small patch at the south-eastern extremity of the Mountain Land, it would appear from available evidence that along the whole eastern edge of the belt correlatives of the Komati formation are poorly if at all developed. To the south in the Forbes Reef area, however, the so-called Magnesian Series of Urie and Jones (1965) is correlated with the Komati formation. The magnesium-rich schists of this area are represented by a wide variety of serpentinous, amphibolitic and talcose rocks which appear to be largely, if not entirely, derived from original ultrabasic rocks. Siliceous horizons are virtually absent, although a few very narrow and not very extensive chert bands have been recorded. An analysis of a magnesium-rich schist from this area is given in column 6, Table I. The similarity with the peridotitic komatiites from the type area is clear and it is contended that the dominant original component of the Komati formation in this area was peridotitic komatiite which occurred to the near exclusion of basaltic varieties.

As with the Sandspruit and Theespruit formations, it is contended that certain of the xenoliths within the granite terrain surrounding the area, and particularly some of the ultramafic and mafic varieties, could represent fragments of the Komati formation. To resolve the question of which of the lower formations a particular xenolith should be ascribed to, however, a detailed consideration of the rock types within the xenolith, their associations, petrology and probably also their chemistry would be necessary.

E. Layered, Differentiated Ultramafic Bodies

Intimately associated with the variety of dominantly extrusive mafic and ultramafic rocks described above and forming an integral part of the Lower Ultramafic Unit are a number of well-layered intrusive, sill or pod-like ultramafic bodies. On the basis of their products of differentiation, three main types have been recognized (Viljoen and Viljoen, 1969). These are:—

(1) The Kaapmuiden type.

(2) The Noordkaap type.

(3) The Stolzburg type.

1. *The Kaapmuiden Type*

This variety is confined to the Komati formation between the towns of Kaapmuiden and Malelane, and has been described in detail elsewhere (Viljoen and Viljoen, 1969). To summarise, three well-layered ultramafic sills, attaining thicknesses of over 610 metres (2,000 ft.) have been emplaced into a broad belt of volcanic rocks with minor interlayered sediments. They are all characterized by broad lower dunitic and peridotitic zones which pass upwards abruptly into rather narrow zones of bronzitite and websterite, followed by very narrow anorthositic gabbro/norite zones. The latter are followed in turn by another broad sequence of dunites and peridotites. Secondary alteration of the dunitic and peridotitic zones has given rise to the extensive development of magnesite and talc which are typical of the Kaapmuiden as well as related types of ultramafic bodies in this area (Paper No. 9). Asbestos mineralization, although present, is generally not well developed.

The bulk chemical composition of these bodies, computed by combining the chemistry of each major rock type in the correct proportions, has been shown to be virtually identical to the average composition of the peridotitic komatiite flows of the Komati formation in the southern portion of the Mountain Land (Paper No. 11).

Two other sill-like dunitic bodies of an identical nature to the dunite or dunite-peridotite assemblages of the Kaapmuiden-type bodies, but without any sign of pyroxenitic or gabbroic rocks, are also present in the area. These, termed the Magnesite and Central Bodies, are generally highly altered and contain most of the important magnesite and talc deposits of the area (see Paper No. 9). Also present in the region is a layered and differentiated mafic body, the Canal Body. All of the layered and differentiated bodies of the area are closely associated with and generally intrusive into an assemblage of magnesium-rich basic rocks (basaltic komatiites) (Fig. 1). The significance of these bodies with respect to an understanding of the genesis of most of the mafic and ultramafic rocks of the lower formations of the Onverwacht Group is discussed in the concluding Paper No. 11 and will not be elaborated upon here.

2. *The Noordkaap Type*

A number of distinctive layered ultramafic bodies termed the Noordkaap type (Viljoen and Viljoen, in press) occur in the eastern portion of the Jamestown schist belt, immediately to the west of the town of Noordkaap. Two main masses, viz. the Handsup and Mundt's Concession bodies, as well as a few smaller disconnected occurrences in the same general area have been recognised (Anhaeusser, 1969). Although the bulk chemical composition of these bodies is closely similar to the bulk chemical composition of the Kaapmuiden type ultramafic bodies (see accompanying Paper No. 11), the style and products of differentiation are somewhat different. The bodies consist of a number of cyclic differentiated units comprising a minimum of four differentiation cycles (Anhaeusser, 1969). Most cycles commence with olivine peridotites, pyroxenites and finally fairly extensive gabbros with small patches of anorthositic gabbros in places (Anhaeusser, 1969). Certain of the rock units mentioned above may be missing in individual cycles. All of these rocks have suffered alteration and now consist essentially of serpentinites and amphibolites with only occasional remnants of primary mineral phases still being present. The bodies are of interest in that they contain small scattered deposits of the rare semi-precious serpentinous rocks verdite and budstone, in addition to small asbestos and magnesite deposits. The significance of the Noordkaap type ultramafic bodies with respect to theories on the genesis and processes of magmatic diversification of the rocks of the lower three formations of the Onverwacht Group has been discussed in an accompanying paper (No. 11) and the mineral deposits associated with them in Paper No. 8.

3. *The Stolzburg Type*

Differentiated ultramafic bodies of the Stolzburg type have been recognised at four localities in the Barberton Mountain Land (Viljoen and Viljoen, 1969, and in press). These include the Stolzburg body itself at the south-western extremity of the Mountain Land, established as the type occurrence, the Kalkloof body within a large xenolith of the lower Onverwacht assemblage in the tonalitic gneisses to the north-west of Badplaas, and several bodies in the Kaapsehoop area at the western extremity of the Jamestown schist belt. All of these bodies are typified by alternating layers of resistant orthopyroxenite (often altered to talc or bastite) and serpentinized dunites. They are thus distinguished from the two other varieties of ultramafic bodies described above by the near total absence of clinopyroxene and felspar-bearing rocks.

Asbestos mineralization is well developed in all of the occurrences and has been economically exploited in numerous places. The mineralization is discussed in more detail in an accompanying paper (No. 9) and the genetic significance of these bodies is considered in Paper No. 11.

All the ultramafic bodies described above as far as can be ascertained from present knowledge have been emplaced into the Komati formation more or less contemporaneously with the mafic and ultramafic vulcanicity typical of the latter.

VI. GEOCHEMISTRY OF THE LOWER ULTRAMAFIC UNIT

A. Introduction

Owing to the fact that the primary mineral assemblages of almost every rock type within the lower formations have suffered varying degrees of metamorphism from the amphibolite or epidote amphibolite facies near granite contacts to the greenschist facies farther away, the chemistry, rather than the mineralogy, has been important in classifying them. All available analyses of ultramafic, mafic and felsic rocks from the lower formations are listed in Tables I to IV. These data indicate that most of the rock classes recognized are totally unlike any widespread and generally accepted rock type yet described, and it is proposed in the present session to discuss their distinctive and unusual chemical features. The reasons for the introduction of a new class of igneous rock (with a number of sub-classes) for which the general name komatiite is proposed are given. The broad chemical characteristics of the mafic and felsic rocks will be discussed in turn, followed by a detailed consideration of the chemistry of the mafic and ultramafic rocks in relation to well-established rock classes.

B. Chemical Characteristics of the Rocks of the Lower Ultramafic Unit

1. *Ultramafic Rocks* (see Figures 1 and 2 for localities and Table I for chemical composition)

A full description of the geology and geochemistry of the relatively well-preserved assemblage of peridotitic rocks from the type area of the Komati formation has been given in an accompanying paper (No. 5). Evidence presented in that paper shows that these peridotitic rocks crystallized mainly subaqueously from a peridotitic magma with a distinctive and unusual chemical composition. The unique chemical character of this magma, the composition of which is presented in column 5 of Table I, led the authors to propose a new class of peridotitic rock, viz. peridotitic komatiite. The main features of this peridotitic rock are a high Ca/Al and Fe/Mg ratio and a low alkali content.

TABLE I

Selected Chemical Analyses of Ultramafic Rocks from the Three Lower Formations of the Onverwacht Group. (Excluding the Layered Differentiated Ultramafic Bodies.)

Column No.	1*	2*	3*	4	5	6	7*
Sample No.	87J	88J	VU33	Average	Average	FOR	UDI
SiO_2	46.29	44.85	43.11	44.72	41.61	43.60	36.84
TiO_2	0.68	0.51	1.35	0.52	0.31	0.08	
Al_2O_3	3.75	2.66	3.41	3.25	2.70	4.60	4.58
Fe_2O_3	6.96	6.90	4.10	6.02	5.63	1.53	3.96
FeO	5.18	4.91	6.46	5.52	4.35	4.14	6.40
MnO	0.19	0.20	0.19	0.19	0.17	0.10	
MgO	21.99	24.96	29.17	25.35	30.58	31.68	31.24
CaO	9.09	7.01	4.84	6.97	4.29	4.82	0.14
Na_2O	0.53	0.41	0.53	0.49	0.15	0.10	0.03
K_2O	0.06	0.05	0.05	0.05	0.03	0.11	0.03
P_2O_5	n.d.	n.d.	n.d.	n.d.	0.02	0.03	
H_2O^+	4.31	6.10	6.31	5.58	8.81	7.19	11.20
H_2O^-	0.10	0.34	0.16	0.21	0.22	0.18	0.24
CO_2	0.07	0.20	0.50	0.26		1.72	
Cr_2O_3	0.36	n.d.	0.36	0.36	0.32	0.37	
NiO	0.09	n.d.	0.19	0.14	0.18	0.10	
TOTAL	99.65	99.10	99.77			100.38	94.66

Analysts and Methods Used.

* National Institute for Metallurgy, Johannesburg. (Standard gravimetric, volumetric, colorimetric and flame photometric.)
Alkali determinations (flame photometric) as well as chromium and nickel determinations (X-ray fluorescence) by the authors.

Columns 1 to 4 Metaperidotites (peridotitic komatiites) from the Sandspruit Formation.
2 Tremolite, chlorite, antigorite and magnetite variety.
3 Tremolite, chlorite, antigorite and magnetite variety with numerous remnants of clinopyroxene.
4 Average chemical composition of 3 metaperidotites (peridotite komatiites) Sandspruit Formation.
5 Average chemical composition of peridotitic komatiites from the Komati Formation (Average of 8 analyses - see Paper No. 5).
6 Olivine-bearing amphibolite from the Magnesian Series (Komati formation) in the Forbes Reef Area of Swaziland (Urie & Jones, 1965).
7 Partly serpentinized peridotite dyke (peridotitic komatiite), Jamestown schist belt.

As has been shown elsewhere (Paper No. 11) the average bulk chemical composition of the layered differentiated ultramafic bodies of the Kaapmuiden and Noordkaap types is closely similar to the average chemistry of the peridotitic komatiite flows of the Komati formation and it is clear, as suggested in Paper No. 11, that they formed from the same peridotitic magma. The bulk composition of the Stolzburg type ultramafic bodies, however, differs considerably from the chemistry of the peridotitic komatiite magma. This is due to the fact that these bodies represent largely the accumulation of the high-melting refractory residues (olivine and orthopyroxene) from a parent peridotitic komatiite magma.

The chemistry of a typical magnesium-rich schist (Table I, column 6) from the Magnesian Series of Swaziland also has close similarities with the chemistry of the peridotitic komatiites of the type area (column 5, Table I) and confirms the geological and petrological conclusion that this series consists essentially of a huge pile of peridotitic komatiites.

As noted earlier, the peridotitic rocks from the Sandspruit formation occur in a similar geological setting to those in the Komati formation. Although they also bear a close mineralogical and chemical similarity to the peridotitic komatiites, they are not as ultrabasic as the latter (see column 4, Table I). They have higher silica, calcium, iron and alkali contents and lower magnesium contents than the peridotites of the Komati formation. Their alumina content is about the same, which means that they have an even higher Ca/Al ratio than the peridotites of the Komati formation. These chemical differences are indicative of an originally higher pyroxene (diopside) content and a substantially lower olivine content of the rock, and could reflect a fundamental evolutionary trend. The unusually high Ca/Al ratio of all classes of peridotitic komatiite in general is due largely to significant amounts of diopsidic pyroxene together with negligible amounts of plagioclase. Chemical data are not available from the peridotitic rocks of the Theespruit formation but their mineralogical composition suggests they are also of the peridotitic komatiite type, bearing a closer similarity to the peridotites of the Komati formation than to those of the Sandspruit formation.

2. *Mafic Rocks*

Although the chemistry of the mafic rocks is far more variable and erratic than the chemistry of the associated ultramafics, it was found, by a consideration of the geological setting, petrology and chemistry, that they could be grouped into four main classes. One of these represents the class of metatholeiites whereas the other three represent the varieties of basaltic komatiite, the geological setting and petrology of which have been described previously. The main chemical characteristics of the group of basaltic komatiites in general are a high magnesium content with fairly high silica and calcium contents and very low aluminium and alkali abundances. The chemical characteristics of each of the main classes of basaltic rock occurring in the area are discussed in turn.

(i) Basaltic Komatiites

The Barberton Type

Analyses of all rocks considered to be representative of the basaltic komatiites of the Barberton type are presented in Table IIA. The average chemical composition of three of the best preserved specimens, yielding the most consistent chemical

TABLE II A

Chemical Analyses of Basaltic Komatiites from the lower three formations of the Onverwacht Group. Basaltic Komatiite of the Barberton Type.

	1*	2*	3*	4	5†	6	7*	8	9*	10‡	11*
	AB9	VB4	AB4	Ave	KJ55	J6	P1		AB21	14J	RV14
SiO_2	52.42	52.78	53.00	52.73	53.35	56.28	51.06	50.74	46.48	46.28	57.92
TiO_2	0.72	0.91	0.92	0.85	0.46	0.73	0.90	0.70	1.43	0.67	0.62
Al_2O_3	8.64	9.43	11.43	9.83	10.13	8.38	8.71	10.97	9.02	10.59	7.77
Fe_2O_3	1.79	1.18	0.73	1.23	10.94	10.77	2.11	1.55	2.50	2.82	0.49
FeO	9.19	9.88	10.03	9.70	0.16	n.d.	8.80	9.65	13.21	9.72	6.20
MnO	0.23	0.22	0.22	0.22	n.d.	0.15	0.26	0.22	0.27	0.22	0.19
MgO	11.42	9.63	9.27	10.10	11.73	9.47	10.70	9.34	10.81	10.50	11.23
CaO	10.63	10.52	8.83	9.99	7.35	8.64	11.49	11.07	10.76	13.49	8.80
Na_2O	2.88§	2.18§	2.91§	2.65	1.46	3.30	0.72	2.66	1.67§	0.96§	2.30
K_2O	0.06§	0.92§	0.40§	0.46	0.10	0.29	0.17	0.34	0.11§	0.11§	0.18
P_2O_5	0.07	0.05	0.05	0.06	0.06	0.07	0.05	0.05	0.10	n.d.	0.04
H_2O^+	1.69	2.00	1.91		L.O.I.	1.00	3.41	1.45	2.35	3.51	2.60
H_2O^-	0.25	0.15	0.07		3.19	0.23	0.12	0.05	0.23	0.34	0.14
CO_2	0.14	0.12	0.16			0.11	1.44	0.92	0.25	0.38	1.60
TOTAL	100.13	99.97	99.93		98.93	99.46	99.94	99.71	99.19	99.59	100.08
Sr ppm	34.4	89.8	200.0				345.0			355.3	
Rb ppm	0.0	9.5	9.6				2.1			0.0	

Analysts and Methods used.

* National Institute for Metallurgy, Johannesburg. (Standard gravimetric, volumetric, colorimetric and flame photometric).
† Dr. J. Gurney, Dept. Geochemistry, University of Cape Town (X-ray fluorescence).
‡ Mr. I. Wright, Dept. Geology, University of the Witwatersrand, Johannesburg (X-ray flueorscence).
§ Alkali determinations by the authors (flame photometric).
L.O.I. Loss on ignition. Rb and Sr determinations by the authors (X-ray fluorescence).

Columns 1-3 Best available analyses of massive felspar-bearing metabasalts from the Komati Formation in the type area.
4 Average of above.
5 Amphibolitized metabasalt - Kaapmuiden-Malelane area.
6 Tremolite-actinolite schist, northern flank of Mountain Land (Anhaeusser 1969).
7 Pillowed and somewhat carbonated metabasalt from the Komati formation in type area.
8 Banded amphibole gneiss ("metasediment") Forbes Reef area Swaziland (Urie and Jones, 1965).
9 Massive amphibolite from the Theespruit formation in the type area (sample from close to granite contact).
10 Pillowed metabasalt from the Komati formation in type area.
11 Spherule from spherulitic pillowed metabasalts.

data (columns 1–3), is presented in column 4. One of the most distinctive chemical features of this type of basaltic rock is that calcium, aluminium and magnesium occur in more or less equal amounts (about 10% for each oxide). In addition, the silica and alkali contents of these rocks are also relatively high.

These chemical pecularities are ascribed to several main mineralogical features including a relatively high original diopside content together with a reasonably high plagioclase and sometimes quartz content. An analysis of a typical spherule from this type of basalt is given in Table IIA, column 11. The main feature is a higher silica content indicating essentially only an enrichment in quartz.

A number of analyses show possible anomalies with respect to one or more of the major element oxides (columns 5–10 of Table IIA) compared to the average of the class given in column 4. The significance of these variations is not fully understood, but it is probable from microscopic evidence that some of these are due to metamorphism, secondary silicification, carbonation and hydration. It is possible that the average chemical composition of this class (column 4) is not entirely representative and further analytical data are probably necessary to define the komatiites of the Barberton type more rigorously.

The Badplaas Type

TABLE II B

Basaltic Komatiites of the Badplaas Type

	1*	2*	3†	4*	5*	6	7
	VB1	VB2	KR35	K262	K188	Average	J9
SiO_2	52.80	50.20	53.02	52.57	52.80	52.22	55.11
TiO_2	0.73	0.66	0.36	0.40	0.66	0.56	0.39
Al_2O_3	5.68	5.20	4.84	5.31	6.05	5.42	4.25
Fe_2O_3	0.32	1.81	9.56	0.74	1.05	0.98	8.14
FeO	9.09	9.12	n.d.	8.05	9.28	8.88	n.d.
MnO	0.20	0.28	0.18	0.22	0.23	0.22	0.15
MgO	15.12	15.40	16.01	15.71	13.99	15.25	16.51
CaO	11.89	13.04	13.78	13.53	11.92	12.83	12.28
Na_2O	1.84	1.58	0.61	1.02	1.02	1.21	0.90
K_2O	0.12	0.06	0.05	0.09	0.13	0.09	0.02
P_2O_5	0.02	0.07	0.08	0.03	n.d.	0.05	0.06
H_2O^+	2.16	1.78	L.O.I.	2.11	2.17	2.05	1.41
H_2O^-	0.05	0.09	1.59	0.12	0.09	0.09	0.14
CO_2	0.16	0.23	0.08	0.24	0.04	0.17	0.11
TOTAL	100.18	99.52	100.16	100.14	99.63		99.52
Sr		51.3					
Rb		1.4					

Analysts and Methods Used.

* National Institute for Metallurgy, Johannesburg. (Standard gravimetric, volumetric, colorimetric and flame photometric.)

† Dr. J. Gurney, Dept. of Geochemistry, University of Cape Town (X-ray fluorescence.)

L.O.I. Loss on ignition.

Columns 1 to 2 Massive tremolite-actinolite metabasalts (metapyroxenites) with occasional diopsidic pyroxene remnants and very minor amounts of felspathic material. Komati formation type area.

Columns 3 to 5 Similar to above, from Kaapmuiden-Malelane area.

6 Average chemical composition of basaltic komatiite of the Badplaas type from best available data (columns 1 to 5).

7 Tremolite-actinolite schist from northern flank of the Mountain Land (Anhaeusser 1969).

Analyses of basaltic komatiites of the Badplaas type from both the northern and southern portions of the Mountain Land are presented in Table IIB, columns 1–5 and their average chemistry in column 6.

The main features are the high silica, magnesium and calcium contents and low aluminium and alkali contents. These rocks have the highest calcium content and the highest Ca/Al ratio of all three classes of basaltic komatiite, chemical features ascribed to an abundance of original diopsidic pyroxene, absence of original olivine and very small amounts of original plagioclase. An analysis of a schist from the northern contact of the Mountain Land which probably belongs to the Badplaas type (Table IIB, column 7) was not used in the computation of the average chemistry of the class because of its somewhat higher than normal silica and magnesium contents. The chemical data from the Badplaas type indicates that these rocks have the most consistent chemistry of all the basaltic rocks from the lower formations.

The Geluk Type

TABLE II C

Basaltic Komatiites of the Geluk Type

	1*	2†	3*	4*	5	6†	7	8	9‡	10†
	AB12	SG3	SG80	KJ39	Ave.	K178	J10	Ave.	42J	KJ40
SiO_2	47.84	46.32	46.85	48.45	47.37	50.02	48.36	49.19	50.61	50.12
TiO_2	0.55	0.38	0.56	0.36	0.46	0.32	0.54	0.43	0.36	0.48
Al_2O_3	7.32	6.82	6.90	6.15	6.79	3.77	3.75	3.76	7.78	8.77
Fe_2O_3	1.22	12.78	1.25	1.09	1.18	9.02	13 97	11.00	1.06	11.43
FeO	8.08	n.d.	8.90	7.26	8.08	n.d.	n.d.	n.d.	5.96	n.d.
MnO	0.18	0.25	0.16	0.17	0.19	0.15	0.19	0.17	0.13	0.22
MgO	19.28	19.61	20.97	21.72	20.39	20.80	19.26	20.03	19.20	15.93
CaO	8.52	8.61	7.43	8.69	8.31	9.92	9.11	9.51	6.30	9.14
Na_2O	0.71	0.27	0.09	0.50	0.39	0.04	0.16	0.10	2.12	1.07
K_2O	0.09	0.03	0.04	0.10	0.06	0.01	0.03	0.02	0.04	0.03
P_2O_5	0.03	0.07	n.d.	n.d.	0.05	0.06	0.04	0.05	n.d.	0.08
H_2O^+	5.08	LOI	5.89	4.86	5.26	3.74	3.55		4.96	LOI
H_2O^-	0.34	3.80	0.29	0.12	0.25	0.56	0.58		0.14	2.77
CO_2	0.90		0.02	0.07			0.03		0.10	
Cr_2O_3			0.30						0.41	
NiO			0.06				0.18		0.12	
Total	100.14	99.28	99.67	99.54		98.43	99.75		99.29	100.04

Analysts and methods used.

* National Institute for Metallurgy, Johannesburg (Standard gravimetric, volumetric, colorimetric and flame photometric).

† Dr. J. Gurney, Dept. Geochemistry University of Cape Town (X-ray fluorescence).

‡ Mr. I. Wright, Dept. Geology, University of the Witwatersrand (X-ray fluorescence).

L.O.I. Loss on ignition.

Column 1 Pillowed metabasalt, Komati formation type area.
2 and 3 Amphibolitized metabasalts, Steynsdorp Valley.
4 Amphibolitized metabasalts, Kaapmuiden-Malelane area.
5 Average chemical composition of the Geluk type (high alumina variety).
6 Amphibolite, Kaapmuiden-Malelane area.
7 Amphibole—talc schist, northern flank of Mountain Land (Anhaeusser, 1969).
8 Average chemical composition of the Geluk type (low alumina variety).
9 Amphibolitized metabasalt, Komati formation type area.
10 Magnesium-rich amphibolite, Kaapmuiden-Malelane area.

Analyses of the most representative specimens of basaltic komatiites of the Geluk type are presented in columns 1–4 of Table IIC and the average chemistry of the class given in column 5 of the same table. As in the case of basaltic komatiites of the Badplaas type, the chemistry is remarkably consistent with distinctive features including a silica content which is considerably lower than for the above-mentioned classes but still relatively high, a very high magnesium content and a very low alkali content. Calcium and aluminium are also fairly low and the majority of these rocks, therefore, have a high Ca/Al ratio. The analyses presented in columns 6 and 7 (average in column 8) are of a variety with a very low aluminium and alkali content and somewhat higher silica and calcium contents compared to the first variety of the Geluk type. The magnesium content is approximately the same as in the first variety. As alumina is the element that shows the greatest change, this variety has been termed the low alumina type to distinguish it from the first group, or what has been termed the high alumina variety. The chemical peculiarities of basaltic komatiites of the Geluk type are ascribed to the presence of significant amounts of original orthopyroxene as well as some olivine in certain instances, and to the originally small amounts of plagioclase, particularly in the low alumina variety. Varying amounts of original diopsidic pyroxene were probably also present.

Analyses of two metabasalts which probably also belong to this group but which contain one or more of the major element oxides in anomalous quantities are given in columns 9 and 10. The excessively high sodium content of the analysis presented in column 9 and the higher than normal aluminium, calcium and sodium, together with lower magnesium of that in column 10, necessitated the exclusion of these analyses from the computation of the class average.

(ii) Metatholeiites

The average chemical composition of meta-tholeiitic basalts from the lower formations of the Onverwacht Group calculated from the most consistent analyses available (columns 1–4 of Table III) are given in column 5 of the same table. The main distinguishing features of these rocks are their higher alumina and alkali contents and lower magnesium contents (compared to the basaltic komatiites) reflecting essentially an increase in the plagioclase content and a decrease in the pyroxene content. The analyses presented in columns 6–11, although clearly tholeiitic, have certain anomalous features due to unknown factors and were therefore excluded from the calculation of the average chemical composition. Microscopic evidence seems to suggest that these anomalous features could be due largely to metamorphism and attendant processes such as silicification rather than to significant primary differences in the basaltic magma. The chemistry of the metatholeiites of the lower formations is comparable to the chemistry of the tholeiites of the upper formations.

TAPLE III

Selected chemical Analyses of Meta-tholeiites from the Lower Formations of the Onverwacht Group:

	1*	2	3*	4*	5	6†	7*	8	9†	10	11
	V11	CAA1	KJ1	KJ57	Ave	K206	KJ16	65	TS1	CAT 79	J7
SiO_2	52.98	52.30	49.19	54.04	52.13	52.68	54.74	54.11	53.76	49.00	49.64
TiO_2	1.08	0.95	1.59	0.75	1.09	1.07	0.75	0.82	0.74	0.70	1.32
Al_2O_3	13.20	14.15	12.99	12.97	13.33	10.50	15.89	14.42	11.82	14.25	12.63
Fe_2O_3	1.61	1.48	3.60	2.26	2.24	0.88	0.73	10.34	10.33	2.12	16.92
FeO	10.07	9.44	12.39	7.86	9.94	11.59	4.70	n.d.	n.d.	8.04	n.d.
MnO	0.18	0.24	0.28	0.14	0.21	0.21	0.08	0.15	0.17	0.18	
MgO	6.25	6.06	6.53	6.55	6.35	6.58	7.16	5.73	8.98	8.33	5.63
CaO	8.30	10.66	8.58	8.27	8.98	8.46	8.94	9.64	9.28	9.88	9.77
Na_2O	3.35	2.15	2.50	3.87	2.97	3.50	4.93	1.93	3.24	1.75	2.69
K_2O	0.48	0.21	0.23	0.17	0.26	0.20	0.13	0.14	0.15	1.63	0.37
P_2O_5	0.09	0.06			0.07	0.13		0.12	0.12	0.05	0.15
H_2O^+	1.96	1.89	1.79	2.25	1.97	LOI	1.44	2.42	LOI	2.60	1.06
H_2O^-	0.10	0.09	0.13	0.12	0.11	1.45	0.10	0.07	1.19	0.17	0.08
CO_2		0.11	0.06	0.05	0.07		0.11	0.53		0.61	0.15
Cr_2O_3				0.03							
NiO								0.11			0.06
TOTAL	99.65	99.79	99.87	99.33		97.25	99.71	100.53	99.78	99.31	100.85

Analysts and Methods Used

* National Institute for Metallurgy, Johannesburg (Standard gravimetric, volumetric, colorimetric and flame photometric).

† Dr. J. Gurney, Dept. Geochemistry, University of Cape Town (X-ray flourescence).

LOI. Loss on ignition.

Column
1. Massive variety, Komati formation type area.
2. Dark contact amphibolite, eastern end of Jamestown schist belt (Anhaeusser 1969).
3. Fine grained massive amphibolite, Kaapmuiden.
4. Massive amphibolite, core of Koedoe fold structure, Malelane area.
5. Average of best analyses (Columns 1 to 4).
6. Massive amphibolite, core of Koedoe fold structure, Melalane area.
7. Coarse grained amphibolitic xenolith in Ship Hill body, Kaapmuiden area.
8. Tremolite-actinolite schist, eastern end of Jamestown belt (Anhaeusser, 1969).
9. Amphibolite from Theespruit formation, Steynsdorp anticline.
10. Dark contact amphibolite, Jamestown Schist belt, Kaap Valley granite contact (Anhaeusser 1969).
11. Contact amphibolite, northern flank of Mountain Land (Anhaeusser 1969).

3. *Felsic Rocks*

(i) Siliceous aluminous tuffs

Chemical analyses of a variety of felsic tuffs occurring in the Theespruit formation are given in columns 1 to 6 of Table IV. The three main varieties recognised, viz. an extremely siliceous type, a very aluminous type lower in silica, and a variety intermediate between these two extremes, are presented in columns 5, 6 and 4 respectively of Table IV.

The most distinctive chemical characteristics of all of these rocks is their high alumina content. It is considered unlikely that a primary tuff would have been so aluminous and it seems probable that the primary felsic pyroclasts were somewhat modified after deposition. This could have been accomplished by processes of weathering similar to those which give rise on surface weathering to aluminous clays from felsic rocks. From the environment of these rocks, however, it is considered more feasible that alteration took place subaqueously, probably

TABLE IV

Chemical analyses of Felsic Rocks from the lower formations of the Onverwacht Group

	1	2	3	4	5*	6†	7†	8	9†
	J13	J14		Ave.	2	AT2	VP5		R17
SiO_2	79.53	78.28	77.50	78.44	90.04	59.93	68.84		71.26
TiO_2	0.53	0.55		0.54	0.05	0.45	0.30		0.21
Al_2O_3	12.76	17.52	16.27	15.52	6.71	26.26	14.60		14.40
Fe_2O_3	0.11	0.39	0.74	0.41	0.23	0.29	0.26		0.73
FeO	0.36	1.01		0.68	0.00	1.64	1.51		1.15
MnO	tr	0.01		0.01	0.01	0.02	0.05		0.03
MgO	0.30	0.46		0.38	0.00	1.26	1.45		1.10
CaO	0.28	0.30		0.29	0.01	0.42	2.80		0.55
Na_2O	0.27	0.53	0.69	0.49	0.12	3.75	5.12	5.46	8.22
K_2O	3.86	1.48	1.65	2.33	2.31	1.63	1.61	1.74	0.27
P_2O_5	0.18	0.11		0.14	n.d.	0.05	0.11		0.10
H_2O^+	1.74	0.88		1.31	1.06	3.72	1.02		0.95
H_2O^-	0.23	0.20		0.21	0.06	0.32	0.11		0.06
CO_2	n.d.				0.00	0.12	2.60		0.31
Total	100.15	101.72			100.60	99.86	100.38		99.24

Analysts and Methods Used.

† National Institute for Metallurgy, Johannesburg (Standard gravimetric, volumetric, colorimetric and flame photometric).

* Mr. I. Wright, Dept. of Geology, University of the Witwatersrand (X-ray fluorescence.)

COLUMNS 1 to 6: Variety of aluminous and siliceous felsic tuffs.

1 Quartz-sericite schist, eastern portion of Jamestown schist belt (Anhaeusser, 1969).
2 Andalusite and chloritoid-bearing quartz-sericite schist from same locality as above (Anhaeusser, 1969).
3 Average partial composition of 5 andalusite- bearing quartz-sericite schists from the eastern part of the Jamestown schist belt (Anhaeusser, 1969).
4 Average chemical composition of siliceous aluminous schists from the eastern part of the Jamestown schist belt.
5 Siliceous, sericite-bearing tuffaceous sediment (Theespruit formation type area.)
6 Water worked, aluminous (pyrophyllitic) felsic tuff (Theespruit formation type area.)

COLUMNS 7 to 9: Intrusive porphyry bodies.

7 Typical widespread type of felspar porphyry intrusive into the Komati formation in the type area.
8 Average alkali values for three porphyry bodies similar to the one above.
9 Variety of soda-rich felspar porphyry intrusive into the Komati formation in the type area.

aided by a process such as hot spring and late volcanic activity of a similar nature to that which in Puerto Rico has given rise to aluminium potassium silicates from andesitic volcanics (Hess, H. H., verbal communication). The large amounts of silica in the environment were probably also generated by late-stage felsic pyroclastic activity. A number of these tuffs are characterized by concentrations of barium (see Paper No. 9).

(ii) Porphyries

The abundant intrusive felspar and quartz porphyries which attain their best development in the Komati formation have a very consistent chemical composition (columns 7 and 9 of Table IV). A distinctive feature is the high soda content of the rocks and, except for a somewhat lower iron and magnesium content, they are similar in composition to the leuco-biotite-quartz tonalties intrusive into the Onverwacht Group in the southern portion of the Mountain Land (Paper No. 7).

C. Chemical Comparison of the Mafic and Ultramafic Rocks of the Lower Ultramafic Unit with other Well-established Classes of Mafic and Ultramafic Rock

1. Introduction

In an accompanying paper the unique chemical characteristics of the extrusive peridotitic magma of the Komati formation (peridotitic komatiite) are illustrated by a comparison with well-established classes of peridotitic and picritic rocks. In the present section the chemistry of all the ultramafic and mafic rocks of the lower formations of the Onverwacht Group are also compared with the chemistry of well-established classes of peridotitic and picritic rocks in addition to basaltic and extra-terrestrial rocks. The classes of peridotitic and picritic rocks used are the same as those employed in the comparison of the peridotitic komatiites of the Komati formation (Paper No. 5). They include the class of so-called high temperature peridotite intrusions (Green, 1967) and the average of the peridotite inclusions in kimberlite pipes and in alkali basalts. The classes of picrite used can be divided into two groups, viz. basaltic picrites and picrites of the picritic minor intrusions (see Paper No. 5). Other classes of igneous rock employed for comparative purposes are limburgites. In addition to the above classes which were all used in the accompanying paper (No. 5) several other additional rock classes have been employed for comparative purposes in the present paper. These include the well-known basalt classes, viz. oceanic tholeiites, continental tholeiites, alkali basalts and olivine basalts (Table V). In addition, the average chemistry of the meta-tholeiites from the upper formations of the Onverwacht Group as well as of Archean metatholeiites from the Superior Province of the Canadian Shield have also been included for comparison. Of the extra-terrestrial material the average chondrite and achondrite as well as the average lunar sample from the Apollo 11 landing site on the Sea of Tranquillity (recalculated to 100% after subtracting 8.3% TiO_2 and 8.3% FeO to eliminate excessive ilmenite) have been included (see Table V).

2. Presentation of Results by Means of Variation Diagrams

(1) CaO vs Al_2O_3

Two of the most distinctive components of both the peridotitic and basaltic komatiites, viz. calcium and aluminium, have been plotted against each other in Fig. 4a. This plot serves to distinguish komatiites of both the peridotitic and basaltic varieties from most well-established classes of basaltic, picritic and peridotitic rocks. The ankaramite picrites of Madagascar and Hawaii have Ca/Al ratios approaching those of the mafic rocks under consideration, with those from Madagascar plotting close to the field of the Badplaas type and those from Hawaii close to the field of the Barberton type (Fig. 4a). The higher alkali and lower silica content of the ankaramites, however, readily distinguishes them from the basaltic komatiites. The calcium-rich (omphacite) eclogites from kimberlite pipes also have a high Ca/Al ratio but are distinguished from basaltic komatiites by their higher alkali (particularly Na_2O) content which is ascribed to the jadeite molecule in the pyroxene. Recent chemical data from the Apollo 11 lunar sample (L.S.P.E.T., 1969) indicates that this material has a Ca/Al ratio approaching that of the basaltic komatiites of the Barberton type. The high iron and titanium content of the lunar material, however, renders it unlike the Barberton or any other earth rocks.

TABLE V

	Komatiites						Ave. perid-otite	Picrites						Lim-burg-ites	Basalts								Extraterrestrial Material		
	Peridotitic		Basaltic																						
	1	2	3	4	5	6	7	8	9	10	11	12	13	14	15	16	17	18	19	20	21	22	23	24	25
SiO_2	47.49	45.94	53.75	53.29	50.02	51.78	44.52	46.00	42.60	46.41	41.26	44.10	45.07	50.23	46.25	45.36	50.24	48.35	49.94	51.50	48.16	49.83	46.86	48.51	49.96
TiO_2	0.55	0.34	0.87	0.57	0.49	0.45	0.15	0.61	0.29	1.98	2.66	2.74	1.30	3.25	0.47	0.75	0.16	2.77	1.51	1.20	2.91	0.94	0.13	0.48	2.46
Al_2O_3	3.45	2.98	10.02	5.50	7.17	3.95	2.75	10.86	8.98	8.53	10.31	11.60	7.85	9.60	15.79	7.49	7.71	13.18	16.69	16.30	18.31	14.64	3.08	13.04	13.17
Fe_2O_3	6.39	6.23	1.25	1.00	1.25	11.58	1.74	3.98	1.50	2.47	5.58	2.84	2.42	2.54	1.06	5.20	1.02	2.35	2.01	2.80	4.24	3.03		1.11	
FeO	5.86	4.80	9.90	9.06	8.53	n.d.	6.74	6.86	8.25	9.82	8.62	9.91	6.44	8.24	9.60	9.72	4.64	9.08	6.90	7.90	5.89	8.77	15.34	15.90	12.79
MnO	0.20	0.18	0.22	0.72	0.23	0.18	0.04	0.14	0.14	0.15	0.14	0.16	–	0.15	–	0.31	0.11	0.14	0.17	0.17	0.16	0.21	0.30		
MgO	26.92	33.79	10.30	15.56	21.53	21.08	40.98	22.15	32.12	20.81	14.00	15.13	18.39	12.97	14.36	15.55	17.79	9.72	7.28	5.90	4.87	7.36	29.37	7.87	9.84
CaO	7.40	4.73	10.18	13.09	8.78	10.01	2.26	7.88	5.48	7.38	11.65	10.69	14.29	7.79	9.12	10.92	16.66	10.34	11.86	9.80	8.79	10.46	2.40	11.00	12.15
Na_2O	0.52	0.15	2.70	1.23	0.41	0.10	0.24	1.07	0.56	1.58	3.19	1.66	1.31	2.40	2.10	2.66	0.97	2.42	2.76	2.50	4.05	2.02	1.21	0.68	0.66
K_2O	0.05	0.03	0.47	0.09	0.06	0.02	0.70	0.04	0.04	0.32	0.88	0.54	1.19	1.92	0.30	0.00	0.11	0.58	0.16	0.86	1.69	0.23	0.21	0.25	0.17

N.B. All analyses re-calculated anhydrous

1 *Peridotitic komatiite* (average of Sandspruit formation).
2 *Perodititic komatiite* (average of Komati formation).
3 *Basaltic komatiite* (average of Barberton type).
4 *Basaltic komatiite* (average of Badplaas type).
5 *Basaltic komatiite* (average of Geluk type).
6 *Basaltic komatiite* (average of Geluk type–low alumina variety).
7 *Peridotite* (average of high temperature intrusions and nodules in basalt and kimberlite, see accompanying Paper No. 5).
8 *Picritic minor intrusion*-chilled margin of dyke – Skye (Drever and Johnson, 1961. p. 81).
9 *Picritic minor intrusion* – centre of sill – Skye (Drever and Johnson, 1967, p. 81).
10 *Oceanite* type of picrite basalt – Hawaii (Mac Donald and Katsura, 1964, p. 19).
11 *Mimosite* type of picrite basalt – Hawaii (Mac Donald, 1949, p. 1571).
12 *Ankaramite* type of picrite basalt – Hawaii (Mac Donald and Katsura, 1964).
13 *Ankaramite* type of picrite basalt – Ankaramy, Madagascar (Johannsen, 1937).
14 *Limburgite* – average from Naunetsi Igneous Province, Rhodesia (Cox et al., 1965, p. 145).
15 *Eclogite* nodule – typical kimberlitic ecolgite from Roberts Victor Mine (new analysis, analysts – N.I.M. = Johannesburg).
16 *Eclogite* nodule with abundant omphacite relative to garnet – Roberts Victor Mine (Williams, 1932).
17 *Eclogite*, Roberts Victor Mine (Wagner, 1928).
18 *Olivine basalt*-average Hawaiian (Mac Donald, 1949).
19 *Oceanic tholeiite* – average given by Engel et al. (1965).
20 *Continental tholeiite*-average given by Mansen (1967).
21 *Alkali Basalt* – average from East Pacific rise (Engel et al., 1965).
22 *Archaean Metabasalt* of Superior Province of Canadian Shield – average given by Wilson et al. (1965).
23 *Chondrite* – average of superior analyses given by Urey and Craig, 1953; recalculated without Fe and FeS.
24 *Basaltic Achondrite* – average given by Urey and Craig, (1953), quoted in Engel et al. (1965).
25 *Lunar Sample* minus excessive ilmenite (8.31% TiO_2 and 8.31% FeO) from Apollo 11 landing site on the Sea of Tranquillity (L.S.P.E.T., 1969).

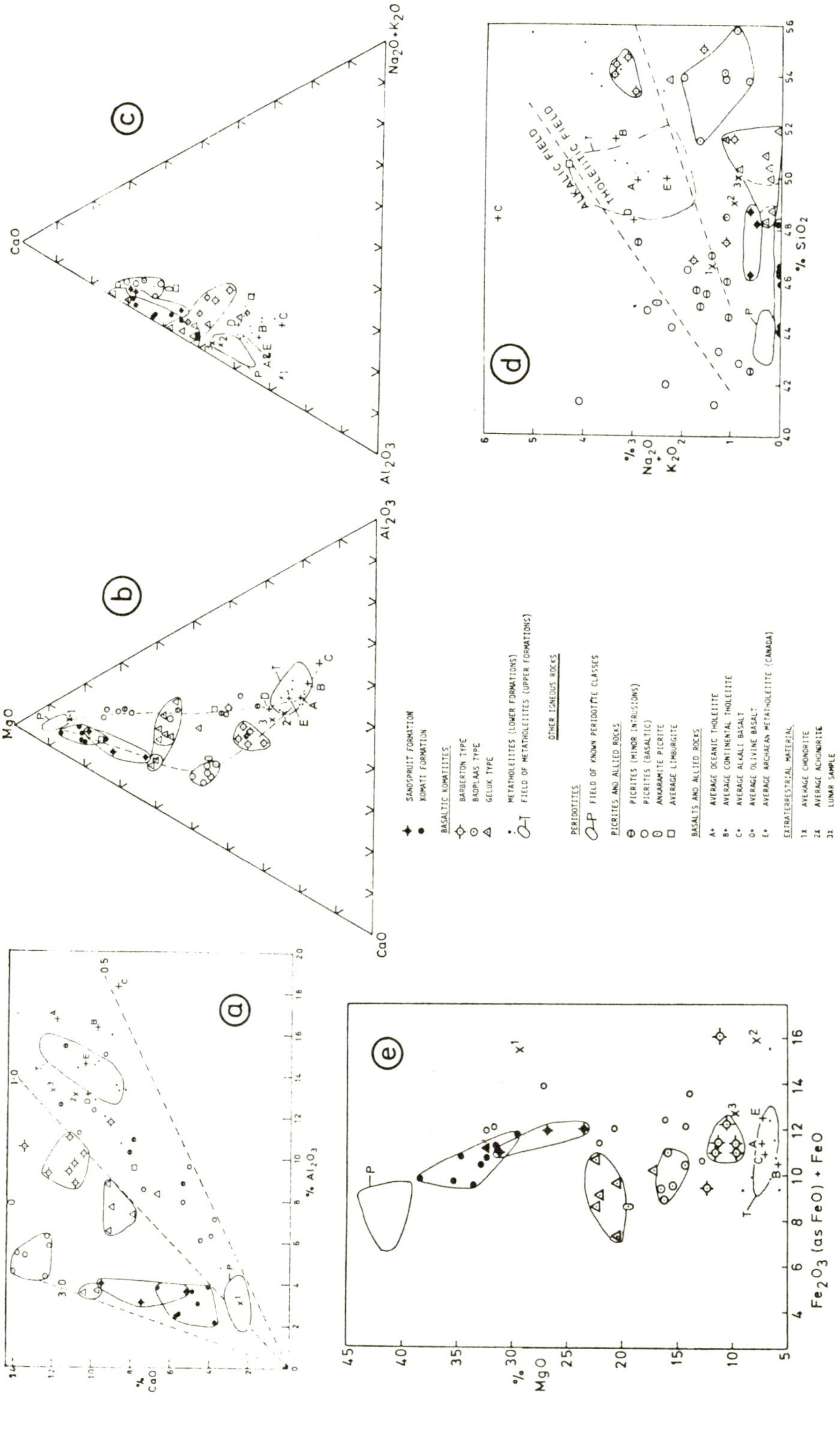

Fig. 4. Chemical comparison of the mafic and ultramafic rocks of the lower formations of the Onverwacht Group with established classes of peridotites, picrites, basalts and allied rocks, as well as extra-terrestrial materials.

The meta-tholeiites of the lower formations of the Onverwacht Group have normal Ca/Al ratios (± 0.70) and are comparable to most classes of tholeiite, including those from the upper formations of the Onverwacht Group, as well as the Archaean meta-tholeiites from the Canadian Shield. Two additional variation diagrams, making use of the other diagnostic variables of the distinctive mafic and ultramafic rocks of the lower formations, viz. magnesium and the alkalies, are depicted in Figs. 4b and 4c.

(ii) MgO—CaO—Al_2O_3

In this plot the peridotitic rocks of the Onverwacht Group fall in a field towards the magnesium corner of the diagram close to the fields of known peridotite classes and olivine-rich picrites (Fig. 4b). They are removed from these fields, however, as a result of their higher Ca/Al ratio, and form an elongate field starting with the peridotites of the Komati formation and extending through the field of peridotites from the Sandspruit formation towards the calcium corner of the diagram (Fig. 4b). Beyond this cluster a trend through the fields of the basaltic komatiites of the Geluk (low-alumina variety) and Badplaas types (at which point it swings towards the aluminium corner of the diagram) towards the field of the basaltic komatiites of the Barberton type and eventually into the field of tholeiitic basalts can be followed (Fig. 4b). A complementary trend joining most of the well-known rock classes employed in this comparison starts at the field of known peridotite classes and the average chondrite at the magnesium corner of the diagram and passes through various types of picrites to the field of basalts. The separation of this and the above trend is mainly a function of calcium and aluminium abundances. When the tholeiite field is reached both trends converge and all the basalt classes mentioned above (except for alkali basalts) cluster in a small, fairly well-defined area (Fig. 4b). The field of one class of rock, viz. basaltic komatiites of the Geluk type, is spread between and partly overlaps the two trends mentioned above, occurring towards the magnesium corner of the diagram (Fig. 4b).

The only rock types falling on or close to the komatiite trend are basaltic kimberlites (discussed in Paper No. 5), ankaramites from Madagascar, certain high calcium (omphacite-rich) eclogites (not plotted) and the average lunar sample. Some genetic connection could exist between the ankaramites and omphacite-rich eclogites and the basaltic komatiites of the Barberton and Badplaas types but, as pointed out previously, other differences clearly separate them. The lunar sample plots on the komatiite trend between the field of the Barberton type and the tholeiite field.

(iii) K_2O+Na_2O—CaO—Al_2O_3

The plot shows a clear separation between the peridotitic and basaltic komatiites and almost all other rock types used in the comparison, including the meta-tholeiites from the lower formations. The peridotitic komatiites from the Komati formation, as well as a number of basaltic komatiites of the Geluk type, lie close to the aluminium-calcium tie line towards the calcium side of the diagram. The peridotitic komatiites from the Sandspruit formation, as well as all of the remaining basaltic komatiites, occur in the same region, but are farther removed from the tie line due to their higher alkali abundances (Fig. 4c). The average ankaramite picrite, basaltic kimberlite and omphacite-rich eclogites plot in the same region with respect to calcium and aluminium, but are generally distinguished by a considerably higher alkali content. The remaining rock types plot on the aluminium side of the tie line but are substantially removed from the latter owing to their higher alkali contents.

(iv) Na_2O+K_2O—SiO_2

In Fig. 4d the alkali abundances have been plotted against the silica abundances following a scheme of classification for the Hawaiian basalts proposed by Macdonald and Katsura (1964). Using this plot the latter authors were able to separate clearly the tholeiitic and alkalic basalts (indicated by means of a diagonal dashed line in Fig. 4d) on the basis of chemical composition. None of the Hawaiian basalts plot below the lowermost dashed line indicated in Fig. 4d. All classes of komatiite, except for the basaltic komatiites of the Barberton type, plot below and well out of the field of all known Hawaiian tholeiites given by Macdonald and Katsura (1964). The only other rocks that plot in this region are the varieties of peridotites quoted, the average achondrite, and the lunar sample. The basaltic komatiites of the Barberton type plot in the lower portion of the tholeiitic field as do the meta-tholeiites from the lower formations, which overlap the field of the basaltic komatiites of the Barberton type (Fig. 4d).

(v) Mgo—$FeO+Fe_2O_3$ (as FeO)

In the final plot (magnesium against total iron) a clear separation is again evident between the different types of komatiite, as well as between the latter and the majority of other rock classes quoted (Fig. 4e). Certain picritic rocks plot close to some of the komatiite fields and the lunar sample plots close to the field of the basaltic komatiite of the Barberton type.

VII. KOMATIITE — A PROPOSED NEW CLASS OF IGNEOUS ROCKS

From the above geochemical comparisons it is clear that the majority of mafic and ultramafic rocks of the lower formations of the Onverwacht Group are not chemically comparable to any well-established class of igneous rock.

In certain respects they bear some similarities to a variety of rocks including basaltic kimberlites, omphacite-rich eclogites, ankaramites, picrites, basaltic achondrites, and the lunar sample. Although some of these similarities may indicate some underlying genetic connection, the bulk chemical composition of the Barberton rocks is clearly different from all of these classes.

Based largely on geochemical data therefore, but also taking into account the petrology, geological setting, development, age and remarkable state of preservation and exposure of the Barberton rocks, it is considered that the main parameters of these are well enough defined to propose the presence of what we consider to be a new and hitherto unrecognised group of igneous rock. The name "komatiite" which has been used throughout this paper and which takes its name from the Komati river which traverses the Scuthern part of the Barberton Mountain Land, including the Komati formation where these distinctive rocks are best developed, is proposed. It is further suggested that the prefix peridotitic or basaltic be used, depending on the particular rock being described. Two varieties of peridotitic komatiite appear to be present, viz. those from the Sandspruit formation and those from the Komati and Theespruit formations (which are more ultrabasic). Three main types of basaltic komatiite, viz. the Barberton, Badplaas and Geluk types have been recognised and although there might be some overlap between their chemistry they appear for the most part to represent well-defined classes.

VIII. THE GEOLOGICAL SIGNIFICANCE OF THE LOWER FORMATIONS OF THE ONVERWACHT GROUP

It will be shown elsewhere (Paper No. 10) that the distinctive formations of the Lower Ultramafic Unit of the Onverwacht Group can be correlated very closely on lithostratigraphic grounds with the lowermost assemblages of many of the Southern African greenstone belts. It is further contended that certain areas within every shield area might well contain similar assemblages, including the widespread development of the distinctive peridotitic and basaltic komatiites. Stronger dynamic and thermal metamorphism, however, coupled probably with extensive elimination of the succession has, it is suggested, resulted in this lowermost unit not having been recognised or appreciated. Examples of this are given in an accompanying paper (No. 10) where it is proposed that the stratigraphy of the Onverwacht Group might well be used as a model which could be of tremendous value as an aid to the elucidation of the geology of the greenstone belts of the shields.

Not only is the Lower Ultramafic Unit well developed within most of the individual greenstone belts of the Southern African shield, but it would appear that it was previously widespread in the areas between discrete greenstone depositories. The granitic rocks, and more particularly the ancient tonalitic varieties, of not only the Barberton region but of the whole of the Kaapvaal craton and certainly the south-eastern sector of the Rhodesia craton, contain rafts and xenoliths which appear to belong almost exclusively to the Lower Ultramafic Unit. This suggests that before its invasion by the early tonalitic magma, the Lower Ultramafic Unit, which probably existed as a succession of reduced thickness in the areas between the large accumulations in the greenstone depositories, covered much of the ancient Rhodesian and Kaapvaal cratons. As the granitic rocks are younger than the rocks of the Lower Ultramafic Unit, the question remains as to what the base of the latter might have been. This has been discussed at length in an accompanying paper (No. 8), where it is concluded that no vestiges of a pre-Onverwacht basement exist and that the whole Lower Ultramafic Unit might well have represented a primitive and primordial earth crust which was more or less transitional into the primitive mantle. It is likely that little or no sialic material existed and in many respects this original earth crust could have been somewhat similar to the present-day oceanic crust. In the areas of the mid-oceanic ridge the oceanic crust consists of a complex assemblage of serpentinized peridotite, pillowed lava, together with numerous sill and dyke swarms, as well as sediments (cherts) and possible tuffs (Hess, 1965, and verbal communication). The emplacement of the widepread ancient tonalitic granites at 3.4 to 3.2 $\times$ 10^9 years could well be a manifestation of the single-cell convective overturn, suggested by Vening Meinesz (1952) and favoured by Hess (1965). This catastrophic event in the earth's early history is regarded by these authors as having resulted in the formation of the nickel-iron core, concomitant with the rising of the low-melting silicates to form a primordial single continent (Hess, 1965).

Data from the Barberton region might well add support to this theory. Assuming a major early convective overturn as outlined above, it is thought that only after or perhaps in part during this catastrophic event did the normal tholeiite basalts of the upper formations develop in abundance and to the near exclusion of the distinctive earlier komatiites. Tholeiitic calc-alkaline assemblages of this type appear to have been confined to much smaller depositories largely within well-defined greenstone successions and could in fact represent the true initial magmatic phase of the greenstone belts developed only locally on a much more extensive crust of the Lower Ultramafic Unit. If the above suggestions are correct, then the geological significance of the Lower Ultramafic Unit is tremendous. It could represent part of the earth's primitive crust developed before the establishment of a substantial sialic plate and nickel-iron core. The lower formations in this case would have developed at a stage when the budget of elements between the crust, mantle, and core was significantly different. Total or near total melting of the uppermost mantle at this stage could well have been expected to produce rock types of a kind not found in later geological times and, it is suggested, afford an explanation as to the unusual and distinctive chemistry of the komatiites which form the bulk of the Lower Ultramafic Unit (Paper No. 11).

The composition of these could clearly assume tremendous importance in providing an insight into such problems as the composition of the earth's primitive mantle as well as the bulk composition of the earth prior to significant differentiation. With respect to the latter a major problem concerns the low alkali abundances of both the basaltic and peridotitic komatiites. If they were derived from a mantle which had not as yet differentiated to form an alkali-enriched crust, then the material derived from this primitive mantle by partial or near total melting might be expected to have had a high alkali content. The authors have no explanation for the observed low alkali characteristic of komatiites except to comment on the fact that there does seem to be a complementary relationship between the widespread mafic and ultramafic komatiites, singularly depleted in lithophile elements, and the immediately preceding generation of the vast alkali (and lithophile) enriched granite event which formed the earliest known continental sial.

[*Editor's Note:* The references cited may be found at the end of the original publication.]

7: SUMMARY OF THE TECTONIC DEVELOPMENT OF THE RHODESIAN ARCHAEAN CRATON

C. W. Stowe

ABSTRACT

The stratigraphy and structure of the Rhodesian Basement Complex craton is described in general outline. Recent observations and ideas regarding the stratigraphy and structures present are reviewed. Relics of an older pre-Bulawayan gneissic basement have been distinguished.

The general tectonic history in some key areas has been shown to be multiphase. Early orogenic type folding was followed by a second phase of vertical tectonics with the formation of domes and rim-synclines. In the last phase the shield behaved as a stable craton surrounded by younger mobile belts.

INTRODUCTION

The Rhodesian Basement Complex craton is 650 km northeast by 350 km southeast. Apart from the northwestern margin it is relatively well exposed. Most of the greenstone schist belts and some of the granitic regions have been mapped in detail by the Rhodesia Geological Survey.

Since Macgregor's (1951) account, a wealth of new information has become available, and a general review of the Rhodesian Basement Complex has been written by Bliss & Stidolph (*in press*). The regional geochemistry was described by Viewing (1968). This paper is to note new information of tectonic significance and to attempt to synthesize this into a provisional outline of the craton structure as a whole.

Radiometric ages within the craton range from 3,400 to 2,300 m.y. A younger limit for deformation within the craton is indicated by the younger limit of 2,530 m.y. for the Great Dyke (Allsopp, 1965), which cuts across all craton structures.

The Archaean craton is bounded by younger metamorphic belts on all sides (Fig. 1).

In the southeast the Limpopo metamorphic belt separates the Rhodesian craton from the Kaapvaal craton in South Africa (Anhaeusser *et al.*, 1969). This belt consists of granulites that have been tightly folded. Some of these are recognisably derived from Basement Complex rock types, re-metamorphosed by 2,600 and 2,000 m.y. events (Holmes & Cahen, 1956; Nicolaysen, 1962).

On the northwestern and northern sides the craton is bounded by the Zambezi paragneiss belt with overprinted metamorphic events at 2,300-2,100 m.y.; 1,900-1,650 m.y.; 1,300-1,100 m.y. and 600-400 m.y. (Vail, 1965; Holmes & Cahen, 1956). This boundary of the craton is not sharply defined, with Proterozoic formations intruded by younger granites. Part of it is covered by Mesozoic and Tertiary sediments.

The eastern boundary of the craton is the Mozambique belt consisting of gneiss and quartzites, thrust up against the craton margin. The youngest metamorphic event was the 600-400 m.y. event (Vail, 1965).

EARLY WORK IN RHODESIA

Macgregor (1932, 1951) drew attention to the characteristic structural pattern of synclinal basement schist belts occupying the spaces between oval "gregarious batholith" domes. He suggested that this structure resembled a thin floating sheet, tightly folded by loading its surface. On a basis of similar lithology and unconformities, Macgregor (1947, 1951) recognized three systems, the Sebakwian, Bulawayan and Shamvaian.

Since 1968 the Rhodesia Geological Survey has adopted a litho-stratigraphic code, so that a chronological correlation between lithologically similar formations of the Sebakwian, Bulawayan or Shamvaian Groups is no longer implied. In the Archaean, it has not yet been possible to apply this code strictly to units larger than local formations. Correlation with

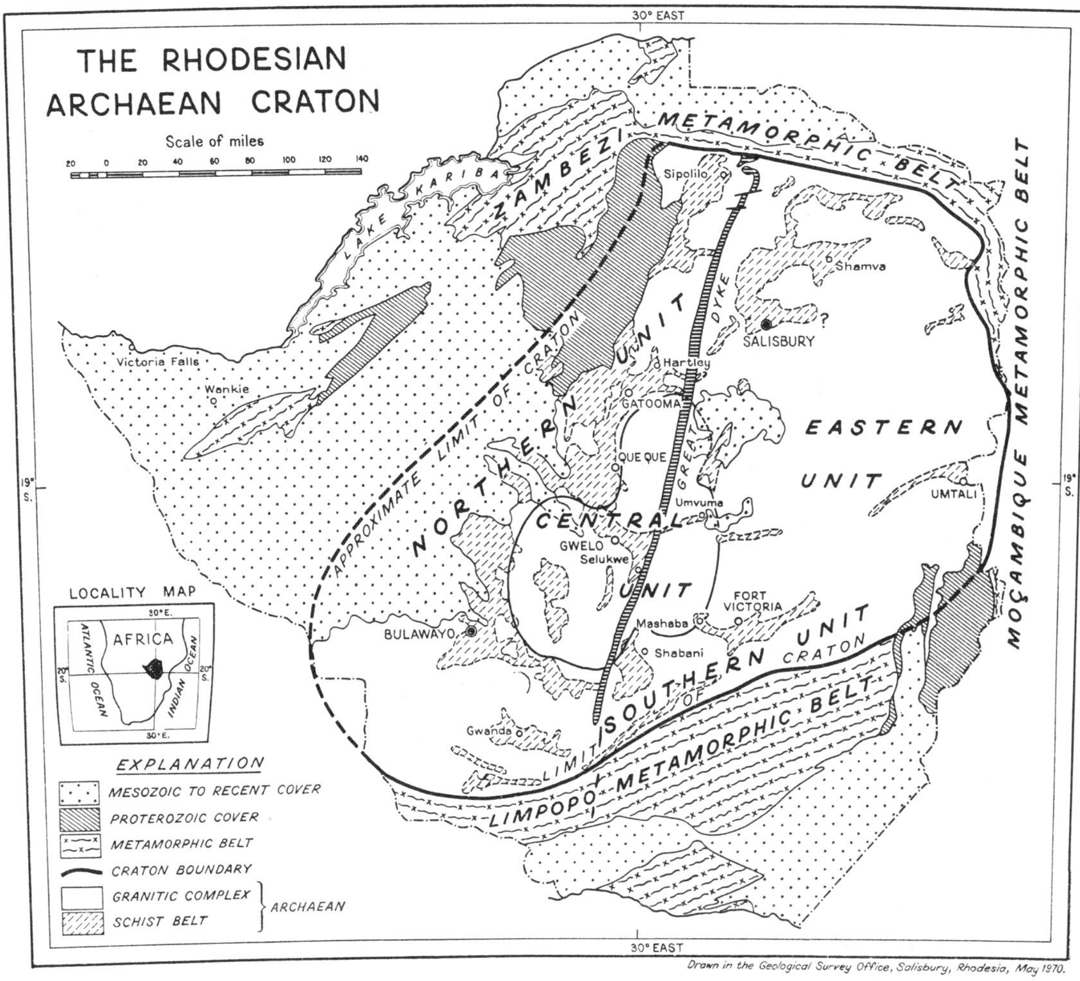

Fig. 1.

the Bulawayan and Shamvaian Groups is still based on lithological similarities with their broadly defined type areas near Bulawayo and Shamva. The Sebakwian Group consists of formations underlying the Bulawayan, but needs to be re-defined. In this paper the divisions such as "Series" or "group", as originally used informally by the authors, have been retained for reference to work published before 1968 and the American 1961 code followed as closely as possible for the rest. The present position is that the Rhodesia Geological Survey have discarded the old terminology and are halfway towards the strict application of the American 1961 code.

The Sebakwian Group consists mainly of jaspilites, magnesia-rich volcanic schists, and granulite relics in gneisses. The Bulawayan Group consists of greenstones (including pillow lavas) with banded ironstones, phyllites and locally derived polymict conglomerates. The Shamvaian Group overlies the Bulawayan Group with an angular unconformity, and consists mainly of poorly sorted, arenaceous rocks derived from gneiss.

RECENT WORK IN RHODESIA

(a) The Que Que area was re-mapped by Harrison (1968) who concluded that the Sebakwian type-area unconformity (Macgregor, 1947) was not regional, and that the "basal Bulawayan" conglomerate was intraformational in the Bulawayan Group. He reinterpreted the large carbonated serpentinite, previously regarded as Sebakwian, as being intrusive into the Bulawayan. The only Sebakwian Group rocks remaining in the area are remnants of sediments, ultramafics and volcanics in the Rhodesdale Gneiss (Fig. 2). K/A ages of 3,400-3,260 m.y. have been determined for this (Poldervaart, 1961).

(b) In the Selukwe area, Stowe (1968*b*, 1965) mapped the Selukwe Schist Belt consisting of 3,000 m of Sebakwian Group rocks, sericitic grit and greenstones intruded by serpentinite, unconformably overlain by a conglomerate and younger greenstones. It is a large nappe structure, thrust over an older gneissic basement. A gravity gliding mechanism was postulated. Mylonite zones were cross-folded, first by northwest and later by east-northeast trending folds. The nappe originated in a 10 km wide, tightly folded crystalline zone of higher metamorphic grade. This may be traced across the older gneissic basement for 60 km in a northwest direction, and is terminated at both ends by younger granites.

The older basement consists of schist and migmatite relics in gneiss, tightly folded on north-trending axes during an earlier episode. These relics may be traced northwards to where they are interrupted by the northwest-trending Gwelo Schist Belt. Similar relics occur on the same line to the north of the Gwelo Schist Belt as the Sebakwian inclusions in the Rhodesdale Gneiss (Fig. 2). The largest relic is the Ghoko Schist Belt 25 km southwest of Selukwe. It is 35 km long, consisting of tremolite schist (derived from volcanics) interbedded with jaspilite and an unconformably overlying banded ironstone and phyllite formation. The whole succession was folded into a tight synclinorium and partly invaded by the enclosing gneiss.

At their intersection, the folding and metamorphism of the Ghoko Schist Belt was shown to be older than the northwest-trending zone of origin of the Selukwe nappe. Therefore there were two deformations, the Rhodesdale-Ghoko (I) and the Selukwe-Gwelo (II) (Fig. 2). Field relations suggest but do not prove, that the Selukwe Schist Belt was probably a new cover, deposited after the first event. Therefore there are probably two divisions of the Sebakwian Group, Sebakwian I and Sebakwian II.

(c) Mapping in the Shabani area, Worst (1955) and Oldham (1968) recognized an "Upper Bulawayan group" of phyllite and banded ironstone overlain by 6,000 m of greenstone, and a "Lower Bulawayan group" consisting of a similar, 4,500 m thick, succession of sediments and greenstones. The lower group was tightly folded on east-northeast trends prior to the deposition of the upper group which was folded co-axially on north-northwest trends during two later events. The gneissic basement beneath the schist was partly re-activated during the intrusion of a large, post-Shamvaian, east-northeast trending granodiorite batholith. Relics of this basement show two earlier episodes of folding and migmatization. Northwest trending relics may be traced to the Sebakwian II in the Selukwe area (Fig. 2).

(d) In the Fort Victoria region, Wilson (1964, 1968) mapped north-northeast trending relics of Sebakwian schist in a gneissic basement, which was shown to be older than the Bulawayan Group in the main Victoria Schist Belt. The Bulawayan Group here consists of a very thick succession of greenstones and phyllitic sediments, folded east-northeast in line with the Lower Bulawayan group in the Shabani Schist Belt. They are overlain unconformably by two Shamvaian formations, consisting of conglomerate, phyllite and banded ironstone. Overlying these is a third formation of felsic volcanics.

(e) The Bulawayo Schist Belt is an important link in the chain of schist belts around the Shangani dome and it is also the type area for the Bulawayan Group. Amm (1940) recognized a threefold division of the "Bulawayan Greenstone Series" (Bulawayan Group); a lower division of pillow lava and banded ironstone; a middle division of andesite-dacite lavas and phyllite; and an unconformable "Upper Greenstone group" of porphyritic andesite and agglomerate. The total thickness of the Bulawayan Group is here about 12,000 m. It was folded about northeast axes and deeply eroded before deposition of the overlying Shamvaian Group sediments. The Lower Conglomerate group was folded on northwest axes before the unconformable deposition of the Upper Phyllite group, which was itself folded about westerly trends.

(f) In recent years there has been renewed interest in mapping the granitic rocks. West of Selukwe (Stowe, 1965, 1968*b*) a large concordant trondjhemite (G2) underlies the older migmatite basement, which was partially homogenized to a permeation gneiss. It was domed up as a mantled gneiss dome structure prior to the east-northeast trending, post-Shamvaian folding. A younger, high-level type of transgressive granite pluton (G3) was intruded after this folding. In the adjacent Shangani region Harrison (1969) mapped another simi-

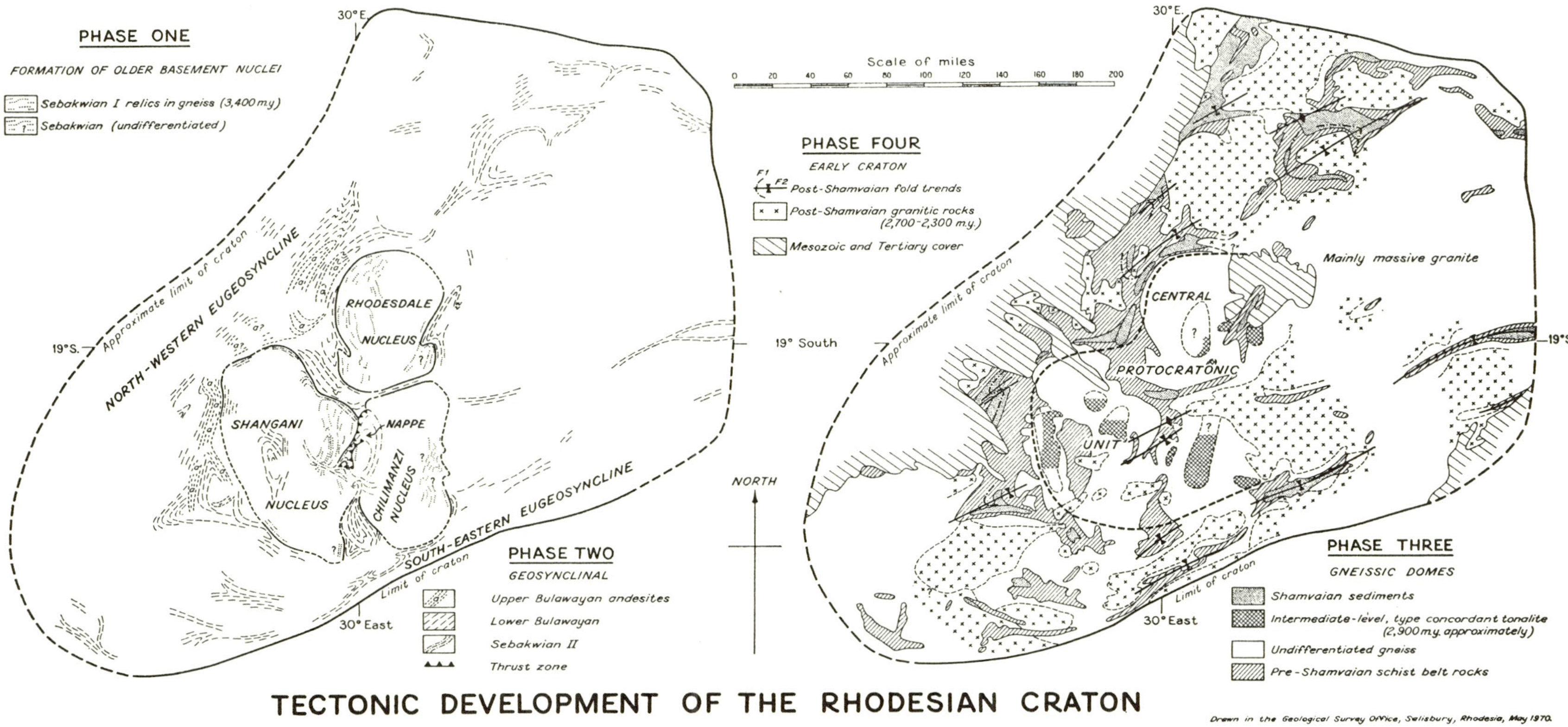
PHASE ONE
FORMATION OF OLDER BASEMENT NUCLEI
Sebakwian I relics in gneiss (3,400 m.y.)
Sebakwian (undifferentiated)
30°E.
19°S.
Approximate limit of craton
NORTH-WESTERN EUGEOSYNCLINE
RHODESDALE NUCLEUS
SHANGANI NUCLEUS
NAPPE
CHILIMANZI NUCLEUS
SOUTH-EASTERN EUGEOSYNCLINE
Limit of craton
30° East
PHASE TWO
GEOSYNCLINAL
Upper Bulawayan andesites
Lower Bulawayan
Sebakwian II
Thrust zone
Scale of miles
0 20 40 60 80 100 120 140 160 180 200
PHASE FOUR
EARLY CRATON
F1 F2 Post-Shamvaian fold trends
Post-Shamvaian granitic rocks (2,700-2,300 m.y.)
Mesozoic and Tertiary cover
19° South
NORTH
30°E.
19°S.
Mainly massive granite
CENTRAL PROTOCRATONIC UNIT
Approximate limit of craton
Limit of craton
30° East
PHASE THREE
GNEISSIC DOMES
Shamvaian sediments
Intermediate-level, type concordant tonalite (2,900 m.y. approximately)
Undifferentiated gneiss
Pre-Shamvaian schist belt rocks
TECTONIC DEVELOPMENT OF THE RHODESIAN CRATON
Drawn in the Geological Survey Office, Salisbury, Rhodesia, May 1970.

Fig. 2.

lar, large, G2-type, concordant tonalite and two phases of the high-level, G3-type plutons.

North of Fort Victoria, Wilson (1964, 1968), mapped a very large post-Shamvaian granite and an older G2-type, concordant tonalite. The post-Shamvaian granite is a high-level transgressive type. This is the extension of the large granite that was mapped by Bliss (1962) in the Felixberg area, 80 km to the north.

(g) Ultramafic rocks of various ages are widespread in the craton (1:000,000 scale geological map of S. Rhodesia 1961). Two Alpine-type serpentinite lenses each about 600 m thick and 15 km long were mapped in the Selukwe and Ghoko Schist Belts (Stowe, 1965). A post-Shamvaian layered ultramafic complex in the Mashaba area was intruded into older gneiss with relics of serpentine (Wilson, 1968). Both contain economic deposits of chromite, although of widely different ages.

TECTONICS

An outline of tectonics within the craton must be regarded as provisional in view of the incomplete state of knowledge. Nevertheless it is hoped that this provisional attempt will pinpoint fields where more work is needed.

Four main tectonic units may be distinguished (Fig. 1). A *central unit* contains the remnants of an older gneissic basement which may be recognized with a general north-northeast fold trend. It constituted an older, central protocraton formed by the coalescence of the Rhodesdale, Shangani, and Chilimanzi nuclei (Fig. 2). A *northern unit* consists of Bulawayan and Shamvaian Groups deposited in an eugeosyncline, which was folded on northeast trends and intruded by post-Shamvaian granites. A *southern unit* of Bulawayan and Shamvaian Groups has been folded on east-northeast axes. Finally, there is an *eastern unit* where relics of schist, partly enclosed in a thin gneiss envelope, were flooded by post-Shamvaian granite.

Recent work and re-examination of older literature shows that many episodes of folding occurred during the formation of the schist belts in an early phase of orogenic-type tectonics. Macgregor (1947) stated "There is evidence of movements at various times during the formation of the Basement Complex, but the movements do not themselves explain the arcuate strikes around so many of the larger granitic intrusions". The arcuate strikes are clearly the result of a second stage of vertical tectonics when gneissic domes were formed.

It is possible to distinguish four phases in the tectonic development of the craton.

The primary phase

The primary phase is the earliest recognizable phase, when the older gneissic basement nuclei with Sebakwian I inclusions were formed. The patchy distribution of the relics makes the recognition of structures uncertain, but they are found only in the older central tectonic unit. In addition to the north-trending Rhodesdale-Ghoko line of Sebakwian I relics, similar lines occur on the east sides of the Rhodesdale and Chilimanzi nuclei (Fig. 2).

A Sebakwian I jaspilite unconformably overlies a pre-Sebakwian I migmatite preserved in the vicinity of Gwenora Dam 20 km west of Selukwe. In this the preponderance of amphibolite and absence of sediments other than chemically deposited, banded ironstones suggest that the crust on which they were deposited may have been amphibolite or basalt (Stowe, 1968*a*).

The 3,400 m.y. Rhodesdale igneous event appears to mark the end of this phase during which the older gneissic basement was formed as the protocratonic nuclei.

The geosynclinal phase

The geosynclinal phase was marked by the development of small, Bulawayan eugeosynclines on either side of, and across the central protocraton (Fig. 2).

The older gneissic basement is crossed by northeast- and northwest-trending narrow fold belts in the area west of Selukwe, which may mark fracture lines. The widespread occurrence of clasts derived from an older gneissic terrain in the sediments interbedded with greenstones, indicates that deposition of the Sebakwian II and Bulawayan groups was in deep, narrow eugeosynclines (Stowe 1968*a*).

Sebakwian II volcanics and sediments fringe the southern margin of the Gwelo Schist Belt where they were distinguished from the overlying Bulawayan rocks by an unconformity and a higher metamorphic grade (Amm, 1946; Tyndale-Biscoe, 1949). In the Selukwe area they were thrust northwards over the older gneissic basement as a large nappe structure (Stowe, 1968*b*).

There are two main eugeosynclinal belts of Bulawayan formations, one northwest and one southeast of the central unit. It is improbable that all the formations included in the Bulawayan by various workers were contemporaneous, but they all belong to this second tectonic phase.

The southeastern belt is preserved as a chain of schist belts from Gwanda to Umtali, folded tightly on an east-southeast trend. The lower formation of the Bulawayan Group in the type area closely resembles formations assigned to the Bulawayan Group in the Shabani, Fort Victoria and Umtali Schist Belts. All consist of thick greenstone successions with pillow lavas and local sedimentary associations of conglomerate and quartzite, or phyllite and banded ironstone.

The northwestern belt, from Bulawayo to Shamva, consists of very thick pillow lava greenstone successions with thin interbedded lenses of banded ironstone, conglomerate and greywacke. In the vicinity of Gwelo a branch extends southeastwards across the central protocraton. This belt was folded on northeast and northwest trends.

The Upper Greenstone group in the Bulawayo area (Amm, 1940), the Maliami River Formation west of Que Que (Harrison, 1968), and the Umniati Group west of Gatooma (Bliss, *in press*) consist of andesites and peculiar "rolled" agglomerate (Amm, 1940, Plate IIIA). They unconformably overlie the Lower Bulawayan near Bulawayo and unconformably underlie the Shamvaian. This constitutes an upper Bulawayan andesite belt along the northwestern protocratonic margin.

The gneissic dome phase

The gneissic dome phase occurred when vertical tectonics were dominant and the older gneissic basement in the protocratonic nuclei was granitized and domed up over the concordant tonalites (Fig. 2). The contact between this older basement and the greenstones was reactivated in many places, either by direct intrusion (Gl granodiorite, Stowe, 1965) into the schist belts, or by a permeation granitization, such as that below the Selukwe nappe, to give a locally intrusive appearance.

Radiometric ages of the tonalites show a wide scatter but Bliss & Stidolph (*in press*) and Macgregor (1951) both tentatively suggest 2,900 m.y.

The rim synclinal basins between the domes may well have been the depositories for the Shamvaian Group sediments and felsic volcanics. These occur within the older protocraton between the domes as well as around it (Fig. 2).

The Shamvaian Group usually includes two or three unconformable formations showing alternating episodes of shallow water deposition and infolding. These usually show a progression from low-energy environment, phyllitic rocks to high-energy, rapidly deposited arenites, derived from gneiss. Along the northeastern margin of the craton the Shamvaian formations occur as a line of local deposits, synclinally infolded in the previously folded Bulawayan greenstones. Shamvaian sediments constitute the bulk of the schist belts along the northern craton margin (*e.g.* Wiles, 1965).

The early cratonic phase

During the early cratonic phase the central unit of the craton was stable, but intense folding on an east-northeast trend, and a flood of post-Shamvaian granites in the marginal tectonic units occurred.

In the Salisbury area, the east-northeast trending folds refolded older, isoclinally folded Shamvaian formations. In the Fort Victoria area Wilson (1964) mapped east-northeast trending, isoclinal folds with axial planes dipping away from the craton to the southeast. At Selukwe, Stowe (1968*b*) noted late, east-northeast trending folds in narrow zones with a tectonic style reflecting a competent basement.

Granites of post-Shamvaian (2,700-2,300 m.y.) age are tentatively shown in Figure 2, but insufficient mapping has been done. Bliss (1969) published a provisional map of the granitic rocks. They vary from small, transgressive, high-level types in the central unit, to intermediate-level types in the northern and southern units, to a vast flood that engulfed most of the eastern unit.

CONCLUSIONS

1. More mapping in the granitic areas is necessary in order to understand the tectonics of the Rhodesian craton as a whole.

2. Similarities with the Barberton Model (Anhaeusser *et al.*, 1968) are obvious, in particular the effects of the diapiric granites in the northern unit. Differences are found in the older central unit where a pre-3,400 m.y. gneissic basement was sufficiently stable to support alpine-type tectonics following the Sebakwian II deposition. Another difference appears to be in the growth of the protocratonic centre by eugeosynclines and orogenesis around its margins.

3. The older gneissic basement of Katarchaean age and the subsequent Archaean eugeosynclines, followed by stabilization of the craton by sialic intrusion may be related to chelogenic cycles (Sutton, 1967).

ACKNOWLEDGMENTS

The author has drawn information freely from publications by staff of the Rhodesia

Geological Survey. He would like to thank the Director and drawing office staff for drafting the text-figures. He would also like to thank Mr Phaup of the Survey and Professor Bond and Dr Wilson of the University College of Rhodesia for the critical reading of this paper.

REFERENCES

ALLSOPP, H. L., 1965: Rb-Sr and K-A age measurements in the Great Dyke of Rhodesia. *J. geophys. Res., 70,* pp. 977-984.

AMM, F. L., 1940: The geology of the country around Bulawayo. *Bull. geol. Surv. Sth. Rhod., 35.*

———, 1946: The geology of the Lower Gwelo Gold Belt. *Bull. geol. Surv. Sth. Rhod., 37.*

ANHAEUSSER, C. R., ROERING, C., VILJOEN, M. J., & VILJOEN, R. P., 1968: The Barberton Mountain Land. A model of the elements and evolution of an Archaean gold belt. *Trans. geol. Soc. S. Afr., 71, Annex.,* pp. 225-254.

ANHAEUSSER, C. R., MASON, R., VILJOEN, M. J., & VILJOEN, R. P., 1969: A reappraisal of some aspects of Precambrian shield geology. *Bull. geol. Soc. Am., 80,* pp. 2175-2200.

BLISS, N. W., 1962: The geology of the country around Umvuma and Felixberg. *Bull. geol. Surv. Sth. Rhod., 56.*

———, 1969: Thermal convection in the Archaean crust. A reply to C. J. Talbot. *Nature, Lond., 222* (5197), pp. 972-974.

———: The geology of the country around Gatooma. *Bull. geol. Surv. Rhod.* [in press].

———, & STIDOLPH, P. A.: The Rhodesian Precambrian Basement Complex: a review. *Trans. geol. Soc. S. Afr.* [in press].

HARRISON, N. M., 1968: A reassessment of the stratigraphy of the Precambrian Basement Complex around Que Que, Gwelo district. *Trans. geol. Soc. S. Afr., 71, Annex.,* pp. 113-124.

———, 1969: The geology of the country around Fort Rixon and Shangani. *Bull. geol. Surv. Rhod., 61.*

HOLMES, A., & CAHEN, L., 1956: Geochronology Africaine. *Mem. Acad. r. Sci. colon. Cl. Sci. nat. med. 8°, 5* (1).

MACGREGOR, A. M., 1932: Batholiths of Southern Rhodesia. *Geol. Mag., 79* (881), pp. 18-20.

———, 1947: An outline of the geological history of Southern Rhodesia. *Bull. geol. Surv. Sth. Rhod., 38.*

———, 1951: Some milestones in the Precambrian of Southern Rhodesia. *Proc. geol. Soc. S. Afr., 54,* pp. xxvii-lxxi.

NICOLAYSEN, L. O., 1962: Stratigraphic interpretation of age measurements in southern Africa; *in* Engel, A. E. J., James, H. L., & Leonard, B. F. (Eds), *Petrologic Studies: a Volume in Honour of A. F. Buddington,* pp. 560-598. Geol. Soc. Am., New York.

POLDERVAART, A., 1961: Letter on file at Rhodesia Geol. Surv., Salisbury.

OLDHAM, J. W., 1968: A short note on the recent geological mapping of the Shabani Area. *Trans. geol. Soc. S. Afr., 71, Annex.,* pp. 189-194.

STOWE, C. W., 1965: Map of the country south and east of Selukwe, to accompany *Bull. geol. Surv. Rhod., 59.*

———, 1968*a*: Ph.D. Thesis, Univ. London.

———, 1968b: Intersecting fold trends in the Rhodesian basement south and east of Selukwe. *Trans. geol. Soc. S. Afr., 71, Annex.,* pp. 53-78.

SUTTON, J., 1967: The extension of the geological record into the Precambrian. *Proc. Geol. Ass., 78* (4), pp. 493-534.

TYNDALE-BISCOE, R., 1949: The geology of the country around Gwelo. *Bull. geol. Surv. Sth. Rhod., 39.*

VAIL, J. R., 1965: An outline of the geochronology of the late Precambrian formations of eastern central Africa. *Proc. R. Soc., Ser. A, 284,* pp. 354-369.

VIEWING, K. A., 1968: Regional geochemistry and the Rhodesian Basement Complex. *Trans. geol. Soc. S. Afr., 71, Annex,* pp. 21-32.

WILES, J. W., 1965: Map of the country around Sipolilo, to accompany *Bull. geol. Surv. Rhod., 63.*

WILSON, J. F., 1964: The geology of the country around Fort Victoria. *Bull. geol. Surv. Sth. Rhod., 58.*

———, 1968: The geology of the country around Mashaba, *Bull. geol. Surv. Rhod., 62.*

WORST, B. G., 1955: The geology of the country between Belingwe and West Nicholson. *Bull. geol. Surv. Rhod., 43.*

C. W. Stowe,
University College of Rhodesia,
Box MP167,
Mount Pleasant,
Salisbury,
Rhodesia.

PAPERS 8 THROUGH 11: COMMENTARY

Literature dealing with the Abitibi Belt, the largest single continuous greenstone belt yet described, is abundant, reflecting its economic importance. The truncation of the belt by later mobile belts of gneissic terrain is sharp and transverse, the older Archean gneiss being reworked in these belts (Paper 23). Radiometric dating has barely penetrated the barrier imposed by the Kenoran event (2550 million years ago), but zircon- and lead-derived values exceeding 3000 million years indicate incorporation of minerals from an older Katarchean terrain. The Kenoran age is likely to be quite close to that of the formation of the volcanic–clastic, sediment–chemical sediment assemblages that comprise the greenstone sequence; the supposition that the Kenoran event largely reworked much older rocks is unwarranted.

Thicknesses of these assemblages are considerable, as in most Archean greenstone belts. The presence of island-arc-type, calc-alkaline eruptives (basalts, andesites, dacites, rhyodacites, and rhyolites) seems to be well established despite counterarguments (Paper 16), at least for this particular province, in which it is thought to account for the extensive copper–zinc mineralization. However, nickel–copper, zinc–silver–lead, molybdenum–bismuth, and lithium mineralization are also prominent, and intense gold mineralization is associated with unique Archean syenites, basanites, trachytes, and leucitites, which accompany the banded ironstones at Kirkland Lake. So the question may well be asked, "Is it the nature of the eruptivity (calc-alkaline, alkaline) that was responsible for the profusion of economic deposits in the belt, or was it a particularly 'pregnant mantle' beneath (in analogy with Arizona–New Mexico; Sillitoe, 1972), or a combination of both?"

The recognition of two sedimentary facies, volcanogenic and flyschoid, is important; both facies reflect unstable depositional environments. The Timiskaming Facies, a flyschoid variant, includes coarse polymictic conglomerates and is believed to mark localized zones of steep sea-floor slope. The Pontiac Facies, truly flyschoid, shows little facies changes, whereas the volcanogenic sediments seldom show much strike continuity of facies. Iron formation is extensively developed, and its degree of continuity indicates that there were extensive basins of deposition. A zonation of oxide, carbonate, and sulfide facies is recognized, indicating shelf-to-basin bathymetry.

Rhythmically banded noritic and anorthositic layered intrusions, both extensive and thick (up to several kilometers), resemble those of the Murchison District of Western Australia (up to 8 km thick; Paper 15). The Canadian examples seem to be more anorthositic, and are economically significant (e.g., association with the Chibougamou gold–copper mineralization).

The granitic rocks include two main types: granodiorite, forming tracts of essentially metasedimentary origin, and discrete plutons of granite. Both types appear to postdate the greenstone sequence. Radiometric ages for the plutons range from Kenoran (2400 million years) to Proterozoic (1700 million years and younger).

The authors postulate two cratonic forelands to the north and south of the belt, but there is a gradation evident across the boundaries, from the predominance of supracrustal to a predominance of granitoid rocks, and the older foreland terrain is by no means clearly demarcated. It is assumed to be present because it seems that a foreland must have existed in both these directions. The interpretation of the marginal granitoid terrain remains controversial, and it is possible that it includes either foreland or only entirely reworked foreland or no foreland rocks.

The regional structure is one of first- and second-order folds, with granite plutons occupying the anticlinal positions and supracrustal rocks the synclinal positions. The anticlines tend to bifurcate and show variable axial plunge. Regional "breaks" have long been recognized; they are not now regarded as abrupt dislocations that destroy all stratigraphic continuity, but as zones of intense shearing following incompetent rock units and lithofacies boundaries. The entire cluster of volcanic edifices seems to have been foreshortened by compression on north–south axes, by a factor of about 50 percent.

The oval, strike-elongated volcanic complexes are sites of low negative gravity anomalies, the granite plutons are sites of high negative anomalies, and the rest of the terrain is without anomaly. This pattern suggests that the granitic plutons are deep-rooted. These plutons can readily be distinguished from the mixed, mostly gneissic terrain of the forelands (?).

Paleogeographic development seems to have been sequential in time from an older shelf sequence (Skead Group, Kirkland Lake) to a more complex unstable basin sequence (Blake River Group) formed toward the center of the basin.

The evidence suggests that the role of sialic reworking or contamination in the development of the acid volcanics may have been slight. Their structural level seems to suggest high-level emplacement and more opportunity to differentiate on their ascent. Tholeiites predominate in the center of the orogen and calc-alkaline suites (including extreme acid differentiates) toward the margin. The possibility that we are seeing the product of spreading of an ancient sea floor is suggested, and an analogy is drawn with modern marginal subduction zones. The mineralization is, in detail, related to individual volcanic and intrusive expressions and to specific sedimentary facies or structural features, but the general pattern forms an integral part of the overall development of the orogen.

The comment in discussion of this paper by Naldrett is important. Later work by Pyke et al. (1973) recorded a sequence of some 60 thin, spinifex-textured ultramafic flows from Munro Township, Ontario, thus confirming that extrusive ultrabasic rocks are represented in Archean greenstone suites, as suggested by Viljoen and Viljoen (Paper 6), McCall and Leishman (1971), and Nesbitt (1971). "Spinifex-textured ultrabasics" are either thin flows, crystallized at 1600–1700°C (Paper 4), that display supercooling textures, vesicles, glass, and flow-top breccias, or else they are near-surface sills of similar character. Both types of occurrence are recognized. These magmas were remarkably fluid, and the "spinifex" material (named after a spikey Australian grass) is a residual crystallization from a melt after the removal of olivine by gravitational sinking (extremely rapidly, in a matter of hours) of olivine phenocrysts to form the underlying cumulate zone of the flow. The process is becoming quite well understood, but the remarkably high temperature of crystallization remains unexplained (Paper 4).

The Slave structural province is characterized by only a minority (15–20 percent) of volcanic content in the supracrustal rock sequence (by area). Volcanic–sedimentary rocks outcrop in areas separated by large batholiths of granite and associated high-grade metamorphics. Although volcanic–sedimentary repetition is locally evident, the sequence is essentially bimodal, comprising a single major cycle: a thick, dominantly volcanic sequence is overlain by a thinner sequence of graywackes and shales. In the transition zone, conglomerates, mature sandstones, impure limestones, and tuffaceous sediments occur, but they may be locally absent. The sediments do not seem to be everywhere underlain by volcanics.

The volcanics are believed to have developed at the margins of basins on the site of deep fracture zones bordering uplifted foreland. Their occurrence marginal to the areas of sedimentary outcrop is not attributed solely to the synclinal nature of such areas. The total thickness of the supracrustal sequence at Yellowknife is estimated at 20,000 m.

The bordering granites may include older basement. Evidence for this is to be found in transecting dikes that may be feeders of the Yellowknife volcanics and in the superimposition of conglomerates carrying boulders derived from the granite. Possible areas of basement occur near the margin of the supracrustal outcrop, which suggests that present boundaries may coincide with those of the ancient basin of deposition. Granite of a later age, intrusive into the sequence, is also recognized, and the differentiation of this intrusive granite from foreland granite remains extremely difficult. Locally, there is unconformity between sediments and volcanics, but this is not considered to represent a major time break. It is a reflection of the steep attitude of the underlying volcanics relative to the flatter sedimentary mantle.

The conglomerates are polymictic and carry both felsitic and basic volcanic phenoclasts, as well as well-rounded granite phenoclasts. This rounding is an indication of a quite distant foreland source (compare angular, locally derived phenoclasts in the Aleutians). Preliminary paleocurrent analyses suggest source areas in marginal granodiorite forelands and support the contention that present synclinal zones of supracrustals closely approximate to the depositories. Primary sedimentary structures are excellently preserved, and the graywackes have the character of turbidites. As at Mt. Belches, Western Australia (Dunbar and McCall, 1970), siliceous magnetic iron formation occurs in the turbidite sequence, but banded iron formation is conspicuous by its absence from the entire Slave Province.

The tremendous volume of siliceous material in the sediments, which are of overall granodiorite composition, requires a foreland source rather than a source in intrabasinal volcanics, but some material may well have been contributed from such intrabasinal sources.

The structure is one of tight, intricate folds within incompetent sediments, and of homoclines within volcanic sequences and thick, competent sediment sequences.

The homoclines face away from the granite areas, which may be either older basement or younger intrusions.

Metamorphism, deformation, and intrusion of granites occurred as an overlapping sequence of events, commencing with low-grade greenschist to amphibolite-grade metamorphism of the supracrustals, followed by syntectonic granite emplacement and associated metamorphism overlapping early folding, and ending with posttectonic, high-level intrusion of granodiorites.

The authors believe that the basins of accumulation of the supracrustals were founded on sialic crust, and favor an origin in marginal sialic contamination for the calc-alkaline volcanics (a concept that has lately lost ground to one of subduction and reworking of oceanic crust). They do not favor an island-arc analogy (compare Papers 8, 10, and 11).

The authors mention Martin's idea that each Archean province bears the geochronologic overprint of the last series of Archean events, of a type never repeated in the history of the earth. In the case of the Slave Province, this overprint is Kenoran, averaging 2490 million years by potassium–argon dating.

In their geochemical treatment, based on 261 analyses, the authors of Paper 10 have drawn an analogy between volcanic assemblages of Archean greenstone-belt associations in Canada and island-arc or continental orogenic calc-alkaline suites in modern systems. A primitive continent composed of typical sialic continental rocks must, they believe, have existed in Archean times. Keewatin greenstones are believed to be remnants of a sequence that covered the entire southern half of the Superior structural province; such remnants are now only preserved in synclinal belts. The standards upon which this comparison is based are various chemical syntheses of classical suites in type areas, of both island-arc and continental calc-alkaline rocks. Niggli alkali–silica and Peacock alkali–lime index plots are compared, and the latter indexes for Archean volcanic suites are found to range between 56.1 and 66.4 (corresponding to tholeiites and calc-alkaline ranges). Analogy with continental flood basalts can be ruled out on account of the different physical nature and mode of occurrence of the Archean suites and the lack of acid differentiates and pyroclastic expressions in the continental association.

More recent data from other regions of the world (South Africa and Australia) suggest that the calc-alkaline suites may be unusually well developed in the Superior Province compared with many other Archean provinces, and that abyssal tholeiites and unique Archean high-magnesia basalts (komatiite equivalents; see Paper 6) have a substantial development in all Archean regions, the latter, however, being more strongly developed than in the Superior Province in those regions that show less development of calc-alkaline rocks.

Discussing the Slave Province of Canada (the subject of Paper 9), the authors of Paper 11 favor the island-arc analogy (Papers 10, 33, and 35), which is not favored by the authors of Paper 9. They stress similarities between components of the greenstone assemblages and the intrusive rocks of this province to the suite of the Fossa Magna, Japan, and the similarity of the calc-alkaline trend of differentiation of the tholeiite, quartz–latite, dacite sequence to the tholeiitic differentiation pattern. Noting the absence of any ages significantly older than Kenoran (as revealed by K/Ar, Rb/Sr, and U/Pb methods), they conclude that, despite the evidence for an older granodiorite basement cited by other authorities (e.g., Paper

9), the granodiorite, which gives ages by the Rb/Sr method of 2500–2600 million years, is either younger intrusive material or older basement completely updated. Their main contention is that the dating by this method of the greenstone and granodiorite components of the province is sufficiently refined to show that the events of the greenstone belt took place over a similarly short period as that known to encompass the development of the eugeosynclinal suite of the Fossa Magna and of the Western Cordillera of Canada.

This argument may be criticized on the grounds that there is still a considerable area of doubt concerning the refinement claimed for the particular age-dating method; it is doubtful if it really can be used to delineate accurately such narrow intervals in greenstone-belt development. There is also the question of whether the authors have unequivocally established the fact that the radiometric number obtained from the volcanics applies to the time of eruption as opposed to the time of superimposed metamorphism. Finally, even if the accumulation did take place within such a short time interval, this fact in itself does not justify drawing an exact analogy. The main basis for the contention must remain the geochemical similarity of the component members and differentiation trends. In that broad sense, the analogy is justified, but doubt remains whether geotectonic patterns and environments of the Archean can be related to extensive lateral dispersion of plates such as is now supposed to have produced island arcs.

References

Dunbar, G. J., and McCall, G. J. H. (1971). Archaean turbidites and banded ironstones of the Mount Belches area, Western Australia. *Sedimentary Geol., 5,* 92–133.

McCall, G. J. H., and Leishman, J. (1971). Clues to the origin of Archaean eugeosynclinal peridotites and the nature of serpentinisation. *Spec. Publ. 3, Geol. Soc. Aust.,* 281–299.

Nesbitt, R. W. (1971). Skeletal crystal forms in the untramafic rocks of the Yilgarn Block, Western Australia: evidence for an Archaean ultramafic liquid. *Spec. Publ. 3, Geol. Soc. Aust.,* 331–346.

Pyke, D. R., Naldrett, A. J., and Eckstrand, O. R. (1973). Archaean ultramafic flows in Munro Township, Ontario. *Geol. Soc. Am. Bull., 84,* 955–978.

Sillitoe, R. H. (1972). A plate tectonic model for the origin of porphyry copper deposits. *Econ. Geol., 67,* 184–197.

8: THE ABITIBI OROGENIC BELT

A. M. Goodwin

Department of Geology, University of Toronto, Toronto, Canada

R. H. Ridler

Department of Geology, University of Western Ontario, London, Ontario, Canada

Abstract

This east-trending tectonic unit, some 500 by 150 miles in dimensions, is the largest single continuous Archean greenstone belt in the Canadian Shield. It is a characteristic Archean orogenic composite belt featuring mafic to felsic volcanics with coeval intrusions, volcanic sediments, both clastic and chemical including banded iron formation, and several large granitic batholiths. Metallogenic patterns conform to lithic distributions. Low- to medium-rank greenschist facies prevail. Supracrustal rocks have been isoclinally folded about east-trending, undulating axes resulting in substantial lithofacies compression. The belt is bounded north and south by granitic-metasedimentary crystalline terrains; it is truncated abruptly east and west by Grenville and Kapuskasing crystalline rocks, both products of younger Precambrian events. Thus the present belt represents only part of an original Archean tectonic entity.

A number of well-recognized mafic to felsic volcanic centres with intercalated volcaniclastics, iron formation and igneous intrusions are present. Despite common features each centre constitutes a semi-independent lithic assemblage of limited stratigraphic continuity. The centres may be confidently ascribed to processes of igneous differentiation and eruption largely by way of central vents. Three main stages of igneous eruption are recognized in the belt. The total time span was probably several hundred thousand years. Major geochronological problems remain.

Recognized mafic to felsic volcanic centres and accompanying clastics with iron formation are concentrated in two main east-trending bands, each 50 miles wide, which respectively cross the northern and southern parts of the belt. In contrast, the intervening or median part of the belt, also 50 miles wide, is underlain, as known, by uniform tholeiitic basalt, fine-grained clastics, and major granitic batholiths.

On this basis, the Abitibi belt is viewed as a remnant of a bilaterally symmetrical intra-tectonic orogen rather than a conventional asymmetrical continental-oceanic tectonic interface (e.g. island arc). Crustal thinning and other orogenic activities may be attributed either to conventional geosynclinal downsinking of thin, supple Archean crust or to spreading apart of Archean crustal blocks in the manner of present-day ocean floor spreading.

INTRODUCTION

Abitibi belt, 475 miles long by 125 miles wide, is the largest continuous Archean (older than 2,500 million years) greenstone belt in the Canadian Shield. Located in the southeastern part of the Superior tectonic province this east-trending belt is truncated east and west by northeast-trending crystalline rocks of Grenville province and Kapuskasing subprovince respectively. The present belt is part only of an originally longer Archean orogen which in Archean time included now-crystalline Grenville rocks immediately on the east and, likely, orogenic rocks of Wawa belt to the west.

[*Editor's Note:* Figure A and the Comments section have been omitted owing to limitations of space.]

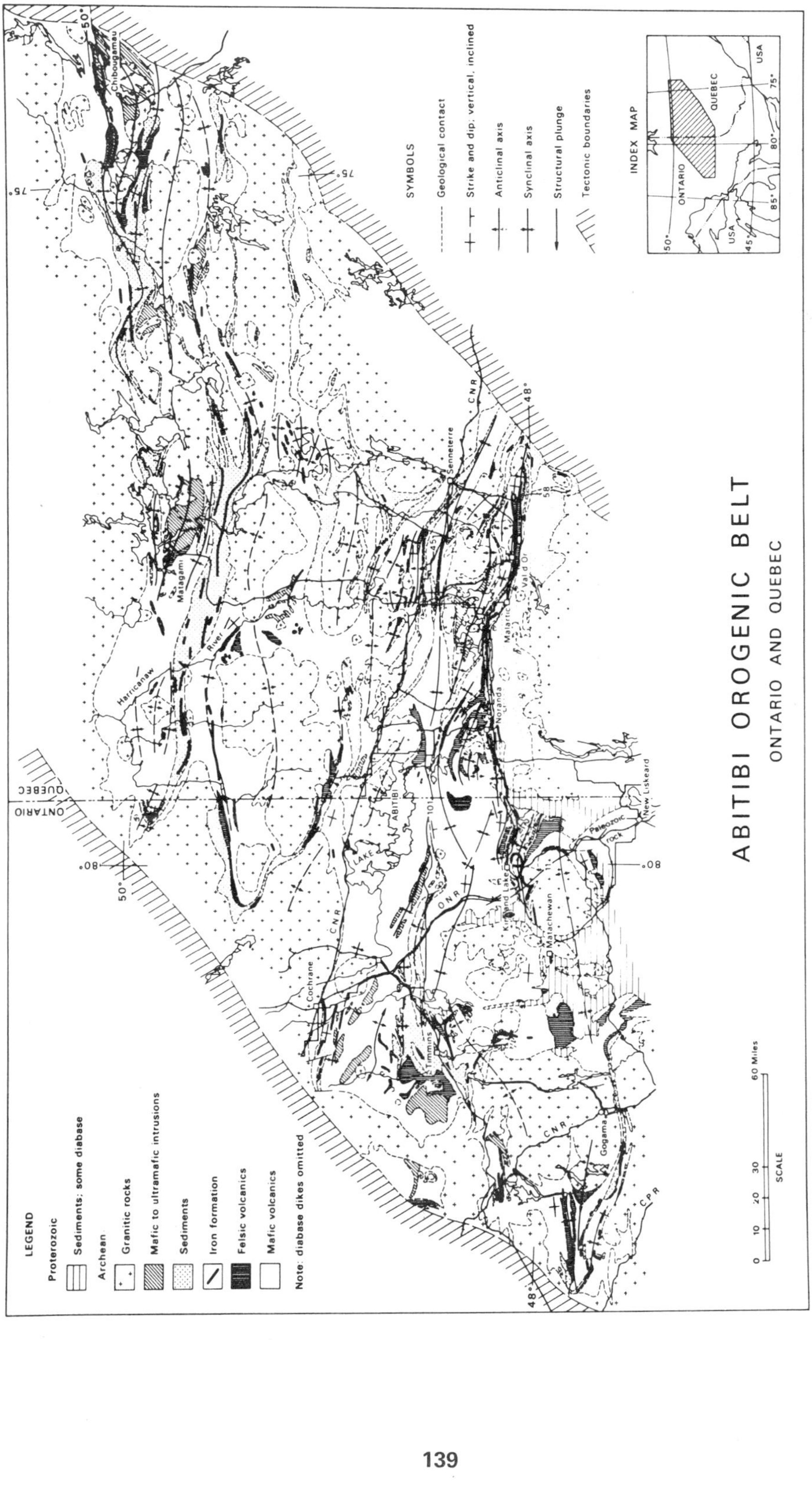

Figure 1. Geologic map of Abitibi orogenic belt. Omitted from the map for cartographic reasons are diabase dikes and some thin pyritiferous carbonaceous (± chert) zones in the Noranda-Val d'Or-Senneterre area. The latter are shown in Figure 2 as sulphide facies iron formation.

Abitibi belt has been geologically studied and explored for more than sixty years. Complete regional air-photo, aeromagnetic and gravity as well as local E.M. surveys are available. Many parts have been geologically mapped in detail. Sustained mineral exploration and mine development over the years have provided a wealth of data there being at least one hundred and fifty producers (past and present) including Au-Ag, Cu-Zn-Au-Ag, Ni-Cu, Fe, Mo-Bi, Li and asbestos deposits. However, extensive drift-covered tracts especially north of the main CN railway obscure many key geologic relationships. Despite the attention received, the major geologic problems including those of stratigraphic correlation and tectonic construction remain largely unresolved. Therefore additional studies, regional and detailed, direct and indirect, will be required for many decades to come.

In common with most greenstone belts of the Canadian Shield the geologic record of this belt doubtless includes a long and complex history of Archean events. So far, age-dating has not penetrated significantly the radiometric barrier imposed by the Kenoran event (at 2,500 million years) (Wanless *et al.*, 1968; see also Roscoe, 1965). This paper presents a brief integrated statement on relationships pertaining to tectonic evolution of this important primitive crustal unit. A principal purpose is to stimulate similar analysis of other Archean orogens in this and other Precambrian Shields of the world.

GENERAL GEOLOGY

The Abitibi lithic assemblage (Fig. 1) is characteristic of many Archean greenstone belts of the Canadian Shield. Older supracrustal and igneous rocks, now substantially deformed and of common greenschist facies, have been intruded by syntectonic to post-orogenic felsic to mafic intrusions ranging up to those of large batholithic dimensions. Numerous younger Precambrian diabase dikes, mostly of the Matchewan swarm, transect. Several tongues of flat-lying Proterozoic rocks and one of Paleozoic rocks protrude from the south. Pleistocene glaciation resulted in both clean-scouring of thick, steeply inclined Archean stratigraphic sections and accumulation of thick glacio-fluvial and lacustrine deposits which obscure bedrock relations particularly in northern parts of the region.

Volcanic Rocks

A variety of volcanic rocks is present. Mafic to felsic flows and pyroclastics of calc-alkaline chemical affinity, all representative of the orogenic suite (Baragar and Goodwin, 1969) greatly predominate. Some alkaline volcanic rocks are present at Kirkland Lake. Basalt flows and associated gabbroic intrusions are common in the lower parts of volcanic assemblages. Andesitic flows and pyroclastics are intercalated with basalt and characteristically increase in proportion upwards. Felsic rocks of dacite to rhyolite composition are generally present in upper stratigraphic parts. Most volcanic assemblages in the region display a generalized mafic to felsic compositional sequence of this type. In some areas the sequence is repeated in whole or in part.

Mafic volcanic rocks comprising basalt and andesite in that order of abundance, are the predominant supracrustal rocks of the region. They are broadly distributed particularly in the central and westen parts. Much remains to be learned about their petrochemistry. Thick mafic (i.e. basalt-andesite) accumulations exceeding 40,000 feet thick are present in Blake River group west of Noranda, and in Skead group south of Kirkland Lake. Similar thicknesses of mafic rocks may be present elsewhere in the region. In the Blake River group a lower zone of low-Al_2O_3 mafic volcanics, approximately 20,000 feet thick, is conformably overlain by equivalent thicknesses of high-Al_2O_3 mafic effusives (Baragar, 1968; Baragar and Goodwin, 1969). Similar low- and high-Al_2O_3 basalt and andesite are present in Swayze area to the west (Goodwin, 1968). In the Skead group south of Kirkland Lake a thick mafic assemblage approximately 28,000 feet thick which includes basanite lava flows exhibits significant upward increase in K_2O content (Ridler, 1970). Volcanic piles north and south of Timmins include tholeiitic basalt and andesite. Preliminary results of a reconnaissance geochemical cross-section of the volcanic rocks between Cadillac, Malartic area in the south and Matagami area in the north indicate a preponderance of basaltic rocks except for the felsic rocks shown in Figure 1 (J. Descarreaux, pers. comm.). Sparse chemical analyses of mafic volcanic rocks in the Matagami area suggest the presence of normal tholeiitic pillow basalt and andesite. In the Chibougamau area to the northeast, low-K_2O tholeiites including basaltic rocks with very low K_2O, high Na_2O and intermediate $K/_{Rb}$ ratios are present (Gunn, 1969).

Mafic lava flows are commonly 50 to 100 feet thick but flows up to 800 feet thick are known. Pillow forms, palagonite, variolites, amygdules and hyaloclastites are widespread and abundant thereby indicating prevailing subaqueous accumulation of the mafic lava. Primary textures are, on the whole, remarkably well preserved (Moorhouse, 1970).

Overlying the great thicknesses of mafic volcanic rocks are numerous concentrations of felsic volcanic rocks, each concentration characteristically representing a felsic eruptive centre (Fig. 1). Principal concentrations occur at Val d'Or, Noranda, Kirkland Lake, Timmins, Matagami and Chibougamau. Rocks of dacitic composition are most common in the felsic concentrations. Rhyolite is present locally as, for example, at Noranda, Timmins, Swayze, Joutel and Matagami but is absent in other centres, such as, for example, Kirkland Lake. Pyroclastic forms are very common in felsic concentrations of Swayze, Val d'Or, Kirkland Lake, Timmins, Matagami and Chibougamau areas. Massive rhyolite lava flows intercalated with andesite flows are common at Noranda. Subaqueous ash-flows recently identified in Noranda area (R.S. Fiske, pers. comm.) may be widespread in other volcanic piles of the region. Trachytic flows and pyroclastics are present at Kirkland Lake.

In addition to specific felsic concentrations, the regional volcanic pile contains many thin discontinuous felsic tuff zones, some notably spherulitic, which are attributed to widespread wind and/or water distribution. Such zones are potentially valuable marker-horizons particularly where intercalated within thick uniform mafic flow accumulations.

Although mineralogically altered, felsic volcanic rocks exhibit a wide range of internal textures including spherules, phenocrysts, vesicles, perlite, pumice and shards, all commonly in excellent state of preservation (Moorhouse, 1970).

Sediments

Clastic zones range from thin discontinuous units to broad regional, east-trending belts of which the principal are situated south of Noranda-Val d'Or, east of Kirkland Lake, near Timmins, northeast of Lake Abitibi, south and southwest of Matagami and west of Chibougamau (Fig. 1). Sediments characteristically occupy higher stratigraphic positions. Most sediments are structurally conformable with associated layered rocks. However, local unconformities have been established at Kirkland Lake and Timmins.

Sediments of the "poured-in" turbidite association predominate thereby suggesting rapid accumulation in tectonically unstable environments. Two principal facies are identical: volcanogenic and flyschoid. The Timiskaming facies, a variation, is present locally. Rocks in the volcanogenic facies, ranging up to 10,000 feet thick include greywacke, shale, lithic sandstone, conglomerate, breccia and iron formation. Identifiable clastic components compare closely in composition with nearby volcanic rocks and have obviously been derived in large part from them by rapid mechanical erosion and subaqueous deposition in nearby troughs and basins. Chaotic textures and polymictic unsorted materials are distinctive features. This facies which is characteristic of strongly active tectonic zones, typically displays intricate soft-sediment deformational structures. Graded bedding and abrupt facies changes are common. Iron formation is present. Carbonaceous and pyritiferous zones are intercalated with finer grained clastic phases.

A prominent belt of more regularly bedded flyschoid facies (Pontiac group) lies south of Noranda-Malartic-Val d'Or. It comprises rhythmically bedded greywacke-argillite sequences of uniform construction and composition which lack marked lateral facies changes. According to J. Holubec (in preparation) the facies is 7,000 to 8,000 feet thick and was developed in a comparatively stable tectonic environment in proximity to a stable crustal domain to the south.

A variant facies, the Timiskaming, is present in narrow, east-trending zones near Kirkland Lake-Noranda and Timmins. This facies is characterized by coarse-grained detritus including conglomerate, abrupt lateral facies changes and erosional unconformities. It occupies narrow zones of pronounced tectonic instability which have been attributed to the local development of steep slopes (J. Holubec). At Kirkland Lake the facies is associated with a unique assemblage of trachytic flows and pyroclastics.

Iron formation is widely distributed (Figs. 1 and 2). Principal bands occur in Swayze, south of Timmins and Kirkland Lake-Val d'Or areas in the south and within a broad east-trending belt south of Matagami in the north. Local bands are present at Chibougamau, in the Barraute-Quevillon area of the central-eastern part and elsewhere. The iron formations are important stratigraphic elements and the closest approach to marker horizons in the regional assemblage. Their continuity across broad parts of the region demonstrates the presence of large extensive basins of deposition in Abitibi time.

Iron formations of the region are classified as oxide, carbonate or sulphide facies depending on the predominant iron mineral. The three facies are transitional across the region (Fig. 2). Oxide (magnetite) facies is common in Swayze, south Timmins, Kirkland Lake and Val d'Or areas in the south and in the principal belt south of Matagami in the north. Generally, oxide facies is transitional eastward and northeastward through narrow carbonate (siderite-ankerite-dolomite) zones to sulphide (pyrite-pyrrhotite) facies iron formation. These facies transitions are considered to define original shelf-to-basin bathymetry. For example in the Timmins area, magnetitic iron formation in the south is transitional northeastward through a narrow zone of sideritic iron formation to thin cherty pyritiferous iron formation. Magnetitic iron formation at Kirkland Lake is transitional eastward to siliceous carbonate iron formation as at Larder Lake and farther to thin pyritiferous zones at Noranda (Ridler, 1970). Similarly, magnetitic iron formation near Val d'Or is transitional northeastward to thin chert-pyrite-carbonaceous zones representing sulphide facies iron formation. Similar transitions are present elsewhere (Fig. 2).

Mafic Intrusions

Numerous mafic sheets, sills and dikes occur in the mafic lava accumulations. They are readily confused with diabasic phases of thick flows and commonly represent intrusive phases of the volcanism. They are also intrusive into felsic volcanic phases as at Noranda.

Differentiated sheets and sills up to several thousand feet thick typically comprise basal zones of peridotite, usually serpentinized, transitional upwards through pyroxenitic or mafic gabbro or norite, to gabbro and quartz gabbro and, locally, to granophyre (micropegmatite). In Chibougamau-Opemiska area, differentiated mafic sills include peridotite-pyroxenite, gabbro-pyroxenite and diorite-quartz diorite transitions. The Chibougamau anorthosite, 30 miles in diameter, comprises coarse-grained greenish white assemblages composed of saussuritized plagioclase and chloritized ferromagnesian minerals.

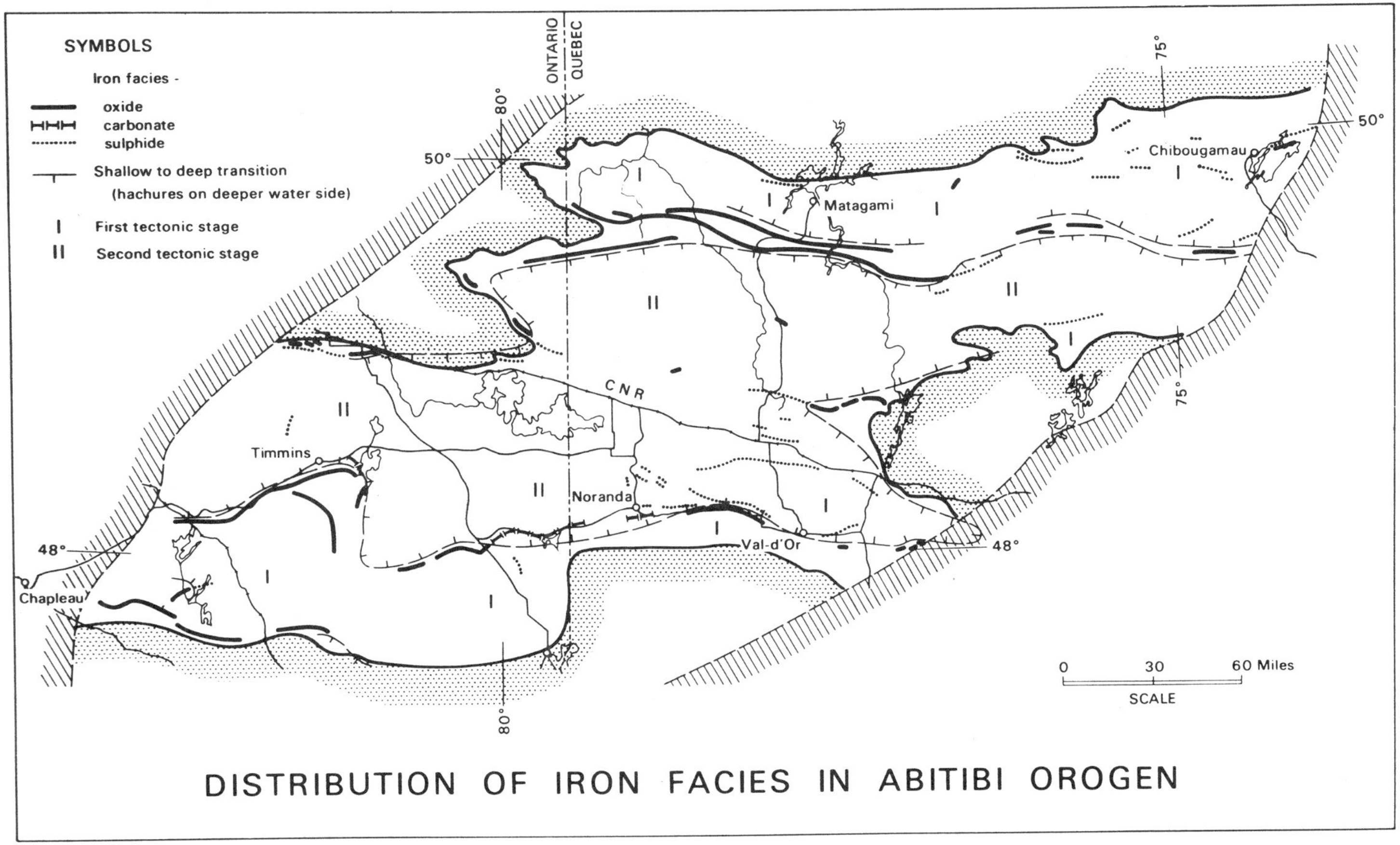

Figure 2. Distribution of facies in Abitibi orogen. Oxide-carbonate-sulphide facies of iron formation transitions delineate original shelf to basin slopes within the orogen. Tectonic stages refer to postulated episodic progressions in construction of the orogen. Presumed younger volcanic-sedimentary rocks are present towards the centre of the orogen.

The Bell River complex at Matagami, at least 15,000 feet thick, comprises strongly and rythmically banded norite, anorthosite, pyroxenite and associated gneissic cataclastic rocks. Another differentiated complex occurs at Kamiskotia Lake north of Timmins.

Granitic Rocks

A great variety of granitic rocks is present in the region but only a brief statement is made on the subject. Foliated to massive granitic rocks commonly underlie the margins of the region. They are associated with more or less metavolcanic and metasedimentary rocks. Where intrusive relations are apparent the granitic rocks have intruded the volcanic and sedimentary rocks and hence post-date them in terms of last emplacement. Large granitic complexes along the north and south boundaries may include primitive crystalline basement. By its nature the demonstration of the reality of such sialic basement is very difficult if indeed possible. The granitic rocks consist in large part of weakly to strongly foliated granodioritic gneiss of possible metasedimentary origin, together with varying proportions of schists, gneiss, migmatites, and undoubted igneous plutons. Based on geochemical reconnaissance of similar crystalline terrain in Ungava region of northern Quebec the average chemical composition of these crystalline rocks approaches that of common granodiorite (Eade, *et al.*, 1966).

In addition granitic plutons are widely distributed within the orogenic belt itself. At least nine major plutons each 40 to 100 miles in diameter are present, most in the northern part of the region. Additional smaller plutons enclosed within volcanic rocks are closely associated with felsic volcanic rocks e.g. Noranda, Malartic, Kirkland Lake and Chibougamau, of which they may represent coeval intrusive equivalents. The Round Lake batholith south of Kirkland Lake contains a southern part of leucrocratic quartz diorite and a northern, presumably younger part of hornblende granite. The Otto stock on the northeast is composed of syenite, nepheline syenite and locally, pegmatitic phases. To the east, in the Malartic area, the Bourlamaque batholith comprises a core of quartz diorite and a border of albite granite. The LaCorne-La Motte group of plutons in the north Malartic area, approximately 40 miles long by 16 miles wide, comprises respectively hornblende-, biotite-, and muscovite-granite phases, the last named associated with spodumene-bearing pegmatites and molybdenite mineralization. To the west the Taschereau, Palmarolle and Flavrian plutons comprise cores of pink granodiorite and borders of grey quartz diorite. Many other types of plutons are present in the belt of which the above provide some example.

Radiometric ages of various intrusions in the Kirkland Lake area, for example, indicate that plutonic activity has occurred there repeatedly from Archean to Mesozoic time. Particular examples are the Round Lake batholith (2,400 million years), the Otto stock (1,700 million years) and the Kimberlite dikes in the Upper Canada Mine which are ascribed to a Jurassic-Cretaceous age (Ridler, 1970). Clearly there are major problems concerning the times and mechanisms of igneous emplacement as well as the origin of sialic material contained in the plutons.

Lithic Proportions

A planimeter survey of the geologic map (Fig. 1) including only those rocks within the orogenic belt (Fig. 3) provides an assessment of lithic composition. Areas of cover rocks of Proterozoic and Paleozoic ages totalling 1,268 sq. mi. have been excluded. The belt has been divided into southern and northern parts with the main CN railway as the dividing line.

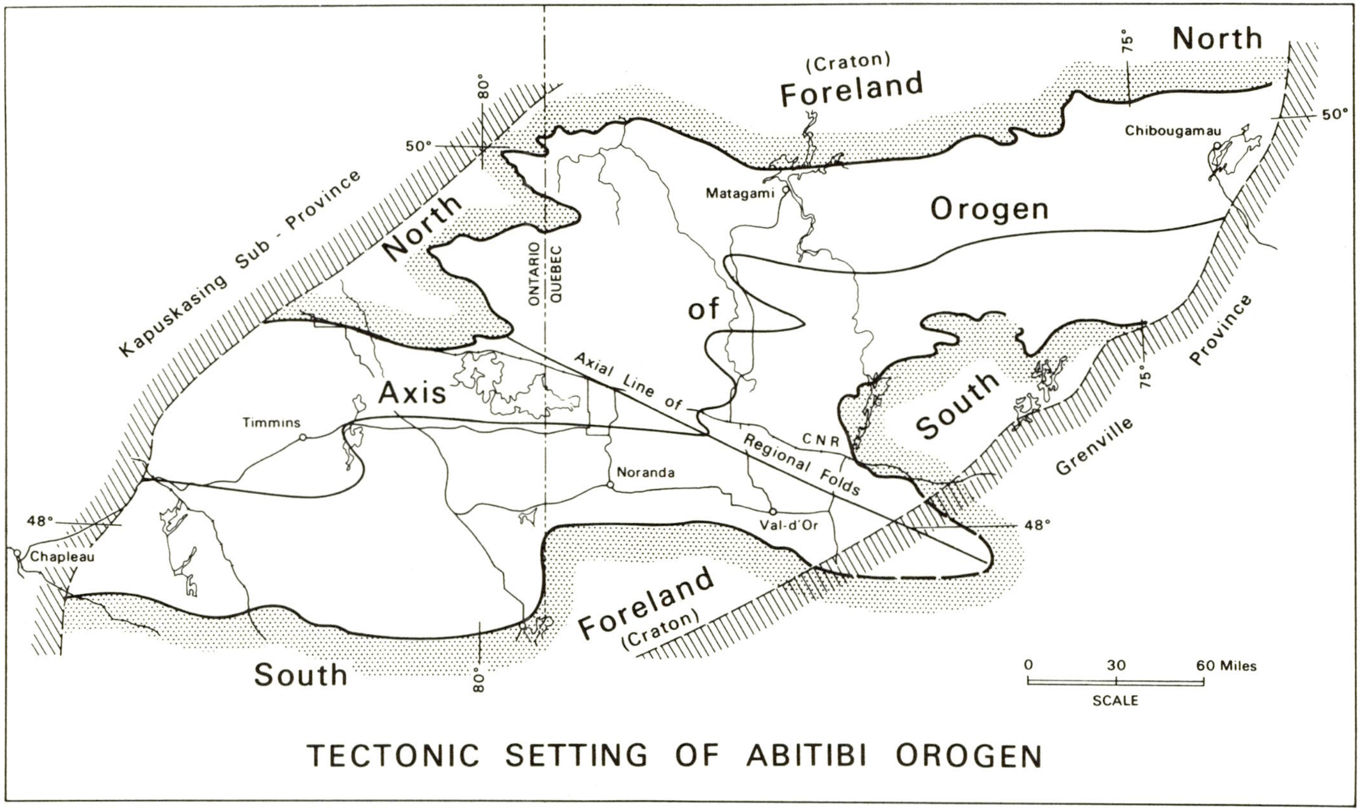

Figure 3. Tectonic setting of Abitibi orogen. Boundaries of the forelands are defined on the basis of preponderant crystalline granitic rocks. S-shaped form of the orogen is thought to be due to a structural deformation.

Abitibi Orogen

	Southern Part	Northern Part	Total
Area	15,948 sq. mi.	20,811 sq. mi.	36,759 sq. mi.
Granitic rocks	26.2 per cent	37.0 per cent	32.3 per cent
Mafic volcanics	47.5 "	44.2 "	45.6 "
Felsic volcanics	5.5 "	2.2 "	3.6 "
Sediments	18.8 "	13.8 "	16.0 "
Mafic intrusions	2.0 "	2.8 "	2.5 "
	100.0 per cent	100.0 per cent	100.0 per cent

Recalculating to Archean supracrustal rocks only (i.e. excluding granitic rocks and mafic intrusions) the following lithic proportions are indicated to be present in the belt:

	Southern Part	Northern Part	Total
Mafic volcanics	66.2 per cent	73.5 per cent	70.0 per cent
Felsic volcanics	7.6 "	3.6 "	5.5 "
Sediments	26.2 "	22.9 "	24.5 "
	100.0 per cent	100.0 per cent	100.0 per cent

Thus the indicated ratio mafic volcanics: felsic volcanics is 13:1. Stated otherwise, felsic volcanic rocks (dacite, rhyodacite, rhyolite and minor trachyte) form an indicated 5.5 per cent of the regional volcanic assemblage. The figures must be viewed with caution considering the extensive drift cover in the northern parts particularly. Past experience suggests that further field studies in the region will reveal additional bands of felsic volcanic rocks, and result in reclassification of some sediments as felsic volcanic rocks. In the meantime the figure of 5.5 per cent felsic volcanics is offered as a working minimum, the true figure possibly approaching 7 to 10 per cent felsic volcanic rocks.

The northern part of the region includes substantially higher proportions of internal granitic rocks in the form of large batholiths and less felsic volcanic rocks. Despite the note of caution expressed in the previous paragraph the stated proportions are considered to reflect a genuine difference in lithic proportions between the northern and southern parts of the region. Major problems remain concerning the origin of the sialic material and the time and mechanism of emplacement of these large northern batholiths.

Relatively more sediments are present in the southern part of the orogen (i.e. 18.8 vs 13.8 per cent). This reflects inclusion within the orogen of Pontiac sediments - the main flyschoid facies - situated south of Noranda-Malartic-Val d'Or (Fig. 1). Present indications are and future studies may prove that this facies was derived from a craton to the south and should more properly be included in the southern foreland (Fig. 3). However in order to be consistent in defining the boundaries of the forelands i.e. preponderant granitic rocks, the Pontiac facies has been included in the orogenic belt for present purposes.

TECTONIC SETTING

The tectonic setting and principal parameters of the Abitibi orogen are illustrated in Figure 3. The orogen lies between two northeast-trending boundaries, the Grenville province on the east and the Kapuskasing subprovince

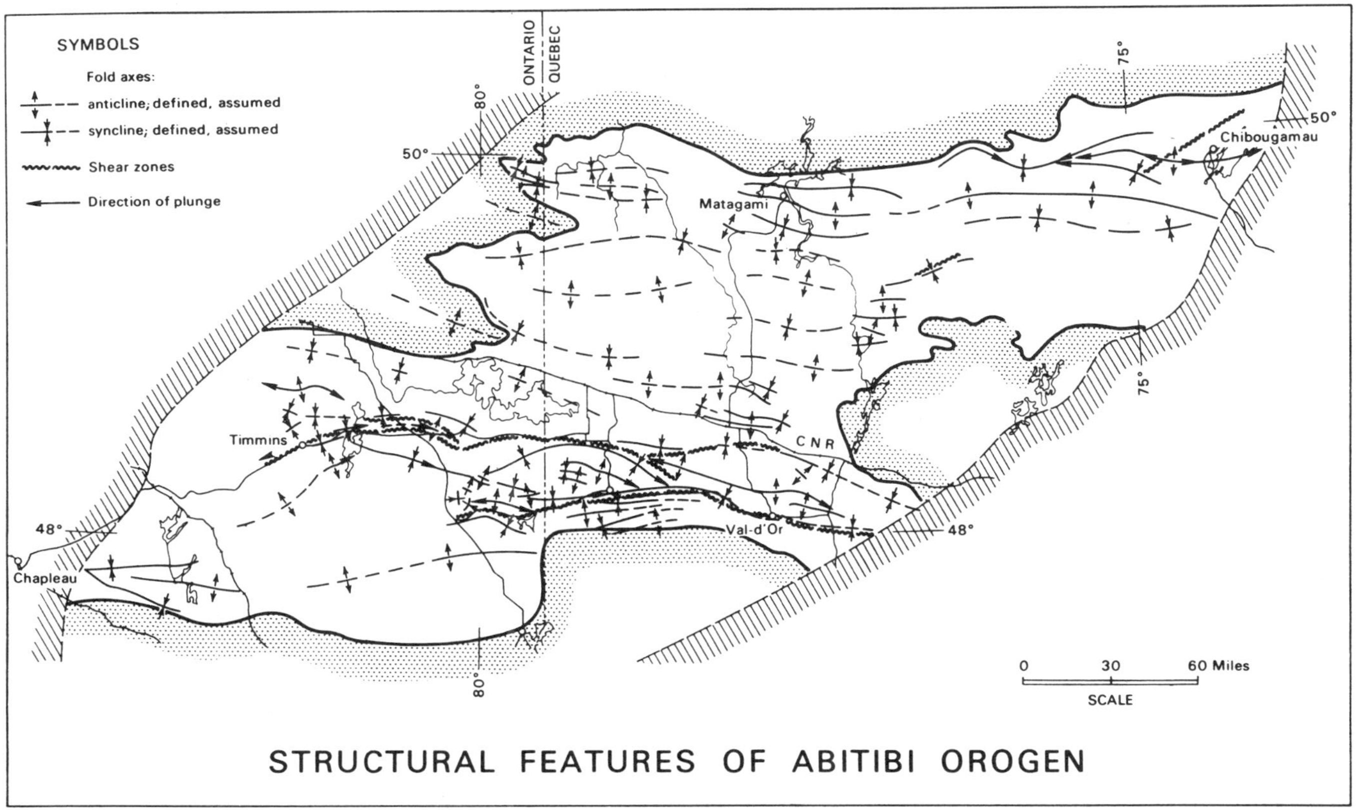

Figure 4. Structural features of Abitibi orogen. Additional folds and shear-zones may be present in drift-covered parts, especially in the north.

on the west. The orogen proper lies between two postulated cratonic forelands, the boundaries of which are defined on the basis of preponderant granitic rocks. Foreland boundaries are in fact gradational from predominantly supracrustal rocks within the orogen to predominantly crystalline complexes without. Also the foreland complexes locally include substantial masses of metavolcanic and metasedimentary rocks. This is particularly so in the north foreland as well as in the western part of the south foreland. The axis of the orogen has been placed midway between the forelands with due regard to internal lithic distribution.

Accordingly the Abitibi orogen constitutes a generally east-trending tectonic unit approximately 500 miles long and 60 to 100 miles wide. Lithologically it features orogenic supracrustal rocks, both volcanics and sediments, together with prominent granitic plutons. It is intracratonically contained between predominantly granitic forelands. The present S-shaped form reflects structural deformation.

STRUCTURE

The first-order structure of the region is an S-shaped regional fold which is delineated by the boundary surface of the orogen (Fig. 3). Thus the north boundary surface extends from Chibougamau vicinity on the east to Bradburn township north of Timmins on the west, and the south boundary surface from Marceau township situated south of Doda Lake on the east to the vicinity of Chapleau on the west. Two conspicuous second order S-folds which form part of the regional fold are present on both boundary-surfaces. Those on the north boundary surface protrude eastward in the area north of Lake Abitibi near the Ontario-Quebec interprovincial boundary. Those on the south boundary surface lie respectively northeast of Val d'Or and southwest of Noranda the latter indicated by the conspicuous southward trend of rock units south of Larder Lake. The axial line of the regional folds trends northwesterly (Fig. 3).

Folding of the compositional layering within the orogen is widespread (Fig. 4). The folds have east-trending axial lines, are characteristically isoclinal, doubly plunging, and commonly bifurcate along strike. Those to the north of the CNR line commonly plunge to the west e.g. Chibougamau and Matagami whereas those to the south commonly plunge to the east e.g. Noranda, Malartic, Kamiskotia. On a regional scale granitic plutons are preferentially distributed along anticlines and supracrustal rocks along synclines. On a local scale synclines commonly bifurcate to enclose an anticline which includes central plutons e.g. Noranda, Malartic, Matagami and Chibougamau. In the west Malartic area, volcanic (Malartic group) and sedimentary (Kewagama) rocks have been folded into a west-facing anticline; these rocks underlie younger volcanic rocks situated to the west and north.

Several prominent, steeply inclined, east-trending, discontinuous shear-zones of undetermined displacement have been identified in the southern part of the region e.g. Porcupine-Destor and Larder Lake "breaks". They follow lithofacies boundaries for the most part, including sedimentary-volcanic interfaces. Other regional shear zones of this type may be present particularly in drift-covered areas to the north. Recent field studies of the Larder Lake "break" cast doubt on its continuity and structural significance (Ridler, 1970). This suggests that a re-examination of the entire problem of the regional "breaks" is in order.

Comparison with modern volcanoes in island arc settings (e.g. Solo zone in East Java, *see* Van Bemmelen, 1949, Fig. 16, p. 26) suggests that the present elliptical outline of the Abitibi volcanic complexes (Fig. 5; *see* page 12) is due to foreshortening in a north-south direction of originally circular volcanic complexes. The present axial ratios of the volcanic complexes average 5:1. This implies that originally circular complexes have been foreshortened by at least 50 per cent. If the complexes were also foreshortened

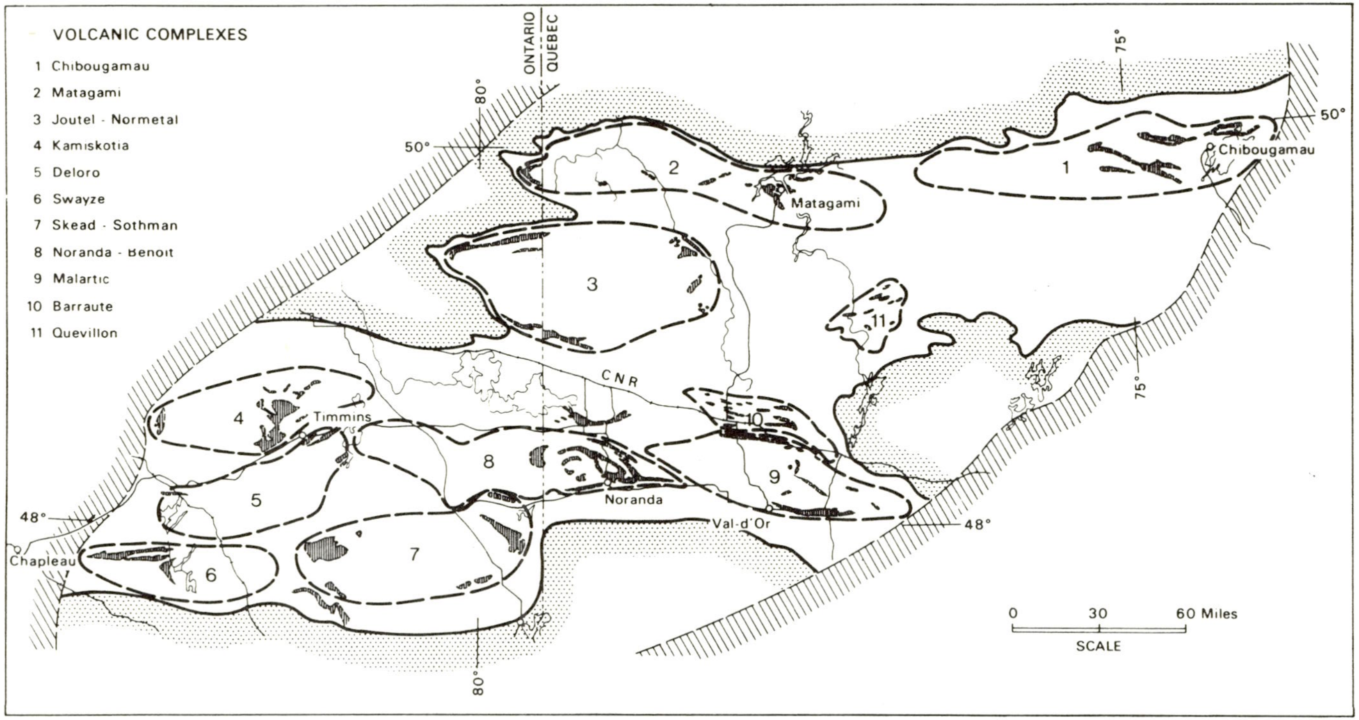

Figure 5. Distribution of volcanic complexes in Abitibi orogen. Felsic volcanic rocks shown by vertically lined pattern. Approximate boundaries of volcanic complexes including cogenetic intrusions and sediments is shown by heavy dashed lines. Present elliptical shape of presumably originally circular volcanic complexes is attributed to compression folding.

in an east-west direction as is suggested by the presence of doubly-plunging east-trending fold axes then the amount of north-south foreshortening may have substantially exceeded 50 per cent.

The present elliptical shape of the volcanic complexes together with closely spaced isoclinal folds suggests that the mechanism of deformation was compression folding. Such deformation would also produce substantial vertical extension of the rocks. Thus Archean orogenic assemblages originally 35,000 feet thick (10 km) may have been increased in thickness to approach that of the existing continental crust (40 km). More definite statements on the mechanics and state of deformation would require additional structural studies in the field.

VOLCANIC COMPLEXES

The regional stratigraphy is dominated by the presence of semi-independent ellipsoidal volcanic-sedimentary domains each of which is termed a volcanic complex. Nine major and two minor volcanic complexes have been delineated in the region (Fig. 5) each with a mafic to felsic volcanic sequence, associated intrusions and sediments.

The best known domain is that of the Blake River group herein called the Noranda-Benoit complex (number 8 in Fig. 5). The lowermost identifiable strata of this group which are of prevailing mafic composition extend from Nighthawk Lake situated 15 miles east of Timmins, 1) eastward to north Noranda area, thence southeastward to Malartic, and 2) southeastward to Kirkland Lake, thence eastward to Malartic, the two boundary lines thus outlining the complex. An identifiable mafic to felsic volcanic sequence with associated intrusions and sediments is contained within this boundary. The single felsic volcanic centre lies at Noranda in the eastern part of the complex. Although lateral extensions of Blake River rocks may fall outside this boundary the bulk of the preserved assemblage appears to lie within it.

Similarly, other volcanic complexes have been delineated with the felsic volcanic concentrations of the region serving in their identification (Fig. 5). Some boundaries have been tentatively defined only e.g. terminations of Skead-Sothman (#7), Matagami (#2), Chibougamau (#1) and Kamiskotia (#4) complexes. Barraute complex (#10) may be a structural extension of Malartic complex (#9). Quevillon complex (#11) is of uncertain definition as are the mutual boundary areas of Deloro (#5), Swayze (#6) and Skead-Sothman (#7) complexes. Other as yet unidentified complexes may be present in the region. Modifications of existing boundaries are to be expected on the basis of future work.

Despite these problems of boundary definition it is suggested that the pattern of complexes illustrated in Figure 5 accurately reflects the style and format prevailing during construction of the Abitibi orogen - a constructional style dominated by development of numerous semi-independent mafic to felsic volcanic piles with coeval intrusions and sediments. Analogy with recent and modern volcanic piles including those of linear tectonic (especially island arc) association suggests that the Abitibi volcanic complexes originally had circular outlines, each apparently in the order of 60 to 100 miles diameter. Their present elliptical outlines are attributed to structural deformation.

The defined volcanic complexes of the region are concentrated along the northern and southern boundaries of the orogen. Thus comparison of Figures 3 and 5 shows that volcanic complexes 1 to 4 lie north of the axis of the orogen and close to the northern boundary, whereas complexes 6 to 11 lie south of the axis of the orogen and close to the southern boundary. This pattern may reflect original linear distribution of strato-volcanic complexes along tectonic structures such as geanticlinal uplifts essentially parallel to the margins of the developing orogen.

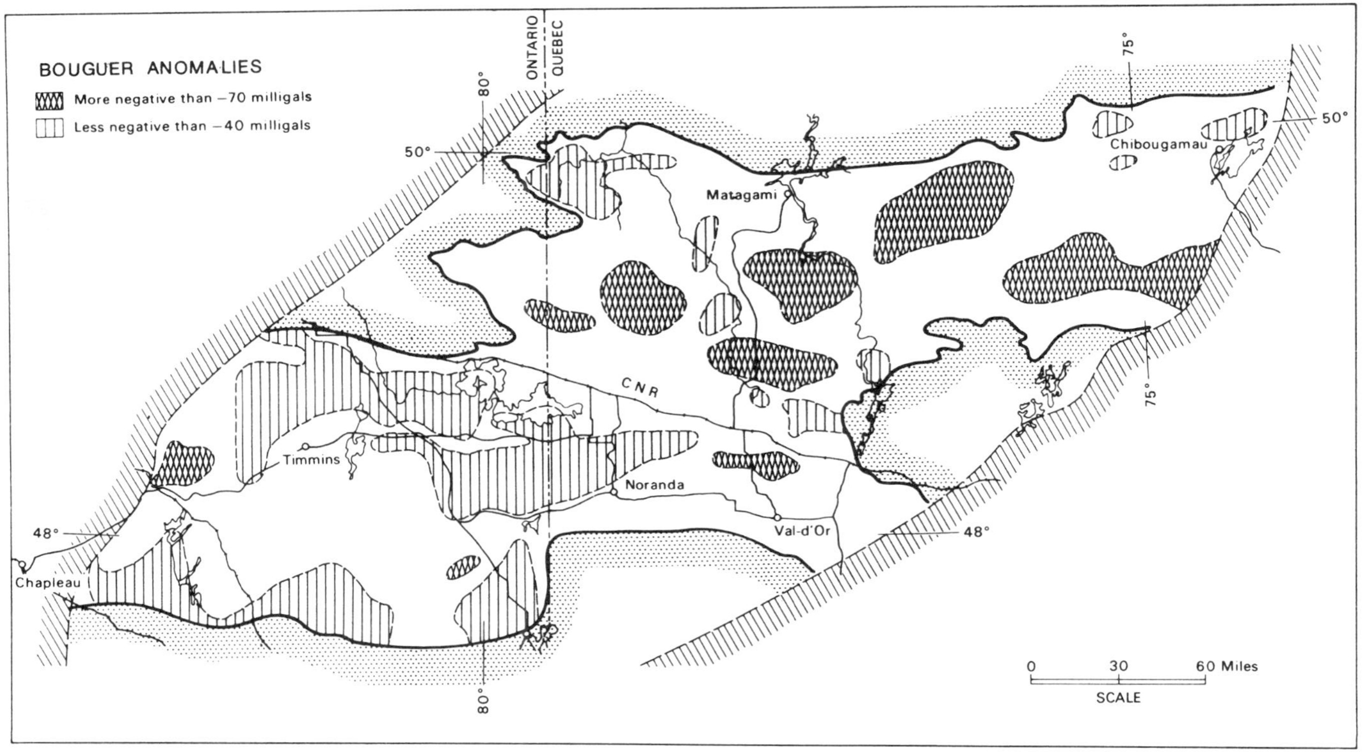

Figure 6. Gravity features in Abitibi orogen. Areas between the designated Bouguer anomalies have intermediate gravity expressions of -70 to -40 milligals. Bouguer anomalies more negative than -70 milligals correspond to granitic batholiths. Anomalies less negative than -40 milligals correspond to large areas of mafic igneous rocks.

GRAVITY

Gravity and geologic features are closely related in Abitibi orogen. This may be seen by comparing the Bouguer anomaly map (Fig. 6, which is based on Gravity Map of Canada GMC 67-1) and the geologic map (Fig. 1). The close correspondence of gravity and geologic features indicates that the regional gravity measurements summarized in the gravity map substantially reflect surface and near-surface crustal features rather than deep-crustal or mantle features.

Six principal high negative (less than - 70 milligals) Bouguer anomalies north of the CN railway coincide remarkably well with the main granite batholiths of the region. By demonstrating the presence in the crust of substantial bodies of low density material, they indicate that the batholiths have deep roots. Three high negative anomalies south of the CN railway coincide with smaller granitic batholiths (near Val d'Or, Kirkland Lake and Timmins). In contrast several granitic batholiths within the orogen do not have comparable high negative gravity expressions e.g. Matachewan, Timmins, Gogama areas. Such batholiths may terminate at shallow depths or alternatively, contain igneous material of intermediate or assorted densities.

The principal low negative (greater than - 40 milligals) Bouguer anomalies coincide with 1) major mafic volcanic-intrusive assemblages and 2) flat-lying Proterozoic diabase along the southwestern margin. For example, thick, extensive mafic assemblages are reflected by major low negative anomalies northwest of Noranda (Blake River group mainly) and north of Timmins. Four local mafic assemblages are reflected by correspondingly small, low negative gravity anomalies northwest of Senneterre, between granitic batholiths near Joutel-Poirier, west of Matagami, and near Chibougamau.

All other parts of the orogen, which yield an intermediate gravity expression (between -70 and -40 milligals) are underlain apparently by mixed lithic assemblages which collectively lack an anomalous gravity expression.

Gravity relationships clearly demonstrate that the part of the orogen north of the main CN railway contains less total mass (i.e. more low-density rock) than the southern part. This is attributed mainly to the presence of numerous granite batholiths with deep roots and proportionately less mafic igneous rock in the north. In addition some undisclosed deeper crustal influence may be reflected.

It is particularly significant that high negative Bouguer anomalies (less negative than -70 milligals) coincide with internal granite batholiths only and not with the regional granitic complexes of the northern and southern forelands as defined in Figure 3. This provides a ready method of differentiating true granitic batholiths from the mixed lithic predominantly granitic assemblages of the forelands. In this connection two high negative Bouguer anomalies, each presumably an expression of a granite batholith, lie outside the orogen; and within the bordering crystalline complexes. One lies in north foreland rocks, 50 miles north of Matagami and the other in Grenville province rocks 50 miles southeast of Chibougamau (*see* Gravity Map of Canada). With the exception of these two high negative areas the forelands have intermediate gravity expressions suggesting the presence of mixed crystalline complexes of assorted gneissic-migmatitic-plutonic composition.

METALLOGENESIS

Type and Distribution of Mineralization

The region contains several thousand known mineral occurrences including 150 producers (past and present) of which 57 lie in Ontario and 93 in

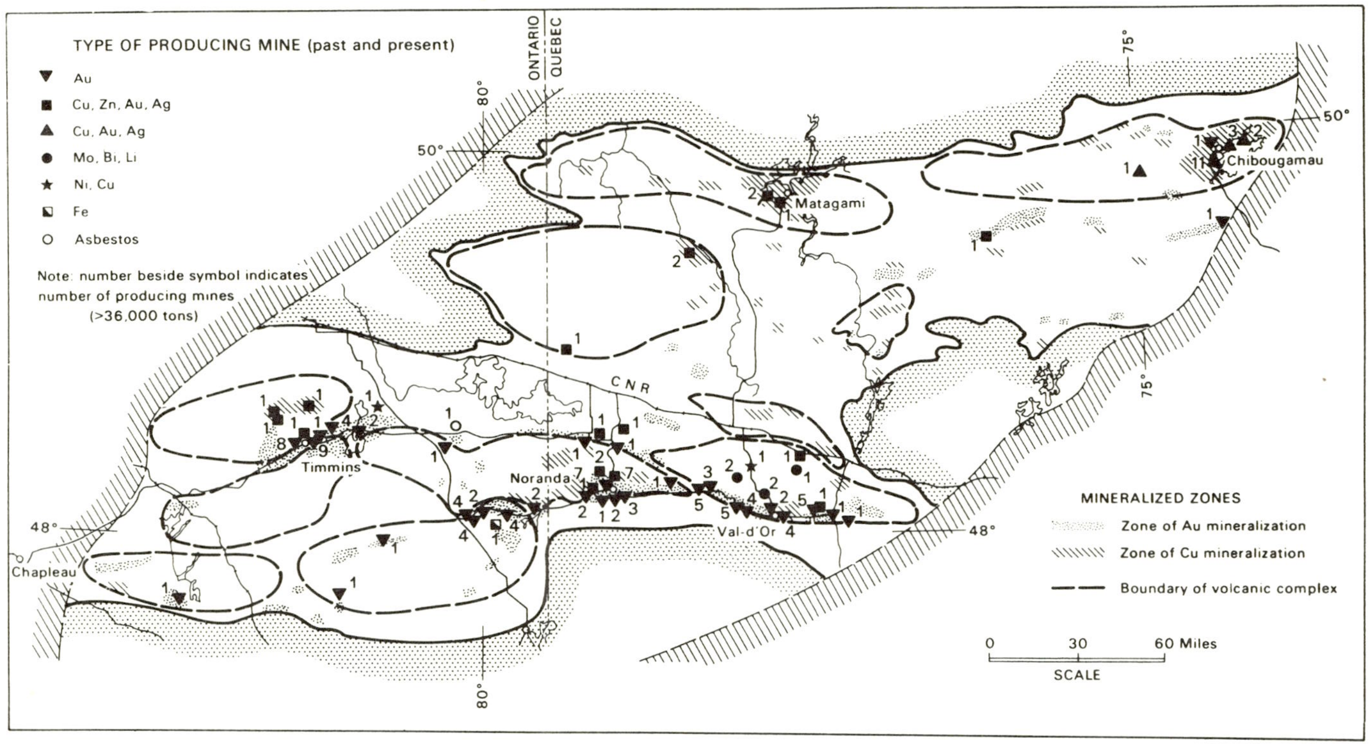

Figure 7. Metallogenic relations in Abitibi orogen. Distributions of main Au and Cu mineralized zones and of producing mines (greater than 36,000 tons; past or present) are shown relative to volcanic complexes.

Quebec. Most occurrences are either gold-bearing quartz veins or disseminated to massive sulphide (Cu-Zn-Au-Ag) concentrations. Others include magnetitic iron formation, asbestos, Ni-Cu, Zn-Ag-Pb, Mo-Bi and Li.

The main mineralized zones based on known occurrences are shown in Figure 7. A principal east-trending Au zone extends intermittently from Sothman-Matachewan area on the west through Kirkland Lake, Larder Lake, Noranda and Malartic to Val d'Or area on the east. A second well-established Au zone extends eastward from Timmins vicinity through Duparquet thence curving southeastward to join the first zone at Malartic. Smaller discontinuous Au zones are widely distributed in the region, for example, in the Swayze, south Timmins, Benoit and north Malartic areas; north of the CN railway in the Doda-Opawica area, and southwest of Chibougamau and Quevillon areas. Principal zones of Cu mineralization, on the basis of producers and known occurrences, lie in north Timmins, Noranda, Malartic-Val d'Or, Joutel, Matagami and Chibougamau areas. Smaller zones lie west and south of Matagami, west of Chibougamau and in the Barraute-Quevillon area. Several other small isolated mineralized patches without obvious pattern are present in the region as illustrated in Figure 7.

Lithic Relationships

Most Au deposits lie in intermediate to felsic volcanic rocks, their intrusive equivalents or associated clastic or exhalative sediments. Included are those at Swayze, Matachewan, Kirkland Lake, Timmins, Cadillac, Malartic and Val d'Or. Sulphide deposits of the Cu-Au and Cu-Zn-Ag types are associated with intermediate to felsic volcanic complexes at Noranda, Joutel, Timmins and Matagami. Mo-Bi and Li deposits are associated with pegmatitic rocks in north Malartic area. A single iron producer lies in banded iron formation enclosed in felsic tuff and mafic lavas south of Kirkland Lake. Asbestos deposits are associated with differentiated ultramafic complexes north of Matheson and west of Timmins. Ni-Cu deposits lie in mafic to ultramafic intrusions near Cochrane, south of Timmins and north of Malartic. Cu-Ag-Au deposits are associated with gabbroic, noritic and anorthositic intrusions of volcanic association at Chibougamau (Duquette, in press) and Cu-Zn-Ag deposits with anorthositic intrusions at Matagami (Sharpe, 1965).

Relation to Volcanic Complexes

As illustrated in Figure 7 most mineralized zones including all but three of the 150 producers lie within or marginal to the designated volcanic complexes. Principal Au zones lie at the margins of complexes and have traditionally been interpreted as being primarily associated with "shear-zones" e.g. Porcupine-Destor and Larder Lake "breaks". For example, one Au zone extends from Matachewan through Kirkland Lake at the northern margin of Skead-Sothman complex (#7 in Fig. 5) eastward along the southern margins of the Noranda-Benoit (#8) and Malartic (#9) complexes; a second Au zone extends from the mutual boundary of Kamiskotia (#4) and Deloro (#5) complexes eastward along the northern and northeastern margins of the Noranda-Benoit (#8) complex. Other, leaner, Au-bearing zones are present in the western parts of Swayze (#6) Skead-Sothman (#7) and Deloro (#5) complexes. In addition, several small Au zones without obvious relation to volcanic complexes occur in the area southwest of Chibougamau and north of the main CN line.

Most Cu-bearing sulphide zones lie in the volcanic complexes directly associated with felsic volcanic rocks. They are most common in the eastern parts of the following six volcanic complexes: Chibougamau (#1), Matagami (#2), Joutel-Normetal (#3), Kamiskotia (#4), Noranda-Benoit (#8) and Malartic (#9). Cu-Au and Cu-Zn-Ag producers are present in Matagami (#2), Joutel-

Normetal (#3), Kamiskotia (#4), Noranda-Benoit (#8) and Malartic (#9) complexes whereas Cu-Au-Ag producers are present in Chibougamau (#1) complex. Other small local zones of Cu mineralization occur in Barraute (#10) and Quevillon (#11) complexes, in the central and western parts of Matagami (#2) and Chibougamau (#1) complexes and elsewhere north of the main CN railway as illustrated in Figure 7.

Mo-Bi and Li occurrences are associated with pegmatitic phases of felsic intrusions in the centre of Malartic (#9) complex. Magnetitic iron formation is associated with volcanic tuff and mafic lavas in northeastern Skead-Sothman (#7) complex. Ni-Cu producers lie in ultramafic intrusions in or near Malartic (#9) and Noranda-Benoit (#8) complexes.

Lithic associations common to sulphide deposits are felsic volcanic concentrations reflecting central vent eruptions, sulphide facies iron-formation reflecting reducing environments, and absence or scarcity of coarse clastics reflecting offshore, deeper water sites of accumulation. Such deeper water environments may be attributed in part to tectonic collapse including cauldron subsidence following felsic volcanic discharge. Conversely some lithic associations are inimical to sulphide occurrences: e.g. oxide facies iron formation and clastic sediments, especially conglomerate. This association reflects topographically higher, shallow water, partially oxidizing shelf-environments. Accordingly, most mineralized sulphide zones in the region are directly associated with felsic volcanic concentrations having sulphide facies iron formation nearby e.g. Noranda, Malartic, Matagami and Chibougamau. On the other hand, many Au occurrences are directly associated with carbonate facies iron formation either at felsic volcanic-intrusive centres or at the margins of volcanic complexes.

Stratigraphic Relations

Cu-Zn-Au-Ag concentrations have the most direct stratigraphic relationship. Many lie at specific volcanic contacts, commonly felsic-mafic transition in upper parts of stratigraphic successions e.g. Kamiskotia, Kidd Creek Noranda, Matagami, Joutel and Malartic areas. Cu-Au-Ag deposits lie along fractures at the northern margin of a volcanic-enclosed, anorthositic complex at Chibougamau. Some Ni-Cu deposits lie in ultramafic intrusions either within stratigraphically lower mafic volcanic rocks or upper felsic volcanic rocks e.g. Marbridge and Alexo mines.

Au deposits have variable degrees of stratigraphic associations. One family of deposits has endogenous relations to felsic alkaline intrusions in the form of subvolcanic sills, discordant plugs or fairly large stocks. The gold occurs disseminated in irregular zones often associated with pyrite or concentrated into various types of vein structures. Although the plutons are believed to be cogenetic volcanic equivalents, and therefore Archean in age, it is possible that post-Archean igneous events have been accompanied by gold mineralization. Examples are the gold mines of Kirkland Lake, Granada, Yonge-Davidson, Eldrich, Powell-Rouyn and Sullivan mines. Flow and pyroclastic equivalents of the above class of plutons may be enriched in gold, either disseminated or in vein form. Examples are Dome, Hollinger and McIntyre mines of the Timmins area and certain parts of Upper Canada mine east of Kirkland Lake. Gold deposits intimately associated with carbonate facies iron formation include Omega and Kerr Addison (Ridler, 1970); those with interflow deposits include Wasamac mine west of Noranda and Dome deposit of Timmins area. Gold deposits of possible placer derivation in whole or in part include Pamour mine, lying in quartzite east of Timmins and McWatters mines in conglomerate south of Noranda, each situated in the upper part of the local stratigraphic succession. In addition, many of the gold deposits lie in dilatant shear and fracture zones into which the precious metal together with accompanying quartz appears to have been concentrated from the foregoing original stratigraphic sites during brittle deformation. This has partly obscured primary stratigraphic distribution and has produced a distribution pattern which has been traditionally related to a regional fracture system.

Finally, magnetitic iron formations are enclosed in volcanic rock in upper stratigraphic parts of local successions e.g. Adams mine at Kirkland Lake, whereas asbestos deposits lie in differentiated mafic sills and complexes towards the base of local stratigraphic successions, e.g. Munro mine north of Matheson.

TECTONIC INTERPRETATION

It has been established that Archean rocks of the Abitibi orogen have been structurally deformed to a considerable degree. Relations indicate that component lithofacies have been tightly folded about east-trending axes, resulting in substantial north-south shortening. Regional faults may be present. Deformations of this type have combined to obscure original tectonic relationships. Despite these reservations and uncertainties which are common to Archean greenstone belts in general, a tectonic reconstruction based on available data is offered for consideration as a working hypothesis.

The nature, distribution and composition of volcanic and sedimentary rocks points clearly to the presence in Archean time of a mobile belt or tectonically unstable, linear basin complex. Archean mafic to felsic volcanic assemblages have their modern analogues in orogenically active, thin-crustal environments such as those of island arcs which are characteristic of continent-ocean interfaces. Archean sediments, notably greywacke, conglomerate and breccia, are of the high-energy, "poured-in" turbidite association. Thus tectonic mobility is a clearly indicated characteristic of the orogen. Further, the present location of the east-trending orogenic belt between adjoining predominantly granitic terrains suggests that the orogen developed between two sialic forelands. The presence of flyschoid clastics (Pontiac group) in Noranda-Val d'Or area derived from a granitic provenance to the south (Holubec in preparation), reinforces this interpretation of the presence of a sialic foreland south of Noranda during Abitibi time. Therefore, we deduce that Abitibi orogen developed intracratonically upon thin mobile crust, between adjoining forelands of predominantly sialic composition. The apparent absence of flyschoid facies along the northern margin of the orogen and along the western part of the southern margin, together with the presence of some volcanic and sedimentary rocks within the adjoining predominantly granitic terrains suggest that comparatively subdued sialic forelands of mixed lithic composition existed in those parts during Abitibi time.

Two elements of regional stratigraphy are used to reconstruct tectonic development of the orogenic assemblage, namely facies transitions of iron and relative ages, as known, of specific volcanic complexes.

The main shelf-to-basin transitions in the orogen, based on iron formation facies-relationships, extend east-west across the northern and southern parts of the region. In addition other local transitions are present. The transitions, in effect, separate common shelf-type oxide facies iron formation, present along the northern and southern parts of the orogen, from deeper water or basinal sulphide facies iron formation towards the centre of the orogen. These relationships, in brief, point to the presence of tectonic basin(s). Successive basins may have been present during progressive development of the orogen. The local presence of sulphide facies iron formation near Chibougamau and Matagami may be due to cauldron subsidence or to some other form of local tectonic collapse.

Detailed stratigraphic studies in Kirkland Lake area (Ridler, 1970) indicate that an older volcanic assemblage (Skead group) mainly to the south of Kirkland Lake proper underlies younger Blake River volcanic rocks to the north (*see* Figs. 1 and 5); that older Skead volcanics constituted a stable shelf

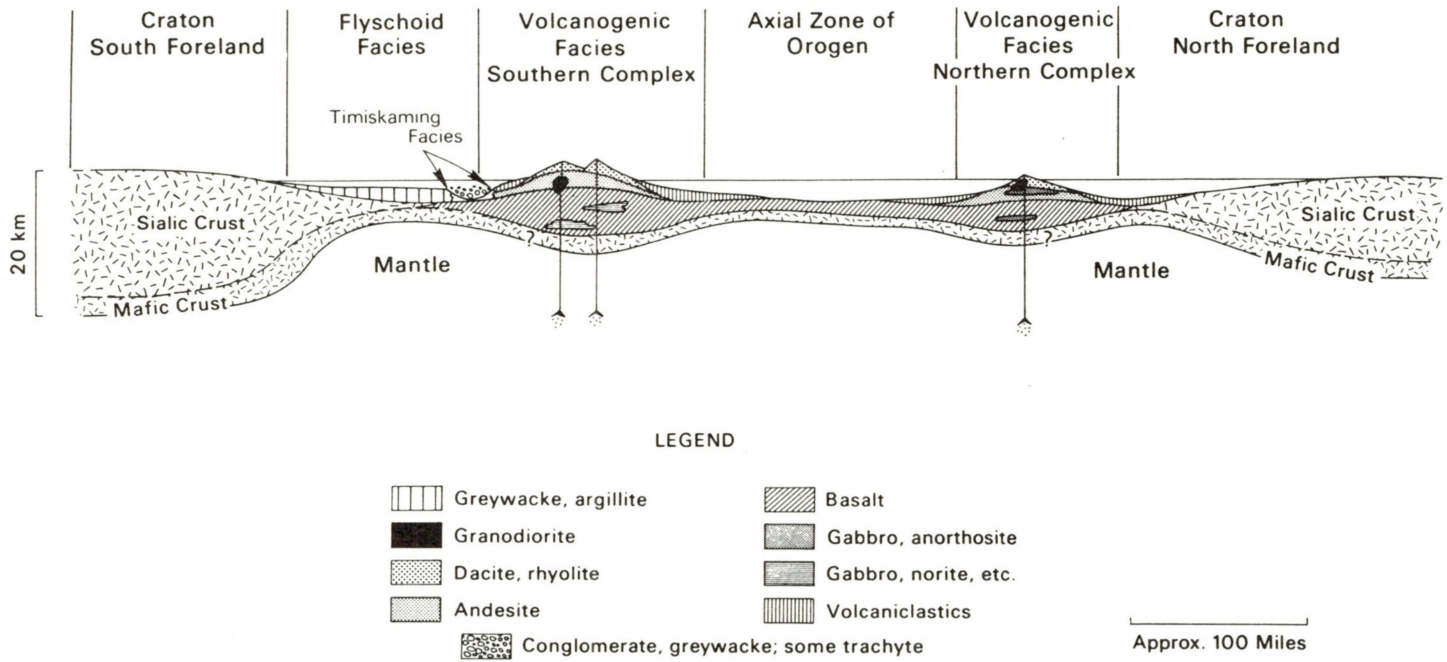

Figure 8. Hypothetical tectonic reconstruction of Abitibi orogen. Vertical cross-section. Length of the orogen is approximately 500 miles. Distance between southern and northern volcanogenic facies may have been in the order of 200 miles. Vertical scale is approximate only. The line of the section is intended to cross Noranda in the south and Matagami in the north. The smaller size and relief of the northern sialic foreland is indicated by the absence of flyschoid facies along its southern margin. The nature and thickness of mafic crust underlying the orogen is conjectural. In the absence of direct evidence of sialic contribution to the volcanic and sedimentary assemblages of the orogen (except for the flyschoid facies, i.e. Pontiac, at the south margin), no subjacent sialic crust is shown as being present, at least during the early development stages of the orogen.

relative to the Blake River basin of mafic volcanic accumulation to the north; and futhermore that the shelf-to-basin transition is marked by oxide- and carbonate-facies iron formation. In the Timmins area the east-trending iron formation lies between underlying Deloro volcanics to the south and what is generally interpreted as younger Tisdale volcanics to the north. Malartic volcanic and sedimentary (Kewagama) rocks stratigraphically underlie volcanic rocks to the west (Blake River) and to the north.

Therefore it is suggested that the main shelf-to-basin transitions of the region, marked by iron formation, delineate volcanic assemblages of slightly different ages. The younger assemblages are interpreted as occupying the deeper water side, towards the basin and hence mainly towards the axis of the orogen. Accordingly, the following volcanic complexes and rock groups appear to possess some degree of contemporaneity: in the south, Deloro (#5 in Fig. 5), Swayze (#6), Skead-Sothman (#7) and Malartic (#9) complexes and Pontiac sediments; in the north, Matagami (#2) and Chibougamau (#1) complexes. Somewhat younger and bearing a similar degree of contemporaneity, on the same basis, are Kamiskotia (#4), Noranda-Benoit (#8) and Joutel-Normetal (#3) complexes together with associated belts of sediments. Thus the postulated sequence of events is one of episodic progression in tectonic development from orogenic margins to the axis of the orogen. This working hypothesis requires detailed testing. In this regard radiometric dating, isotopic and paleomagnetic studies, and detailed stratigraphic studies may be fruitful.

Most of the volcanic complexes with their felsic concentrations lie close to the boundaries of the orogen. All felsic concentrations lie within 40 miles of a foreland boundary and all but two (Kamiskotia and Joutel) lie within 25 miles. Thus the main bulk of defined felsic volcanic rock is situated near the margins of the orogen. Because most mineral deposits (excluding asbestos and Ni) are closely associated with felsic rocks, it follows that metallogenic patterns bear a similarly close spatial relationship to foreland boundaries. The two exceptions referred to above are in 1) the Timmins area where major Au and Cu-Zn deposits lie 40 and 30 miles respectively from the northern foreland boundary and 2) the Joutel Cu-Zn deposit approximately 40 miles east of a folded spur of the same boundary. However, rock assemblages in both of these areas are highly deformed and extensively drift-covered so that their true stratigraphic positions relative to original foreland boundaries are difficult to interpret.

Within the limitations imposed by the existing state of deformation, drift-cover and knowledge of stratigraphic relations the main tectonic features of the orogen have been reconstructed and presented diagrammatically in Figure 8. Thus the orogen is considered to have developed intracratonically between two sialic forelands. Accumulation of mafic to felsic volcanic complexes proceeded, presumably under deep-fracture control, in proximity to sialic forelands i.e. at thick-thin crustal interfaces. Igneous differentiation produced effusive products ranging from basalt through andesite to rhyolite, in that general order of abundance. The role of sialic contamination in development of felsic differentiates has not been adequately tested but may have been slight. Within the orogen and remote from sialic forelands, tholeiitic basalt was apparently the predominant volcanic effusive, to the exclusion of felsic differentiates. This apparent restriction of felsic differentiates to the vicinity of sialic forelands may reflect, in addition to deep-fracture control at thick-thin crustal interfaces, a more advanced stage of igneous differentiation in high-level magma chambers at topographically higher, near-shore sites. Conversely, the central part of the orogen with predominant tholeiitic basalt and fine-grained sediments may reflect deeper water basinal sites at lower elevation resting upon thinner mafic crust.

Construction of volcanic complexes apparently proceeded in epidsodic progression, the axis of successive volcanic accumulation shifting away from forelands and towards the axis of the orogen. Accordingly younger assemblages appear to lie towards the axial zone. This distribution pattern may reflect progressive tectonic spreading of sialic forelands possibly a manifestation of crustal floor spreading in Archean time. Accordingly the apparent preference of Abitibi volcanic complexes for the margins of the forelands may correspond to the world-wide distribution of currently active volcanoes which are mainly in areas where the moving sea-floor turns down under the continents (Heirtzler, p. 9).

As previously stated, metallogenic patterns are intricately related to volcanic patterns in the orogen. Many deposits, e.g. asbestos, Cu-Ni, Mo-Bi, Li, Cu-Zn-Au-Ag and iron are directly enclosed in original lithic hosts either intrusive or extrusive. Other deposits, e.g. Au-Ag, have been involved in structural deformation with more or less migration of components to favourable sites. On a regional scale, there is little doubt that mineralization represents an integral part of the tectonic evolution of the orogen.

CONCLUSIONS

The east-trending Abitibi orogen, 500 miles long by 60 to 100 miles broad, is bounded east and west by northeast-trending crystalline rocks of the Grenville province and Kapuskasing subprovince respectively. The orogen is intracratonically contained between northern and southern predominantly granitic forelands. Its present S-shaped form reflects severe structural deformation, dominated internally by tight isoclinal east-trending folds. The original orogen constituted a much longer and in all likelihood, broader mobile belt.

Based on planimeter survey of the geologic map, the present orogen is 37,000 square miles in area. It is underlain by 32.3 per cent granitic rocks, 45.7 per cent mafic volcanics, 3.6 per cent felsic volcanics, 16.0 per cent sediments and 2.5 per cent mafic intrusions. The proportion of granitic rocks, in the form of large batholiths, is higher in the northern parts of the orogen, a fact supported by regional gravity anomalies.

Archean supracrustal rocks, of predominantly orogenic type, are products of tectonically unstable, thin-crustal, mobile environments. Mafic to felsic volcanic sequences of predominant calc-alkaline affinity correspond in significant degree to those of modern island arcs situated at continent-ocean interfaces. Accompanying clastic sediments including volcanogenic and flyschoid facies have the essential characteristics of turbidite associations including greywacke-conglomerate assemblages, graded bedding, soft rock slump structures and other evidence of rapid, high energy, "poured-in" accumulation histories.

Iron formations, widely distributed in the regional assemblage, comprise oxide, carbonate and sulphide facies. The facies are arranged across the orogen in shelf to basin transitions corresponding to original basin bathymetry.

Regional stratigraphy is dominated by the presence of numerous semi-independent, elliptical volcanic masses, each with a mafic to felsic extrusive sequence, cogenetic intrusions and sediments. Of eleven such volcanic complexes within the orogen four lie close to the northern foreland and six close to the southern foreland.

Mineral occurrences and ore deposits, mainly Au-Ag, Cu-Zn-Ag, Cu-Au-Ag, Ni-Cu, Mo-Bi-Li, asbestos and iron, preferentially lie within volcanic complexes, where they are commonly associated with felsic rocks. Cu-Zn-Au-Ag deposits favour deeper water lithic associations marked by felsic volcanic concentrations, sulphide facies iron formation and scarcity or absence of coarse clastics. Many Au-Ag occurrences are associated with shear zones in shelf facies with felsic concentrations, clastic sediments and carbonate facies iron formation. Asbestos and Ni deposits, are directly associated with mafic intrusions.

Abitibi orogen is considered to have been developed intracratonically between spreading sialic forelands. Igneous differentiation leading to development of mafic to felsic volcanic sequences occurred within the orogen under presumed deep-fracture control in proximity to sialic forelands which represented the thick-thin crustal interfaces of Abitibi time. The role of sialic contamination in development of felsic differentiates is uncertain but may have been minimal. Accumulation of supracrustal assemblages apparently proceeded in episodic progression, with the younger rocks being distributed towards the axis of the orogen.

Tectonic development may be attributed to either, 1) intracratonic ocean-floor spreading in response to mantle convection in the manner of modern plate tectonics, or, 2) withdrawal of deep-crustal support with consequent downsinking and progressive filling of the resulting mobile belt or basin complex in the manner of conventional geosynclinal development. Analogy with modern crustal architecture lends some support to the concept of Archean ocean-floor spreading. However a great deal of additional work would be required to test the hypothesis.

ACKNOWLEDGMENTS

We have drawn freely upon the following published maps in compiling the geologic map and preparing other illustrations of the Abitibi orogenic belt: Ontario Department of Mines - Timmins Kirkland Lake Sheet (Map 2046), Coral Rapids-Cochrane Sheet (Map 2161), Chapleau-Foleyet Sheet (Map 2116); Department of Natural Resources, Quebec-Noranda, Matagami, Val d'Or and Chibougamau area (Maps 1600-iii, iv and v); Geological Survey of Canada - Mineral Deposits of Canada (Map 1252A); and Observatories Branch of Department of Energy, Mines and Resources, Ottawa - Gravity Map of Canada (Map GMC 67-1).

We gratefully acknowledge the value of discussions on specific topics with G. Duquette, P. Sauvé, W.M. Schwerdtner, A.O. Sergiades, J I. Sharpe and M. Latulippe.

All illustrations were prepared by F. Jurgeneit, University of Toronto.

REFERENCES

Baragar, W.R.A.
1968: Major element geochemistry of the Noranda volcanic belt, Quebec-Ontario; *Can. J. Earth Sci.*, vol. 5, pp. 733-790.

Baragar, W.R.A., and Goodwin, A.M.
1969: Andesites and Archean volcanism of the Canadian Shield; *Proc. of the Andesite Conf., Intern. Upper Mantle Project*, Sc. Rept. 16, pp. 121-142.

Duquette, G.
in press: Archean stratigraphy and ore relationships in the Chibougamau district; Ministère Richesses Naturelles, Quebec.

Eade, K.E., Fahrig, W.F., and Maxwell, J.A.
1966: Composition of crystalline rocks and fractionating effects of regional metamorphism; *Nature*, vol. 211, No. 5055, pp. 1245-1249.

Goodwin, A.M.
1965: Mineralized volcanic complexes in the Porcupine-Kirkland Lake-Noranda region, Canada; *Econ. Geol.*, vol. 60, No. 5, pp. 955-971.

1968: Archean protocontinental growth and early crustal history of the Canadian Shield; *XXIII Intern. Geol. Cong. Proc.*, vol. 1, pp. 69-89.

Goodwin, A.M., and Shklanka, R.
1967: Archean volcano-tectonic basins: form and pattern; *Can. J. Earth Sci.*, vol. 4, pp. 777-795.

Gunn, B.M.
1969: Geochemistry of Archaean metavolcanics, Chibougamau; *in* Report on Volcanology, *Can. Geophys. Bull.*, vol. 21.

Heirtzler, J.R.
1968: Sea-floor spreading; *Sci. Am.*, vol. 219, No. 6, pp. 60-70.

Holubec, J.
in prep.: Lithostratigraphy, structure and deep crustal relations of Archean rocks, Rouyn-Noranda area, Quebec.

Moorhouse, W.W.
1970: Atlas of textures in Archaean and modern volcanic rocks; *Geol. Assoc. Can.*, SP 8.

Ridler, R.H.
1970: Relationship of mineralization to volcanic stratigraphy in the Kirkland-Larder Lakes area, Ontario; *Proc. Geol. Assoc. Can.*, vol. 21, pp. 33-42.

Roscoe, S.M.
1965: Geochemical and isotopic studies, Noranda and Matagami areas; *Can. Inst. Mining Met., Trans.*, vol. LXVIII, pp. 279-285.

Sharpe, J.I.
1965: Field relations of Matagami sulphide masses bearing on their disposition in time and space; *Can. Inst. Mining Met., Trans.*, vol. LXVIII, pp. 265-278.

Wanless, R.K., Stevens, R.D., Lachance, G.R., and Edmonds, C.M.
1968: Age determinations and geological studies; *Geol. Surv. Can.*, Paper 67-2, Pt. A.

Van Bemmelen, R.W.
1949: The geology of Indonesia; Government Printing Office, The Hague, 732 pp.

Comments from: A.J. Naldrett,
University of Toronto, Toronto

Re: Ultramafic and related mafic rocks of the Abitibi orogen

Ultramafic rocks and their related mafic rocks in the Abitibi orogen fall into three broad classes:

1) large (20-50 miles long, over 20,000 ft. thick) igneous complexes

2) smaller (1-5 miles long, up to 4,000 ft. thick) layered intrusions

3) lenses of peridotite and pyroxenite, 10 x 200 ft. up to ½ mile x 4 miles in area.

The large complexes of class 1 include, the Dore Lake complex of the Chibougamau area (Allard, 1970), the Bell River complex of the Matagami area and possibly the large body of gabbroic rocks located 15 miles west of Timmins. The first two of these consist largely of alternating stratiform layers of anorthosite, pyroxenite and gabbro, successive layers becoming gradually more quartz-rich upwards in the sequence. Anorthositic layers are particularly common towards the base of each complex and both complexes are notably rich in titaniferous magnetite and ilmenite. Sharp (1965) comments that multiple intrusion has played a role in the development of the Bell River Complex. However, the main feature of these complexes is their stratiform nature and the conformity of this stratification with that in the enclosing volcanic rocks. This indicates that cooling and fractionation of the complexes were essentially complete when folding first affected the orogen.

Phinney (1970) has pointed out that the differentiation trend of the early Keewenawan lavas along the north shore of Lake Superior can be explained if they are regarded as successive extrusions of magma, decanted from the main body of the Duluth complex which was undergoing fractionation at this time. There are some close similarities between the anorthositic gabbros of the Duluth complex and the lower portion of the Dore Lake complex as described by Allard. We should seriously entertain the possibility that the large igneous complexes of the Abitibi orogen are actually examples of magma chambers, similar to the Duluth complex, in which differentiation was occurring to give rise to the felsic portions of the volcanic pile. The close correlation between the igneous complexes and centres of felsic volcanism at Timmins, Matagami and Chibougamau as shown on Goodwin and Ridler's (1970) geologic compilation is certainly remarkable. If these complexes are in fact the source of the felsic volcanism, it is possible that they are also the source of the economically important fumerolic activity associated with the felsic volcanics.

The small layered intrusions of class 2) have been described by MacRae (1969) and Naldrett and Mason (1968). Many of them are restricted to a narrow belt, approximately 20 miles wide, extending 90 miles, from just south of Cochrane, east-southeast across the orogen to the Ontario-Quebec border. Although the proportion of mafic to ultramafic rocks varies from intrusion to intrusion, most of them are classic examples of a differentiated sill, with a basal zone of cumulus olivine passing upwards through an augite cumulate and then a plagioclase-augite cumulate to a capping of granophyric gabbro. The igneous layering in these bodies coincides very closely with stratification in the enclosing volcanic rocks, indicating that these intrusions also crystallized before folding had affected this part of the orogen. The only economic mineralization known to be associated with this class of intrusion is the Munro asbestos deposit near Matheson, Ontario, mined by the Canadian Johns-Manville Co. Ltd., from 1950 to 1964.

The lenses of class 3 are best developed along and to the south of the axis of the orogen, particularly south and east of Timmins, Ontario and north of Malartic, Quebec. The lenses are important because all of the important-looking nickel sulfide deposits of the orogen are associated with them.

Typically, the lenses are conformable with the surrounding volcanic rocks and consist of a core of closely packed equant olivine crystals with interstitial hypersthene and augite and a marginal zone, 10 to 100 feet wide, over which the proportion of olivine to pyroxene gradually decreases. It is thought that many of the lenses were intruded as a suspension of olivine crystals within a pyroxenitic liquid. As is usual in cases of the flow of suspended solids (Gibb, 1969), some of the interstitial liquid appears to have migrated to the margins and served to lubricate the intrusions.

Naldrett and Mason (1968) have described some lenses of class 3 type from Dundonald township, east of Timmins, in which the marginal portions contain large skeletal crystals of olivine, pyroxene and chrome spinel (Fig. A). Pyke (1970) has reported similar rocks from south of Timmins and MacRae (personal communication) has observed them in the Matheson area. Except that they are much larger, these skeletal crystals are identical with those produced on quenching experimental charges of appropriate compositions and it seems certain that they are also the result of very rapid chilling. The rocks in which the crystals occur are pyroxenites containing up to 50 modal per cent olivine, which indicates that the liquid from which they have formed was ultramafic rather than mafic in composition.

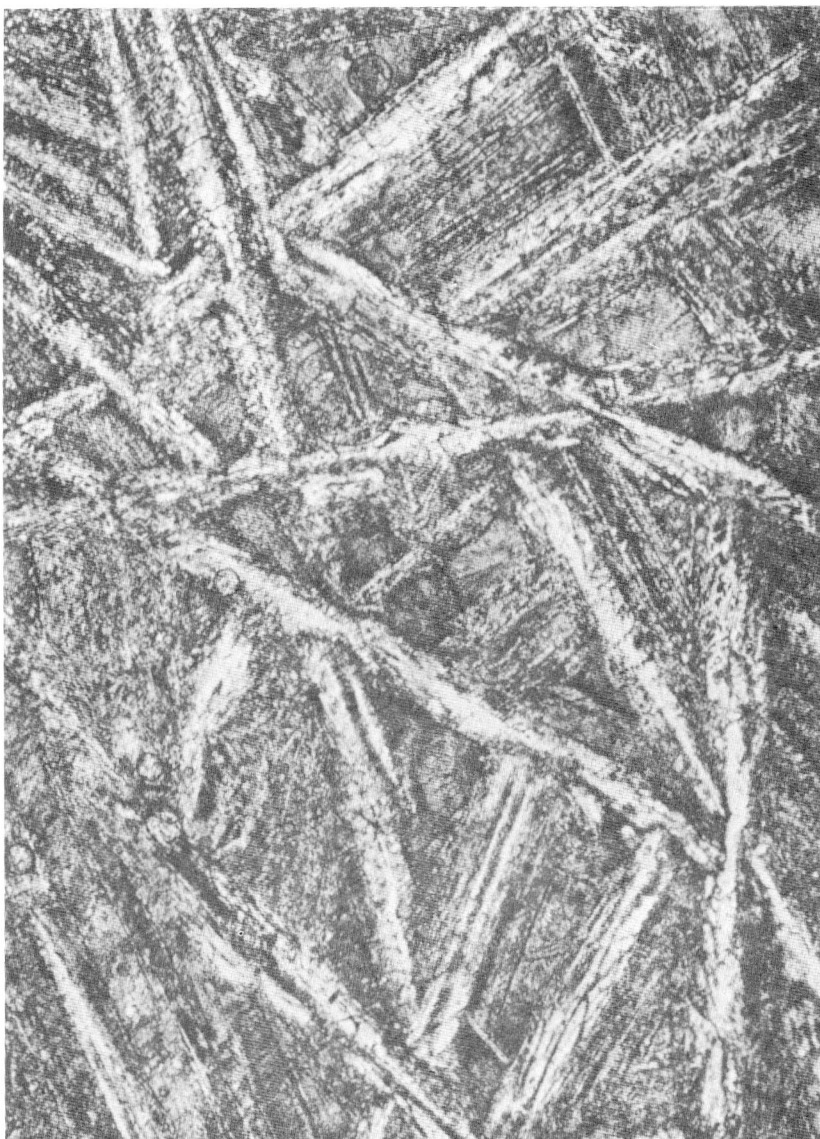

Figure B. Thin section of rock containing skeletal olivine (light grey) from the Komati formation of the Onverwacht group, Barberton Mountainland, Republic of South Africa. X10.

The rapid cooling indicated by the skeletal crystals raises the question of whether these lenses are extrusive or not. So far, poor outcrop has prevented our solving this problem with observations on the Abitibi rocks (most of our information has come from drill core) but Viljoen and Viljoen (1969) have described rocks with very similar skeletal olivine (Fig. 2) from the Barberton Mountainland of South Africa. The rocks of the Mountainland are part of a well preserved Archean geosyncline that has been dated as 3.4 billion years old. The exposure at Barberton is much better than in Canada. The skeletal rocks are closely associated with peridotite lenses, up to two miles long, that are interstratified with pillowed basalts in the lower portion of the volcanic pile. Pillow structures are also developed in the ultramafic rocks and the Viljoen brothers regard all of this sequence as extrusive. It seems very likely that some of the Canadian examples are also extrusive. Whether this is true or not, the Canadian rocks clearly crystallized in a very cool environment, close to the top of the volcanic pile, showing that they are part of the eugeosynclinal volcanism.

In conclusion, I stress that all of the ultramafic and associated mafic rocks of the Abitibi orogen crystallized at an early stage in the history of the orogen, before much folding had occurred. This indicates that the magmas involved may have a common origin with those forming the volcanic rocks. The large anorthositic complexes may represent the magma chambers in which differentiation occurred to produce some of the more felsic volcanic rocks. Petrographic and geochemical work on the complexes and surrounding volcanic rocks should be able to prove or disprove this suggestion.

References

Allard, G.O.
1970: The Dore Lake complex of Chibougamau, Quebec, a metamorphosed Bushveld-type layered complex. Symposium on the Bushveld Igneous complex and other layered intrusions; *Geol. Soc. S. Afr.*, Spec. Publ., vol. 1, in press.

Gibb, F.G.F.
1968: Flow differentiation in the xenolithic ultrabasic dykes of the Cuilins and the Strathard peninsula, Isle of Skye, Scotland; *J. Petrol.*, vol. 9, pp. 411-443.

Goodwin, A.M., and Ridler, R.H.
1970: Abitibi orogenic belt, this volume.

MacRae, N.D.
1969: Ultramafic intrusions of the Abitibi area, Ontario; *Can. J. Earth Sci.*, vol. 6, pp. 281-303.

Naldrett, A.J., and Mason, G.D.
1968: Contrasting Archean ultramafic igneous bodies in Dundonald and Clergue townships, Ontario; *Can. J. Earth Sci.*, vol. 5, pp. 111-143.

Phinney, W.C.
1970: Chemical relations between Keewenawan lavas and the Duluth complex, Minnesota; *Bull. Geol. Soc. Am.* (in press).

Pyke, D.
1970: The geology of Langmuir and Blackstock townships; *Ontario Dept. Mines Rept.* (in press).

Sharp, J.I.
1965: Field relations of Matagami sulfide masses bearing on their disposition in time and space; *Trans. Can. Inst. Mining Met.*, vol. 68, pp. 265-278.

Viljoen, M.J., and Viljoen, R.P.
1969: Evidence for the existence of a mobile extrusive peridotitic magma from the Komati formation of the Onverwacht group; *Symp. on the Upper Mantle, Pretoria, South Africa, 1969.*

9: ARCHEAN VOLCANISM AND SEDIMENTATION IN THE SLAVE STRUCTURAL PROVINCE

J. C. McGlynn and J. B. Henderson
Geological Survey of Canada, Ottawa, Canada

ABSTRACT

Archean volcanic and sedimentary rocks of the Yellowknife Supergroup within the Slave Province occur in a number of discontinuous northerly trending belts separated by large areas underlain by granitic rocks. Volcanic sequences that in part border these belts are composed of mafic lavas with both calc-alkali and tholeiitic differentiation trends and are commonly capped by more silicic volcanic and pyroclastic rocks. The central parts of the belts are filled with volumetrically much more important greywackes and mudstones that have been interpreted as turbidites. The composition of the greywackes suggest derivation from a sialic source. Such paleocurrent evidence that is available indicates transport into the basins more or less perpendicular to the basin margins from areas now occupied by large granitic batholiths. With a few exceptions the contact between the sediments and the volcanics is conformable and in places is marked by a conglomerate composed mainly of angular volcanic clasts but commonly with well rounded granitic boulders. A few instances of apparent granitic basement to the Yellowknife rocks have been documented. The supracrustal rocks have been complexly folded, metamorphosed at a relatively low grade and intruded by granitic rocks of various ages.

It is suggested that the Yellowknife rocks were deposited on a continuous, highly mobile sialic crust. The mafic volcanics were extruded from linear fractures paralleling contacts between positive and negative areas of the crust. Sediment derived mainly from the positive sialic areas was poured into the intervening negative basinal areas. Ultimately, extensive deformation and diapiric granitic intrusion at the margin of the basins has resulted in the situation seen today.

INTRODUCTION

The Slave Structural Province underlies some 75,000 square miles of the northwestern part of the Canadian Shield. The rocks of the province are for the most part Archean in age. To the northwest, southeast and northeast, the Archean rocks are unconformably overlain by Aphebian sediments of the Snare and Epworth Groups, the Great Slave Supergroup and the Goulburn Group respectively.

Of the total area, somewhat less than two-thirds of the Slave Province is composed of Archean sedimentary and volcanic rocks or their metamorphic equivalents (Davidson, 1967). The remainder consists of extensive granitic batholiths. Between 15 and 20 per cent of the stratified rocks are volcanic in origin while the remaining 85 to 80 per cent are of sedimentary origin. In this respect, the Slave Province differs markedly from the much larger Superior Province where Archean volcanic rocks are considerably more abundant than the associated sedimentary rocks.

The Archean stratified rocks of the Slave Structural Province all belong to the Yellowknife Supergroup (J.B. Henderson, 1970b). On a map of the province (Fig. 1), the Yellowknife rocks can be seen to occur in three major areas: (1) along the western margin of the province, (2) in a belt extending north from north of Great Slave Lake to the central part of the province and (3) in the large, more or less equidimensional area in the northeastern part of the province. These areas of volcanics and sediments are separated by large

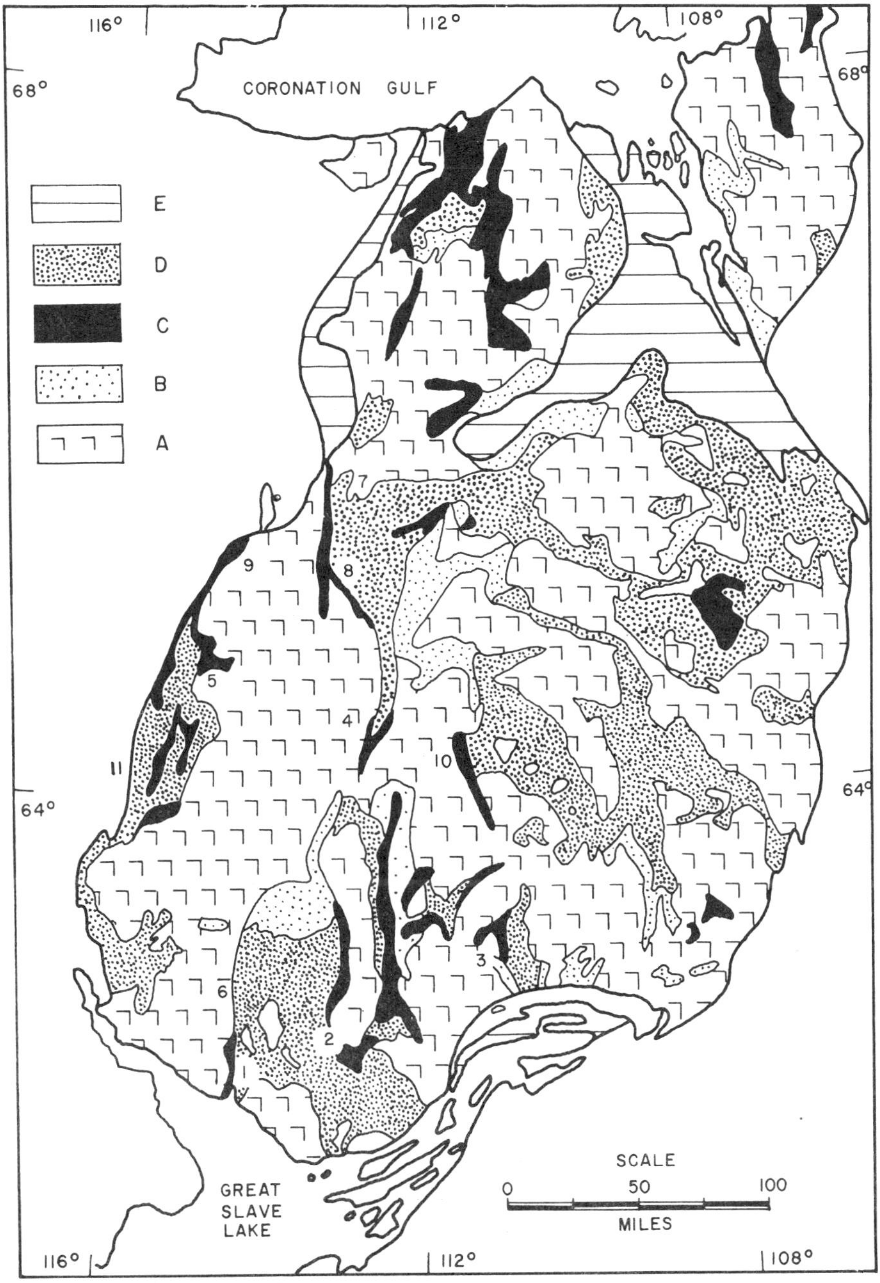

A - Predominantly granitic rocks.
B - Mixed granitic and metasedimentary and metavolcanic rocks.
C - Volcanic rocks of the Yellowknife Supergroup.
D - Sedimentary rocks of the Yellowknife Supergroup.
E - Aphebian cover on the Slave Province.

1. Yellowknife area.
2. Ross Lake, Cameron Lake, Bealieu River areas.
3. Benjamin Lake area.
4. Winter Lake area.
5. Arseno Lake area.
6. Giauque Lake area.
7. Itchen Lake area.
8. Point Lake area.
9. Mesa Lake area.
10. Courageous Lake-Matthews Lake.
11. Ingray Lake area.

Figure 1: Generalized geological map of the Slave Structural Province

granitic batholiths and high grade metamorphic rocks. In addition there are several smaller occurrences of Yellowknife sediments and volcanics, notably north of the East Arm of Great Slave Lake and south of the Arctic Coast.

With a few exceptions, geological work in the Slave Province has been primarily of a reconnaissance mapping nature (Lord, 1942; J.F. Henderson, 1944; Fortier, 1949; Folinsbee, 1949; Fraser, 1969). Recently, however, more attention has been given to detailed studies of the geochemistry of the volcanics (Boyle, 1961; Baragar, 1966), geochronology (Green *et al.*, 1968), sedimentology (J.B. Henderson, 1970a), structure and metamorphism (Davidson, 1967) in particular areas of the province.

STRATIGRAPHY

The stratigraphy of the Yellowknife succession is broadly similar throughout the province and indeed is similar to the rather restricted assemblage that characterizes Archean stratified rocks throughout the world. This consists of a thick sequence of predominantly mafic volcanics followed by a sedimentary series of typically greywackes and shales. In the transition zone between these two dominant lithologies, minor amounts of conglomerate, more mature sandstones, limestones and tuffaceous sediments may or may not occur.

The Yellowknife rocks have almost always been divided into at least two units: a predominantly mafic volcanic unit and a sedimentary unit. In the past they were rarely named, but were commonly referred to by map-legend number, or as the Yellowknife sediments or volcanics. In some areas the sequence was divided into "divisions" (Jolliffe, 1942). For example, at Yellowknife the mafic volcanic rocks were referred to as Division A, while the more felsic flows and sediments were placed in Division B and C.

In some of the areas studied in greater detail it was found that the basic sedimentary and volcanic units could, in turn, be further subdivided into mappable lithological units. Thus, in some of the volcanic sequences, where the top of the sequence is more silicic in composition or contains a greater abundance of sedimentary material, one or more additional units were added. Similarly, with the sediments, conglomeratic or shallow-water crossbedded sandstones were separated from the predominant greywacke-mudstone unit. In the Yellowknife area, where the stratigraphy has been recently formally described, the Yellowknife Supergroup can be subdivided into six formations - three of volcanic origin and three of sedimentary origin (Henderson, 1970b).

It is generally accepted that the major basic volcanic units of the Yellowknife Supergroup are older, on the whole, than the sedimentary formations. Wherever the sedimentary-volcanic contact is observed, the sediments overlie the volcanics except at Mesa Lake (Ross, 1962) where the major basic volcanic unit is found between two thick sedimentary sequences. The sediments are typically conformable with the volcanics and, commonly, minor flows and tuffaceous beds are found immediately above the contact, suggesting that there was little or no time lapse between volcanism and sedimentation. At Yellowknife, however, sediments overlie the basic volcanics with angular unconformity. It is felt that this unconformity does not represent a major time break, but possibly a tilting of the volcanic pile on the thin relatively unstable Archean crust.

Where conglomerates occur in the sedimentary sequence, they almost invariably are at the contact between the volcanic and sedimentary units. This is certainly true at Yellowknife where the conglomerate fills large depressions on the surface of the unconformity. In most other areas where a conglomerate is reported, it lies with apparent conformity between volcanics and finer grained sediments. Examples of this relationship can be seen in the Arseno Lake (McGlynn and Ross, 1963), Beaulieu River (J.F. Henderson, 1938) and Winter Lake (Fraser, 1969) map-areas. In the Courageous-Matthews Lakes map-areas (Moore, 1956) conglomerate is reported within the major basic volcanic unit. At Point Lake conglomerate is interstratified with basic volcanics and in part overlies an older granitic body (Stockwell, 1933).

It is interesting to note the map distribution of the various units of the Yellowknife Supergroup within the Slave Province. As already mentioned, rocks of the Yellowknife Supergroup occur in three major zones with, in some cases, rather irregular diffuse borders. The sedimentary units underlie by far the greatest area and occupy the central parts of these zones. The volcanics, where present, are restricted to the margins of the zones. This is best exemplified by the southern part of the central zone (Fig. 1), where the volcanic rocks occur along the margin of the basin at Yellowknife in the west and along the Cameron River to the east. This suggests that these zones of Archean stratified rocks may represent basins of accumulation that were, in part, bordered by volcanic piles.

Table 1 shows estimated comparative thicknesses of the various units of the Yellowknife Supergroup in several different parts of the Slave Province.

Table 1

Comparative Unit Thicknesses in the Yellowknife Supergroup

Area	Basic Volcanics	Intermed. & Felsic Volc.	Cong. & Assoc. Seds.	Sediments
Yellowknife	40,000' ±	4,000'	0-800'	15,000' +
Benjamin Lake	2,000'	1,000'	-	?
Courageous Lake-Matthews Lake	12,000 - 2,000'	0-8,000'	0-400'	5,000'
Arseno Lake	18,000 - 1,500'	-	0-1,500'	2,000-3,000'
Mesa Lake	1,000'	-	-	2,000-3,500'

Estimates of thicknesses are often very difficult to make, due to extreme structural complexities. This is particularly true for sediments, as the top of the section is rarely defined.

BASEMENT TO THE YELLOWKNIFE SUPERGROUP

Despite the large areas of the province underlain by rocks of the Yellowknife Supergroup, very few observations have been made of what can be considered basement to the Yellowknife. Most contacts with non-Yellowknife rocks appear to be intrusive. However, a few possible documentations of basement have been made.

Baragar (1966) points out that a large zone of amphibolitic dykes in the granodiorite east of Ross Lake, on the eastern margin of the central area of Archean strata may represent feeders to the Yellowknife volcanics in the area. The dykes cut both the granodiorite and the volcanic strata but apparently not the overlying Yellowknife sediments. Thus, despite the fact that other small bodies of intrusive granitic rocks are present in the contact area, it is possible that much of the granodiorite in this particular area might represent basement.

Davidson (1967) using a similar argument describes a small granitic body exposed in the core of an anticline in the Benjamin Lake area, that may also be basement. Metamorphosed basic dykes that cut the granitic body and the possible pre-Yellowknife sediments that overlie it, were not traced into the overlying Yellowknife basic volcanic flows.

Another possible area of basement is located in the Beaulieu River area (J.F. Henderson, 1938) where a small, oval, highly altered and granulated chlorite granite body is surrounded by basic Yellowknife lavas. Along one contact between the granite and the volcanics is a fragmental zone that could represent a granitic conglomerate above the basement, although Henderson also suggested that it may represent a tectonic breccia zone.

The first discovered and best documented case for a basement to the Yellowknife Supergroup is described by Stockwell (1933) at Point Lake, in the north-central part of the Slave Province. Again, the rock in question is a highly altered chloritized granite that is intruded by other granitic bodies that cut both it and the overlying Yellowknife lavas and sediments. Immediately overlying the chloritized granite is a conglomerate that dips away from the granite body and whose cobbles are identical to it. Within the conglomerate there is a cobble size gradation away from the granite source and there appears to be no metamorphic gradient, as would be expected if the granite was intruded into the conglomerate. As well, basic dykes, bearing a close resemblance to some of the overlying lavas interstratified with the conglomerate, cut the older chloritized granite but were not observed to cut the younger granites in the area. It is interesting that this possible basement granite contains inclusions of a "dark green schist", possible evidence of the crustal material into which the Yellowknife "basement" granite intruded.

In all instances cited, the basement material is a strongly altered granitic rock. In addition these granitic rocks are all part of, or at least very close to the large batholithic areas of granodiorite that separate the various areas of Archean lavas and sediments. Although the number of instances of possible basement is very meager, the fact that they all are at, or at least close to, the margins of the supracrustal outcrop areas supports the suggestion that these zones do indeed represent basins of accumulation and are not just fortuitous downfoldings of an originally much more extensive area. If this is true, the present day margins are a fairly close approximation to the original margins.

VOLCANIC ROCKS OF THE YELLOWKNIFE SUPERGROUP

Volcanic rocks of the Yellowknife Supergroup underlie about five per cent of the Slave Structural Province. They occur for the most part in 18 or 19 discreet belts that trend north to northeast. Most of the volcanic belts occur along the margins of areas underlain by sediments and are most abundant in the western half of the province.

The volcanic rocks lie at the base of the Yellowknife succession as discontinuous belts, except at Mesa Lake (Ross, 1962) where sediments occur below volcanic rocks. In some areas, thin layers of volcanic rocks occur within the overlying sediments and may represent resumption of volcanism or may be contemporaneous with volcanism occurring in other parts of the volcanic complex.

The volcanic strata comprise massive mafic lavas (basalt or andesite), pillowed basic lavas and fragmental rocks, and intermediate to acidic lavas and tuffaceous rocks that include dacites, latites and quartz latites. The intermediate to acidic strata occur in the upper part of individual piles. Sills, dykes, and irregularly shaped masses of gabbro and diorite occur in the volcanic rocks and, less commonly, in overlying sediments. They are thought to be intrusive equivalents of the lavas. Dykes of acidic porphyry also occur in the lavas, and probably represent intrusive equivalents of the acidic lavas.

The stratigraphy or anatomy of only a few complexes is known in any detail. At Yellowknife (J.F. Henderson and Brown, 1966; Baragar, 1966) 40,000 feet of volcanic strata are exposed. The succession comprises massive, pillowed and variolitic basalts and andesites, with intermediate and acidic layers at about mid point, and at the top of the sequence. Latite, dacite and quartz latite comprise 15 to 20 per cent of the strata (Baragar, 1966). In the Cameron River belt (Baragar, 1966) where about 7,000 feet of volcanic rocks are exposed, massive and pillowed basalts and andesites form the bulk of the succession and a discontinuous layer, up to 1,400 feet thick, of latite, dacite and quartz latite flows and fragmental strata occurs in the upper part of the unit. At Arseno Lake (McGlynn and Ross, 1963) 18,000 to 1,500 feet of lava are composed of basic massive and pillowed basalts and andesites with minor amounts of more acidic material. In the Benjamin Lake area, 2,000 feet of massive and pillowed flows are overlain by about 1,000 feet of latites and acidic or intermediate tuffs and breccias. Available data on other volcanic belts in the Slave Structural Province suggest that relative proportions of basic and acidic lavas, and stratigraphic position of the acidic layers are similar to the successions known in greater detail.

Baragar (1966) in his study of the geochemistry of volcanic rocks at Yellowknife and along the Cameron River recognizes two differentiation trends, namely a calc-alkali trend, involving enrichment in alkaline (plus quartz) components, and a tholeiitic trend, characterized by enrichment in iron at the expense of magnesium. The lavas, therefore, have mixed tholeiitic and calc-alkaline characteristics. Baragar postulates that tholeiitic basalts represent the fundamental magma type, with calc-alkaline trends resulting from sialic wallrock contamination.

The thickness of the numerous volcanic belts is variable. The thickest measured section is about 40,000 feet and as much as 18,000 feet has been estimated in other belts; however, maximum thicknesses of 5,000 to 10,000 feet are more representative. In most belts, the thickness varies considerably and appears to decrease along strike. In longitudinal section, a pinch and swell effect would be evident, with the belts thinning out away from centers of volcanism. In addition, it is thought that the lavas wedge out across their strike under the sediments, and that the sedimentary basins are not everywhere underlain by volcanic rock. The volcanic rocks were extruded from centers along linear zones of weakness in the crust, and thinned out in all directions from these centers.

SEDIMENTARY ROCKS OF THE YELLOWKNIFE SUPERGROUP

Archean sedimentary rocks or metamorphic rocks derived from sedimentary rocks underlie a large proportion of the province, and certainly greatly exceed the area underlain by volcanic rocks. Yet despite their great abundance, they are quite restricted as to type, being predominantly immature greywackes and mudstones with only minor amounts of conglomerate, limestone and more mature sandstones. These sediments show for the most part a remarkable similarity throughout the province, both within and between the main basins of accumulation.

For discussion purposes the sediments will be broken down into three groups: (1) conglomerates, (2) greywacke-mudstones, and (3) other sediments.

Conglomerates

Conglomerates where present, are characteristically found in rather thick sections between the thick basic volcanic sequences and the greywacke-mudstone facies. In most areas, the transition from volcanism to sedimentation is not marked by a conglomerate. Good examples of the conglomerate occur in the Arseno Lake and Winter Lake areas and at Point Lake, Ross Lake and Yellowknife.

Conglomerates occur in elongate lenses ranging in thickness from only a few feet to over 1,500 feet in some cases, and can be traced laterally for over ten miles. Some occurrences, however, are much more discontinuous, with other sediments occurring between the pockets or lenses of conglomerate. In some instances, conglomerates occur in areas where the associated mafic volcanic sequence thins. At Ross Lake, the Cameron River volcanic belt thins out from a maximum thickness of 8,000 feet and disappears over a distance of about four or five miles. It is in this area of thinning and south of it that the conglomerate is found. A similar situation is seen at Point Lake. This may suggest that the conglomerates tend to be deposited on the flanks of the volcanic piles.

Boulders in the conglomerates are mainly of volcanic origin. Their angular nature and compositional similarity to the underlying mafic volcanics suggest local derivation. The deposits commonly grade upwards from coarse angular mafic blocks, often several feet in length, to finer and commonly more felsic varieties. Cobbles and pebbles of greywacke, shale, quartz and chert are minor components but locally can be abundant. In addition, many of the conglomerates contain well rounded pink to grey granitic cobbles - a minor but quite spectacular component. The size of the clasts varies from less than an inch to several feet in the case of the angular basal boulders. The average size ranges between three inches and one foot, and except for the basal clasts, the cobbles, in particular the granitic boulders, show considerable rounding.

The matrix of the conglomerates is quite variable, ranging from a mixture of quartz and finely comminuted basic volcanic rock fragments through quartz and silicic volcanic rock fragments, to greywacke and, in one case, a calcareous and quartzose mixture. Again, the preponderance of volcanically derived material is of note. In many of the conglomerates thin discontinuous lenses or patches of sediment identical to the matrix of the conglomerates is found. These frequently show evidence of vigorous current activity with abundant cross-bedding, scours and ripples.

The conglomerates for the most part are conformable with the underlying volcanics and at Point Lake and Courageous-Matthews Lake the conglomerate is interbedded with volcanic flows. One exception is the conglomerate at Yellowknife which overlies the mafic volcanics with angular unconformity and fills depressions in the erosional surface that, in some cases, are several hundred feet deep. The general consensus of opinion is that although there was some erosion (most of the clasts in the conglomerate are locally derived), the conglomerate does not represent a long period of erosion after the major volcanic event prior to the principal period of sedimentation. The presence of a few granitic boulders in the conglomerate does not necessarily require an orogeny involving the intrusion and subsequent unroofing of great tracts of granitic rock prior to sedimentation. The contrast between the hard, very well rounded granitic cobbles and the relatively soft, angular, apparently locally derived, volcanic clasts suggests derivation of the granitic cobbles from a more distant source. This would preclude derivation from local leucocratic stocks in the volcanic pile as is seen in some recent volcanic sequences such as in the Aleutian Islands (Coats, 1956). There, conglomerates contain angular volcanic and granitic clasts both of local origin. A more likely source for the granitic cobbles in the Archean would be a granitic basement area.

Greywacke-Mudstone Facies

The greywacke-mudstone couplets make up by far the largest proportion of the sediments and, indeed, the greatest part of all the Archean stratified rocks. They typically overlie or apparently overlie the thick sequences of mafic volcanics and, where present, the conglomerates. The contacts are conformable and are commonly gradational; the volcanics giving way through a series of interbedded flows and tuffaceous beds to the typical greywackes and mudstones. Where associated with conglomerates, gradation is through a decreasing number of beds and lenses of ever finer grained conglomerate and cleaner sandstones.

Regionally these sediments occupy the large central areas of basins rimmed in part by the mafic volcanic sequences, and surrounded by granitic areas. Because of structural complexity, lack of good marker horizons, and difficulty in defining the true top of the unit, thicknesses are difficult to determine. Estimates range from one to two thousand to in excess of 15,000 feet.

This facies is made up of interbedded greywacke and mudstone with greywacke greatly predominating in most cases. The individual greywacke-mudstone couplets range in thickness from less than an inch to thick massive units measurable in tens of feet. It is commonly found, however, that many of the excessively thick units are composite in nature, with two or more greywacke units occurring together without the intervening mudstone.

The less metamorphosed sediments present a magnificent array of sedimentary structures. A high proportion of the greywacke beds are graded - a very useful feature for determining the complex structure of these areas. In addition the Bouma sequence of sedimentary structures (Bouma, 1962), so commonly noted in turbidites is also present in these rocks. This sequence includes a basal graded division followed by a parallel laminated division above which is a current ripple-laminated division. A second or upper division of parallel lamination is next, followed finally by the uppermost pelitic division.

The base of each individual bed is invariably sharp but is not necessarily straight or even. Many beds show evidence of minor scouring due to current activity or small channels that cut down through underlying beds. In addition, the bases of some beds made very irregular by the development of load and flame structures due to post-depositional soft sediment deformation.

Individual beds are laterally continuous. Pinching out of beds is rarely observed, although in places a pair of greywackes can be traced laterally into a single amalgamated bed. Even the thinnest of beds can be traced across very large outcrops (up to several hundred feet). Tremblay (1952) in the Giauque Lake area reported that when groups of thicker greywacke beds were traced along strike over several thousand feet, they sometimes are seen to grade into much more thinly bedded greywackes with a higher proportion of mudstone.

The sediments have been classified according to the thickness and proportion of greywacke to mudstone. Although the various types are commonly mixed, they do sometimes occur in groups of predominantly one type. In some sections, a zone of thin beds with a high shale to sand ratio will be followed by a group of very thick massive greywackes.

The greywackes are composed of a sand-sized fraction of quartz, rock fragments and feldspar in a matrix of chlorite, muscovite, and very fine grained quartzo-feldspathic material. The mudstones have a composition similar to the greywacke matrix. Depending on the relative proportions of the three major components, the sandstones have been variously called impure quartzites, greywackes and arkoses. At Yellowknife, the greywackes have a relatively high proportion of silicic and intermediate volcanic rock fragments which suggest that silicic volcanics were an important part of the source area. In addition, a minor but persistant component of granitic rock fragments is present. The compositional make-up of the greywackes is similar at Gordon Lake, suggesting again that a silicic volcanic source was a very important contributor to these sediments.

At Yellowknife and in the Mesa Lake area, paleocurrent measurements have been made in these sediments. For the most part measurements have been made on small-scale cross-laminations of the ripples or small cross-laminated scour fills which, when rotated back to overcome orientation errors caused by folding, give a good approximation of the depositional current sense. Considering the extreme degree of deformation in these Archean sediments, too much significance should not be placed on the exact current azimuths that result as there are bound to be considerable errors in the unfolding of the beds. The results are probably significant if not taken to indicate more than the general quadrant from which the currents came. At Yellowknife, the sediments appear to have been derived from a source to the west - a direction more or less perpendicular to the present-day margin of the basin, suggesting that the sediments were being shed from a land mass in the position now occupied by the western granodiorite. In the Mesa Lake area (Ross, 1962) in another Archean basin west of the same large granitic area, the paleocurrent trend is towards the southwest, away from the granodiorite area, but at more of an angle than was the case at Yellowknife. Ross suggests the direction may indicate the axis of a trough along which the currents were flowing. More information of this type would certainly be useful in outlining the shape of the original Archean basins.

The great proportion of the Yellowknife sediments are greywackes, believed to be deposited as a result of turbidity currents. The fact that the sediments are mainly greywackes composed predominantly of quartz, rock fragments and feldspar strongly suggests a source area of sialic composition. Indeed the few chemical analyses of these greywackes that are available show close similarities to those of a granodiorite. Greywackes being immature sediments, have undergone relatively minor sorting and concentration and probably represent a close approximation of the composition of their source area. Such paleocurrent evidence that is available indicates that the greywackes are derived from areas now occupied by granodiorite masses.

Other Sediments

Although the greywacke-mudstone assemblage makes up by far the greatest proportion of the sediments, other relatively minor sediment types are present.

Many of these other sediments occur close to the contact between the volcanics and the greywackes. Tuffaceous beds are most prevalent in this area and mark the transition from predominantly volcanic activity to sediment deposition. Thick tuffaceous units may also occur higher in the sedimentary section as is seen at Yellowknife and Arseno Lake.

A unit which is frequently reported as an impure quartzite is commonly found in the contact area. This rock occurs in well defined beds that are either massive or exhibit large-scale tangential crossbedding. In some cases this unit is interbedded with the normal greywacke-mudstone facies. At Giauque Lake (Tremblay, 1952), a thick section of this impure quartzite is interbedded with minor amounts of greywacke while at Benjamin Lake tangentially crossbedded units between six inches and five feet thick are a minor occurrence in the greywacke-mudstone section (Davidson, 1967). At Yellowknife a sandstone composed of quartz and silicic volcanic rock fragments together with a patchy conglomerate overlies the mafic volcanic pile.

Where present, calcareous sediments are also found near the contact with the volcanics. Lord (1942) in the Ingray Lake area reports a transition from mafic volcanics through tuffaceous beds into ninety feet of crystalline siliceous limestone. In the Ross Lake area, the conglomerate has a calcareous matrix and a unit of impure limestone occurs below the transition to the normal greywacke-mudstone facies.

One notable feature of Archean sediments in the Slave Province is the general lack of iron formation. This is in strong contrast to the Superior Province where iron formation is a common, although usually volumetrically small facies in the Archean sediments. An exception to this paucity of iron formation is the Itchen Lake area (Bostock, 1967) where siliceous magnetic iron formation occurs locally within the greywacke facies.

STRUCTURE OF THE SLAVE PROVINCE

Rocks of the Yellowknife Supergroup have been folded, metamorphosed and intruded by granitic rocks. This activity is considered by Stockwell (1964) to result from an orogeny that has been assigned to the Kenoran orogeny. Most of the K-Ar age determinations on minerals formed during this orogeny fall in the range of 2,300 to 2,600 m.y. and their mean value is 2,490 m.y.

Rocks of the Yellowknife Supergroup were deformed into northerly to northeasterly trending similar folds that plunge moderately north and south. The more northerly trends are common in the western half of the area whereas the more easterly trends occur in sediments of the eastern part of the province. These folds in many areas were mofidied by steeply plunging cross-folds. The sedimentary rocks are generally tightly and intricately folded but volcanic rocks, probably because of their greater competency, behaved differently. Thin volcanic successions are folded into broad folds parallel to those in the sediments but thicker successions occur as homoclinal sequences that face away from bounding granitic rocks that have been presumed to be younger. An alternative interpretation is that the homoclinal successions are in contact with granitic rocks that were part of the basement and that have been somewhat remobilized during deformation of the supracrustal rocks.

The Yellowknife strata, particularly the volcanic sequences, are cut by chloritic shear-zones that may be contemporaneous with folding and also by later high-level faults that strike north, northwest and northeast and seem to be of several ages.

About two-thirds of the Slave Structural Province is underlain by granitic rocks of which possibly as much as half are migmatite, mixed gneiss, banded gneiss and granitic gneiss, with the remainder consisting of batholiths of massive or slightly gneissic granitic rocks of various compositions. In a number of areas, Yellowknife sediments are separated by zones of mixed gneisses that were derived, at least in part, from the supracrustal rocks. It seems likely that they were parts of single larger basins.

Yellowknife volcanic and sedimentary rocks have been metamorphosed to degrees varying from low greenschist to amphibolite facies.

In the areas where detailed work has been completed it can usually be shown that deformation, metamorphism and intrusion of granites were elements of a sequence of overlapping events that include folding and low grade metamorphism followed by granitization, intrusion of syntectonic granitic rocks and associated metamorphism that overlaps in time with the early folding, later cross-folding and intrusion of high-level late tectonic granodiorites and pegmatites.

CRUSTAL MODEL FOR THE SLAVE STRUCTURAL PROVINCE

The authors' model of Archean deposition in the Slave Structural Province has been strongly influenced by the work of Martin (1969) and Anhaeusser *et al.* (1969) among others.

A model to explain evolution of this province during the Archean must take into account the following features:

(1) The areas of Archean volcanic and sedimentary rocks are separated by large granodioritic areas.

(2) Where volcanics are present they tend to occur at the margins of areas underlain by stratified rocks.

(3) The sedimentary rocks, which greatly exceed the volcanics in volume, have on the average a granodioritic composition.

(4) All areas that have been suggested as possible basement to the stratified rocks are granitic in nature.

(5) Areas of sedimentary and volcanic rocks have a dominant north to north-northeasterly trend, particularly in the western, central and southern parts of the province.

The zones of Yellowknife supracrustal rocks probably represent basins of accumulation and the present-day borders to these areas are a close approximation to the original margins. That is to say these areas do not merely represent preserved "keels" of large synclines that originally covered a much more extensive area. Between these negative basinal areas must have been positive areas that are now occupied by extensive granitic batholiths. Volcanic activity occurred in large part at the contacts between positive and negative areas in the crust. Only rarely is there any evidence of volcanic activity in the central part of the basins.

The meager paleocurrent evidence that is available suggests that the sediments were derived from the interbasinal positive areas. The sediments are predominantly greywackes and mudstones; chemically they approximate a granodioritic composition indicating that the source area was sialic. Further evidence that sialic material was available is indicated by the scattered, well rounded granitic cobbles found in the conglomerates at the margins of the basins. It is possible that the granitic cobbles could have been derived from local minor intrusions into the mafic volcanics that typically underlie them. However, the high degree of rounding of these granitic boulders, in contrast with the rather angular, locally derived, volcanic clasts would suggest a somewhat more distant source. In this regard it is significant that in all places where possible basement to the Yellowknife Supergroup has been described, basement rocks are granite.

The marginal volcanic sequences are not regarded as a major contributor of sediment although erosion of the comparatively thin silicic top of some of the volcanic piles is of local importance. The composition and volume of the sediments would require tremendous quantities of silicic material - far beyond the capabilities of the essentially mafic volcanic belts.

It appears likely that a positive terrain of sialic composition did exist during Archean times and from the evidence of the relict basement and granitic cobbles in the Yellowknife Supergroup that at least some of the basement was of plutonic or hypabyssal aspect. This terrain could conceivably have consisted of intruded and deformed remnants of pre-Yellowknife sedimentary and/or older volcanic sequences. With the present information available it is not possible to precisely define such areas but it is felt that the granitic terrain west and north of Yellowknife up to the Arctic coast and the area of granitic rocks between the Cameron River and Benjamin Lake north of the East Arm of Great Slave Lake may be two examples.

It is believed that the Yellowknife basins were completely underlain by sialic crust throughout their history and hence contrast with many of the more recent geosynclines that have developed at continental-oceanic crustal junctions. The relatively small size of many of the zones that are surrounded by granitic batholiths makes it difficult to conceive of them being underlain by simatic crust if the batholiths represented sialic areas at the time of basin filling. The fact that the volcanic sequences show evidence of sialic contamination would also suggest intrusion through a sialic crust. It seems unlikely therefore that the Archean sequences in the Slave Province are analogous to island arc systems.

In Archean time the Slave Structural Province is thought to have been underlain by a continuous highly mobile sialic crust. This mobility may have been a result of the crust being much thinner than present day continental areas or possibly due to a much steeper thermal gradient resulting from a higher rate of radioactive decay (Urry, 1949). Parts of the crust were uplifted, forming ridges and concomitant negative areas that have been preserved in the northerly trends seen in the province today. Deep fractures or zones of weakness developed parallel to the positive areas and along these the volcanic rocks were extruded. In response to the uplift of the positive areas, silicic debris was eroded and deposited in the basinal areas between the positive zones. Volcanic activity was concentrated along parts of the margins of the basins while the sediments flowed into the basin between volcanic areas in a series of large fans (Fig. 2).

With continued accumulation in the basins these areas were depressed into the mobile crust, resulting in the eventual deformation of the supracrustal rocks. Diapiric granitic bodies were intruded preferentially between positive and negative regions of the crust, which resulted in intrusive relationships between the Yellowknife rocks and the batholithic areas (Fig. 3). Deformation and intrusion took place at relatively shallow depths in the crust (possibly as a result of its mobility) as shown by the low pressure metamorphic effects (Davidson, 1967). In some instances deformation and intrusion were much more extensive, resulting in the partial obliteration of parts of basins. Thus the depositional areas may have been much more extensive than is indicated now.

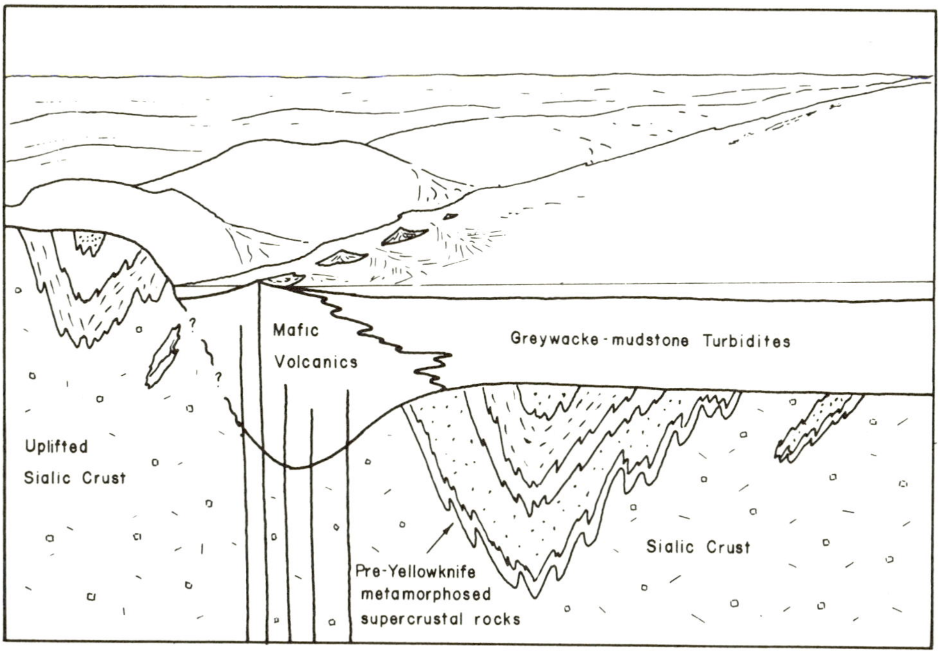

Figure 2. Cross-section through part of an Archean basin prior to deformation.

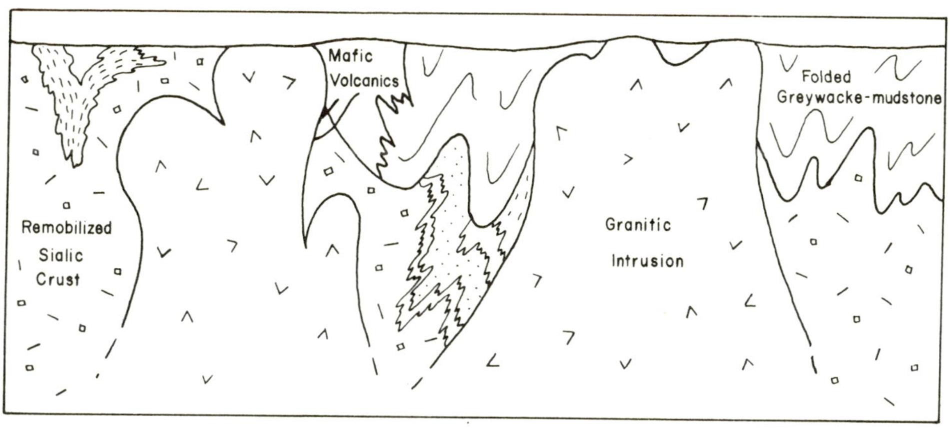

Figure 3. Cross-section through the same basin after deformation.

This cycle may have been repeated over a long period of time with basins being filled, deformed and eventually cannibalized to fill newly developing basins. As suggested by Martin (1969) the 2,500 m.y. number that keeps appearing in Archean terrains may reflect the achievement of stability for unknown reasons by the up-to-then mobile crust. What we see preserved in the Archean is the last of a long series of cycles that operated under conditions that have never again existed on the earth.

The exercise of developing this model or synthesis of the Archean geology of the Slave Province, based as it is on rather scattered evidence, emphasizes the need for further work in this area. Particularly the granitic areas in the province need re-studying to test for the presence of basement rocks, to outline their extent, and to study the nature of younger granitic rocks and thereby place them in an orogenic sequence. The composition and stratigraphy of many volcanic complexes needs to be determined; and sedimentological and paleocurrent studies in the sedimentary rocks of the Yellowknife Supergroup are required to define the extent and nature of the original basins.

REFERENCES

Anhaeusser, C.R., Mason, R., Viljoen, M.J., and Viljoen, R.P.
1969: A reappraisal of some aspects of Precambrian Shield geology; *Geol. Soc. Am. Bull.*, vol. 80, pp. 2175-2200.

Baragar, W.R.A.
1966: Geochemistry of the Yellowknife volcanic rocks; *Can. J. Earth Sci.*, vol. 3, pp. 9-30.

Bostock, H.H.
1967: Geological notes Itchen Lake map-area, Dist. of Mackenzie, part of 76E and 86H; *Geol. Surv. Can.*, Paper 66-24.

Boyle, R.W.
1961: The geology, geochemistry and origin of the gold deposits of the Yellowknife District; *Geol. Surv. Can.*, Mem. 310.

Bouma, A.H.
1962: Sedimentology of some Flysch Deposits, Elsevier, Amsterdam.

Coats, R.R.
1956: Reconnaissance geology of some Western Aleutian Islands, Alaska; *U.S. Geol. Surv. Bull.* 1028E.

Davidson, A.
1967: Metamorphism and intrusion in the Benjamin Lake map-area, Northwest Territories, Unpubl. Ph.D. thesis, Univ. British Columbia.

Fraser, J.A.
1969: Winter Lake; *Geol. Surv. Can.*, Map 1291A.

Folinsbee, R.E.
1949: Lac de Gras, District of Mackenzie, N.W.T.; *Geol. Surv. Can.*, Map 977A.

Fortier, Y.O.
1949: Indin Lake map-area (east half), Northwest Territories; *Geol. Surv. Can.*, Paper 49-10.

Green, D.C., Baadsgaard, H., and Cumming, G.L.
1968: Geochronology of the Yellowknife area, Northwest Territories, Canada; *Can. J. Earth Sci.*, vol. 5, pp. 725-735.

Henderson, J.B.
1970a: Petrology and origin of the sediments of the Yellowknife Supergroup (Archean), Yellowknife, District of Mackenzie, Canada; Unpubl. thesis, The Johns Hopkins Univ.

1970b: Stratigraphy of the Yellowknife Supergroup, Yellowknife Bay-Prosperous Lake area, District of Mackenzie; *Geol. Surv. Can.*, Paper 70-26.

Henderson, J.F.
1938: Beaulieu River area, Northwest Territories; *Geol. Surv. Can.*, Paper 38-1.

Henderson, J.F., and Brown, I.C.
1966: Geology and structure of the Yellowknife Greenstone Belt, District of Mackenzie; *Geol. Surv. Can.*, Bull. 141.

Jolliffe, A.W.
1942: Yellowknife Bay, District of Mackenzie, Northwest Territories; *Geol. Surv. Can.*, Map 709A.

Lord, C.S.
1942: Snare River and Ingray Lake map-areas; *Geol. Surv. Can.*, Mem. 235.

Martin, H.
1969: Problems of age relations and structure in some metamorphic belts of South Africa; *Geol. Assoc. Can.*, Spec. Paper No. 5, pp. 17-26.

McGlynn, J.C., and Ross, J.V.
1963: Arseno Lake map-area, District of Mackenzie 86B/12; *Geol. Surv. Can.*, Paper 63-26.

Moore, J.C.G.
1956: Courageous Lake-Matthews Lake area, District of Mackenzie, N.W.T.; *Geol. Surv. Can.*, Mem. 283.

Ross, J.V.
1962: Deposition and current direction within the Yellowknife Group at Mesa Lake, N.W.T., Canada; *Geol. Soc. Am.*, Bull. 73, pp. 1159-1162.

Stockwell, C.H.
1933: Great Slave Lake-Coppermine River area, Northwest Territories; *Geol. Surv. Can.*, Sum. Rept. 1932, Pt. C, pp. 37-63.

Tremblay, L.P.
1952: Giauque Lake map-area; *Geol. Surv. Can.*, Mem. 266.

Urry, W.D.
1949: Significance of radioactivity in geophysics-Thermal history of the Earth; *Trans. Am. Geophys. Union*, vol. 30, pp. 171-180.

10: ARCHAEAN VOLCANISM IN THE CANADIAN SHIELD

H. D. B. Wilson, Peter Andrews, R. L. Moxham, and K. Ramlal
Department of Geology, University of Manitoba, Winnipeg, Manitoba, Canada

ABSTRACT

Chemical compositions of Archaean volcanic rocks in the Superior province of the Canadian shield have been determined from 261 new analyses from 10 volcanic belts. The analyses are compared with those of the various volcanic associations. This comparison shows that the volcanic rocks of all belts belong to the basalt andesite rhyolite association typical of continental orogenic belts or island arc systems. A primitive continent with continental rocks must, therefore, have existed in early Archaean time.

The Keewatin lava sequences in the various belts are remnants of a great volcanic sequence that covered the southern half of the Superior province of the Canadian shield.

INTRODUCTION

The Archaean strata of the Canadian shield have usually been considered to "occur as scattered troughs and basins of severely deformed and altered volcanic and sedimentary rocks". (Stockwell 1957, p. 5). The term Keewatin was first applied to the lavas of these Archaean troughs in the Lake of the Woods and Rainy Lake areas in northwestern Ontario. The name has subsequently been applied to nearly all similar lavas throughout northern Ontario and northwestern Quebec that were thought to be the oldest strata of the particular region being mapped (Stockwell 1957, p. 29). The composition and regional structure and relationships of these Keewatin lavas are the subjects of this paper.

The Keewatin-type lavas have been described in numerous published reports and papers. Descriptions usually indicate that the volcanic rocks have the composition of basalt or andesite but rhyolite and trachyte have been described, and locally may attain considerable volume. The volcanic rocks commonly consist of metamorphic minerals so that it is not possible to use the mineralogical classification to apply names to the lavas. Primary volcanic structures such as pillows, flow lines, and amygdales are preserved and separate flows may be recognizable. Pyroclastic structures are also abundant in all the volcanic series.

"One of the notable features of the Keewatin assemblages is their great thickness. M. E. Wilson gives an average of 25,000 feet in the Noranda area, and similar thicknesses are reported from many other areas" (Stockwell 1957, p. 30). Sedimentary rocks are commonly interbedded with the volcanic rocks. Much of the sedimentary material is tuffaceous but greywacke, slate, conglomerate, chert, and iron formation are also common. Many larger sedimentary areas were commonly included with or not distinguished from the Keewatin formations but it is now possible to separate many of these gneissic sedimentary areas from the true Keewatin (Wilson and Brisbin, unpublished).

In the past, few attempts have been made to establish the nature of the volcanism because of the lack of original igneous minerals and the scarcity of chemical data. The Keewatin lavas are shown as the oldest formation on most geological maps of the Superior province of the Canadian shield. Thus, they appear to be amongst the oldest known rocks. A knowledge of the nature of Keewatin volcanism is obviously necessary if we are to understand the structure and formation of the continental nucleus. Compositions of lavas are commonly diagnostic of the nature of the volcanic action; hence a study of the composition of the Keewatin volcanic rocks was undertaken so that they could be compared with more recent volcanic piles in various tectonic settings. The project was conceived as part of the Canadian Upper Mantle Project and was supported by grants from the National Research Council. This paper represents the findings to date, based upon suites of rocks from nine volcanic belts in the Superior province of the Canadian shield.

METHOD

Sample Collection

Samples of eight Keewatin volcanic belts were collected by the authors. The sample collecting for the initial study was designed to obtain as many composition types as possible rather than to obtain a statistical sample of the overall composition of a belt. The least altered samples available were collected for analysis. All Keewatin lavas are regionally metamorphosed to some degree but sheared samples, silicified samples, and samples with secondary veinlets and alterations were avoided as far as possible. A clear distinction was made in the field between flow rocks and fragmental rocks, but both types were collected. The abundant exposures of the recently glaciated bedrock surface on which textures, structures, and alteration are clearly etched out partly compensate for the low and medium grade metamorphism of all samples.

The ninth set of samples from the Miminiska area was obtained from Dr. R. C. Holman of the Geological Survey of Canada. A few samples from the Cross Lake belt in Manitoba were collected by D. H. Rousell. A total of 261 new rock analyses were made for this paper. The location of volcanic areas sampled and discussed in this paper is shown on Fig. 1.

Method of Analysis

The project was started using the wet rapid analysis method for silicate rocks but an A.R.L. multichannel vacuum X-ray quantometer became available early in the program. This instrument was then used for all analyses because of its speed and accuracy.

Sample Preparation and Analysis

Samples for analysis by X-ray fluorescence spectrometry are prepared using an adaptation of the method of Rose, Adler, and Flanagan (1963). In this method 0.5000 g of sample (−200 mesh), 0.5000 g lanthanum oxide (La_2O_3), and 1.000 g lithium tetraborate ($Li_2B_4O_7$), are thoroughly mixed in a platinum crucible using a platinum spatula. This mixture is fused at 1000 °C for 30 minutes in a graphite crucible. Enough boric acid (H_3BO_3) is added to the

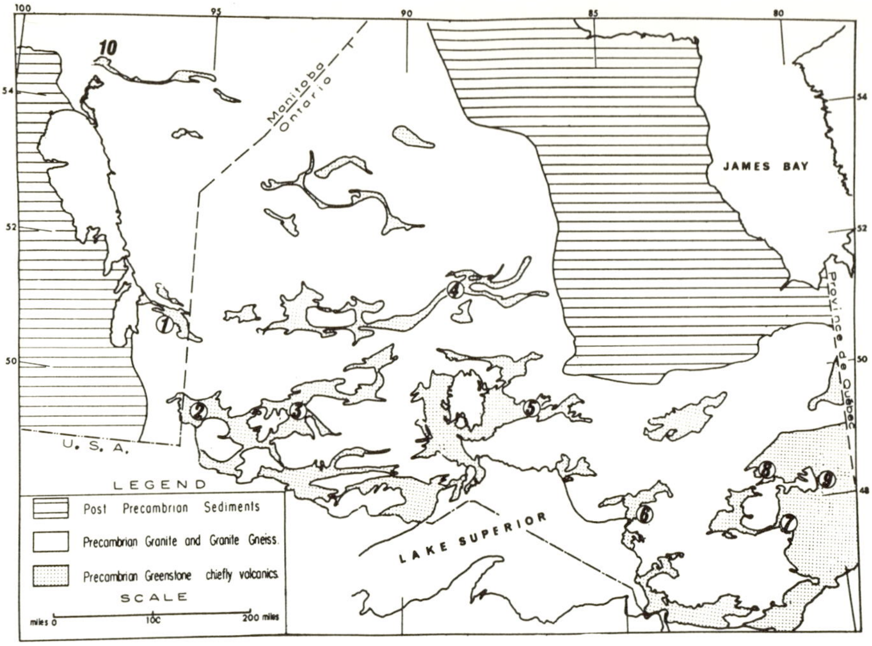

FIG. 1. Location of volcanic belts. 1. Beresford Lake. 2. Kenora. 3. Dryden. 4. Miminiska. 5. Longlac. 6. Goudreau. 7. Shiningtree. 8. Timmins. 9. Kirkland Lake. 10. Cross Lake.

resulting bead to make the total weight of bead plus boric acid equal to 2.1000 g. The bead plus boric acid are ground to −200 mesh in a Bleuler rotary grinder (about 1 minute) then compressed at 50 000 p.s.i. using boric acid as a backing material. Table I shows the results of the analyses of four pellets of syenite rock (CAAS)[1] prepared in this manner.

The pellets prepared in this manner when tested over a period of several months did not experience the intensity loss observed with some methods of sample preparation.

Working curves for the eight elements determined simultaneously were prepared using U.S.G.S. samples G1 and W1 as primary standards. Figure 2 shows the SiO_2 curve for a number of standard samples ranging from 6.0% to 95.96% SiO_2. Samples used in preparation of this curve were G1, W1, NBS[2] 69A, NBS 77, NBS 78, NBS 88, NBS 91, NBS 93, NBS 97, NBS 98, NBS 99, NBS 102, NBS 199, CAAS sulphide ore, and CAAS syenite rock. The curve extends down to 6.0% SiO_2 although none of the rocks analyzed so far has had less than 30% SiO_2. Experience with a variety of silicate rock types has not shown any noticeable matrix effects in the analyses of elements listed in Table I.

Certain constituents cannot be determined by the X-ray fluorescence method; hence total moisture (H_2O) was determined using the rapid method of Shapiro and Brannock (1956), carbon dioxide was determined using an absorption

[1]CAAS = Canadian Association of Applied Spectroscopy.
[2]NBS = National Bureau of Standards.

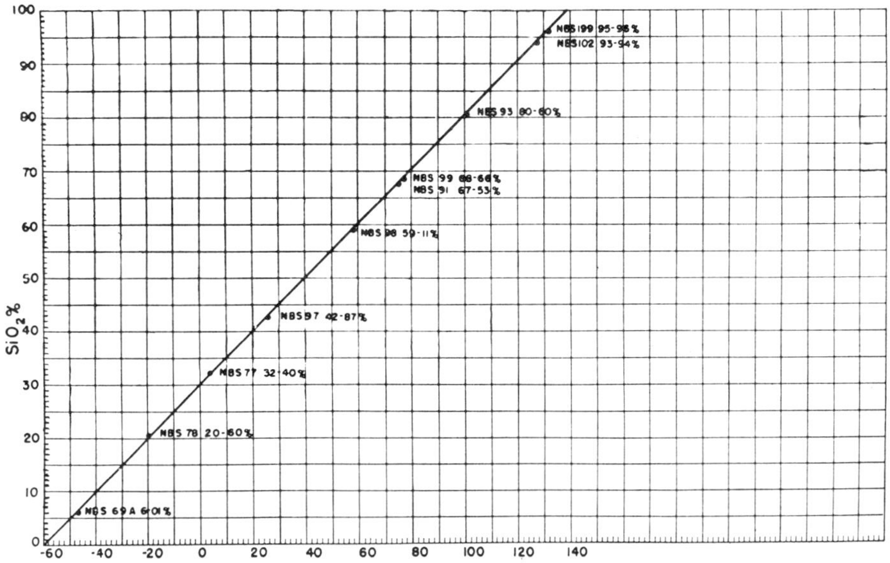

FIG. 2. X-Ray fluorescence calibration curve for SiO_2.

TABLE I

Analysis of syenite rock (CAAS)

Pellet No.	SiO_2	Al_2O_3	Total Fe as Fe_2O_3	MgO	CaO	K_2O	TiO_2	MnO
1	59.0	9.3	8.1	4.0	10.0	2.62	0.46	0.42
2	58.8	9.4	8.3	4.0	10.1	2.58	0.45	0.39
3	58.9	9.1	8.2	3.9	10.2	2.53	0.44	0.41
4	59.2	9.2	8.2	3.75	10.2	2.59	0.46	0.41

train collecting CO_2 on Caroxite, and soda was determined using a Coleman flame photometer.

Processing of Analytical Results

The results of the analyses were tabulated on Fortran cards and the following data were determined on a 1620 I.B.M. computer.

1. Niggli molecular equivalent normative minerals.
2. Ritmann sigma index.
3. Larsen variation diagram, values in percent.
4. Femalk metal values in atom percent.
5. Alkali metals values in atom percent.
6. SKM diagram position.
7. ACF diagram position
8. The Niggli values.
9. Various metallic ratios.
10. The name of the rock using normative minerals.
11. Nature and amount of alteration.

Chemical analyses were also tabulated by the machine for purposes of record.

The above data were plotted and the arrays of points on the diagram were used to evaluate the chemical nature of the lavas. Analytical data for other lava suites from recent environments were programmed in the same manner and the results were used as controls for comparison and contrast with the Archaean lavas.

The main control groups used were as follows.

Oceanic association
- Truk Islands (Stark and Hay 1963)
- Honolulu series, Hawaiian Islands (Winchell 1947)
- Kerguelen Islands (Turner and Verhoogen 1960, p. 190)
- Tahiti, Society Islands (Turner and Verhoogen, 1960, p. 187)

Continental alkali association
- Alkaline olivine basalts from continental sources (Turner and Verhoogen 1960, p. 183)

Orogenic association (continental and island arc)
- Volcanic rocks of orogenic regions (Turner and Verhoogen 1960, p. 281)
- Fuji Province (Turner and Verhoogen 1960, p. 281)
- Cascade Province (Waters 1961)
- Marianas Islands (Schmidt 1957)
- Izu Group (Ono 1962)

METAMORPHISM AND ALTERATION

The petrographic examination of the samples will not be complete for some time but it is apparent that most samples are in the greenschist or epidote amphibolite metamorphic facies with some in the amphibolite facies. Igneous and fragmental textures are still clear in some belts, but in others the original mineral texture has been lost.

The Larsen variation diagrams and silica variation diagrams of the Archaean lavas, although not shown in this paper, are useful in assessing the possible migration of cations during metamorphism. In general the Larsen diagrams for individual areas show that the cations behave in a manner similar to those of fresh lavas. The spread of points on the variation diagrams for the Precambrian lavas is similar to that of the lavas from recent volcanic fields such as those of Truk.

The least altered volcanic rocks were selected during field sampling as mentioned earlier, but limited petrographic work as well as the normative calculations showed that many have large amounts of carbonate minerals and a few have been silicified. Lavas containing more than 4% normative calcite and 42% quartz were considered too altered and were eliminated from the calculation of average lava composition. These arbitrary cutoff levels were selected after a study of histograms of the calcite and quartz abundances in the rocks. Figure 3 is a histogram showing the frequency distribution of normative calcite in all samples of Archaean volcanic rocks used in this study. The calcite

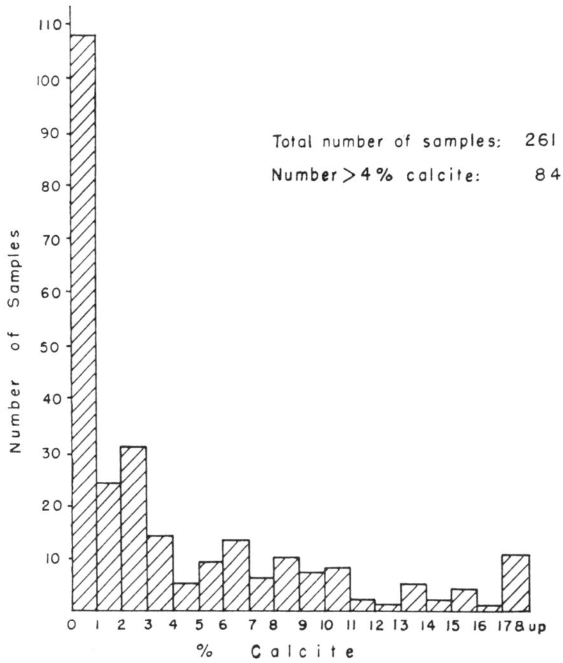

FIG. 3. Frequency distribution of normative calcite in volcanic rocks of the Superior province.

occurs as a fine-grained alteration product of the calcium silicate minerals and not as introduced calcite in veinlets.

Lavas and fragmental volcanic rocks are included in the samples but each type was identified in the field permitting later correlation of genetic types. It turns out that the fragmental rocks tend to be somewhat more altered than the lavas; however, analyses of the fresher fragmental types plot close to those of the lavas in the various diagrams, suggesting that the fragmental rocks were representative of the other extrusion products at the time of volcanism.

The evidence of the variation diagrams shows that cation migration is not very great in most specimens and that the analyses are reliable representations of the compositions of the original lavas.

AVERAGE COMPOSITION OF CANADIAN ARCHAEAN LAVAS

The average composition of Canadian Archaean lavas of the Superior province of the Canadian shield is shown in Table II. Metamorphism has destroyed the igneous minerals which are used for rock classification; thus the rock names of the lavas in the table were derived by computer from the proportions of the normative minerals. The Niggli molecular equivalent normative minerals were used according to a slight modification of the Classification of Igneous Rocks used by Nockolds (1954).

The validity of naming rocks from the molecular equivalent normative minerals was checked by computing the names of Nockolds' (1954) rock types

TABLE II

Average composition of Canadian archaean lavas

Lava No. samples	Basalt 53	Andesite 20	Dacite 3	Rhyodacite 3	Rhyolite 24	Na-rhyolite 12	K-rhyolite 12
SiO_2	49.83	51.83	61.74	67.16	74.11	72.43	75.78
Al_2O_3	14.64	14.53	15.17	16.73	13.59	14.23	12.95
Fe_2O_3	3.03	2.96	1.87	.89	.64	.84	.44
FeO	8.77	8.46	5.74	2.77	1.27	1.54	.99
MgO	7.36	6.22	2.30	1.86	1.04	1.47	.62
CaO	10.46	8.42	4.80	3.29	.60	.88	.31
Na_2O	2.02	3.40	3.57	3.61	2.74	4.67	.81
K_2O	.23	.29	.54	1.31	3.82	1.32	6.32
H_2O	1.81	1.56	1.87	.86	.90	1.04	.76
CO_2	.33	.29	.47	.24	.51	.58	.43
TiO_2	.94	1.11	1.22	.46	.14	.16	.11
MnO	.21	.18	.14	.02	.02	.02	.02
P_2O_5	.19	.22	.32	.12	.12	.20	.03
	99.82	99.52	99.73	99.33	99.50	99.38	99.57

from the chemical analyses and comparing these names with those derived from the modal minerals. The names of the control groups of recent lavas were checked in the same manner. The only appreciable discrepancy appeared in intermediate members because K_2O is assigned entirely to normative orthoclase, whereas in the mode it is distributed between orthoclase and other minerals, such as the sheet silicates. After the various rock groups were scanned, it was found that only the boundary between dacite and rhyodacite was appreciably affected. Trial and error tests showed that when the boundary between rhyodacite and dacite was changed from 10% to 18% of the total feldspar being potassium feldspar, the computed name from normative minerals was similar to the rock name from the modal minerals.

Normative minerals cannot be used to name the Archaean lavas with a large proportion of normative and modal calcite because much of the CaO in calcite was probably derived from plagioclase. Therefore, names of volcanic rocks which contained more than 4% normative calcite were not used in any compilation.

The analyses used in Table II represent analyses of flow lavas which have minimal carbonate alteration. The original 261 analyses were thus reduced to 103 analyses by eliminating all fragmental rocks and any flow rocks which contained more than 4% normative calcite. Low normative calcite is deemed necessary in the rocks used in Table II because carbonatization may have affected cation distribution in the lavas in addition to introducing errors in the rock name.

The rhyolites of column 5, Table II, have been separated into soda rhyolites and potash rhyolites in columns 6 and 7. Figure 4 is a histogram showing the frequency distribution of the soda:potash ratio in the rhyolites. The bimodal distribution indicates the presence of two types of rhyolite.

High-potash and high-soda rhyolite lavas commonly occur in the same greenstone belts, both types occurring in the Miminiska, Kenora, Timmins, and Shiningtree belts. Acid fragmental rocks analyzed in this study turned out to

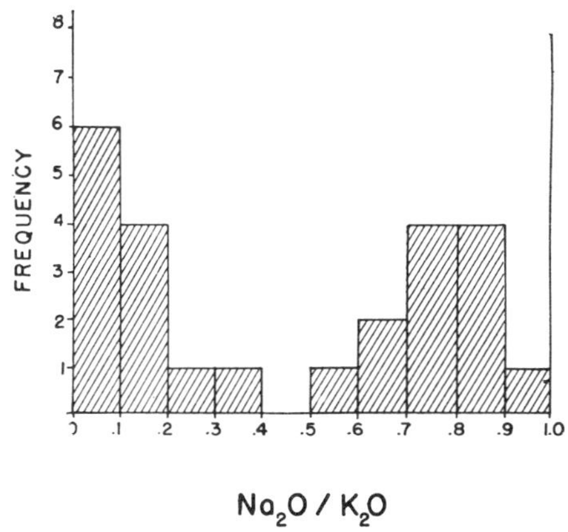

FIG. 4. Frequency distribution of the soda: potash ratio of rhyolites in the Superior province.

be dominantly sodic, but potassium varieties occur. The significance of the two types is not yet understood. Further field mapping and sampling of rhyolite flows are necessary before we speculate about the origin of the two types of lava.

Table II shows a scarcity of lava specimens in the intermediate silica range occupied by the dacite and rhyodacite groups indicating a possible compositional gap between basic and acid lavas. This possibility was tested by using the analyses of the fragmental rocks in addition to those of the flow rocks, but excluding the more altered varieties containing more than 4% normative calcite or 42% normative quartz. The results of the compilation are shown in Table III.

TABLE III
Frequency of rock types (flows and fragmentals)

Rock name	No. specimen	% of specimen
Basalt	86	48.6
Andesite	23	13.0
Dacite	8	4.5
Rhyodacite	23	13.0
Dellenite	3	1.7
Rhyolite	33	18.6
Trachyte	1	0.6

The compositional gap is no longer apparent and it appears that the intermediate varieties tend to form fragmental deposits rather than flows. Table III should not be taken to represent the average amount of the various rock types in the Archaean volcanic belts because samples did not represent unit areas.

NATURE OF ARCHAEAN VOLCANISM

The primary purpose of the study of the composition of the oldest volcanic rocks of North America was, as stated earlier, to determine the nature of the volcanism. Compositions and nature of the lavas were therefore compared and contrasted with the compositions and nature of the five (generally accepted) volcanic associations of the crust; the alkaline olivine–basalt association of the ocean basins, the flood basalt quartz diabase association of non-orogenic continental areas, the leucite–basalt potash–trachybasalt association of non-geosynclinal regions, the spilite keratophyre association of geosynclines, and the basalt andesite rhyolite association of continental orogenic belts.

Comparison with Alkaline Olivine–Basalt Association of the Ocean Basins

The strongly alkaline composition of the alkaline olivine–basalt association may be demonstrated by certain diagrams and indices derived from the chemical composition of the lavas. The most useful diagrams and indices for this purpose are as follows:

(1) the Niggli values for alk plotted against si,
(2) the alkali–lime index as proposed by Peacock (1931), and
(3) the Ritmann sigma value and suite index.

A chemical division between oceanic lavas and continental orogenic or island arc lavas was established on the alk–si diagram by plotting the values from analyses of three typical oceanic suites, Truk (Stark and Hay 1963), Kerguelan (Turner and Verhoogen 1960), and Tahiti, (Turner and Verhoogen 1960); and the values from three typical continental orogenic or island arc suites, Izu (Ono 1962), Fuji (Turner and Verhoogen 1960), and Marianas (Schmidt 1957). Figure 5 shows that a smooth curve can be drawn separating the two

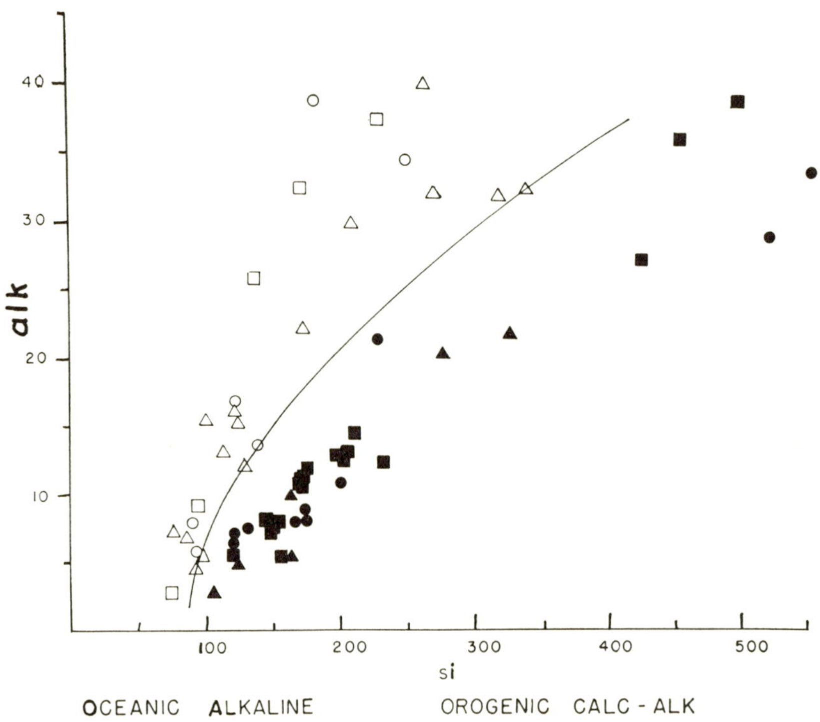

FIG. 5. Plot of Niggli alk–si values for control groups of lavas. Oceanic alkaline: △ Truk, □ Tahiti, ○ Kerguelen; orogenic calc-alk: ▲ Fuji, ■ Izu, ● Mariana.

lava associations. The diagram was checked by plotting analyses from other oceanic islands in the Pacific and Atlantic, and analyses of continental orogenic lavas from American Cordilleran areas such as the Cascades. The division between the lava types is apparently valid. It is recognized, however, that both types may occur on some oceanic islands such as Hawaii.

The alk–si plots for the Archaean lavas from the various belts in the Superior province are shown in Fig. 6. It is apparent that the Keewatin lavas have

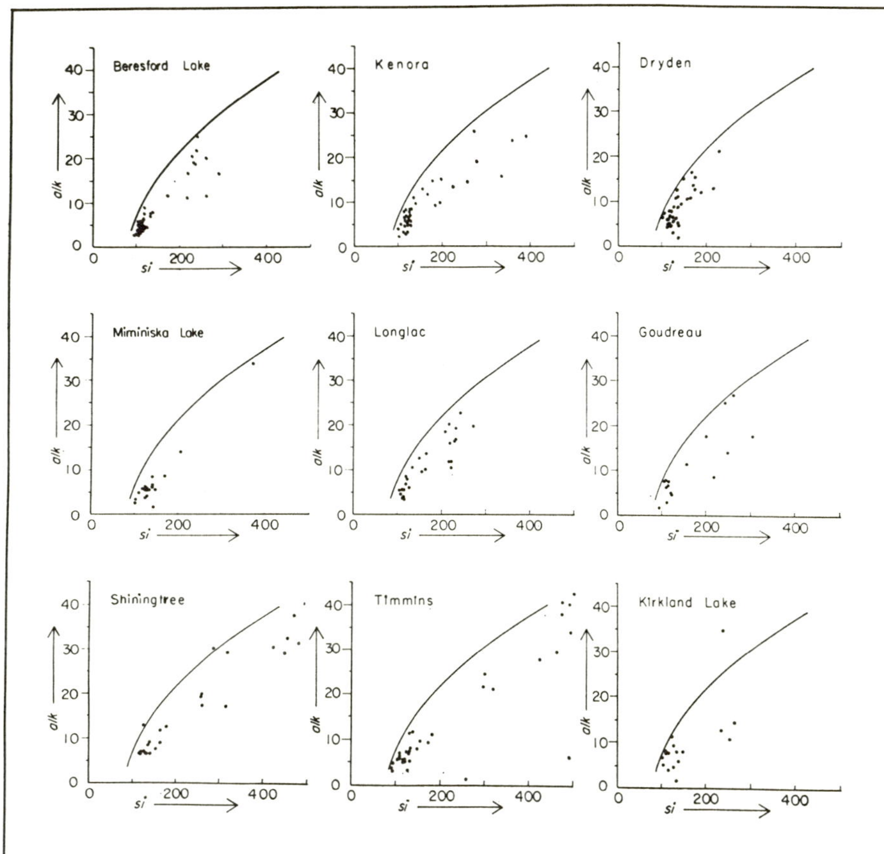

FIG. 6. Plot of Niggli alk–si values for Archaean lava belts in the Superior province.

alk:si ratios similar to those of continental orogenic or island arc lavas and that they are not similar to oceanic alkaline lavas. A single sample from the Kirkland Lake area is alkaline trachyte.

Alkali:lime indices of the Canadian lavas and control groups of lavas of known associations were determined by the method proposed by Peacok (1931). The lime:alkali index of the typical oceanic type from the Truk islands is 51.5 indicating an alkali:calcic character. The continental Izu group from Japan has a lime:alkali index of 62.5 indicating a calcic nature. The lime:alkali indices of the Keewatin lavas are shown in Fig. 7. The range of the indices is 56.1 to

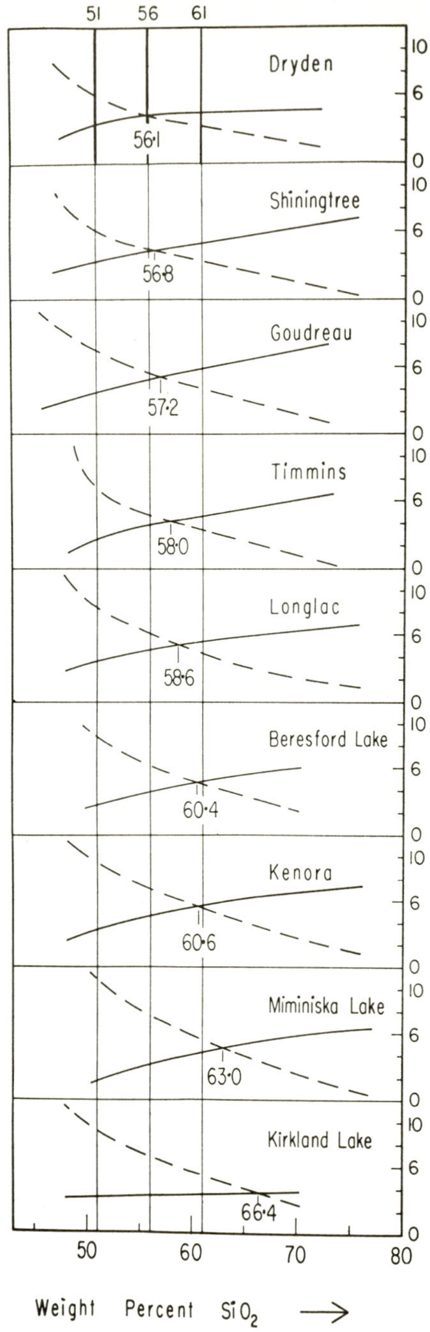

FIG. 7. Lime–alkali indices of lavas in volcanic belts in the Superior province. Weight percent SiO_2 is plotted against weight percent CaO (broken line) and $Na_2O + K_2O$ (solid line).

66.4 indicating that all are calc–alkalic or calcic. These indices are derived from the analyses of all specimens regardless of the flow or fragmental nature or degree of alteration of the specimens, so they may be slightly modified when only fresh lavas are considered. However, the general calc–alkalic or calcic nature of the Keewatin lavas is unmistakable.

Rittmann (1962, p. 110) has used the soda, potash, and silica contents of lavas to derive a suite index and suite type. Statistical examination of these indices and types shows that various petrographic provinces may be differentiated from one another. Table IV shows the suite indices and suite types for

TABLE IV
Rittmann suite index

Area	No. specimens	Suite index	Suite type
Miminiska	16	0.6	Extreme calc–alkaline
Dryden	13	0.7	Extreme calc–alkaline
Beresford Lake	34	0.9	Extreme calc-alkaline
Kenora	41	1.1	Strong calc–alkaline
Cross Lake	8	1.1	Strong calc–alkaline
Timmins	22	1.2	Strong calc–alkaline
Kirkland Lake	13	1.3	Strong calc–alkaline
Shiningtree	15	1.4	Strong calc–alkaline
Longlac	23	1.4	Strong calc–alkaline
Algoma	23	2.1	Average calc–alkaline

the Keewatin areas after elimination of altered specimens containing more than 4% calcite. It is again apparent that the Keewatin volcanic rocks are strongly calc–alkaline. The Peacock and Rittmann computations of suite character each demonstrate the calc–alkaline nature of the Archaean volcanic belts, but a comparison of Fig. 7 and Table IV shows that exact calc–alkalinity indices are not comparable when computed by different methods using different oxides.

Comparison with Flood Basalts

Flood basalt extrusion occurs in non-orogenic areas. It is characterized by thick series of extensive flows of uniform composition. Fragmental rocks are not prominent. Examples of such volcanism include the Columbia plateau in northwestern United States and the Deccan basalts in India.

The Keewatin lavas occur in very thick formations but they are not uniform in composition. The present sampling is not representative of the average composition of the volcanic belts, but it is apparent that the complete range of compositions from basalt to rhyolite is present in almost all belts. Andesite may be approximately one-third as abundant as basalt. Dacites, rhyodacites, and rhyolites are abundant and probably form much more of the volcanic section than usually supposed because dacites are not distinguished from andesites and basalts on most geological maps of the Superior province. Compositions of the Keewatin volcanic rocks are therefore more closely comparable to lavas associated with orogenic volcanism than to non-tectonic flood basalts.

Fragmental rocks are abundant in Keewatin belts and may even be predominant in some belts, such as the Kenora belt. This fact would also lend support to an orogenic association for the Keewatin volcanic belts.

Comparison with Leucite–Basalt, Potash–Trachybasalt Association of Non-geosynclinal Regions

This association occurs at widely scattered places on all continents, although individual areas are restricted. Chemically these lavas are remarkable because of their highly potassic character, which is particularly apparent in the basic members of the lava series, lavas in the 50% SiO_2 range containing from 2.5% to 9.5% K_2O. Turner and Verhoogen (p. 235, 1960) list several examples of this association.

Keewatin lavas have an extremely low potassium content in the basic members. Andesites and basalts average 0.25% and all have less than 1% K_2O. The usual Keewatin basalt andesite rhyolite association with very low potash content bears little resemblance to the potash-rich leucite–basalt association. The potash-rich rhyolites in the Keewatin series do not have any corresponding basic members and do not have the composition of differentiates from a leucite–basalt series.

Comparison with the Spilite Keratophyre Association of Geosynclines

The geosynclinal accumulation phase of the tectonic cycle is usually characterized by eruption of dominantly basic lavas (Turner and Verhoogen 1960, p. 257). Rittmann (p. 164, 1962) discussing geosynclinal volcanicity uses the geosynclinal concept in its original sense as vast elongated troughs filled with thick sedimentary deposits. The characteristic volcanic association is that of serpentinite flows and basic pillow lavas. The composition of the effusives is basaltic or picritic (Rittmann p. 165, 1962).

The spilite association of lavas, tuffs, and associated intrusions is prominent in and almost confined to the geosynclinal environment (Turner and Verhoogen 1960, p. 258). Other lavas of normal basalt composition, or the basalt andesite rhyolite suite, however, may also occur in geosynclines. The most characteristic feature of spilites is that they are highly sodic. Other chemical characteristics include a relatively high TiO_2 content. Spilites also have characteristic mineralogical and textural features, but these are not likely to be retained in the metamorphosed Keewatin lavas.

Figure 8 shows the contrast between the soda:silica ratios in spilites and the Keewatin lavas. The average Keewatin lava has only approximately one-half of the soda content of the average spilite, and even the most sodic specimens of Keewatin basalts and andesites contain less soda than the average spilite.

The contrast in TiO_2 content between spilites and Keewatin andesites and basalts is as marked as that of soda. Turner and Verhoogen (1960, p. 262) show the average spilite contains 51.22% SiO_2 and 3.32% TiO_2. The average of 73 unaltered Archaean andesites and basalt flow rocks is 50.3% SiO_2 and 0.99% TiO_2. Only one Keewatin specimen contains more than 2.0% TiO_2, and only six specimens contain one-half of the average amount of TiO_2 in spilites.

"The spilitic association, whatever its relation to the other basaltic kindreds

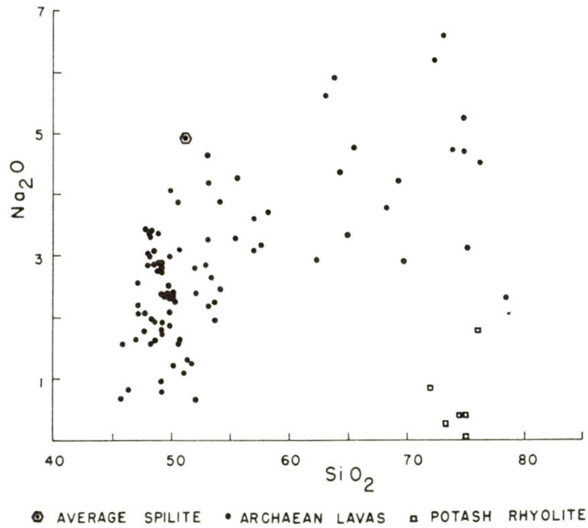

FIG. 8. Contrast between Na_2O content of average spilite, Archaean lavas of the Superior province, and potash–rhyolite of the Superior province.

may be, is one of striking individuality, maintained in widely scattered provinces of all ages and recognized wherever the rocks of geosynclinal terranes have been petrographically investigated" (Turner and Verhoogen 1960, p. 263). The Keewatin volcanic rocks certainly are not spilites. Sixty analyses of basic intrusive rocks in the Keewatin show the intrusions have soda and titania contents similar to those of the lavas.

It is apparent that positive compositional evidence relating the Keewatin volcanism to a geosynclinal facies is completely lacking. None of the characteristic geosynclinal lava compositions are present in any of the 261 specimen analyzed from 10 different areas.

Comparison with the Basalt Andesite Rhyolite Association of Continental Belts

This type of association is typical of continental and island arc areas of the earth's crust and is characteristically associated with orogenic regions. Present-day volcanic districts of this type include Japan, Indonesia, the Andes, and the Antilles. It is the type of volcanism that occurs all around the Pacific Ocean, in the Cordillera of North and South America, and in the east Asian island arcs.

Tables II and III show that the lavas are typical of the basalt andesite rhyolite association. Only one specimen of the 261 samples analyzed and named by the computer has a strong alkaline affinity. This specimen is a trachyte from Boston township in the Kirkland Lake area. The calc–alkaline nature of the lava suites has been shown by all tests so it may be concluded that the Keewatin volcanism is continental and not oceanic. Thus a primitive continent or island system was present during the time of the Keewatin volcanism.

CONCLUSIONS

The composition of the Archaean lavas shows a similar type of volcanism in the 10 areas studied. They appear to be differentiated series and are amenable to treatment as such.

The compositions of the volcanic rocks are typical of continental orogenic associations or island arc systems.

The Keewatin lavas are remnants of a great volcanic sequence that covered the southern half of the Superior province of the Canadian shield. The time span of lava extrusion may have been as long as that of the present-day orogenic belts, a number of which have been active from the early Tertiary to the present time.

Great thicknesses of pillow lavas suggest that much of the Keewatin lava consisted of underwater flows, but interlayered sedimentary rocks show that land areas occurred nearby. Many of the massive flows and the volcanic breccias and lapilli tuffs are typical of subaerial volcanism.

Work remains to be done on areas of high-potash rhyolites and on the extensive carbonate alteration of the Archaean Lavas.

The term Keewatin seems applicable to the volcanic sequences of the Superior province of the Canadian shield because the volcanic association in each belt is similar and because most of the belts are interconnected through more highly metamorphosed volcanic equivalents. It is thus logical to examine the geochemistry of the volcanic rocks of the whole Superior province because, for example, areal variation such as that of the alumina content of the basalts may be meaningful in working out the continental history. Sequences in individual belts cannot be correlated but the present type of investigation and an age-dating program in the Superior and neighboring shield provinces may lead ultimately to a picture of the stratigraphic correlation in the various areas and provinces.

REFERENCES

NOCKOLDS, S. R. 1954. Average chemical composition of some igneous rocks. Geol. Soc. Am. Bull. **65**, 1007.

ONO, KOJI. 1962. Chemical composition of volcanic rocks in Japan. Geological Survey of Japan.

PEACOCK, M. A. 1931. Classification of igneous rock series. J. Geol. **39**, 54.

RITTMANN, A. 1962. Volcanoes and their activity. *Transl. by* E. A. Vincent (Interscience Publishers).

ROSE, H. J., JR., ADLER, I., and FLANAGAN, F. J. 1963. X-Ray fluorescence analysis of the light elements in rocks and minerals. Appl. Spectry. **17**, 81.

SCHMIDT, R. G. 1957. Geology of Saipan, Marianas Islands, petrology of the volcanic rocks. U.S. Geol. Surv. Profess. Paper No. 280B, 127.

SHAPIRO, L. and BRANNOCK, W. W. 1956. Rapid analysis of silicate rocks. U.S. Geol. Surv. Bull. 1036-C, 19.

STARK, J. T. and HAY, R. L. 1963. Geology and petrography of volcanic rocks of the Truk Islands, East Caroline Islands. U.S. Geol. Surv. Profess. Paper No. 409.

STOCKWELL, C. H. 1957. Geology and economic minerals of Canada. Can. Dept. Mines Tech. Surv. Geol. Surv. Can. Econo. Geol. Ser. No. 1, 4th ed.

TURNER, F. J. and Verhoogen, J. 1960. Igneous and metamorphic petrology. 2nd ed. (McGraw-Hill).

WATERS, A. C. 1961. Stratigraphic and lithological variation in the Columbia River basalt. Am. J. Sci. **259**, 583.

WINCHELL, H. 1947. Honolulu series, Oahu, Hawaii. Geol. Soc. Am. Bull. **58**, 1.

11: A VERY ANCIENT ISLAND ARC

R. E. Folinsbee, H. Baadsgaard, G. L. Cumming, and D. C. Green

Department of Geology and Physics, University of Alberta, Edmonton, Alberta, Canada

Abstract. In the early Precambrian of the Slave Province, Yellowknife area, N.W.T., Canada, mafic volcanic piles accumulated to heights of 2000 to 12000 meters, presumably as ancient island arcs. The pattern of eruption at Yellowknife commenced with the outpouring of a thick sequence of tholeiitic basalts and terminated with a thinner sequence of quartz latites and dacites, following in general the differentiation patterns noted by Kuno. A rubidium-strontium isochron suggests that these volcanics poured out in submarine eruptions 2.6 billion years ago. If, as has been suggested, the volcanic piles rest on basement gneisses (granodiorites), these gneisses must be amongst the most ancient crustal rocks known in North America. However, Rb/Sr and K/Ar dates of 2.5 to 2.6 b.y. for the granodiorites suggest that either the granite rocks are for the most part younger than the volcanics or that they were completely updated during the general period of Kenoran orogeny and intrusion at the close of the Archean.

The Slave Province of the Northwest Territories was one of the first in Canada to yield dates greater than 2 billion years by the K/Ar and galena methods [*Shillibeer and Russell,* 1954; *Russell et al.,* 1954]. The area has since been subject to many investigations by radiometric dating [*Folinsbee,* 1955; *Burwash and Baadsgaard,* 1962; *Leech,* 1966] and has been encompassed by the Geological Survey of Canada project to produce a new tectonic map of the Canadian shield incorporating radiometric dates [*Stockwell,* 1962]. It is now clearly recognized that the Kenoran orogeny (yielding K/Ar dates of 2500 ± 150 m.y.) was the last period of intensive folding, metamorphism, and intrusion undergone by the Yellowknife group of volcanic and sedimentary rocks that were developed in an island arc environment of the Slave Province—an area of about 20,000 km^2, lying north of Great Slave Lake, N.W.T. (Figure 1).

GEOLOGIC SETTING

In the Archean of the Yellowknife area, mafic volcanic piles, characterized by marvelously well developed pillows in many of the basalt flows, accumulated to a thickness of 2000 to 12000 meters in a structural setting that seems to have been remarkably similar to that of a classic island arc, the fossa magna of Japan, as described by *Matsuda* [1962]. The volume of tholeiite and alkaline basalt magma in the Slave Province is of the same order as that of the fossa magna, 20,000 km^3. It is clearly demonstrable that thick sequences of volcanic rocks followed by diorite intrusion and graywacke type sedimentation were extruded, intruded, eroded, and buried within a short period of time during the Miocene in Japan. It was felt that if the Yellowknife and Japanese volcano-sedimentary sequences were found to yield a similar age pattern, the island arc analogy would be firmly established. The Yellowknife rocks are chemically variable and offered attractive possibilities for dating by the application of the Rb/Sr whole rock method (Figure 1).

The pattern of eruption in the Yellowknife area is one of a succession of lavas of slowly increasing acidity, capped by a layer of acidic volcanic rocks.

[*Editor's Note:* Material has been omitted at this point.]

The Yellowknife volcanics were extruded in an island arc environment 2625 (2480) m.y. ago, as is evidenced by a well defined Rb/Sr isochron derived by mass spectrometric analysis of 6 rocks from the volcanic sequence, ranging in composition from basalt to dacite. The range in rubidium values is particularly

[1] A Canadian contribution to the International Upper Mantle Project.

TABLE 4. Comparison of Geologic Setting and History of the Yellowknife Area and the Fossa Magna of Japan

Feature	Fossa Magna, Japan	Yellowknife, Canada
		Prosperous pegmatite and other granite intrusions.
Narrow crustal folding	Middle and upper Miocene early Pliocene Fujikawa series of rhythmically alternating sandstones, mudstone and conglomerate, minor intermediate volcanics.	Upper Yellowknife graywackes (alternating sandstone and mudstones), conglomerate, minor intermediate volcanics.
Intrusion and uplift	Potash-deficient Miocene quartz diorites intrude Misaka, occur as boulders in Fujikawa.	Oligoclase granodiorites and diorites intrude Yellowknife volcanics, occur as boulders in Yellowknife conglomerates.
Broad crustal downwarping	Extrusion lower Miocene Misaka pillow basalts and tuff, volcanic breccia agglomerate intercalated shales and limestones, extrusives becoming dacitic.	Extrusion lower Yellowknife pillow basalts, tuff, volcanic breccia, agglomerate, intercalated cherts, extrusives becoming dacitic to rhyolitic.
Basement	Lower Tertiary and Mesozoic (unseen).	Primary crust or basement gneisses (unseen).
Volume	20,000 km³ volcanics	20,000 km³ volcanics

(Side annotations: 15 m.y. spanning narrow crustal folding to broad crustal downwarping in both columns; Thickness: 5000 meters / 3200+ meters (folding), 10000 meters / 12000 meters (downwarping).)

favorable for defining an isochron, and the volcanics have been little disturbed by later intrusions. All the granitic rocks belong to a closely related granite series ranging in composition from quartz diorite through granodiorite to granite and occur as plutons intruding the volcanics and sediments. The western granodiorite gives a younger Rb/Sr date of 2610 (2460) m.y., confirmed by potassium-argon dates; however, these potassium-argon dates may represent only the culminating orogeny and intrusion of the later granites rather than the true age of intrusion of the earlier members of the granite series, the quartz diorites and granodiorites. The writers advance the possibility (Table 4) that, as in the fossa magna of Japan, the oligoclase granodiorites were emplaced into the volcanics of the Yellowknife group and unroofed before the laying down of the Yellowknife conglomerates which contain boulders of oligoclase granodiorite [*Folinsbee et al.*, 1956].

The whole pattern of extrusion of basic to acid volcanics, graywacke type sedimentation, and intrusion of a granite series seems to be remarkably similar to the pattern in both time and space for the fossa magna of Japan, and other island arc areas of the world, though the Yellowknife volcanics are remnants of a particularly ancient volcanic pile.

Acknowledgments. The lead isotope dating of Yamba Lake monazite was done by Lorin Stieff of the U. S. Geological Survey; three of the rock analyses were carried out in the Rock Analysis Laboratory of the University of Minnesota by Thaemlitz and Kane, the fourth at the University of Alberta by *Van Breemen* [1965]. Work in radiometric dating at the University of Alberta is supported by the National Research Council of Canada, the Geological Survey of Canada through the National Advisory Committee for Research in the Geological Sciences, and by the Boreal Institute of the University of Alberta. This assistance is gratefully acknowledged.

REFERENCES

Baadsgaard, H., R. E. Folinsbee, and J. Lipson, Potassium-argon dates of biotites from Cordilleran granites, *Bull. Geol. Soc. Am., 72,* 689–702, 1961.

Baragar, W. R. A., Geochemistry of the Yellowknife volcanic rocks, *Can. J. Earth Sci., 3,* 9–30, 1966.

Boyle, R. W., The geology, geochemistry, and origin of the gold deposits of the Yellowknife district, *Geol. Surv. Can., Memoir 310,* 70–76, 1961.

Burwash, R. A., and H. Baadsgaard, Yellowknife-Nonacho age and structural relations, in *Tectonics of the Canadian Shield, Roy. Soc. Can. Spec. Publ. 4,* pp. 22–29, 1962.

Edie, R. W., Petrography and petrology of the Cameron River volcanic belt, N.W.T., unpublished M.Sc. thesis, geology, Univ. of Alberta, 1949.

Folinsbee, R. E., Archean monazite in beach concentrates, Yellowknife geologic province, Northwest Territories, Canada, *Trans. Roy. Soc. Can., Sect. IV,* [3]*49,* 1955.

Folinsbee, R. E., J. Lipson, and J. H. Reynolds, Potassium-argon dating, *Geochim. Cosmochim. Acta, 10*(1, 2), 60–68, 1956.

Geological Survey of Canada, Age Determinations and Geological Studies, *Papers 61-17, 62-17, 63-17, 64-17,* and *65-17* (J. A. Lowdon, G. R. Lachance, R. D. Stevens, C. H. Stockwell, R. Y. H. Rimsaite, H. W. Tipper, R. K. Wanless), 1961–1965.

Henderson, J. F., Gordon Lake South, *Map 645A,* Geol. Surv. Can., 1941.

Jolliffe, A. W., Yellowknife Bay, N.W.T., *Map 709A,* Geol. Surv. Can., 1942.

Jolliffe, A. W., Prosperous Lake, N.W.T., *Map 868A,* Geol. Surv. Can., 1946.

Kuno, H., Fractionation trends of basalt magmas in lava flows, *J. Petrol., 6,* 302–321, 1965.

Leech, A., Potassium-argon dates of basic intrusive rocks of the District of Mackenzie, N.W.T., *Can. J. Earth Sci., 3,* 389–412, 1966.

Matsuda, T., Crustal deformation and igneous activity in the South fossa magna, Japan, in *Crust of the Pacific Basin, Geophys. Monograph 6,* pp. 140–150, American Geophysical Union, Washington, D.C., 1962.

McIntyre, G. A., C. Brooks, W. Compston, and A. Turek, The statistical assessment of Rb-Sr isochrons, *J. Geophys. Res. 71,* 5459–5468, 1966.

McMullen, C. C., K. Fritze, and R. H. Tomlinson, The half-life of Rubidium-87, *Can. J. Phys., 44,* 3033-3038, 1966.

Moore, E. S., Porphyries of the Porcupine area, Ontario, *Trans. Roy. Soc. Can., Sec. IV,* [3]*48,* 1954.

Novikova, A. S., On some aspects of Precambrian tectonics of the East European Platform, in *Symposium on Precambrian of the Commission for Absolute Age Determination in Geological Formations,* Div. Earth Sciences, Academy of Sciences, Kiev, USSR, Oct. 1966 (in press).

Purdy, J., and D. York, A geochronometric study of the Superior Province near Red Lake, Northwestern Ontario, *Can. J. Earth Sci., 3,* 277–286, 1966.

Russell, R. D., R. M. Farquhar, G. L. Cumming, and J. T. Wilson, Dating galenas by their isotopic constitutions, *Trans. Am. Geophys. Union, 35,* 301–309, 1954.

Shillibeer, H. A., and R. D. Russell, The potassium-argon method of geological age determination, *Can. J. Phys., 32,* 681–693, 1954.

Stockwell, C. H., A tectonic map of the Canadian Shield, in *Tectonics of the Canadian Shield, Roy. Soc. Can. Spec. Publ. 4,* pp. 6–15, edited by Stevenson, 1962.

Thornton, C. P., and O. F. Tuttle, Chemistry of igneous rocks, 1, Differentiation index, *Am. J. Sci., 258*(9), 664–684, 1960.

Tilton, G. R., Comparison of isotopic dating methods, Summer course on Nuclear Geology, Varenna, unpublished lectures, pp. 331–345, 1960.

Van Breemen, O., Thermally induced relocation of strontium and rubidium in a granodiorite, unpublished M.Sc. thesis, geology, University of Alberta, 1965.

White, W. H., Cordilleran tectonics in British Columbia, *Bull. Am. Assoc. Petrol. Geologists, 43,* 60–100, 1959.

PAPERS 12 THROUGH 14: COMMENTARY

The results of mapping by a research group of the University of Western Australia in a key area west of Lake Lefroy, characterized by remarkably fresh and extensive exposure along the west shore of the lake, are summarized in Paper 12. Such continuous, fresh, low cliff and pavement exposure is seldom seen in Canada (where the very fresh rocks form island exposures in glacial terrain) or in South Africa (where such outcrop is restricted to river valley sections such as that of the Komati Gorge). In contrast, the inland exposure away from Lake Lefroy (a salt-pan lake) is extremely degraded and sparse, and the east shores are covered by gypsum dunes.

The method of mapping used was quite different from the regional approach of Williams (Paper 13). Key areas of small extent were studied in detail, and linking mapping was carried out to couple up through poor terrain. The various units mapped were given the status of "beds," and were defined on local criteria. No attempt was made to introduce regional subdivisions, as has been attempted at Kurnalpi (Paper 13); but Gemuts and Theron (1973) have extended these subdivisions to cover this area in the context of the entire greenstone tract south of Kalgoorlie. Although previous reports in this series had been issued covering adjacent areas, Williams's report is exceptional and outstanding, for in it he made one of the first attempts to derive a realistic stratigraphic and structural analysis of the volcanic–sedimentary sequence of the Kalgoorlie area. The text is unsuitable for facsimile reproduction, being only available in cyclostyled provisional edition; the printed edition is as yet unissued. Thus, only a summary is given here, prepared by the volume editor and checked by Williams. This paper marks a milestone in the study of the vast tract of Archean terrain in Western Australia. It is to be hoped that the remainder of that tract will be similarly treated, and later subjected to studies of key areas and specialized aspects by academic geologists, like Williams basing their work on field mapping, and also to regional synthesis in the manner of Gemuts and Theron (1973).

Comments that may be advanced are, first, that too much uncritical extension should not be given to the *association units* throughout the Yilgarn Block. Regional geologists should, in the first place, erect local stratigraphies and then suggest possible correlations with other areas, rather than use these association names in their stratigraphy, possibly erroneously. The use of local stratigraphies (see Paper 12) allows subsequent revisions of regional correlations, whereas use of the five Kurnalpi association names leads to problems in revision.

Williams's suspicion that the paleogeographic trend may have been orientated north–south poses the question: "Could not the sinuosity owe something to the effect of this paleogeographic lineament interfering with the dominant north–northwesterly tectonic trends?

There is remarkably little modern coverage of the Katarchean of the Pilbara, an important tract of little modified rocks comparable with the Barberton Mountain Land (McCall, 1972).

Ingram, in this privately issued text, has demonstrated the cyclicity of the three predominantly volcanic lower associations, and that the later sedimentary associations closely resemble the Fig Tree–Moodies sequence of Swaziland (Papers 5 and 26) and the Shamvaian of Rhodesia (Paper 7). Innumerable fields for detailed, academic research await study in this region; the volcanic, intrusive, and sedimentary rocks (some of the earth's earliest major clastic sedimentary formations) all deserve attention, as does the possible expanse of even more ancient gneisses. Ingram has treated the eastern half of this hostile terrain, unbearably hot in summer; older reports are available covering the West Pilbara (Ryan, 1965, Kriewaldt, 1964), which, like the eastern division, give ages of around 3050 million years on granites and metamorphics (see Paper 24: Commentary), an age midway between that of the Barberton, Sebakwian, and Bulawayan (Papers 5, 7, and 35) and Superior, Slave, Kalgoorlie, Dharwar, Iron Ore age groups of Canada, Australia, and India (Papers 8, 9, 12, and 13). It seems that this intermediate age could represent yet another maximum of activity (Gastil, 1960; Paper 27).

Ingram has covered most aspects of this ancient terrain, but a subject that calls for further study is the relationship between acid intrusives (the plutons) and the acid volcanics. Only in Canada (Paper 8) has there been some attempt to relate these; too commonly they tend to be regarded in isolation. In the original text the word "caldera" was applied to the volcanotectonic central depressions; at the editor's instigation this term has been removed by the author from the summary, for such structures, well known from the Noranda–Rouyn area of Quebec, appear to be depressions formed on major submarine volcanic centers, broadly analogous to calderas or surface cauldrons, but outside the definition of either.

12: THE ARCHEAN SUCCESSION TO THE WEST OF LAKE LEFROY

G. J. H. McCall

ABSTRACT

This paper summarises and correlates work carried out on the early Precambrian Kalgoorlie System (c. 2700 m.y. metamorphic age), in the area to the west of Lake Lefroy, by the author and by five students whose work he supervised. It includes a map of the entire area, a statement of the stratigraphy derived, and general accounts of the physiography and geology, including notes on the overall structural pattern and possible correlations with better known sequences to the north and south, at Kalgoorlie, Coolgardie and Norseman. While the synthesis is based on the results obtained by all six geologists, the conclusions drawn are entirely the responsibility of the author named. It is intended to publish further accounts giving the results of the investigations of certain sub-areas in more detail.

The stratigraphy deduced involves seven sedimentary units, together with a number of ophiolite belts: of the seven sedimentary units, three are exposed to the east and four to the west of a dislocation, the Yilmia dislocation, which appears to have a significant sinistral transcurrent displacement. The three eastern units are believed to be equivalent to the lower three of the four western units. Any stratigraphic resolution of this area must involve the acceptance of significant lateral thinning and thickening—that is, an essentially lensed sequence, whether by tectonic or sedimentological agencies or a combination of both. Another dislocation separates the eastern three units from the Kambalda ophiolites; no detailed stratigraphy is given for these, which are much better known by the geologists of the Western Mining Corporation. However, the major structure, a synclinorium, deduced from unequivocal facing indications, seems to require that the nickel sulphide bearing ophiolites of Kambalda are, broadly speaking, stratigraphically equivalent to similar nickel sulphide bearing ophiolites, associated with only thin, dominantly pelitic metasediment intercalations, that extend from Widgiemooltha southwards through Higginsville.

SUMMARY

What emerges is that there are a series of belts of ultramafic and mafic igneous rocks, strips that are separated by clastic sediments. Porphyries are locally developed in the mafic strips and in the clastic sequences overlying them. The clastic sediments include thick lenses of coarse, polymictic conglomerates (McCall et al.,

 The summary was prepared by the volume editor.

1970), and these occur throughout the entire succession; they are minor components low down, but appear in force high up. One lense of conglomerate and graywacke pebble beds is more than 5000 m thick, with only minor siltstone and no igneous intercalations. It is difficult to accept Glikson's term "molasse" applied to such deposits (Paper 34), for a molasse is by definition a terminal accumulation *after postorogenic uplift;* these conglomerates seem to have suffered the full effects of orogenesis, including the folding, and display no evidence of terminal relationship to postorogenic uplift. They are both metamorphosed and deformed.

Facing indications are excellent, and it is evident that a 16-km-wide synclinorium is the major structural element. This is complicated by second- and third-order folds and major north–northwest trending strike faults, which locally impose second-order, complex, and tight crenulations of steep plunge on the first-formed fold limbs. Such structures are like the "breaks" of the Superior Province (Paper 8), and this study represents a significant advance in our understanding of the Kalgoorlie region, in showing that the stratigraphy could be mapped across such breaks. This was demonstrated at the same time by Williams (Paper 13) and has been confirmed by Gemuts and Theron (1973).

In this sequence, argillites and thinly bedded siltstones have accommodated by extreme crumpling, whereas thick sequences of volcanics, graywackes or conglomerates, or thick sills have adopted steep homoclinal patterns (compare Paper 9). The latter units are capable of flexing into broad folds, open or isoclinal, but not into small-scale concertina structures, like the incompetent beds. Thin ultramafic–mafic layers behave in an incompetent manner when enclosed by argillites.

There are at least four successive strips of mafic and ultramafic rocks separated by clastic sediments. Locally, such a strip or clastic belt may tail out, bringing two like strips into conjunction, but downstrike the full stratigraphy reappears. The mafic–ultramafic strips may widen to as much as 4 km; such widenings are the site of thick, coarse-textured conformable bodies that may be sills or ponded flows (e.g., Yilmia Hill, McCall and Doepel, 1969; McCall, Paper 16b).

Ultrabasic and basic rocks are intimately associated in all mafic strips, as are pillowed and unpillowed metavolcanics, effusions and hyaloclastite (pillow) breccias, and layered or unlayered sills and ponded flows, mainly of gabbroic overall character. The lowest strip of all in this area displays a thick development of mafic and ultramafic rocks, and includes spinifex-textured lenses associated with nickel sulfide deposits. The association of potassium-poor abyssal tholeiites (Paper 16), ultrabasics, pyroxenites, and high-magnesia basalts (Papers 16a and 16b) is repeated in successive strips, and high-magnesia basalts are not restricted to the basal strip. The amount of basic and ultramafic material present does show a general decrease through the succession, in proportion to clastic sediments.

The mafic–ultramafic igneous rocks carry thin intercalations of sediments, commonly argillaceous, and, more rarely, cherts, psammites, and rudites. This particular part of the Kalgoorlie Belt is lacking in banded ironstones of primary origin, and acid volcanics are extremely scarce. The tendency to develop spinifex-textured ultrabasics where banded ironstones are locally absent has been noted by Naldrett (personal communication, 1973) from Canada.

The entire sequence is estimated at 13,000 m; in a later extension to the south, additional units were added at the base within the lowermost strip (McCall and Hudson, 1970), and a summary of this revision is given in Table 1. Gemuts and Theron believe that two complete major cycles, each equivalent to that of Barberton (Paper 5), are represented in this succession.

TABLE 1 Stratigraphic succession of the Lake Lefroy synclinorium

Sequence	Part	No.	Unit	Thickness	Lithology
		10	Kurrawang Conglomerates (occur on both limbs)	ca. 7500 ft	
East-facing sequence (west limb)		9	Beds above the Abattoirs Line on the east limb	(thickness)	Metasandstones Metasiltstones Pelites (fine)
	(upper part)	8		?	
	(lower part)	8	Cave Rocks Beds[2]	ca. 200 ft	Pelites (fine) Metasandstones
		8b	Cave Rocks Ophiolite Belt (≡ Abattoirs Line)		
		7b	Merougil Beds (possible gap or overlap)	ca. 16,000 ft	Polymictic Metaconglomerates Metasiltstones Metasandstones
		7a	Triangle Island Beds	ca. 2500 ft	(as above)
			Kambalda Ophiolites (≡ 2,3,4 ?		Metabasalt pillow lavas Serpentinites Pelites (fine)
			Yilmia Dislocation		
West-facing sequence (east limb)		9	Lake Dam Beds	c. 300 ft.	Metasandstones (fresh sandstones) Metasiltstones Pelites (fine) Cherty or quartzose breccias
		8	Wanda Wanda Beds[3]	c. 7500 ft	Metasiltstones Pelites (fine) Metasandstones ("grits") Polymictic metaconglomerates (Actinolitic metasediments)
		8b	Yilmia Main Ophiolite Belt Stratiform sills Pillow lavas Serpentinites		
		8a	Yilmia Lower Ophiolite Belt (Pillow lavas		
		7	Mandilla Beds	c. 5000 ft	Metasandstones Polymictic metaconglomerates Metasiltstones (Actinolitic metasediments)
		7a	Vesicular Metadolerite Sill		

TABLE 1—*Continued*

Unit	Name	Thickness	Lithology
6	Causeway Beds	ca. 7500 ft	Pelites (fine) Metasiltstones Metasandstones Polymictic metaconglomerates Actinolitic metasediments
6a	Town Dam Ophiolite Belt (tapers out north and widens south into Big Island Ophiolite Belt (ca. 4000-ft pillow lava sequence)		
5	Cardiff Castle Beds[2]	ca. 200 ft	Metamorphosed breccias Metaconglomerates Metasiltstones Pelites (fine)
4*	Widgiemooltha-Dordie Ultramafic Belt [2]	ca. 1000 ft	Serpentinites, tremolite rocks, actinolite rocks, metabasalts Pelites (fine)
3*	Mt Morgan metabasalts[2] (≡ Mt Eaton Metabasalts)[2] tapers southwards from 8000-ft pillow lava sequence at Mt Morgan)	max. 8000 ft	(Metabasalts
2*	Dordie South Ultramafic Belt[2]	max. 3500 ft	(Tremolite rocks
	(* These constitute the Widgiemooltha Ophiolite Belt)		
1	High-grade aureole metasediments at Dordie[1]	ca. 500 ft	
	(Contact transgressive through Dordie South ultramafics and Mt. Morgan metabasalts)		
	Dordie Rocks Granite		

[1] Reported by J. Sofoulis (1966).
[2] Provisional name.
[3] Units 5–9 ≡ Black Flag Beds.
[4] The pelites (fine) are metamorphosed argillites or claystones (commonly carbonaceous).

References

Gemuts, I., and Theron, A. (1973). Stratigraphy and mineralisation of a portion of the Archaean between Norseman and Coolgardie, Western Australia. Abstract Vol., Sec. 3, ANZAAS, 45th Congr., Perth, p. 88.

McCall, G. J. H., and Doepel, J. J. G. (1969). Stratiform basic bodies within the Yilmia ophiolites, Western Australia. *Proc. Aust. I.M.M., 231,* 67–75.

____, and Hudson, D. R. (1970). Excursion guide, Symposium on Archaean Rocks, Geological Society of Australia, Perth, 1970.

____, Braybrooke, J. C., and Middleton, D. D. (1970). The Merougil sub-area. *J. Roy. Soc. W. Aust., 53,* 9–32.

13: EXPLANATORY NOTES ON THE KURNALPI 1:250,000 GEOLOGICAL SHEET, WESTERN AUSTRALIA

I. R. Williams

SUMMARY

The layered sequence of the greenstone belt* is intruded by granites dated at 2615 million years (∓), and it includes acid and basic intrusives dated between 2615 and 2715 million years (all dated by Rb/Sr method; Turek, 1966). The last two may be metamorphic ages. An upper limit to the Archean sequence is provided by unmetamorphosed basic dikes of the Widgiemooltha Suite (Sofoulis, 1966) dated at 2420 million years (∓) (Turek, 1966). Williams established a stratigraphic succession and defined *lithological associations,* which later were afforded the status of *formations.* These associations are groups of lithological units that are either genetically related or else show consistent spatial relationships as a result of geological environment. Such associations are of two types: (1) acid volcanic–clastic, and (2) basic volcanic.

Five associations numbered I to V were defined in the following sequence: b, a, b, a, b. Boundaries between associations I and II, and III and IV, are transitional, whereas those between associations II and III, and IV and V, are disconformable or unconformable. A pattern of binary cycles emerges, each cycle comprising a basic lower part and an acid upper part. Three binary cycles were recognized, the uppermost incompletely represented in the Kurnalpi area. The associations are named as follows:

Cycle 1:
- I Morelands: basic
- II Gindalbie: acid–clastic

Cycle 2:
- III Mulgabbie: basic
- IV Gundockerta acid–clastic

Cycle 3:
- V Kalpini: basic

Morelands. This association is comprised of basic lavas, some intermediate lavas, minor fine-grained clastic sediments with a high content of pyroclastic material. Base not exposed, but at least 17,000 ft thick. Pillow lavas displaying varioles and pillow breccias are prominent. The sediments are schistose derivatives

*The editor uses the term in the internationaııy accepted sense, to denote the entire tract of stratified rock.

This summary, from *Geological Survey of Western Australia,* 37p. (1970), was prepared by the volume editor.

of black, carbonaceous, pyritic shales, siliceous and dolomitic shales, minor graywackes, and thin chert beds. Acid porphyry intrusives and rare dacite–rhyolite extrusives are recognized, the latter mainly in a transition zone at the top of the association.

Gindalbie. Acid volcanic complexes and clastic sequences, contemporaneous but spatially separate, comprise this association. Minor components are included siltstones, sandstones, and thinly banded conglomerates. Most of the pebbles in these are of acid volcanic or intrusive types. Locally, thinly banded cherts are interbedded with pyroclasts and magnesitic black shales. This position may alternatively be taken by banded iron formation or jaspilite.

This association is referred to an environment of increasing tectonism and explosive acid volcanic emission, the latter declining as erosive agencies became dominant. The bulk of the clastic detritus is ascribed to volcanic sources internal to the basin. The chert, jaspilite, and banded iron formation at the top are equated with quiescence between cycles.

Mulgabbie. This association is similar to the first, but includes finer-grained, clastic sediments, mainly derivatives of carbonaceous shales and siliceous mudstones. Layered basic intrusions are less numerous in this association, the thick extrusive basalt sequences containing few such bodies. Thin, coarse-grained basic intrusions locally follow sedimentary horizons in the basic extrusive piles. Ultrabasic bodies and albitic acid porphyry intrusions are also represented. The thickness of the association is consistent along strike at about 14,000 ft.

The Kalgoorlie sequence of the Golden Mile forms part of this association, as does the Kambalda nickel-bearing sequence. The clastic content ranges from 1 percent to nearly 50 percent of the total assemblage. Fuchsite-bearing schists and quartzites seem to occur overlying ultrabasic rocks, although not in contact with them. The basal units are the banded cherts, referred to above, and the assemblage is overlain unconformably by conglomerates of the Gundockerta association.

Gundockerta. This association is similar to the second, but acid volcanics only comprise about 13 percent by area. The acid complexes also are developed on a less extensive scale. There is a high proportion of clastic sediments. The conglomerates at the base locally cut out association III entirely, and thus rest on association II, but they still contain basic phenoclasts attributable to sources in association III. One large acid volcanic pile is recognized at Steeple Hill–Gooragoogine Lake, and consists of rhyodacite lavas, tuffs, and basic to intermediate amygdaloidal lavas. The acid volcanic complexes are characteristically overlain by banded pyritic cherts, thinner and less continuous than those of association II, and also differing in being overlain by thick beds of coarse-grained clastic sediments. Polymictic conglomerates are prominent in this association and contain phenoclasts of acid volcanic, porphyry, chert, jasper, shaley sediment, quartz, basic rocks, and granite porphyries. They pass laterally and vertically into well-bedded shales, siltstones, and minor, thin conglomerate bands, which commonly exhibit grading and slumping structures.

Turbidites form a thick sequence south of Randell (Dunbar and McCall, 1971), and include graywackes, siltstones, and shales. Rhythms of sedimentation and sedimentary structures are prolifically developed in this sequence, believed to be a deeper-water equivalent of the shallow-water Gundockerta facies to the north. Where acid volcanics predominate, which accounts for two thirds of the total area of the formation, the thickness is up to 25,000 ft; where clastic sediments predominate, it is up to 15,000 ft. The acid volcanics form distinct centers, and include lavas, tuffs, and agglomerates. Rhyolite breccias at Kanowna include flow-banded perlite and felsite. "Welded tuffs" and "tufflavas" are also recognized. Associated quartz–muscovite, quartz–muscovite–chlorite–dolomite, and chloritoid-muscovite schists may represent pyroclasts. Amygdaloidal and vesicular intermediate volcanics have also been recognized. Oligomictic conglomerates form thick deposits marginal to the acid volcanic complexes, exhibiting graded and cross bedding. They are believed to result from a combination of direct pyroclastic accession to water with detritus eroded from the volcanic pile.

The possibility that these conglomerates stem from the breakdown of lavas and sills, the rounding occurring during bulk flow, is entertained. Other clastic types are prolific in this sequence, which is believed to be a deeper-water equivalent of the shallow-water Gundockerta facies to the north. In general, the environment was similar to that of association II, but marked by less intense acid igneous activity and more intense erosive activity.

Kalpini. Like the preceding association, this one tends to be preserved only in synclinal troughs. Pillowed carbonated basic lavas and pillow breccias are intercalated with fine-grained mudstones and shales. There are some coarse-grained basic intrusive bodies, and also ultramafics, serpentinized peridotites, and pyroxenites, which occur near the base of the association. Banded cherts are prominent and continuous along strike near the base. This association is characterized by less alteration, regional metamorphism, and acid intrusive development than associations I and III.

The total thickness of the five associations is at least 91,000 ft (maximum), and Williams sees this great thickness as having resulted from more or less continuous volcanic activity and sedimentation; he attributes most of the clastics to local sources in the volcanic piles themselves. He describes the immense Bulong Complex, within assemblages I and II, a boat-shaped multiple intrusion consisting of several sills of dunite, peridotite, pyroxenite, gabbro, and diorite that displays layering and cumulate textures. It is an emplacement of cycle 2 or cycle 3.

The granitic rocks that make up 33 percent of the area are intrusive into the sequences described. The dominant petrographic type is a fine-grained biotite granite that is porphyritic in about 5 percent of the area of outcrop. Field relationships suggest more than one age of emplacement, but Rb/Sr dating gives a consistent age. Fine-grained leucocratic, commonly alaskitic granites, form late plutons. The margins of the major granite bodies are foliated parallel to the regional foliation. There are some developments of gneissic granites and migmatites.

The folding is dominantly on north northwest–south southeast trends, but sinuosity suggests interference patterns, a northeast–southwest en echelon pattern apparently being superimposed. Similar, nearly isoclinal folds predominate, with steep to vertical bedding dips and shallow to locally steep and even overturned plunges. Major north trending faults and minor faults of varying trend complicate the structure. The major dislocations appear to be strike faults with dominantly vertical movements on near-vertical surfaces. It is suggested that the area comprises two paleogeographic divisions, the western being characterized by volcanics and shallow-water sediments that underwent more or less continuous instability and tectonism, whereas the eastern was characterized by turbidites that underwent only intermittent tectonism. The tectonism is believed to have attained a maximum at the commencement of acid volcanic activity in midcycle and dropped to a minimum at the end of the cycles. The paleogeographic trend may well have been north–south, and not the trend of the superimposed tectonic structures.

References

Dunbar, G. J., and McCall, G. J. H. (1971). Archaean turbidites and banded ironstones of the Mt. Belches area, Western Australia. *Sedimentary Geol., 5,* 93–133.

Gemuts, I., and Theron, A. (1973). Stratigraphy and mineralisation of a portion of the Archaean between Norseman and Coolgardie, Western Australia. Abstract vol., Sec. 3, ANZAAS, 45th Congr., Perth, p. 88.

Sofoulis, J. (1966). Explanatory notes on the Widgiemooltha Sheet, Western Australia. Bureau of Mineral Resources, Geology, and Geophysics, Canberra.

Turek, A. (1966). Rubidium/strontium isotopic studies in the Kalgoorlie-Norseman area. Ph.D. thesis, Australian National University, Canberra.

14: A SUMMARY OF THE GEOLOGY OF A PORTION OF THE PILBARA GOLDFIELD, WESTERN AUSTRALIA

P. A. J. Ingram

SUMMARY

The Archean of the East Pilbara Goldfield consists of a series of cyclical volcanic formations and two major sequences of sedimentary rocks. These are structurally underlain by a gneissic complex, which probably represents the original basement upon which the volcanic–sedimentary sequence was deposited, and are overlain by a thick sequence of volcanic and sedimentary strata of the Lower Proterozoic Nullagine Succession. All these rocks are included in the Pilbara System (Ryan, 1965).

Within the Archean, a subdivision into two groups of rocks has long been recognized–the predominantly volcanic *Warrawoona Succession* (Noldart and Wyatt, 1962) and the predominantly sedimentary *Mosquito Creek Succession.* In Paper 14, the former is further subdivided into four formations, three dominantly volcanic and one dominantly sedimentary.

In the Marble Bar–Nullagine area of the Pilbara Goldfield (the central core of the Archean), large granite complexes have intruded the geosynclinal rocks under the influence of dominantly vertical forces. Erosion has now uncovered these domal granites, leaving the greenstones distributed in sinuous synclinal belts wrapped around them. It is within these essentially symmetrically disposed synclines of steeply dipping rocks that the cyclical nature of the deposition can best be observed. This cyclicity was first observed by the writer during the course of a regional mapping program in 1968. Later, similar developments in Archean greenstone belts were described (Anhaeusser, 1971a, 1970b; Viljoen and Viljoen, 1971a, 1971b) and confirmed the writer's tentative recognition of cyclical development in the Katarchean of the Pilbara.

Summary of Stratigraphy

Noldart and Wyatt (1962) initially divided the Archean of the East Pilbara into two successions, the lower Warrawoona Succession and the upper Mosquito Creek Succession. They further subdivided the latter into four formations, including the Gorge Creek Formation. The present writer has placed the Gorge Creek Formation back in the Warrawoona Succession, and has not been able to confirm the remaining three formations of the Mosquito Creek Succession from very broad regional ground traverses and air photo interpretation in the main outcrop area east of Nullagine. (Three new formations have been informally named and defined within the Warrawoona Succession.) Figure 1 shows the stratigraphic succession as mapped by the writer.

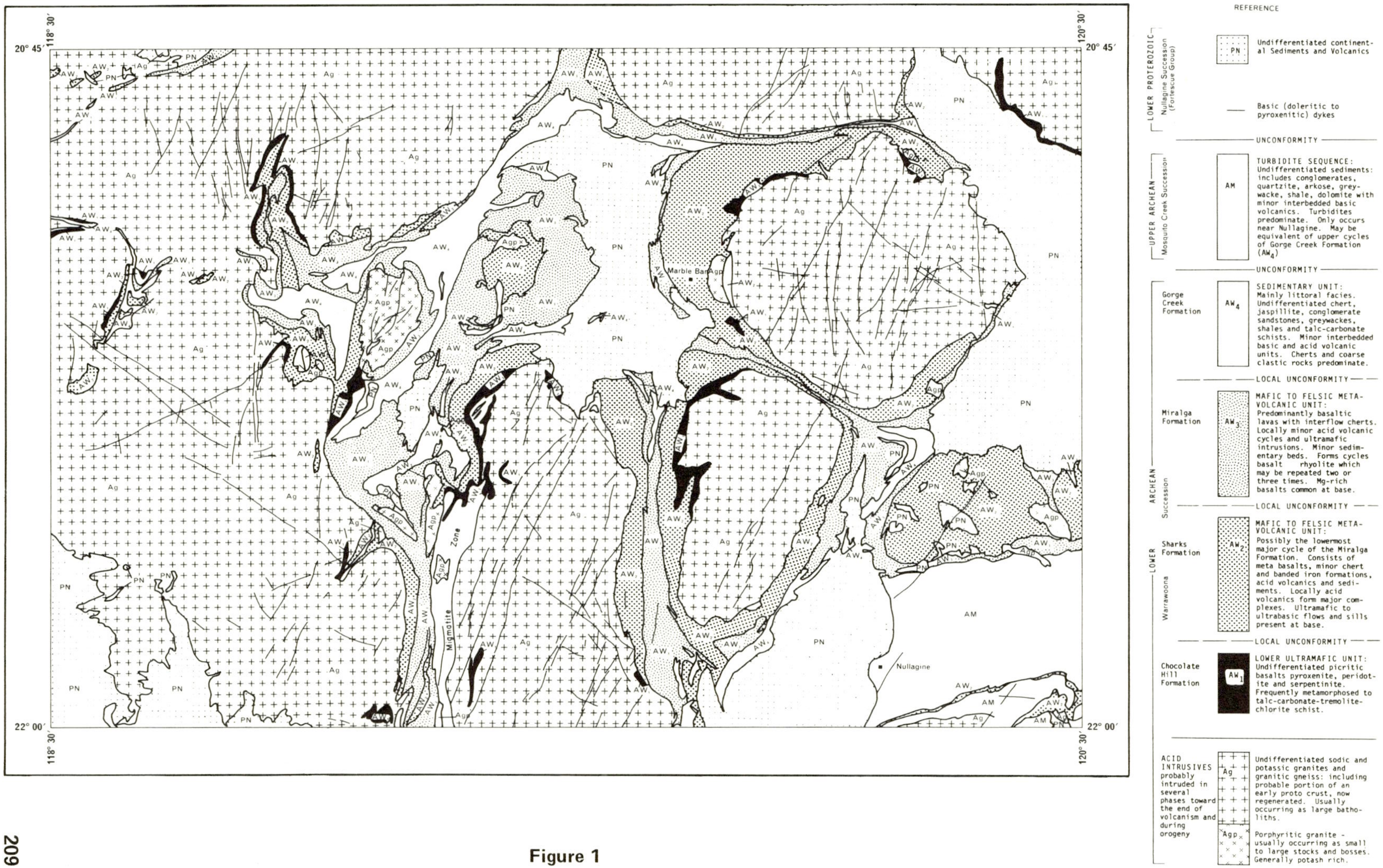
20° 45′
118° 30′
120° 30′
22° 00′
Marble Bar
Nullagine
Migmatite Zone
REFERENCE
LOWER PROTEROZOIC
Nullagine Succession (Fortescue Group)
PN
Undifferentiated continental Sediments and Volcanics
Basic (doleritic to pyroxenitic) dykes
UNCONFORMITY
UPPER ARCHEAN
Mosquito Creek Succession
AM
TURBIDITE SEQUENCE: Undifferentiated sediments: includes conglomerates, quartzite, arkose, greywacke, shale, dolomite with minor interbedded basic volcanics. Turbidites predominate. Only occurs near Nullagine. May be equivalent of upper cycles of Gorge Creek Formation (AW4)
UNCONFORMITY
LOWER ARCHEAN
Warrawoona Succession
Gorge Creek Formation
AW4
SEDIMENTARY UNIT: Mainly littoral facies. Undifferentiated chert, jaspillite, conglomerate sandstones, greywackes, shales and talc-carbonate schists. Minor interbedded basic and acid volcanic units. Cherts and coarse clastic rocks predominate.
LOCAL UNCONFORMITY
Miralga Formation
AW3
MAFIC TO FELSIC METAVOLCANIC UNIT: Predominantly basaltic lavas with interflow cherts. Locally minor acid volcanic cycles and ultramafic intrusions. Minor sedimentary beds. Forms cycles basalt rhyolite which may be repeated two or three times. Mg-rich basalts common at base.
LOCAL UNCONFORMITY
Sharks Formation
AW2
MAFIC TO FELSIC METAVOLCANIC UNIT: Possibly the lowermost major cycle of the Miralga Formation. Consists of meta basalts, minor chert and banded iron formations, acid volcanics and sediments. Locally acid volcanics form major complexes. Ultramafic to ultrabasic flows and sills present at base.
LOCAL UNCONFORMITY
Chocolate Hill Formation
AW1
LOWER ULTRAMAFIC UNIT: Undifferentiated picritic basalts pyroxenite, peridotite and serpentinite. Frequently metamorphosed to talc-carbonate-tremolite-chlorite schist.
ACID INTRUSIVES probably intruded in several phases toward the end of volcanism and during orogeny
Ag
Undifferentiated sodic and potassic granites and granitic gneiss: including probable portion of an early proto crust, now regenerated. Usually occurring as large batholiths.
Agp
Porphyritic granite - usually occurring as small to large stocks and bosses. Generally potash rich.

Figure 1

Chocolate Hill Formation (formerly Mulgandinnah). This is the lowermost recognizable unit in the Warrawoona Succession. It occurs as remnants along granite margins and occasionally in the course of anticlines within greenstone belts and as roof pendants within granites and gneisses. Highly magnesian basalts of the komati type predominate, being particularly evident at the base of the sequence. They are interbedded with tholeiitic basalts, flows or sills of metaperidotite, some acid volcanic flows, and minor tuffaceous sediments (quartz-sericite schists). Cherts are rare, except at the top of the formation, which represents the first cycle of the geosynclinal period. The high magnesia basalt developments possibly reflect tapping of virtually uncontaminated and undifferentiated mantle-derived material. The formation has a maximum true thickness of 2500 m.

Sharks Formation (formerly "Split Rock") (AW_2). Unconformable to conformable on the above formation, this occurs around margins of large intrusions of granitic rocks and in the anticlinal cores. Tholeiitic basalts predominate, and acid volcanics come in strongly near the top of the formation. Ultramafic (including ultrabasic) intrusions comprise 5 percent of this formation, being developed mainly near its base. Cherts are rare developments within the acid volcanic sections. A very thick banded chert, the Marble Bar Chert, terminates the formation, particularly where there is a prominent development of acid volcanic rocks. The maximum thickness of the formation is 3800 m.

Miralga Formation (AW3). This formation overlies the *Sharks Formation* unconformably. It outcrops in the central parts of the synclinoria. Pillowed basalts form the most prominent members of this formation, accounting for 80 percent of the sequence. Interflow cherts are abundant and sediments associated with these are more in evidence than in the lower two formations. In the type area, this formation attains a thickness of 5000 m.

Gorge Creek Formation (AW4). This formation consists of conglomerate, grits, and arkosic sandstones, with minor interbedded shales, acid tuffs, and other volcanics. Banded jaspilites, cherts, and banded iron formation are common toward the bottom of this formation, which attains a thickness of at least 2000 m, and may well attain a much greater thickness.

Noldart and Wyatt (1962) removed this formation from the Warrawoona Succession, within which it was placed by Noldart (1960), placing it within the Mosquito Creek Succession on account of similarities to the Eastern Creek and Dromedary Formations within that succession. The writer believes that the apparent interbedded relationship displayed by this formation at its contact with the Miralga Formation beneath and its obvious volcanic affinities, evidenced by contained tuffs, banded cherts, and the like, justify returning it to the Warrawoona Succession. The purely sedimentary Mosquito Creek Succession may possibly be a lateral equivalent of the entire Warrawoona Succession, rather than a discrete superimposed entity.

Rock Types

Mafic Volcanics. There is a great variety of mafic volcanics within the Warrawoona Succession, ranging from tholeiitic to picritic composition. The basalts within the Chocolate Hill Formation are characteristically magnesia-rich, highly altered, and metamorphosed to talc–tremolite–carbonate–chlorite–actinolite schists. The basalts of the Sharks and Miralga formations are mostly tholeiites, although some magnesia-rich basalts are present in the Miralga Formation, amounting to about 10 percent of the outcrop. On the margins of the area mapped, especially the western margin, basalts of the Sharks Formation are not uncommonly found to be magnesia-rich. Alteration to carbonate–chlorite assemblages is more evident in the basalts of the Miralga Formation. No basalt flows have been recognized in the Gorge Creek formation.

Acid Volcanics. The major acid volcanic piles all occur within the Sharks Formation; they consist of a variety of sodic rock types, including rhyolites, dacites, quartz latites, and andesites, and include flows, tuffs, agglomerates, and both explosive and flow breccias. The flows may be porphyritic, vesicular, scoriaceous, or massive. Two of the most common rock types are the following:

1. Lapilli breccia: a complex association of both flow and pyroclastic breccias derived from a porphyritic sodic-rhyolite magma. Agglomerate and vent breccias are common. These rocks are described as quartz-feldspar porphyry intrusions in the older literature.

2. Welded tuff: also previously described as quartz-feldspar porphyries, these rocks are also of varied texture. The Copper Hills Porphyry is reinterpreted as an association of some true porphyritic flows, possibly some porphyry intrusions, and a large amount of crystal tuff, welded tuff, and ignimbrite, probably of glowing avalanche type. The composition of these rocks is predominantly dacitic–rhyolitic.

Other acid rocks present are bedded tuffs. Quartz-sericite schists, widespread in the area, are mainly derivatives of tuffs or tuffaceous sediments. Shearing and metamorphism close to granite margins have, however, locally produced similar schists from quite coarse pyroclasts.

Two occurrences only are known of acid volcanics in the Chocolate Hill Formation, both metamorphosed, recrystallized, fine-grained (almost glassy) rhyolitic rocks grading into porphyritic rhyolites of medium grain size.

The Miralga Formation is characterized by only minor acid volcanic developments of tuffs and tuffaceous sediments and restricted occurrences of lapilli breccias.

Mafic Intrusions. Mafic intrusions include the following:

1. Gabbroic dikes and sills of normal character, and not requiring further discussion.

2. Large layered sills: there are several such sills of considerable size, of bulk composition approximating to that of basalt or perhaps picrite. They tend to be late intrusives in the overall geosynclinal sequence. They are known from the Miralga and Gorge Creek but not the Chocolate Hill or Sharks formations. Typical examples are those at the Lional Asbestos Workings and north of the Soansville Asbestos Workings. Rock units include thin layers of cumulate-textured peridotites at the base, through alternating bands of pyroxenite and gabbro, into thick gabbro sequences at the top. It is not unusual for the total thickness of these sills to exceed 1000 m.

3. Gabbroic plugs: plugs of gabbro, dolerite, or basalt are common in or near volcanic vents. Textures vary within the limits of single plugs, and features are of both typical intrusive and typical extrusive character (chilled margins, coarse grain, vesicles, pillows, etc.). Such plugs are common in the McPhee Creek volcanic depression sructure.

Acid Intrusives (Other Than Granites). Acid intrusives vary from sodic feldspar–porphyry sills and plugs (in the lower Chocolate Hill and Sharks stratigraphy) to quartz-feldspar and quartz-porphyries, which form very small bodies. Many of these intrusions are closely associated with acid volcanic piles and porphyritic granite stocks or bosses.

Cherts. There are two types of cherts:

1. Primary cherts: volcano-sedimentary, banded cherts of large areal extent and thickness occurring at the top of individual flows and of individual formations. In the Miralga and Gorge Creek formations these may develop into banded iron

formation sequences. Fumarolic (or sinter) cherts are banded gray, white, and black, and are less thick and areally extensive. They occur in interflow positions within acid volcanic piles, frequently in association with small, scattered domes and vents, and are commonly associated with stratabound base metal sufide deposits.

2. Replacement cherts: silica replacements in the Pilbara commonly take the form of chert rather than crystalline quartz veins, although there are many true quartz veins, too. Such cherts show similar banding to the stratiform, primary cherts. Some replacement cherts are stratiform, replacing black shale, tuff, siltstone, ultrabasic, and acid volcanics. Some are similar to the "cherty tuffites" referred to in Canadian literature, which are of significance in the search for gold-pyrite and copper–lead–zinc pyrite deposits.

Ultrabasic Rocks. Three classes of ultrabasic rocks have been distinguished on the basis of stratigraphic and structural setting and physical and chemical characteristics. The three classes can be compared with those of Wilson et al. (1969) and Thayer (1971).

1. *Volcanogenic ultrabasics* are the oldest, even though they may also occur high in the stratigraphy. They are typicaly stratiform in habit, not structurally controlled, of fine to medium grain (except where metamorphosed), dark green to black in color, of blocky outcrop, and have the composition of serpentinized dunites and peridotites, altered in part or entirely to talc–tremolite–carbonate assemblages. Locally sulfide bearing, they are but rarely asbestos bearing. They are commonly associated with magnesia-rich basalts, and may display possible pillow structures. Micro- and macro-spinifex textures are recognized in several occurrences. Kunkar and superficial calcareous cappings typically overlie them.

Volcanogenic ultrabasics are extrusives or subvolcanic sill-form intrusives. Related feeder dikes can occasionally be detected beneath them. Later ultrabasics commonly intrude them. The primary olivine has been serpentinized and then replaced by talc–tremolite–carbonate–chlorite assemblages. The oldest of this class of ultrabasics within the Chocolate Hill Formation are the most altered and metamorphosed, having a very high carbonate content.

2. *Layered intrusions* are less common than bodies of the other two classes. Near the western margin of the geosynclinal rock outcrop at Soansville and Mt. York, however, there are several examples. They vary in composition from dunite to pyroxenite, with dunite and peridotite predominant and gabbroic material very subordinate. They carry asbestos, but are not significantly productive. McCall (1971) has given a brief description of one of these sills at Soansville, a composite sill 1200 m thick made up of three separate layered sills, one on top of the other, and showing a differentiation trend through the successive intrusions, which are separated by a chert of sedimentary (replacement) origin. Subvolcanic textures (glass, skeletal crystals of microspinifex texture; McCall and Leishman, 1971) were observed in the top 300 m.

Such sills intrude at many stratigraphic levels, from low in the Sharks to high in the Miralga Formation; they predominate in the Miralga Formation, being essentially a late magmatic suite in the geosynclinal sequence. The dunites and peridotites of this class are similar to those of the orogenic class and could be comagmatic with them.

Nowhere have these ultrabasics been found to yield significant amounts of sulfides.

3. *Orogenic ultrabasics* are readily distinguished from the volcanogenic ultrabasics, but show physical characteristics and pyroxenite–gabbro associations similar to the layered intrusions, from which they are not so readily distinguishable. They

appear, however, to be differentiated by multiple intrusion rather than gravitational layering.

Their characteristic feature is obvious structural control (intrusion along major faults) rather than stratigraphic control. Most are discordant or subconcordant. Despite their structural control, they are mostly concentrated within the Miralga Formation.

Medium to coarse grained and apple green to pale brown in color, orogenic ultrabasics are generally serpentinized dunites or peridotites, little altered or modified by metamorphism. The texture is that of a cumulus of rounded to subhedral olivines set in a variable, fine groundmass. The olivines are unstrained and show no signs of rheid or plastic flow (mosaic or alignment of elongate grains). Olivine cumulus accounts for up to 80 percent of these rocks.

Orogenic ultrabasics are associated with economic to subeconomic chrysotile deposits, but sulfides are not characteristically associated with them.

The geochemistry of the ultrabasics has been studied in a limited analytical program; despite difficulty in obtaining unaltered samples, results suggest that the orogenic class are MgO rich (37.0–38.8 percent). The volcanogenic and layered classes are less MgO rich (30.9–37 percent). Calcium oxide values for unmetamorphosed ultrabasics from the Pilbara are lower than those from South Africa, Canada, or the Kalgoorlie area of the Yilgarn Block, and Al_2O_3 values are similar to those obtained from other provinces.

Granitic Rocks. Granitic rocks have not been studied in detail by the writer. Noldart and Wyatt (1962) provide general summaries of the granites, and Blockley is at present carrying out a fuller study. There are three distinct provinces:

1. The extensive granite–gneiss "platforms" to the west and north of the geosynclinal tracts.
2. The smaller, but still areally extensive, "batholithic" granite domes.
3. The quite small, commonly porphyritic, granite stocks and bosses intruded into the Warrawoona and Mosquito Creek successions and granite of the batholithic types.

Migmatites and granitization aureoles are most commonly associated with the batholithic granites. The magmatic granites appear to have been emplaced in three or four stages.

The grouping of the granites in the Pilbara resembles that applied to the Barberton Mountain Land (Anhaeusser, 1971a).

Metamorphism

The overall metamorphic grade is low greenschist facies throughout the province. Metamorphism is accentuated in zones of shearing and intense deformation and close to granite contacts, where it locally attains amphibolite grade due to contact effects and granitization processes.

Alteration

Carbonate–chlorite alteration of basic rocks, with minor talc formation and tremolitization, and similar alteration of ultramafic rocks of the Chocolate Hill Formation, with more pronounced tremolite and talc formation, are attributed to interaction of carbon dioxide–rich gases, seawater, and lavas (or subvolcanic sills) on the sea floor during the late fumarolic stages of volcanism. Chlorite and/or sericite development and silicification are developed near stratabound mineralization and appear to have a similar origin. Most silicification in the area is the result, however, of chemical activity related to Tertiary weathering.

Structures

Folding in the Warrawoona Succession is complex; major axes trend about 20° across the greater part of the area. Major granite intrusions occupy the cores of doubly plunging anticlines following this trend. Subsidiary folds, mainly synclinoria, tend to wrap around the granitic rocks, with axes of adjacent synclinal volcanic belts tending to point intersections. These intersections form major structural basins, which are reflected in overlying Proterozoic volcanic and sedimentary rocks. Cross-folding on an east–west to 120° trend is also common.

All the fold elements interact to produce the pattern of linear domal granites, linear synclinal greenstone belts curving around them, and larger expanses of greenstones within the major structural basins.

In the Mosquito Creek Succession a major central syncline follows an east–west dominant trend, bisecting the area of sedimentary outcrop. Several minor folds flank this major structure. Folding is open to isoclinal and complicated by minor developments of chevron folding. Intraformational slumping is evident. The east–west trend in this succession approximates to the main trend in the Warrawoona Succession in the area immediately to the north of the sedimentary outcrop area.

In the eastern part of the area mapped there is little evidence of major flexure, although a strong syncline trending east–northeast is bordered to the southeast by a weak anticline. Dips are quite shallow in this anticlinal zone.

Vertical stresses during granite intrusion are thought to have only produced slight modification of the structure. The concentric patterns are thought to be primary features related to a major volcanic structure, the McPhee Creek structure, and a smaller, similar structure at Reedy Creek.

Faulting is developed in a variety of styles in the Archean terrain; strike faulting with unknown but probably extremely large displacements is prominent; the displacements possibly amount to thousands of meters. Transcurrent faults on north–south trends are prominent to the east, and one of these has displaced the Reedy Creek structure at least 4500 m relative to the McPhee Creek structure. Ultrabasic bodies of the orogenic type appear to be locally related to some of these faults; these major disruptions may have tapped the primary mantle material in Archean times.

Mineralization and Economic Geology

Nickel sulfides have been recorded from four locations in the Pilbara; that at Cooke Creek may have economic significance, but none of these occurrences appears to be of major dimensions. The volcanogenic ultrabasics provide the most likely host. The ultrabasics of this group do carry disseminated sulfides. Many of the ultrabasics of the Pilbara are low in copper relative to nickel.

There is a potential for stratabound base metal deposits (Cu, Pb, Zn, Ag, Au) in association with acid volcanics, especially in the Sharks Formation. These are associated with agglomerate, fumarolic chert and cherty tuffite in acid volcanic piles. None as yet has been proved to be of economic significance, however, in the East Pilbara.

Other mineralizations include alluvial tin (and tantalite) and some minor lode working associated with pegmatite developments in the late granites, stratabound gold-pyrite deposits associated with volcanic rocks of the Miralga Formation, and antimony–gold–pyrite deposits in the Mosquito Creek Succession controlled by strike fault intersections. Of these the Blue Spec is important. The antimony mineralization shows a preference for sandstone beds. Pyrite and quartz are associated with the antimony. These deposits are believed to be hydrothermal, but their genesis remains uncertain.

Chrysotile deposits in the orogenic ultrabasics (and some layered ultrabasics) have been worked on a small scale but none are now operative.

Geological History and Summary of Events

1. Formation of a primitive crust of unknown composition. If this was mafic or ultramafic, the Chocolate Hill Formation could represent remnants of such a crust (cf. Barberton Mountain Land, Engel, 1969; Anhaeusser, 1971a). There is no evidence to support such a hypothesis in the case of the Pilbara, and it could equally be sialic and represented by the gneisses.
2. Downwarping of this crust to form primitive geosynclinal basins, with long axes trending west northwest–east southeast. These are not analogous with geosynclines in the modern sense of plate tectonics, but must rather be regarded as protogeosynclines.
3. Tectonism associated with the downwarping–tapped magma chambers in the thin primitive crust and the mantle producing the first phases of basic volcanism, possibly limited to scattered centers along the depository, and probably aligned on north northeast–south southwest trends.
4. Volcanism continued in spurts, the products becoming less magnesian as time progressed.
5. Each major cycle of volcanism commenced with ultramafic to mafic extrusion and terminated with explosive acid volcanism.
6. Subsidence continued under the influences of gravity, and the granite bodies were thrust up into the volcanic pile, again under the influence of gravity.
7. The early volcanic piles spread and eventually overlapped to form one vast volcanic province.
8. Minor sedimentation occurred toward the end of each volcanic cycle. This process increased in importance in the last cycles of the volcanic complex, culminating in the Gorge Creek Formation composed of thick sediments, littoral grits, and conglomerates, as well as banded ironstones.
9. Folding and faulting increased in intensity with continued subsidence of the depository as the granites made their final uprise.
10. Subsidence continued with turbidite sedimentation extending to cover portions of the volcanic province, forming the Mosquito Creek sediments.
11. Subsidence and tectonism continued until the geosyncline became stabilized into a semirigid crust, with continued tectonic adjustments producing late fracturing of granite areas and the tapping of new magma reservoirs. This produced the earliest Proterozoic dikes that traverse the Archean craton and fed the continental volcanics of the Proterozoic.

References

Anhaeusser, C. R. (1970a). Cyclic volcanicity and sedimentation in the evolutionary development of Archaean greenstone belts and shield areas. *Spec. Publ. 3, Geol. Soc. Aust.*, 57–70.

_____. (1970b). The Barberton Mountain Land, South Africa: a guide to the understanding of the Archaean geology of Western Australia. *Spec. Publ. 3, Geol. Soc. Aust.*, 103–120.

Gastil, G. (1960). Continents and mobile belts in the light of mineral dating. *21st Int. Geol. Congr., 9,* 162–169.

Kriewaldt, M. (1964). Explanatory notes on the Dampier and Barrow Island sheets, 1:250,000, SF 50/1,SF 50/2. Geol. Surv. Western Aust., 13p.

McCall, G. J. H. (1971). Some ultrabasic and basic igneous rocks in the Archaean of Western Australia. *Spec. Publ. 3, Geol. Soc. Aust.*, 429–442.

_____. (1972). The nickel-sulphide bearing ultramafic rocks and their environment in the Archaean of Western Australia. *24th Int. Geol. Congr., 1,* 354–362.

_____, and Leishman, J. (1971). Clues to the origin of Archaean eugeosynclinal peridotites and the nature of serpentinisation. *Spec. Publ. 3, Geol. Soc. Aust.*, 281–300.

Noldart, A. J. (1960). Summary progress report on a reconnaissance survey of portion of the Pilbara Goldfield, Western Australia. *Geol. Surv. Western Aust., Bull. 114.*

_____, and Wyatt, J. D. (1962). Geology of portion of the Pilbara Goldfield, covering Marble Bar and Nullagine 4-mile map sheets, *Geol. Surv. Western Aust., Bull. 115.*

Ryan, G. R. (1965). The Geology of the Pilbara Block. *Proc. Aust. I.M.M., 214.*

Thayer, T. P. (1971). Authogenic, polygenic and allogenic ultramafic and gabbroic rocks as hosts for magmatic ore deposits. *Spec. Publ. 3, Geol. Soc. Aust.*, 239–252.

Viljoen, M. J., and R. P. (1971a). The Barberton granite-greenstone model as an aid to interpretation of part of the West Australian shield. *Spec. Publ. 3, Geol. Soc. Aust.*, 153.

Viljoen, R. P., and M. J. (1971b). Geological and geochemical evolution of the Onverwacht volcanic group of the Barberton Mountain Land, South Africa. *Spec. Publ. 3, Geol. Soc. Aust.*, 133–150.

Wilson, H. D. B., Kilburn, L. C., Graham, A. R., and Ramlal, K. (1969). Geochemistry of some Canadian nickeliferous ultrabasic intrusions. *Econ. Geol. Monogr.* 4, 294–309.

PAPERS 15 THROUGH 16C: COMMENTARY

Paper 15 provides a summary of the principal features of the various types of extrusion, effusion, and pyroclastic expression that comprise the ultrabasic–basic assemblages of the Yilgarn and Pilbara Blocks, the emphasis being on the most fully studied Kalgoorlie belt of the Yilgarn Block. Treatments of this nature are important in Archean studies. There is a need to get into the field and obtain detailed information concerning the true nature of the various components of greenstone belts and granite areas.

Other papers deal with this subject matter in more detail, for example, McCall and Doepel (1969), Bartram and McCall (1971), McCall and Leishman (1971), Nesbitt (1971), Williams (1971), and the three papers of a geochemical nature in this volume (Papers 16A, 16B, and 16C). Similar studies from Canada include Naldrett and Gasparini (1971) and Pyke et al. (1972) on the ultramafics, and Allard (1970), Macrae (1969), and Goodwin and Shklanka (1967) on the basic rocks. Despite this quite long list, there is a need for continuing programs of specialized study of small selected areas of igneous rocks in Archean regions, augmenting the many excellent regional studies, the instrument-oriented, laboratory-based studies that are appearing in great numbers, and the review treatments.

Basic and ultrabasic* occurrences of the Yilgarn Block of the Archean include (1) gabbroic sills, (2) serpentinite sills and flows, and (3) basic volcanics, flows, and hyaloclastite breccias. The latter include high-magnesia basalts (Papers 16B and 16C), broadly equivalent to the komatiites (Paper 6). Serpentinized peridotite and dunite and metapyroxenite (now chlorite–tremolite) lenses associated with the high-magnesia basalts probably include surface flows (see Commentary, Paper 8); studies by Nesbitt et al. (1972) and J. D. Lewis (personal communication) of drill core from Scotia and other areas north of Kalgoorlie confirm this and reveal similarities to the flows described by Pyke et al. (1972) from Canada. The point is made that all these rocks occur in successive greenstone strips (Paper 12) as one passes through the alternating stratigraphy (Paper 13). Facing indications, including pillow facings and differentiation trends in the igneous rocks, establish thick successions composed of a complex intermingling of ultrabasic, ultramafic, and basic rocks occurring at several stratigraphic levels. The cyclicity is a multiply hiatal cyclicity of ultrabasic–basic and clastic–acid volcanic (porphyry) units (Papers 13 and 34), not a smoothly progressive cyclicity such as that described from the Abitibi Belt of Canada (Paper 33). The coarse gabbro bodies and layered intrusions

*The reason for the retention of the terms ultrabasic and ultramafic is given in McCall (1972).

derived from magmas of gabbroic composition are penecontemporaneous with the volcanics, almost invariably lying concordantly within a greenstone strip. The concept of penecontemporaneity of the greenstones replaces that of older and younger greenstones (Prider, 1961), which is quite incompatible with the observed distribution.

The serpentinites have two modes of occurrence, as discrete lenses and as the serpentinized lower layers to large stratiform intrusions. The chemistry of the former is characteristically magnesia-rich and alumina-poor. Such chemistry is rare in the second type, which tends to picritic composition and higher iron and alumina contents. Spinifex textures are manifestly supercooling textures. Near-monomineralic olivine "cumulus" aggregates contrast with the spinifex zones, which were formerly glassy, marginal zones in the discrete lenses. Spinifex texture may be present in supercooled zones of flows or near surface sills but is not found in the basal serpentinite zones of large, layered intrusions.

A number of differentiated sills are described. Particularly interesting are the folded but almost unmetamorphosed Mission Sill, near Norseman (McCall and Hudson, 1970), which shows macrorhythms, and the Mount Thirsty sill complex immediately beneath it stratigraphically, a type of sill repetition also noted at Soansville in the Pilbara Block (Paper 14). Such sill complexes show internal layering in a succession of closely spaced sills, with a progressive differentiation trend also evident through the group overall. A peculiar occurrence is the Bluebird Gabbro Sill at Norseman, a spectacular manifestation of the little understood "cat-rock" phenomenon; it is a giant feldspar phenocrystic dolerite.

Suites elsewhere in the Yilgarn Block are also mentioned. The suite at Mt. Singleton, in an enclave in the western granite batholiths, contains ultrabasic and ultramafic (pyroxenitic, now tremolitic) pyroclasts, including agglomerates and eutaxites. The Coobina ultramafic dike, set in an enclave of Archean within the Proterozoic of the Hamersley Range, is an immense, ramifying dike of ultramafics and minor gabbro that contains nodular chromite. This east–west trending dike provides the only possible known analogue of the true Alpine chromiferous ultramafic (Thayer, 1971) in the Western Australian Archean, despite the fact that this Archean terrain carries more ultrabasic occurrences than any part of the globe.

Enormously thick layered intrusions characterized by hundreds of internal macrorhythms and in some cases extensive granophyre developments occur in the Murchison District of the Yilgarn Block within the extensive granite terrain. These are of lopolithic dimensions, the Narndee Complex being a synclinally folded body more than 8 km thick.

Papers 16A–16C deal with geochemical aspects of the mafic and the ultramafic rocks of the same well-studied tract of Archean greenstones. Hallberg's is the most exhaustive geochemical study of a restricted area within a single greenstone belt yet undertaken. More than 500 rock samples, all basic or ultramafic, were analyzed by XRF methods for major and minor elements. The monotonous nature of the chemistry of the thick pillow lava sequences at Norseman and Coolgardie is so convincingly demonstrated that there can be no doubt that these submarine eruptives are ancient equivalents of abyssal tholeiites found on the present ocean floor. Only abyssal potassium-poor tholeiites are recognized at Norseman (more recent studies by Gemuts and Theron, 1973, suggest that this great pile is the lateral equivalent of the basal mafic–ultramafic group that to the north contains high-magnesia basalts and nickel sulfide deposits). Hallberg found no evidence of calc-alkaline, island-arc-type chemistry. He considers that the abundance of intermediate volcanics (andesites) may have been exaggerated in the case of the Archean of the Superior Province, Canada (Papers 8, 10, and 11), and that the analogy with modern island arcs may have been overstated. He attributes this to analysis of not completely unweathered or otherwise unaltered samples and the resultant spuriously high silica values. He asserts that the Barberton sequence (Paper 5) is, in its upper part, one of tholeiites of similar type overlain by acid volcanics.

Despite Hallberg's reservations, it does not now seem likely that the calc-alkaline suites of island-arc affinity have been erroneously recognized in Canada. Both abyssal tholeiites and island-arc-type, calc-alkaline eruptives occur in Archean greenstone assemblages, and the proportion varies from region to region. The area studied by Hallberg represents one extreme case; only high-magnesia basalts, abyssal tholeiites, and sodic porphyries (and rhyolites) are represented. The Abitibi belt of Canada, near Noranda–Rouyn, represents the opposite extreme; high-magnesia basalts, abyssal tholeiites, and sodic acid eruptives are overshadowed by prolific developments of island-arc-type, calc-alkaline suites in central edifices. The Barberton suite may well represent an intermediate case.

Hallberg correctly points out the limitations of the analytical data available. It is only by mounting programs in other key areas on his model, utilizing fresh drill-core samples as far as possible, that we shall obtain really reliable analytical data on greenstone terrains. Goodwin's definition of andesite (Paper 33) *is* based solely on silica percentage, but the alkali–lime index plots are convincing (Paper 10) evidence of calc-alkaline, island-arc chemistry. Nevertheless, the results obtained would in the case of tholeiitic andesites be vitiated by using nonfresh samples, and this could affect the aklali–lime indexes.

Hallberg recognizes an unrelieved thickness of pillow lavas of up to 9000 m at Norseman, and this is real, for gold mine workings traverse and reveal the consistent pillow facings. Such developments are in the nature of lenses formed on the sea floor, and are not maintained in thickness along strike. The term *basic plate* has been suggested for these by Ridler (personal communication, 1972). Rare examples of spilite chemistry only were noted (recorded in Hallberg's thesis, 1970). These are attributed to interaction between seawater and pillows.

References

Allard, G. O. (1970). The Dore Lake complex, Chibougamou—a metamorphosed Bushveld-type layered intrusion. *Spec. Publ. 1, Geol. Soc. South Africa,* 477–491.

Bartram, G. D., and McCall, G. J. H. (1970). Wall-rock alteration associated with auriferous lodes in the Golden Mile, Kalgoorlie. *Spec. Publ. 3, Geol. Soc. Aust.,* 191–199.

Gemuts, I., and Theron, A. (1973). Stratigraphy and mineralisation of a portion of the Archaean between Norseman and Coolgardie, Western Australia, Abstract vol., Sec. 3, ANZAAS, 45th Congr., Perth, p. 88.

Goodwin, A. M., and Shklanka, R. (1967). Archaean volcano-tectonic basins; form and pattern. *Can. J. Earth Sci., 4,* 777–795.

Hallberg, J. (1970). The petrology and geochemistry of metamorphosed Archaean basic volcanic and related rocks between Coolgardie and Norseman, Western Australia. Ph.D. thesis, Univ. Western Aust., 191p.

Macrae, N. D. (1969). Ultramafic intrusions of the Abitibi area, Canada. *Can. J. Earth Sci., 6,* 281–304.

McCall, G. J. H. (1972). The nickel-sulphide bearing ultramafic rocks and their environment in the Archaean of Western Australia. 24th Int. Geol. Congr., *1,* 354–362.

_____, and Doepel, J. J. G. (1969). Stratiform basic bodies within the Yilmia ophiolites, Western Australia. *Proc. Aust. I.M.M., 231,* 67–74.

_____, and Hudson, D. R. (1970). Excursion guide, Symposium on the Archaean Rocks. Geological Society of Australia, Perth, 1970.

_____, and Leishman, J. (1971). Clues to the origin of Archaean eugeosynclinal peridotites and the nature of serpentinisation. *Spec. Publ. 3, Geol. Soc. Aust.,* 281–299.

Naldrett, A. J., and Gasparini, E. L. (1971). Archaean nickel sulphide deposits in Canada: their classification, geological setting and genesis with some suggestions as to exploration. *Spec. Publ. 3, Geol. Soc. Aust.,* 201–226.

Nesbitt, R. W. (1971). Skeletal crystal forms in the ultramafic rocks of the Yilgarn block, Western Australia: evidence for an Archaean ultramafic liquid. *Spec. Publ. 3, Geol. Soc. Aust.,* 331–347.

_____, Purvis, A. C., Barry, J., and McKay, K. G. (1972). The Archaean greenstone peridotites of the Eastern Goldfields: an example of the peridotite nickel sulphide association. Abstract vol., Joint Specialist Groups Meeting, Geological Society of Australia, Canberra, p. H3–H6.

Prider, R. T. (1961). The greenstones of south-western Australia. *J. Roy. Soc. Western Aust., 44,* 1–9.

Pyke, D. R., Naldrett, A. J., and Eckstrand, O. R. (1973). Archaean ultramafic flows in Munro Township, Ontario. *Am. Bull. Geol. Soc., 84,* 955–978.

Thayer, T. P. (1971). Authigenic, polygenic and allogenic ultramafic and gabbroic rocks as hosts for magmatic ore deposits. *Spec. Publ. 3, Geol. Soc. Aust.,* 239–251.

Williams, D. A. C. (1971). Determination of primary mineralogy and textures in ultramafic rocks from Mt. Monger, Western Australia. *Spec. Publ. 3, Geol. Soc. Aust.,* 259–268.

15: SOME ULTRABASIC AND BASIC IGNEOUS ROCK OCCURRENCES OF THE ARCHAEAN OF WESTERN AUSTRALIA

G. J. H. McCall

ABSTRACT

An attempt is made to summarise information concerning the forms of intrusive and extrusive bodies, and the nature of the pyroclastic rocks, in the basic and ultrabasic eruptive suites of the Archaean in Western Australia. Sills, both undifferentiated and regularly stratiform (derived from basic parental melts) are described. Intrusions of lopolithic dimension showing prolific macro-rhythmic layering are also recognised. In the Eastern Goldfields and Pilbara there are sill complexes that show progressive differentiation trends through a series of individually differentiated sills. There are also lenses individually composed of basic, ultramafic, pyroxenitic, or ultrabasic peridotitic-serpentinitic rocks that form intimate spatial associations and show no regular upward trends in differentiation: they exhibit super-cooling textures, vesicles and devitrified glass, and are probably mostly shallowly emplaced sills, with some surface effusions. These associations include the nickeliferous serpentinites. Pillowed volcanic sequences associated with hyaloclastic breccias, basic and ultramafic agglomerates, basic shardy tuffs and basic laminar eutaxites of tufflava or ignimbrite affinities are among the other occurrences described.

INTRODUCTION

A complete review of basic and ultrabasic bodies in the Archaean of Western Australia cannot be given by any one geologist. This account deals with occurrences known to the author, to set up a model for the whole. Unforunately there is no space for inclusion of accumulated geochemical data, some of which were presented at the Perth Symposium.

THE KALGOORLIE-YILGARN SYSTEM

The basic rocks of the Eastern Goldfields, the best known tract of the Kalgoorlie-Yilgarn System (Fig. 1), were separated by Prider (1961) into older and younger greenstones. The older greenstones were regarded as mainly of basic volcanic origin, and were believed to be spilites: the younger greenstones were mainly intrusive sills. Subsequent detailed study of many greenstones belts has shown that both types of greenstone occur in linear belts in which metasediments (predominantly fine pelites) are subordinate to basic and ultrabasic rocks. Between such belts, which form residual ridges standing above the laterite surface, are considerable thicknesses of predominantly clastic sediments, including conglomerates. The constant spatial association of sills and volcanics in linear belts can only be reconciled with penecontemporaneity of all the rocks in these belts, except for the later acid porphyries that may occur within them (McCall & Doepel, 1969).

The components of these greenstone belts, which can be broadly equated with the Alpine ophiolite belts (Aubouin, 1965; McCall *et al.*, 1967; Glikson, 1968; McCall, 1969) are:

(c) *Gabbroic sills*: undifferentiated or stratiform (formerly regarded as younger greenstones),

(b) *Serpentinite sills*: Lenticular (may include some surface flows),

(a) *Basic volcanics*: basalts and picrite basalts.

In addition there may be irregularly distributed pyroxenites, and basaltic dolerites closely associated with the serpentinites, as parts of the same sills or separate sills.

All these rock types appear again and again in successive greenstone belts as one ascends through the stratigraphic succession, and it is difficult to avoid the conclusion that volcanics and intrusives in any one belt were almost always emplaced before the clastic sediments that overlie it and separate it from the belt above. Minor differences between the overall character of successive belts are thus to be

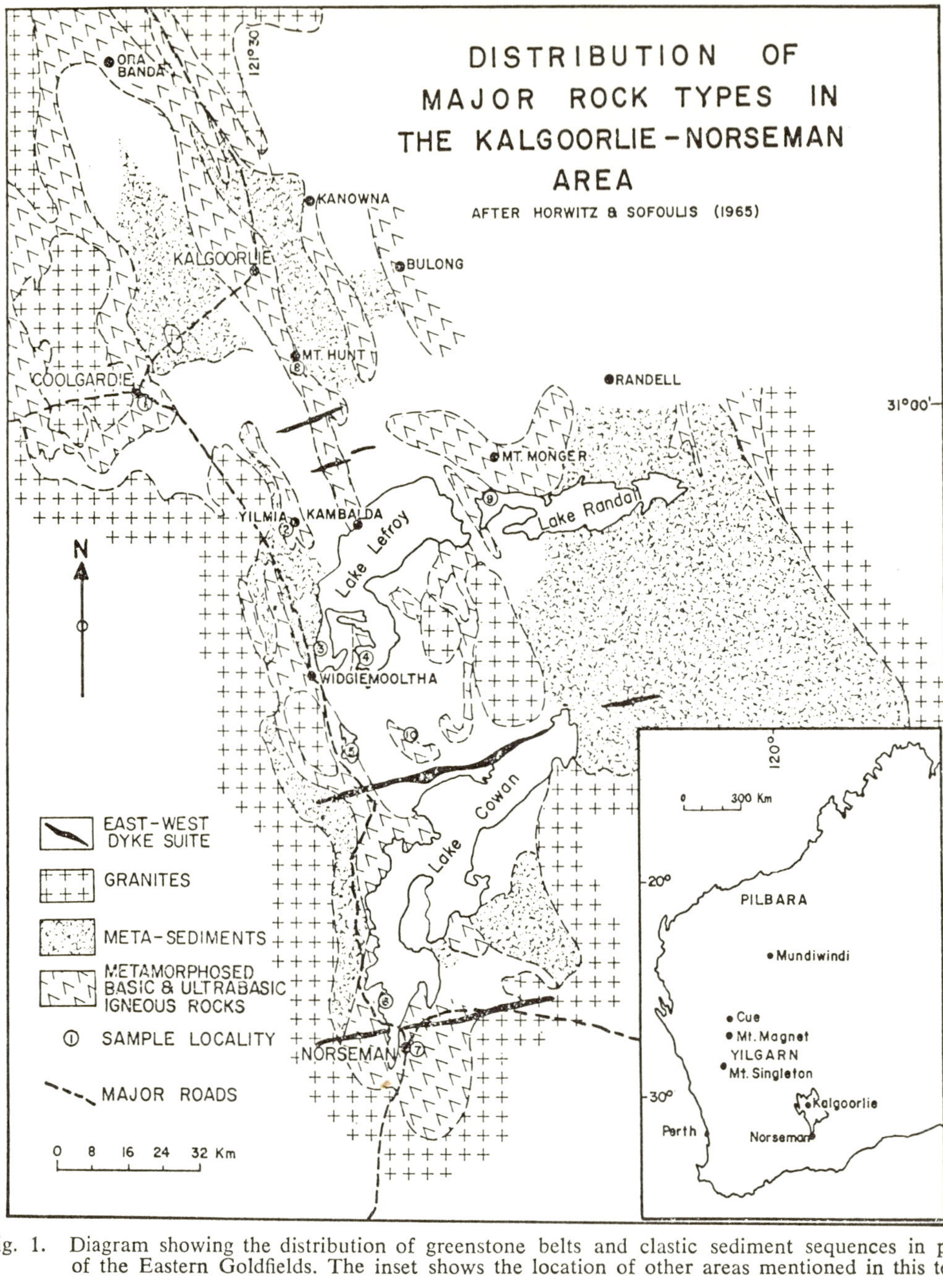

Fig. 1. Diagram showing the distribution of greenstone belts and clastic sediment sequences in part of the Eastern Goldfields. The inset shows the location of other areas mentioned in this text.

expected. Some greenstone belts in the area south of Kalgoorlie are impersistent, tailing out as the enclosing sediments change along the strike into clastic sediments and coming back with the reappearance of pelitic sediments. The concept of penecontemporaneity, replacing the dualistic interpretation in which all sills were taken to be later than all basic volcanics, accords well with Hallberg's conclusion that the magma of the differentiated sills was of the same stem as that of the lavas (Hallberg, 1970).

THE VOLCANIC ROCKS

The volcanic rocks form thick sequences of pillowed metabasalts, which may be intercalated with massive, unpillowed flow material or may form unrelieved pillowed sequences thousands of metres thick (Hallberg, 1970). Fine-textured tremolite-chlorite-(talc) rocks are associated closely with the metabasalts, and are probably partly of volcanic origin. Some may have originally been picrite-basalts (which have an olivine-pyroxenite chemistry), but most seem to have been pyroxenites devoid of olivine. Such rocks show a close spatial association with basalts of highly magnesian chemistry (akin to the basaltic komatiites of Barberton type, of Viljoen & Viljoen, 1969). These magnesian basalts contrast with the K-poor oceanic tholeiite type that characteristically forms monotonous pillowed sequences, typical examples of which are the Paringa Basalts at Kalgoorlie and the Desirable Pillow Lavas at Norseman, west of the Causeway. The highly magnesian basalts at Yilmia Hill, however (McCall & Doepel, 1969: Hallberg, 1970), seem to grade into tholeiites: here, as in the Widgiemooltha-Higginsville greenstone belt and the Carnilya Hill belt (McCall & Leishman, 1970), the highly magnesian basalts are, spatially, intimately related to the tremolite-chlorite rocks (metapyroxenites, broadly equivalent to the basaltic komatiites of Geluk type of Viljoen & Viljoen, 1969), and both are intimately interleaved with serpentinites.

The tholeiitic pillow lavas, which show little geochemical variation (Hallberg, 1970), do show considerable textural variation: they are essentially actinolite-plagioclase rocks, with or without a visible quartz mesostasis and carrying leucoxene and sphene after accessory ilmenite (or titanomagnetite). They show basaltic textures in palimpsest, many were clearly originally hemihyaline, and both vesicles and varioles are characteristic—the latter being found wherever pillow structure is developed. Epidote and chlorite are generally scarce, except in vesicle fillings (a locus of hydroxyl ions). Spilitic rocks are recorded tentatively by Hallberg (1970), from Mt Hunt (the remarkable 'wheel rock' pillows), though they are rare and do not form the predominant type of the Goldfields as Prider (1961) supposed. Interaction of magma and sea water or of magma and deuteric agents has produced a zonation of soda within individual pillows (Hallberg, 1970). The variole patterns are consistent with sea-floor depths at the time of effusion of not more than 120 metres. Varioles are known to grow progressively smaller in pillow lavas down to about 1200 metres, at which depth they disappear, because the hydrostatic pressure of the overlying water inhibits their development. Gash-like vesicles are, however, common in the cores of pillows, appearing like parallel arrays of bayonet gashes, and this structure is supposed to indicate a considerable depth of water (C. J. Annhaeusser, *verb. comm.*). These gash patterns have not so far been used as facing indications as in South Africa. The pillows show wide variations in shape and size, and include 'tailed' pillows, bolster-like elongate sacks without tails, oval forms without tails, and as already noted, the 'wheel rock' variant with spherical form and pronounced radial cracking. Very large pillows up to 3 metres long and very small forms not more than 0.2 metres long are recorded, but most are of moderate size.

It is not easy to distinguish separate flows of basalt even in mine sections, for though it is customary to demarcate 'flow-top' breccias, most basic breccias associated with the lavas are not of this type, which is a feature of subaerial eruption. The breccias are mainly hyaloclastite breccias, and may have a rather irregular distributional relationship to the massive flow rocks. Such rocks, very similar to those described by Rittman (1962) and Solomon (1969), are exposed on the Lefroy Peninsula (McCall, 1969): here interpillow glassy material (peperite) has a pronounced local development, and forms the matrix to a breccia consisting of dark, once glassy fragments, pillow fragments and even whole pillows. Other breccias exposed at Red Lake (Glikson, 1968) and on Hampton Block 53 to the south, have the characteristics of flood breccias.

Some unusual breccias at Bayley's Workings and near Emu Well (Muhling, 1965; McCall, 1969) contain acid rock enclaves, commonly spherulitic, set in a basaltic matrix. Basic pyroclasts showing a finer fragmental texture, exposed on South Island, Dordie Lake, form a mélange of basaltic material and fine, grey pelitic material, strongly sheared out. These are interpreted as peperitic breccias composed of a mixture of fine hyaloclastite material and interpillow sedimentary material. They may be derived from the breakdown of a layer of small pillows set in a copious shale matrix, the whole mass slithering down a seabed slope as a mélange. The pyroclastics here form a belt about 30 metres thick beneath a great thickness of pillow lavas. Many of the pillow-lava sequences show scanty interpillow material, the pillows being tightly packed. Where present, the interpillow material consists of sedimentary material and/or devitrified glass, converted to fine actinolitic aggregates. Pockets of sulphide occur in the interpillow material at Kambalda and at Norseman (Desirable Pillow Lavas, west of the Causeway). Whether these pockets, which include a range of sulphide minerals, are of primary origin is unknown: they could be derived from a deuteric magmatic residuum or be metasomatically introduced at a later date. Such sulphide occurrences are of interest in connection with a possible sulphurisation control on nickel mineralisation.

FINE-TEXTURED ULTRAMAFIC ROCKS

The fine-textured ultramafic rocks, completely converted to colourless needles of tremolite, subsidiary chlorite and in some cases talc, have, like some of the fine serpentinites, been allocated a volcanic origin by some geologists, whereas others regard them as sill-rocks. Their chemistry, characteristic of a picrite-basalt or pyroxenite, has already been noted. The magnesia content commonly ranges between 20% and 25%, and the rocks range from slightly olivine-normative to slightly quartz-normative, though seldom displaying a quartz mesostasis. Their association with the discrete serpentinite lenses, alongside unusually magnesian basalts, suggests that they have the same origin as the serpentinites, which are believed to be mostly sill rocks (see below, and McCall & Leishman, 1971). The relative proportions of serpentinite, tremolitic rocks and basaltic rocks in any one belt or lens of such a triple association—for these ultramagnesian rocks commonly form distinct belts or lenses within greenstone belts (Fig. 2)—are highly variable. For instance, at Yilmia, only a thin band of serpentinite occurs within abundant tremolitic and slightly less abundant basaltic rocks: at Carnilya Hill, near Mt Monger, serpentinite predominates, tremolitic rocks are also abundant, but basalts are scarce. At Carnilya Hill, too, the basalts show "slaggy" microtextures akin to those in the "spinifex" serpentinites nearby, and it is almost certain that all three types form entire shallow-level lenticular sills or parts of such sills. The effect is to produce a composite lens of mainly ultramafic rocks containing a number of smaller lenses of the three rock types. There are certainly a number of individual sills in any such a composite lens, for continuous fine pelite intercalations form screens between them.

SERPENTINITES AND SERPENTINISED PERIDOTITES

Two types of serpentinite occurrence have been distinguished, the *discrete lenses* and the *serpentinised lower layers of large stratiform intrusions*. These correspond to the two types of Naldrett (1964). The discrete lenses characteristically show high magnesia and low alumina chemistry. Serpentinisation is complete or partial. The parent rocks were forsteritic peridotites ranging towards dunite composition. Primary forsteritic olivine may be found preserved. There is evidence (McCall & Leishman, 1971), that both wehrlitic and harzburgitic magmas were represented among the discrete intrusions. 'Spinifex' or 'bird track' textures are common near margins of these bodies. Most of these textures can be shown to be derived from congealing rapidly under supercooling conditions. The original rocks were glass-based. Such textures are typical of small shallow intrusions rather than flows. Though some geologists have suggested that flows are numbered among the serpentinite lenses, these textures are atypical of flows, and the evidence, in the form of supposed pillow lavas (for instance on the west shore of Dordie Lake), is not completely convincing. These appear rather to be networks of joints defining rounded bodies within the rock mass, and infilled with secondary minerals. The bulk of the serpentinite bodies consists of olivine pseudomorphs in antigorite: these close-to-monomineralic olivine accumulations are believed to be due to excessive olivine growth *in situ* instead of mechanical concentration (McCall & Leishman, 1971).

It is likely that some spinifex rocks are derived from amphibolitic pegmatoids, rather than supercooling. These are not common, but are evident in the Dordie Lake area. Many rocks showing the typical supercooling textures of spinifex rocks are tremolitic: such antigorite-free rocks occur about 16 m south of Coolgardie, close to the Coolgardie-Norseman Highway on Hampton block 53, and are probably metamorphic derivatives of serpentinites in a raised metamorphic environment, close to the granites. This process is perhaps more correctly termed metasomatic for it must involve addition and subtraction of material in an open system.

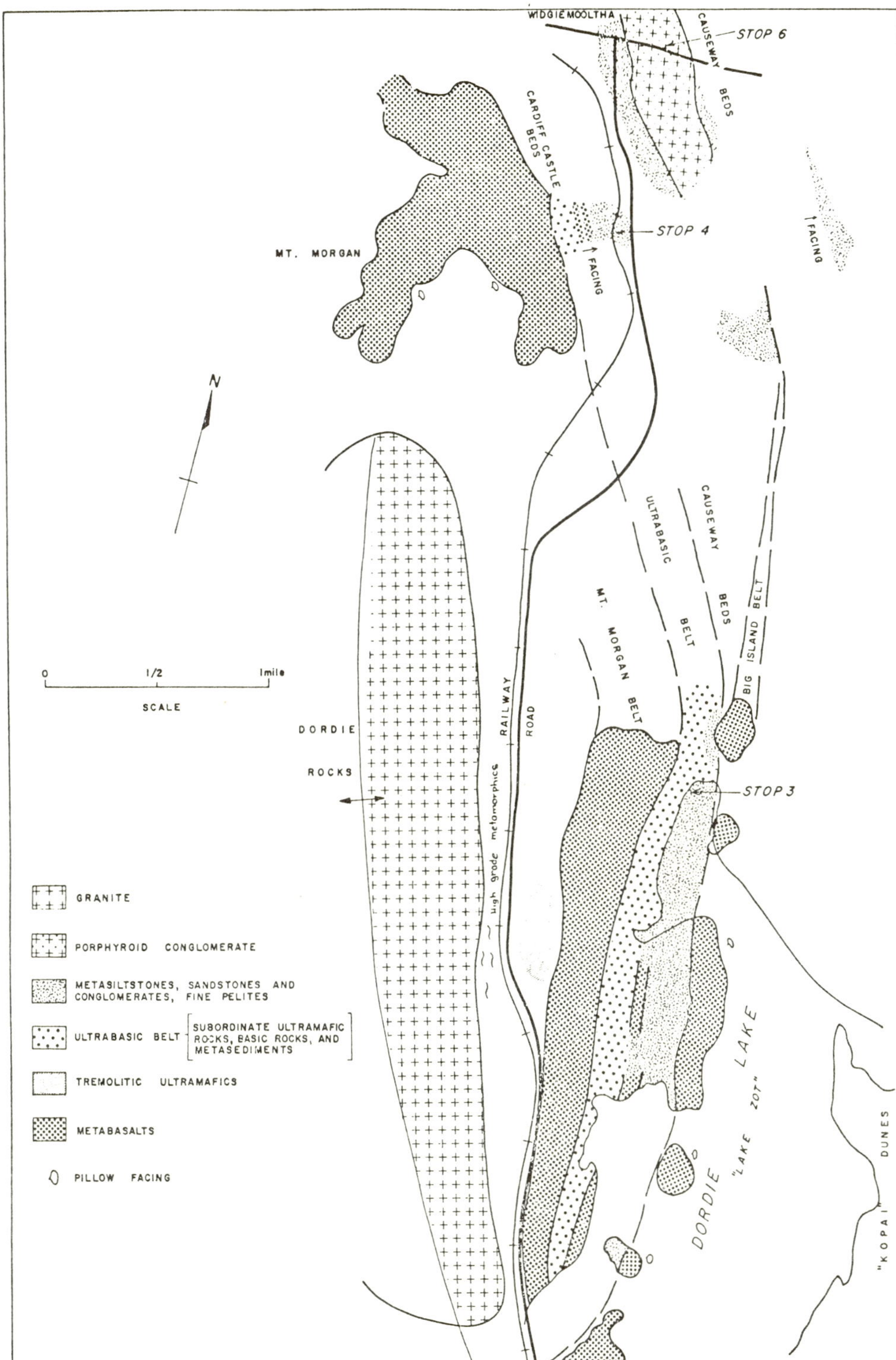

Fig. 2. Diagrammatic sketch of an area south of Widgiemooltha showing how high magnesian ultramafic rocks (serpentinites, tremolitic rocks, highly magnesian basalts) outcrop in a linear zone within the greenstone succession.

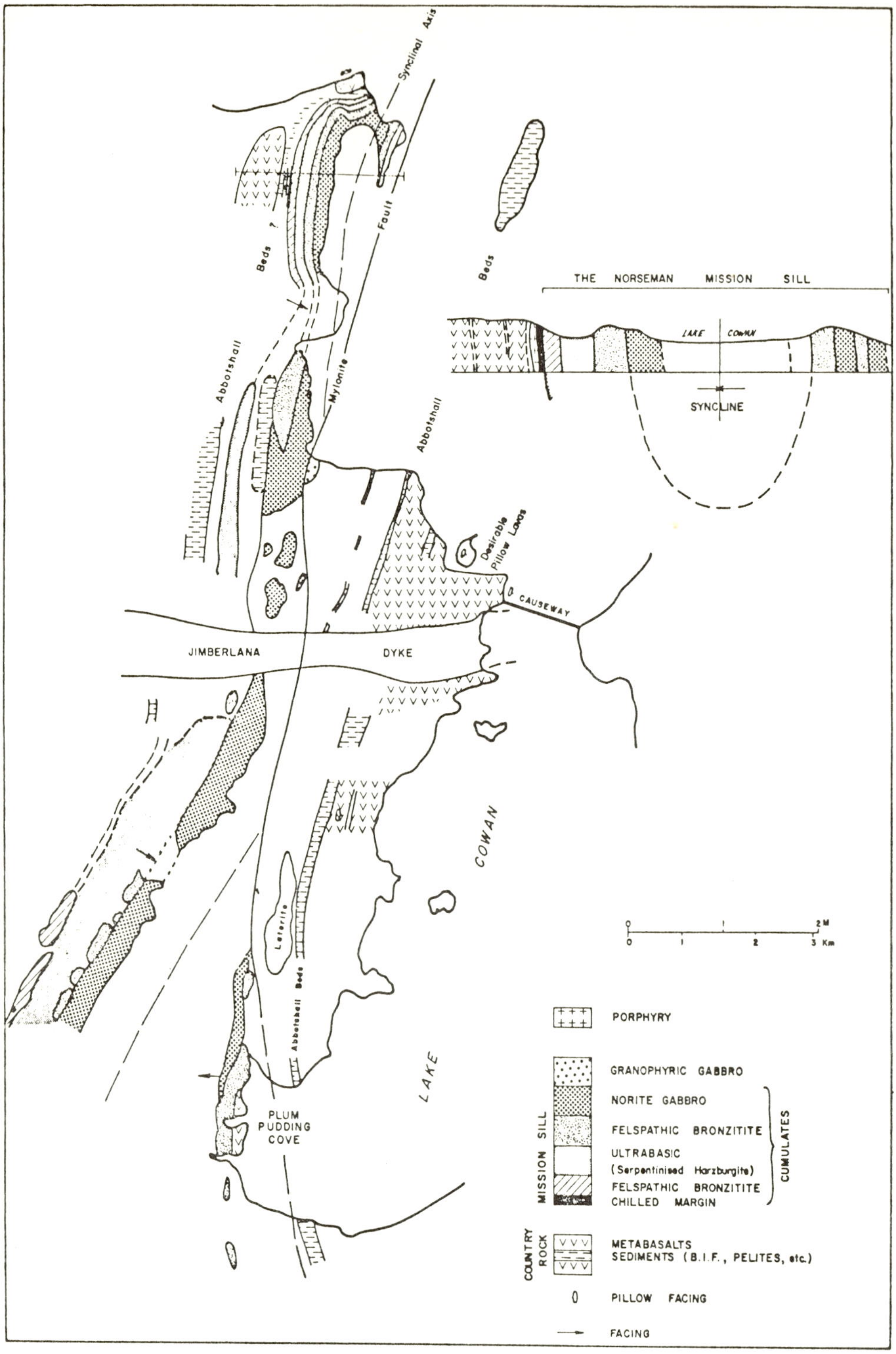

Fig. 3. The Mission Sill. Based on mapping by G. R. Hudson, G. J. H. McCall, C. T. Williams, and D. Harley.

A large cross-cutting body of ultrabasic rocks has been reported from Bulong (I. R. Williams, *verb. comm.*), but other geologists believe it to be concordant. Apparent concordance certainly seems to be the general rule. Though minor discordances at contacts have undoubtedly been obscured by shearing effects, all the evidence of repeated triple associations of clearly genetically related highly magnesian rock types, previously noted, indicates that the serpentinites are more or less *in situ*, and not exotic material mechanically transferred in a cold state, long after emplacement, from a quite different magmatic emplacement environment. The cold emplacement theory, though accepted in other parts of the world, owes much to the lack of metamorphic effects in the form of contact aureoles to the ultrabasic bodies. Contact effects are certainly not always evident in the Eastern Goldfields, but this argument ignores the fact that many large and obviously very hot emplacements of basic magmas produced negligible contact effects—for example the Jimberlana and Binneringie Dykes. The scale of development of contact aureoles is not, apparently, directly proportional to the temperature of the magma and the size of the emplaced mass—the involvement of volatiles is probably also a major factor.

Serpentinisation was in many cases followed by steatitisation, productive of talc-carbonate rocks.

THE GABBROIC SILLS

Differentiated sills studied in some detail include those of Kalgoorlie (Williamstown Dolerite, Golden Mile Dolerite: Bartram & McCall, 1971) the two Yilmia sills (McCall & Doepel, 1969) the Mission Sill (Fig. 3), Mt Thirsty Sill (Fig. 4), (really a number of sills), the Ora Banda Sill (Williams, 1967), and the Mt Monger sills. Others are known. In addition, there are many sills of undifferentiated gabbro which may form swarms of large lenses. The differentiated sills may tail off into gabbro. The Ora Banda Sill shows serpentinite at its base, whereas the Mt Thirsty Sill shows partly serpentinised dunite and harzburgite. Other layered sills show only harzburgite at their base. The phase-layer above the ultrabasic layer is of feldspathic bronzitite, and above this layer come norite, norite-gabbro and dolerite. The top of the sill may be composed of granophyric gabbro. The ultrabasic/bronzitite layer boundary is invariably abrupt, as is also the boundary between the bronzitite and the over-

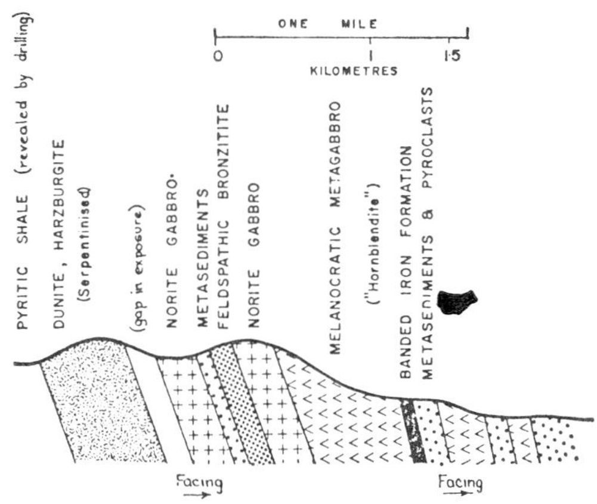

Fig. 4. Diagrammatic section, Mt Thirsty, showing the grouping of sills (after G. Whitfield).

lying gabbroic rocks. The harzburgite/dunite boundary is obscure and not a distinct phase-layer boundary in the Mt Thirsty Sill, and the noritic, gabbroic and dolerite rocks are likewise not marked by clear phase-layer boundaries. The granophyric gabbro occurrences are ill defined, having the form of irregular patches near the top of the gabbroic sequence, which, despite its lack of abrupt internal breaks, must include more than one phase-layer boundary corresponding to the incoming of clinopyroxene and the outgoing of orthopyroxene.

The rocks are mostly cumulates: the dunites and harzburgites are mostly olivine cumulates, and thus the dunite/harzburgite boundary need not be a true phase-layer boundary. The bronzitites are bronzite cumulates, the norites and norite-gabbros pyroxene-feldspar cumulates (clinopyroxene coming in as a cumulus mineral after othopyroxene and feldspar).

The state of metamorphic modification is variable. The feldspathic bronzitites are commonly fresh, preserving both primary minerals. Elsewhere, the bronzitite has been made over to tremolite, but the cumulate texture is preserved. The cumulate texture of the ultrabasic rocks is almost invariably preserved, despite serpentinisation or metamorphic modification to tremolite and chlorite. The gabbroic rocks may preserve their original feldspars or both original feldspars and pyroxenes, though commonly the pyroxenes have been made over to tremolite and actinolite. Again, cumulate textures are preserved. Some gabbroic rocks have been modified to coarse plagioclase amphibolites. The Golden Mile Dolerite is unusual (Bartram, 1969; Bartram & McCall, 1971), in that it consists of ten layers including two chilled margins. The most mafic layer above the basal chilled margin is a melanocratic 'dolerite', showing traces of original cumulate texture. The sill formed by roof-downward as well as floor-upwards crystallisation, and suffered two later injections of magma sandwiched within the resultant layered series. The later injections were the extreme differentiates, ferrodolerites and granophyres.

Though the ultrabasic rocks basal to the stratiform sills are commonly picritic and therefore aluminous, some highly magnesian and alumina-poor ultrabasics occur within the stratiform intrusions. An analysis of the picrite from the Mission Sill shows the same composition as the typical serpentinite of the discrete intrusions. The ultrabasic layer in this sill is not at the base, being sandwiched between two feldspathic bronzitite layers (though it is conceivable that there are two sills here). The sill shows a chilled margin beneath the lower feldspathic bronzitite layer composed of small bronzites, set in a fine, very altered base. If this unusual rock represents the parental melt, its composition will be significant. The Mission Sill shows repetition of layers at its northern end: this may be a form of macro-rhythmic layering, or could be due to tectonic folding or imbrication. The sill is certainly tightly folded, despite its massive character.

Rhythmic layering, including feldspar/pyroxene grading occurs in the gabbroic rocks, but is rare. In the norite-gabbro of the Mission Sill, the layering shows feldspar concentrated at the base of the grades. It has been suggested that this is a mineral grain size rather than density grading pattern, as in the Jimberlana Dyke (Campbell *et al.*, 1970).* Cryptic and phase layering are evident in all these sills. Hallberg (1970) suggests that these sills stem from the same K-poor oceanic-type tholeiite melt as the monotonous pillow lavas, though the post-metamorphic dykes stem from a typical continental, more potassic tholeiite melt. The geochemical evidence (McCall & Hallberg, unpublished data) suggests that already differentiated liquids gained access to the sites of sill emplacement, and were further differentiated *in situ*. Yet it broadly confirms the close genetic relationship between the parent melts of the stratiform sills and of the basaltic volcanics.

An unusual type of sill is represented by the Bluebird Gabbro at Norseman (Bekker, 1962). Varying from gabbro to dolerite and ranging from 66 to 100 metres thick, this sill displays a remarkable coarsely feldspar-phyric basal zone. This zone is quite thin, however, and seems to be an extreme development of the "cat-rock" phenomenon, reflected in outcrops of mega-porphyritic dolerite throughout the Eastern Goldfields. Similar developments are seen in dykes at Kelmscott, near Perth. The calcic plagioclases, several centimetres long, now metamorphically modified to a sodic variety, occur in basalts beyond the Gee Cee Slate horizon (Bekker, 1962) in the country rock beneath the sill, and one may thus justifiably question their primary igneous nature. The feldspars weather out abundantly on an island in Lake Cowan and show a proliferation of faces, some of which have defied identification in terms of known crystal faces for feldspars.

Pre-metamorphic cross-cutting basic intrusions occur as dykes at Norseman (Bekker, 1962; Hudson, 1966), closely following the east-dipping lode shears (laminated auriferous quartz reefs). The dykes appear to exert some broad control on the development of these shears. They are petrographically similar to many of the non-stratiform tholeiitic gabbro and dolerite pre-metamorphic sills. Other small pre-metamorphic dykes outcrop at Dordie Lake and within turbidites of Mt Belches (Dunbar, 1966). These cross-cutting pre-metamorphic bodies are, however, comparatively rare. A markedly vesicular sill of dolerite occurs within the Mandilla Beds near Widgiemooltha (McCall, 1969): I. R. Williams (*verb. comm.*) reports such vesicular dolerite bodies are not uncommon within clastic sediment/acid volcanic formations of the Kurnalpi area. The vesicles (Plate 18B) display quartz, chlorite and epidote, reflecting the local concentration of water, but the rock elsewhere has the usual chlorite- and epidote-poor character of the basic igneous rocks of this region, not, as in the Golden Mile (Bartram, 1969: Bartram & McCall, 1971), being affected by a superimposed hydrothermal metasomatism.

THE MT SINGLETON ENCLAVE

The interesting enclave of Archaean metamorphic rocks within the Archaean granite north of the Wheat Belt contains unusual types of basic and ultrabasic igneous rocks

* See also Campbell (1966) for an account of the Jimberlana Norite.

(Plate 17 A-F). Low in the sequence, the ultrabasic rocks form narrow concordant layers (sills?) similar to those in the Eastern Goldfields: however, both the non-feldspathic ultramafic and feldspathic basic rocks include among their number an unusual abundance of diverse pyroclastic rocks. Porphyritic rocks of intermediate composition have also been identified. The ultrabasic rocks are serpentinised wehrlites containing fresh augite: they display the same textural range from closely packed aggregates of antigorite pseudomorphs of olivine to looser olivine aggregations, the pseudomorphs being set in a once glassy base which also contains the pyroxenes (compare rocks from Marshall Pool, McCall & Leishman, 1971). Brown amphiboles are an abundant accessory. Clinopyroxenites (metamorphosed to tremolitic rocks but still displaying the outlines the original pyroxene accumulation (?) in palimpsest) are closely associated, spatially, with the serpentinites, either forming thin, separate sills or else parts of the same differentiated sills. The field relationships are obscure, but, as in the Eastern Goldfields, there seems to be a lack of regular stratiform layering and a lack of feldspathic rocks in close association with the serpentinites and clinopyroxenites. The basic rocks are mostly pyroclastics, and include shardy tuffs, eutaxitic tuffs, stratified tuffs and coarse volcanic breccias. Pillow lavas are represented, but these display a weakly developed (possibly incipient) structure, marked by much autobrecciation along the pillow margins. The finer shardy tuffs megascopically resemble dark green cherts, but reveal the outlines of shards, indicating their true character, in thin sections. These, and all the other basic volcanic rocks, are mostly ultramafic, the only major mineral component being pale tremolite, in minute needles. The coarse volcanic breccias include feldspathic, basaltic varieties, and also rare ultramafic varieties, the latter being close to the silica content range of true ultrabasic rocks. Antigorite pseudomorphs of olivine are abundant in these breccias. Much of the pyroclastic rock content of this eruptive sequence may well have originated as flood-tuffs (tuff-lavas, ignimbrites, ash-flows). The laminar eutaxite has obvious affinities of this sort, and the shardy tuffs and volcanic breccias may well have been derived as overland pyroclastic flows.

The chemistry of these rocks indicates that there are andesine basalts and oligoclase basalts represented. The intermediate rocks have been loosely termed andesites in the field, but they are not true andesites. They are glomeroporphyritic feldspar-phyric flows, marked by abundant and irregular epidote-filled vesicles. The chemistry of one analysed rock is that of sodic keratophyre, but because nearly half of the volume is vesicle filling, and because of the metamorphic alteration, its exact primary composition remains obscure.

Sediments are rare in this sequence, being limited to thin horizons of fine ferruginous metasediments (pelites?), and some obscure chloritoid-andalusite rocks.

The pyroclasts become more abundant towards the top of the sequence, which is in the form of a broad open syncline, with gentle opposing dips. It is possible that about 650 metres of strata are exposed in Mt Singleton and Wyalocopping Hill to the west, in a hillside section—a rare thing in this part of Western Australia. East-west dykes of gabbro and some small bodies of granophyre, residual to the tholeiitic magma, cut the entire sequence. These dykes are extensively uralitised, but not metamorphosed, unlike the country rocks. There is little doubt that the Mt Singleton sequence is Archaean, and that the dykes belong to the "Early Proterozoic" —*circa* 2420 m.y. suite of Turek (1966)—that is, they are equivalent to the Widgiemooltha Dyke Suite of the Goldfields. The chemistry of the metamorphic rocks is consistent with their being Archaean eruptives.

Base metal associations render this area economically interesting.

CUE—HYALOCLASTITE BRECCIAS

Close to Cue there are excellent outcrops of hyaloclastite or pillow breccias, described by R. S. Chaturvedi, who supplied Plate 18A. The amoeboid form of the pillows left floating in the fine fragmental matrix (a mélange of volcanic glass fragments) and not, as is common, pressed on to and moulded by adjacent pillows, is revealing, and could have implications to theories of pillow genesis.

COOBINA—CHROMIFEROUS ULTRAMAFICS

Studies have been initiated by S. M. Bye (under the author's supervision) on the Coobina area, near Mundiwindi (Fig. 5). A great branching dyke of serpentinised peridotite, pyroxenite and subordinate gabbro cuts an enclave of Archaean banded ironstone, quartzites and schistose plagioclase amphibolites, the whole being hemmed in on either side of

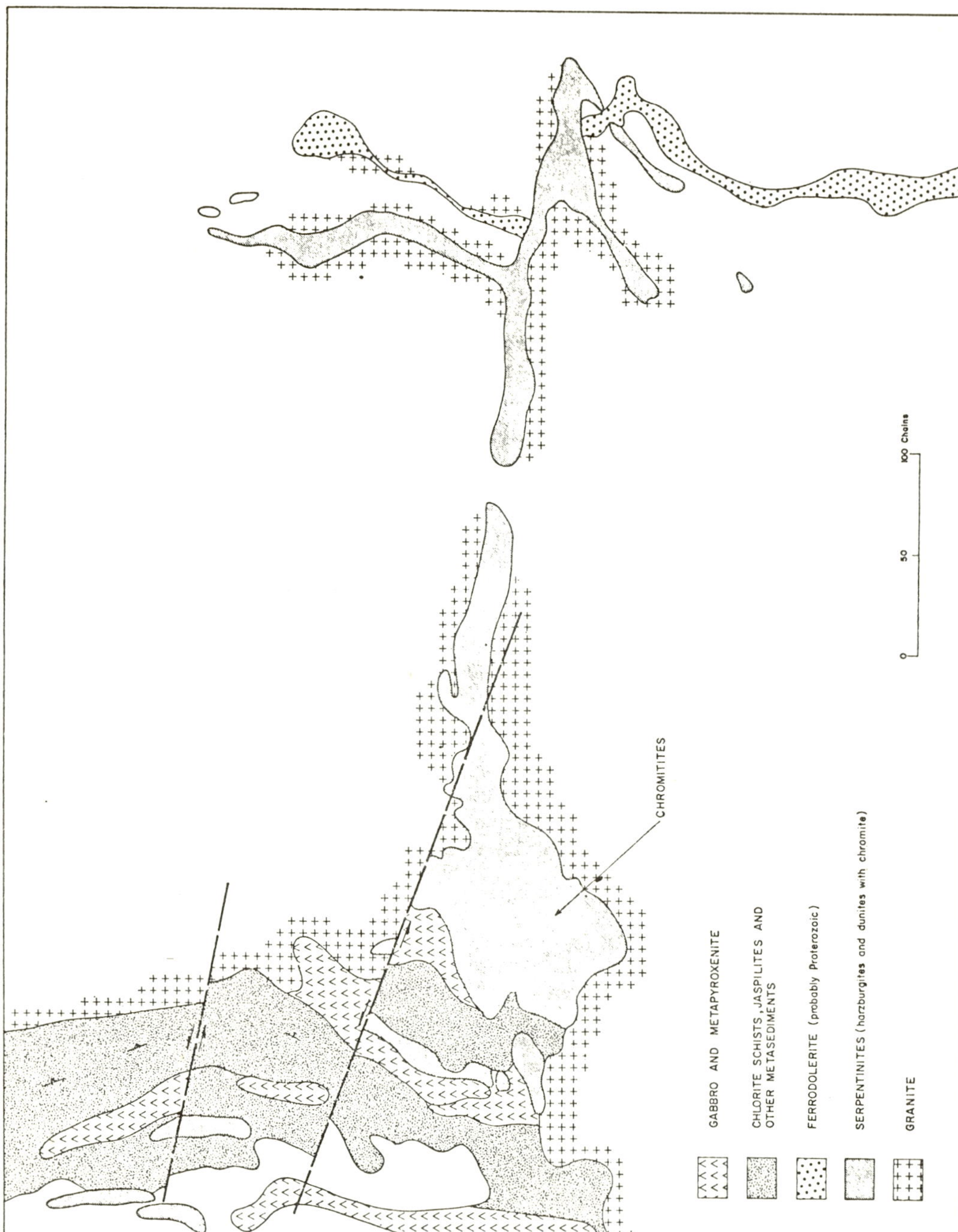

Fig. 5. The Coobina chromiferous dyke (after S. M. Bye).

the dyke by granites, which lap up against the dyke and post-date both dyke and its country rocks. The rocks of the dyke are metamorphosed, so it is unlikely to belong to the east-west suite of the Wheat Belt and Goldfields. It swells to a pod at its west end, where it abuts on a band of quartzite and banded ironstone, which may have prevented westward extension of the magma filling the dyke. The pod contains smaller lenses of chromite, showing schlieren and nodular structure. This chromite is of economic interest, though not of metallurgical grade. Further west, pyroxenites reappear on the same line. Galena veins border the dyke at its eastern end.

This occurrence is of great academic interest, for it closely resembles the Alpine or ophiolitic type of ultramafic body extensively studied by Thayer (1967), who considers that only such nodular chromite-bearing ultramafics can be equated with the Alpine ophiolites (*pers. comm.*). Such occurrences are rare in very ancient Precambrian Systems, but are known from southern Africa. The Coobina occurrence has the petrological character of the ophiolites (*sensu stricto*) but not the typical form or palaeogeographic association: the Goldfields ultrabasic lenses have the typical eugeosynclinal palaeogeographic association and concordant form, but not the characteristic petrology, and perhaps should be designated Archaeogeosynclinal serpentinites, not ophiolites. They carry abundant magnetite, and chromite only in accessory quantities.

PAYNES FIND-SANDSTONE—MAJOR INTRUSIONS

Of great interest are the immense Archaean layered intrusions isolated as residual enclaves within the granites east of the Paynes Find-Mt Magnet highway. The best known of these is the Windimurra Complex, a vast layered complex extending over hundreds of square kilometres.

These complexes show repeated ultramafic/gabbro macro-rhythms extending over immense tracts of steeply dipping strata. A thickness of 5,550 metres of such strata, devoid of country rock intercalations, has been recorded in the synclinal outcrop of the Narndee Complex, a smaller complex that contains harzburgite, pyroxenite and gabbro layers. Fine rhythmic banding is only very rarely detected in these complexes.

Besides their academic interest, these complexes of Stillwater or Bushveld Complex dimensions could prove of considerable economic significance in respect to nickel, platinoids and vanadium.

THE PILBARA SYSTEM

Though Pilbara rocks are superficially similar to the sequences of the Eastern Goldfields, rubidium/strontium dating on rhyolites and granites (Arriens, 1971) is tending to establish an older, Katarchaean (⩾ 3000 m.y.) age for them. The apparent strike continuity across the Proterozoic gap into the Pilbara Block cannot be taken as evidence of exact contemporaneity between the Kalgoorlie-Yilgarn and Pilbara Systems, because these systems may incorporate rocks laid down over a very considerable time interval. The progressively older ages being obtained outward west and north from the Eastern Goldfields bear no relation to distinguishable structural boundaries, and there may be no distinct structural boundary between the Kalgoorlie-Yilgarn and Pilbara Systems, despite the Pilbara System's much greater apparent age.

The discrepancy between the apparent ages of the Kalgoorlie-Yilgarn and Pilbara Systems and the progressive increase in maximum ages from east to west away from the Eastern Goldfields could, however, be reconciled with the Kalgoorlie-Yilgarn and Pilbara deposition being mostly of like age. It is possible that the lower rubidium-strontium ages obtained for the Kalgoorlie-Yilgarn rocks are due to a metamorphic superimposition, related to the granite uprise, which took place about 2615 million years ago (Turek, 1966). This smoothing out might not have affected the Pilbara rocks, for the granites invading them appear to be more than 3000 million years old (Arriens, 1971). We can not at present entirely discount the possibility that varying ages of metamorphic superimposition, related to varying ages of granite uprise, are the controlling influences on the rather strange pattern of maximum rubidium-strontium ages in the Archaean of Western Australia.

Pillowed metabasalts of the Pilbara System resemble those of the Eastern Goldfields, but there does not seem to be the same definition of greenstone belts in the Pilbara. There is, however, a profusion of ultrabasic-basic sills, and the sill complex pattern, similar to that at Mt Thirsty, is of particular interest.

In the Soanesville area, southwest of Marble Bar, there is a succession of sills, some more than 1000 metres thick: there is a pattern of overall differentiation trends upwards through the succession of sills, which are defined by enclosing cherts of sedimentary origin: and there is also a pattern of internal differentiation within the individual sills (produced *in situ* after emplacement by convective/gravitational influences or possibly some other differentiation process). These sills form sill-complexes as at Mt Thirsty, groups of sills apparently successively injected from a magma reservoir progressively differentiating in depth. coarser, holocrystalline rocks display spectacular crystal accumulation textures. These rocks are little modified by metamorphism. The parental magma appears to have been ultramafic and highly magnesian, and though chemical analyses are not yet available, this association of sills seems to come from a similar stem to the komatiites (Viljoen & Viljoen, 1969)—it might well be a stem of similar composition to that of the basaltic komatiites of Geluk type. If chemical analyses support

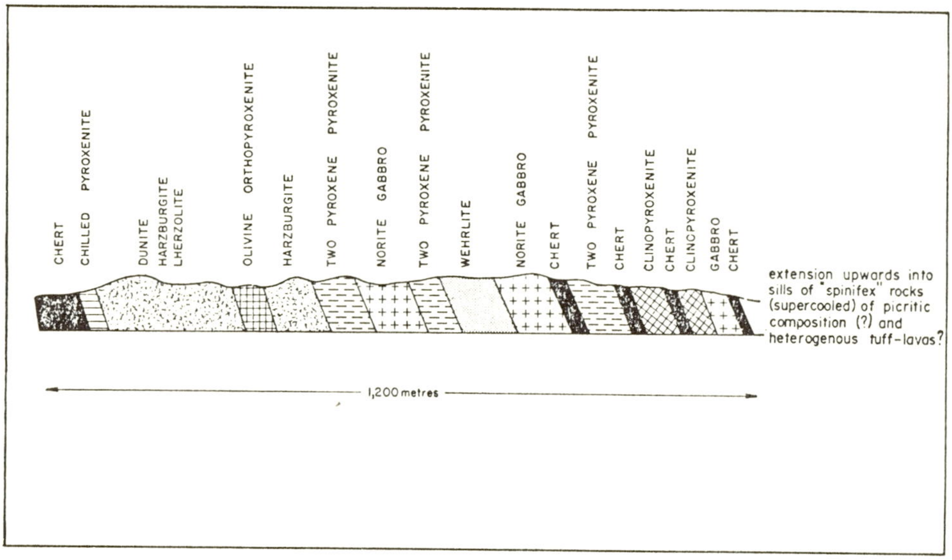

Fig. 6. Diagram of a typical sill complex in the Pilbara (this is a generalisation based on a composite section across a sill complex in the Soanesville area, southwest of Marble Bar).

One of these complexes is shown in diagrammatic section in Figure 6. The rocks forming the layers on the sills include dunite, harzburgite, wehrlite, lherzolite, clinopyroxenite, two-pyroxene pyroxenite, gabbro and diorite. Besides these coarsely holocrystalline rock types, there are thicknesses of several hundred metres near the top of the complexes, consisting of once glassy, spinifex-textured rocks of probable picritic or peridotitic composition. Further chilled phases are recognised near the base of the complex, and appear to be of pyroxenitic rather than basaltic composition. Indeed, feldspar is quite rare in rocks of this association, the gabbroic and dioritic rocks being subordinate to ultramafic rocks. The chilled basal rocks display chlorite-filled vesicles (*cf.* McCall & Leishman, 1971). The this tentative conclusion, and if the Pilbara rocks are really older than the Kalgoorlie-Yilgarn rocks, the ultramagnesian (komatiitic) eruptives would have had access to upper levels of the crust throughout a long period of Archaean time, and therefore the eruption of these unusual magmas would not have been, as Viljoen & Viljoen suppose, a one-burst affair.

There are gneissic rocks in the Soanesville area bordering small granitic stocks: among these are garnetiferous rocks and clearly the metamorphism adjacent to the granites was of high grade. However, away from the granite, the metamorphic effects diminish abruptly, and one passes almost immeditely onto almost unmetamorphosed rocks. This fact, coupled with the association of cumulate rocks and overlying sub-volcanic spinifex rocks in the same sill complex, and the presence of vesicles in chilled margin rocks, might indicate a very steep fall in background geothermal gradient in these ancient times. If the above indication is correct, it might be that the pattern of high-grade metamorphic rocks closely related spatially to the granite margins, also seen in the Eastern Goldfields, reflects a similar but less marked ancient steepening of the thermal gradient. There is a very interesting field for research into the metamorphism of these ancient systems in Western Australia in relation to the uprise of the granites.

REFERENCES

ARRIENS, P. A., 1971: The Archaean geochronology of Australia. *Spec. Publs geol. Soc. Aust., 3*, pp. 11-23.

AUBOUIN, J., 1966: *Geosynclines*. Elsevier, Amsterdam.

BARTRAM, G. D., 1969: Wall-rock alteration associated with auriferous lodes at Kalgoorlie. Ph.D. Thesis, Univ. West. Aust. [unpublished].

———, & MCCALL, G. J. H., 1971: Wall-rock alteration associated with auriferous lodes at Kalgoorlie. *Spec. Publs geol. Soc. Aust., 3*, pp. 191-199.

BEKKER, C., 1962: The greenstones of the Norseman District. M.Sc. Thesis, Univ. West. Aust. [unpublished].

CAMPBELL, I. H., 1966: The petrology of the Jimberlana Norite, near Norseman, Western Australia. B.Sc. Hons. Thesis, Univ. West. Aust. [unpublished].

———, MCCALL, G. J. H., & TYRWHITT, D. S., 1970: The Jimberlana Norite, Western Australia—a smaller analogue of the Great Dyke of Rhodesia. *Geol. Mag., 107*, pp. 1-11.

DUNBAR, G. J., 1966: The geology of the Mt Belches area, east-south-east of Kalgoorlie, Western Australia. B.Sc. Hons. Thesis, Univ. West. Aust. [unpublished].

GLIKSON, Y. A., 1968: Metamorphic and igneous petrology and structural geology of the early Precambrian rocks of the Mungari-Kurrawang Area, Kalgoorlie, Western Australia. Ph.D. Thesis, Univ. West. Aust. [unpublished].

HALLBERG, J., 1970: The petrology and geochemistry of metamorphosed basic volcanic and related rocks, between Coolgardie and Norseman, Western Australia. Ph.D. Thesis, Univ. West. Aust. [unpublished].

HORWITZ, R. C., & SOFOULIS, J., 1965: Igneous activity and sedimentation in the Precambrian between Kalgoorlie and Norseman. *Proc. Aust. Inst. Min. Metall., 214*, pp. 45-59.

HUDSON, G. R., 1966: The geology of part of the Norseman succession west of the mining area, Norseman, Western Australia. B.Sc. Hons. Thesis, Univ. West. Aust. [unpublished].

LEISHMAN, J., 1969: The geology of an area near Marshall Pool, Mt Margaret Goldfield, Western Australia. B.Sc. Hons. Thesis, Univ. West. Aust. [unpublished].

MCCALL, G. J. H., 1969: The Archaean succession to the west of Lake Lefroy. *Proc. R. Soc. West. Aust., 52*, pp. 119-128.

———, BRAYBROOKE, J. C., MIDDLETON, D. D., & MUHLING, P. C., 1967: Sedimentology of some eugeosynclinal rocks of the early Precambrian Kalgoorlie System, Western Australia. *Proc. 7th int. Sedimentol. Congr., Reading & Edinburgh.*

———, & DOEPEL, J. J. G., 1969: Stratiform basic bodies within the Yilmia ophiolites. Western Australia. *Proc. Aust. Inst. Min. Metall., 231*, pp. 67-74.

———, & LEISHMAN, J., 1971: Clues to the origin of Archaean ophiolitic peridotites and the nature of serpentinisation. *Spec. Publs geol. Soc. Aust., 3*, pp. 281-299.

MUHLING, P. C., 1965: The geology of the Bayley's Workings area, west of Lake Lefroy, Western Australia. B.Sc. Hons. Thesis, Univ. West. Aust. [unpublished].

NALDRETT, A. J., 1964: Ultrabasic bodies of the Porcupine and related nickel deposits. Ph.D. Thesis, Queens Univ., Ontario [unpublished].

PRIDER, R. T., 1961: The greenstones of southwestern Australia. *J. Proc. R. Soc. West. Aust., 44*, pp. 1-9.

RITTMAN, A., 1962: *Volcanoes and Their Activity* (Trans. E. A. Vincent). Wiley, New York.

SOLOMON, M., 1969: The nature and possible origin of pillow lavas and hyaloclastite breccias of King Island, Australia. *Q. Jl geol. Soc. Lond., 124*, pp. 153-170.

THAYER, T. P., 1967: Chemical and structural relations of ultramafic and felspathic intrusive complexes; in Wyllie, P. J. (Ed.), *Ultramafic and Related Rocks*, pp. 222-239. Wiley, New York.

TUREK, A., 1966: Rubidium-strontium isotopic studies in the Kalgoorlie-Norseman area. Ph.D. Thesis, Aust. Nat. Univ. [unpublished].

VILJOEN, M. J., & VILJOEN, R. P., 1969: Evidence for the existence of a mobile extrusive peridotitic magma from the Komati formation of the Onverwacht group; *in* Upper Mantle Project. *Spec. Publs geol. Soc. S. Afr., 2*, pp. 87-112.

WILLIAMS, D. A. C., 1967: The geology of the Ora Banda Norite. B.Sc. Hons. Thesis, Univ. West. Aust. [unpublished].

16A: GEOCHEMISTRY OF ARCHAEAN VOLCANIC BELTS IN THE EASTERN GOLDFIELDS REGION OF WESTERN AUSTRALIA

J. A. Hallberg
Division of Mineralogy, CSIRO, Wembley, Western Australia

ABSTRACT

Volcanic belts in the Kalgoorlie–Norseman district of the Eastern Goldfields region of the Western Australian Shield are dominated by monotonous sequences of metamorphosed tholeiitic pillow basalts showing little vertical or lateral variation. These meta-basalts are chemically similar to Archaean tholeiites from Canada and South Africa and are characterized by their low content of K, Rb, and Sr and low Fe_2O_3/FeO ratios. There is little evidence of major cyclic compositional trends within the volcanic belts, with both major and trace elements showing limited dispersion. An undifferentiated, uncontaminated source of basaltic material was available throughout the development of thick volcanic piles in the region.

INTRODUCTION

THE paucity of geochemical data pertaining to Archaean volcanic systems has been aptly emphasized by Condie *et al.* (1969). The following paper presents major and trace element characteristics of Archaean volcanic and related rocks in part of the Eastern Goldfields region of the Western Australian Shield. The Kalgoorlie–Coolgardie–Norseman area has been the focus of numerous local studies involving the Archaean which have provided the groundwork for a regional investigation. Although outcrop within the area is generally poor, intensive exploratory drilling has provided a source of unweathered (and unmineralized) material for study. Sections exposed on the western shores of the numerous salt lakes in the area have also provided samples.

The Yilgarn Block of the Western Australian Shield is a 614 000 km^2 division which lies in the southern half of the State of Western Australia (Daniels & Horwitz, 1969). Within, intrusive granites are numerous N.-to-NNW.-trending lenticular areas of metamorphosed basic volcanic rocks and geosynclinal sediments representing the remnants of Archaean successions. The Kalgoorlie–Coolgardie–Norseman area abounds in basic volcanic sequences, acid porphyries, pre-metamorphic basic dykes and sills, and geosynclinal sediments, all of which have been regionally metamorphosed, steeply folded on NNW.-trending axes, and mildly cross-folded about east–west axes. The Archaean succession has been intruded along the east–west structural element by swarms of post-metamorphic, late Archaean basic dykes, the youngest basic igneous rocks in the area. Overlying the Archaean rocks are scattered patches of flat-lying sediments of Proterozoic and Tertiary ages.

[*Editor's Note:* Figure 1 and the Acknowledgments have been omitted owing to limitations of space.]

Volcanic belts within the area are composed principally of basic pillow lavas; flows without pillows can account for a maximum of 15 per cent of the basic volcanic material. Acid volcanics are rare, but when present form sequences up to 1220 m thick. Individual volcanic belts cannot be traced for great distances because of insufficient exposure, and most correlations are tentative. Stratigraphic successions have been established near major mining centres in the Eastern Goldfields (Hall and Bekker, 1965; Woodall, 1965; Glikson, 1970; McCall, 1969; Williams, 1970).

Volcanic belts in the Kalgoorlie–Norseman area have been loosely referred to as 'ophiolites' (McCall and Doepel, 1969; McCall, 1969; Glikson, 1970), a term which is unwarranted in view of the nature and composition of these belts. Thayer (1967) has outlined characteristic features of alpine ophiolite complexes. Unlike alpine complexes, volcanic belts in the area studied do not contain podiform chromite deposits, have simple relationships between gabbroic and ultramafic rocks and lack granophyric rocks and soda-rich diorites. Quench textures rather than cumulate textures dominate many of the more mafic and ultramafic rock types throughout the Eastern Goldfields region. Furthermore, ultramafic intrusions are pre-orogenic and stratigraphically controlled rather than orogenic and tectonically controlled.

Samples were taken at 30 m intervals throughout a 9000 m volcanic sequence in the Norseman succession and across the better exposed portions of volcanic belts to the north of Norseman; only sediments, acid volcanics and acid porphyries were excluded from sampling. In all, over 500 samples from the localities shown in Fig. 1 were analysed.

MINERALOGY AND PETROGRAPHY

The Kalgoorlie–Norseman region has undergone a low grade regional metamorphism of the Abukuma (Japan) type (Miyashiro, 1958, 1961). Under conditions of low water and CO_2 pressure the stability field of calcic plagioclase in the basic assemblages has been increased. Basalts are composed of slightly altered relicts of primary plagioclase (euhedral phenocrysts An_{65-80} and microlites An_{45-60}) in a sea of dense uralite to feathery actinolite which has replaced a largely glassy mesostasis; relict ilmenite and magnetite are common and quartz rare. Carbonate, chlorite, and clinozoisite are restricted to amygdale and secondary vein fillings. Pillow basalts are commonly amygdaloidal and rarely variolitic and are encased in a skin composed of dense uralite replacing glass. Pre-metamorphic gabbros and dolerites have ophitic to subophitic textures with large relict lathes of plagioclase (An_{55-85}), relict skeletal ilmenite and interstitial quartz associated with a metamorphic amphibole (tremolite or actinolite) which has formed from clinopyroxene and a partial breakdown of the plagioclase.

A discussion of these rocks in relation to their igneous parentage rather than their metamorphic history hinges upon the assumption that metamorphism has been isochemical. Miyashiro (1950) has commented upon the narrowness of the scope of diffusion in the Abukuma-type metamorphism. The restriction of secondary biotite and carbonate to contact aureoles, the limited dispersion of highly mobile potassium and the retention of the physical and chemical integrity of diverse rock types suggests that metamorphism has, indeed, been largely isochemical.

ANALYTICAL METHODS

SiO_2, Al_2O_3, total iron as Fe_2O_3, MgO, CaO, K_2O, TiO_2, P_2O_5, and MnO were determined by X-ray fluorescence spectrography (XRF) using a sample preparation technique described by Norrish and Hutton (1969). Standardization was based on USGS natural standards W–1, BCR–1, AGV–1, PCC–1, DTS–1, GSP–1, and G–2 (Fleischer, 1969; Flanagan, 1969) and a series of synthetic standards in the ranges of interest prepared from specpure chemicals. Matrix corrections given by Norrish and Hutton (1969) were applied by computer.

FeO was determined by titration of a sample solution against standardized ceric sulphate using n-phenyl anthranilic acid as an indicator, and Na was determined by flame photometry. CO_2 and H_2O (+) were determined by a modified loss on ignition method based on a technique described by Riley (1958). Cu, Ni, Zn, and Li were determined by atomic absorption spectrophotometry and Co, Cr, Rb, Sr, V, Y, and Zr by XRF on pressed powder discs using W–1 and a series of synthetic standards for calibration.

CHEMISTRY

Both the basalts and related dolerites and gabbros are quartz-normative tholeiites (Table 1). *AFM* and *FMC* plots (Fig. 2) show that the dolerites and gabbros are more variable in composition than the basalts, especially in the distribution of alkalis, MgO, and total iron. Compared with the basalts, the dolerites and gabbros are enriched in Fe, Ti, Rb, and K and depleted in Ca. However, the similarity in chemistry, especially in trace element distribution (Table 2), between the two groups indicates that they are comagmatic, the dolerites and gabbros representing 'penecontemporaneous, but slightly later phases from the same magmatic source' as suggested by McCall and Doepel (1969).

Individual basic pillows show only a limited chemical variation from the inner surface of their chilled skins to their cores (Table 3). Few pillows are strongly zoned, suggesting that with the initial chilling of the outer skin, pillows have acted as chemically closed systems, with relatively free exchange between individual portions of the pillows, but little or no exchange with the external environment. Individual pillows from widely separated localities are similar in bulk composition and the average composition of pillowed sequences varies little from locality to locality (Table 1).

Major and trace elements show limited dispersion in the volcanic piles (Table 2). Although Rb has a very high dispersion in the basalts (relative standard deviation = 110 per cent), Rb has, in some instances, been introduced before metamorphism with the development of secondary biotite in extensive contact aureoles near large basic and acid intrusive bodies. The median for Rb (4 ppm) is probably more indicative of the true content of the basalts,

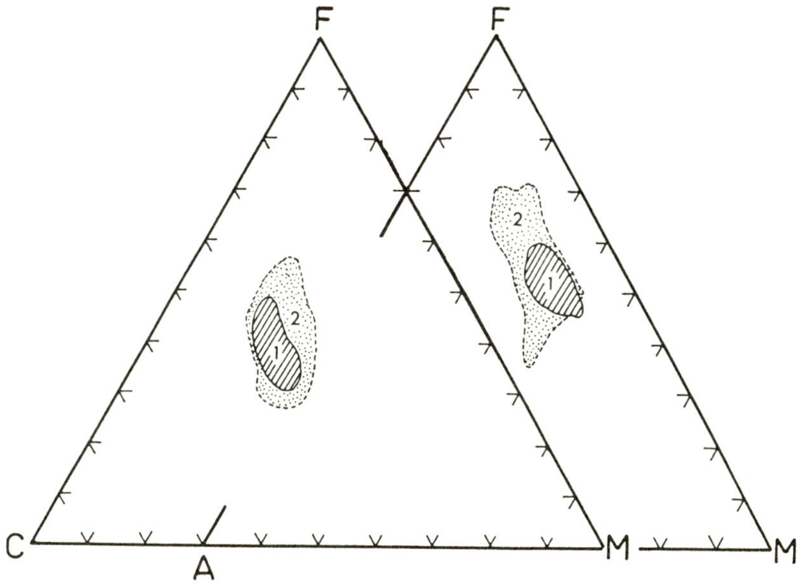

FIG. 2. *AFM* and *FMC* plots for basalts, gabbros, and dolerites from the Eastern Goldfields region. 1 = field occupied by 141 basalts; 2 = field occupied by 101 dolerites and gabbros, $A = Na_2O + K_2O$, F = total iron as FeO, M = MgO, C = CaO.

although the actual value may be lower since all samples containing traces of biotite have not been excluded from the average. Copper has a moderate dispersion, reflecting the variable sulphide content of the basalts, while Co, Cr, Ni, Sr, V, Y, Zn, and Zr have low dispersions.

The limited dispersion of major and trace elements is reflected in the distribution of rock types in the volcanic belts. Basalts (normative An/An+Ab < 50) predominate in the volcanic piles almost to the exclusion of intermediate rocks. In over 300 analyses from these volcanic piles, only 7 approach a tholeiitic andesite in composition, and these are variolitic pillow margins. The paucity of intermediate extrusives is further shown in Fig. 3, a plot of element distribution incorporating data on acid volcanic and related rocks from O'Beirne (1968). In over 400 analyses of volcanic and related rocks from the area, not one lies in the 55 per cent to 60 per cent silica range and only 9 lie between 55 per cent and 65 per cent silica. Total iron, MgO, and CaO also reflect the scarcity of intermediate values. It becomes apparent that an uncontaminated source of tholeiitic basalt was available throughout the development of thick, basic piles

TABLE 1

Chemical composition of Archaean tholeiites from Western Australia

	1	*2*	*3*	*4*	*5*	*6*	*7*	*8*	*9*	*10*	*11*	*12*	*13*	*14*	*15*	*16*	*17*	*18*	*19*
SiO_2*	51·2	51·9	51·4	52·6	51·7	49·0	50·49	51·3	51·4	50·8	51·1	49·3	49·3	51·3	51·5	49·8	50·4	49·64	49·7
Al_2O_3	15·2	14·1	13·9	13·9	13·9	13·6	15·26	14·6	14·8	14·5	16·2	17·0	14·6	18·0	16·3	14·6	15·4	15·12	14·3
Fe_2O_3	1·4	2·3	2·3	1·5	2·4	12·1†	2·06	2·4	1·5	1·6	3·1	2·0	3·2	3·4	2·8	3·0	1·6	2·88	2·7
FeO	9·4	8·9	9·4	8·3	9·3		9·51	8·0	9·1	10·3	7·6	6·8	8·5	5·7	7·9	8·7	9·6	9·07	9·0
MgO	6·4	7·1	7·7	7·9	6·0	5·6	3·89	5·6	6·7	6·9	6·2	7·2	7·4	6·0	5·9	7·3	6·0	7·49	7·0
CaO	10·7	10·3	10·1	10·1	10·2	10·2	12·69	11·2	10·7	9·9	9·9	11·7	10·6	10·0	9·8	10·4	11·7	11·69	10·1
Na_2O	2·8	2·3	2·8	2·8	2·8	2·4	2·28	2·1	2·7	2·7	2·5	2·7	2·2	2·7	2·5	2·0	1·7	1·60	2·4
K_2O	0·20	0·14	0·11	0·17	0·24	0·07	0·20	0·13	0·18	0·25	0·70	0·16	0·53	0·82	0·86	0·23	0·57	0·18	0·22
$H_2O(+)$	1·0	1·1	0·8	0·9	1·1	3·0	1·88	1·4	1·0	1·0	0·7		0·8	0·4	0·8	1·8	1·1	1·32	2·0
CO_2	0·1	0·1	0·0	0·1	0·0		0·42	0·2	0·1	0·1						0·3			
TiO_2	0·96	1·01	1·33	0·65	1·70	1·10	1·03	0·96	0·92	1·16	1·60	1·49	2·40	1·10	1·20	0·94	0·89	0·88	0·80
P_2O_5	0·14	0·13	0·10	0·08	0·15		0·12	0·05	0·13	0·17	0·22	0·16	0·26	0·16	0·21	0·19		0·13	0·08
MnO	0·22	0·18	0·17	0·19	0·26	0·35	0·22	0·26	0·21	0·21	0·17	0·17	0·17	0·16	0·17	0·21		0·20	0·18
Totals	99·72	99·56	100·11	99·19	99·75	97·42	100·05	98·00	99·44	99·59	99·99	98·68	99·96	99·74	99·94	99·47	98·96	100·20	98·48
Co*	60	54	64	46			48	70	59	57	39	32				48			
Cr	400	407	493	387	375		242	259	395	314	162	297				200			
Cu	111	41	78	55	18		134	64	98	111	127	77				100			
Li	5	4	7	7	2		1	1	5	6	10	9							
Ni	166	150	143	82	149		160	190	161	145	85	97			124	150			
Rb	10	3	6	4			11		9	9	30	10			36				
Sr	102	97	122	140			101		105	91	400	130			428	465			
V	323	309	350	254			253		320	307	245					250			
Y	22	17	25	20			10		22	22	32	43				25			
Zn	115	86	116	74	142		103		112	107						100			
Zr	57	53	85	78			67		60	54	100	95			224	110			

C.I.P.W. Norms calculated water and CO_2 free

Q	0·3	4·6	0·9	1·7	3·2	2·4	4·0	6·6	0·8	0·2	3·5		2·0	2·7	3·2	3·4	3·0	2·9	1·7
or	1·2	0·8	0·6	1·0	1·4	0·4	1·2	0·7	1·0	1·5	4·1	0·9	3·1	4·8	5·1	1·3	3·4	1·0	1·3
ab	24·1	19·7	23·9	23·9	24·5	20·9	19·7	18·0	23·5	23·3	21·3	22·8	18·7	22·9	21·3	17·2	14·5	13·7	20·7
an	28·7	28·1	25·1	25·1	24·8	26·9	31·1	30·5	28·0	26·9	31·1	34·3	28·6	34·7	30·9	30·7	33·0	33·9	28·1
di { wo	10·2	9·5	10·2	10·4	10·7	10·5	13·3	10·7	10·4	9·1	7·0	9·6	9·4	5·8	6·9	8·5	10·6	10·0	9·3
di { en	5·1	5·2	5·8	5·9	5·6	4·7	5·4	5·6	5·4	4·6	4·2	6·0	5·9	3·8	3·8	4·8	5·2	5·6	5·1
di { fs	4·8	3·9	4·0	4·0	4·8	5·6	8·0	4·7	4·6	4·3	2·4	2·9	2·9	1·5	2·7	3·3	5·2	3·9	3·9
hy { en	10·9	12·6	13·5	13·9	9·5	9·5	4·3	8·5	11·6	12·8	11·3	9·4	12·6	11·1	10·9	13·6	9·8	13·2	12·6
hy { of	10·2	9·3	9·5	9·4	8·1	11·3	6·5	7·1	9·9	12·0	6·6	4·6	6·4	4·5	7·8	9·2	9·7	9·2	9·6
ol { fo												1·7							
ol { fa												0·9							
mt	2·1	3·3	3·3	2·2	3·5	2·2	3·0	3·5	2·2	2·4	4·5	2·8	4·6	4·9	4·0	4·4	2·3	4·2	3·9
il	1·8	1·9	2·5	1·2	3·2	2·1	1·9	1·8	1·7	2·2	3·0	2·8	4·5	2·0	2·2	1·8	1·7	1·6	1·5
ap	0·3	0·3	0·2	0·2	0·3	—	0·3	0·1	0·3	0·4	0·5	0·3	0·6	0·3	0·5	0·4		0·3	0·1
Totals	99·7	99·2	99·5	98·9	99·6	96·5	98·7	97·8	99·4	99·7	99·5	99·0	99·3	99·0	99·3	98·6	98·4	99·5	97·8
Fe_2O_3/FeO	0·14	0·25	0·24	0·18	0·25		0·21	0·3	0·16	0·15	0·40	0·29	0·37	0·59	0·35	0·34	0·16	0·31	0·3

* Major oxides in weight %, trace elements in ppm.

1. Average of 83 basalts, Norseman, W.A.
2. Average of 13 basalts, Coolgardie, W.A.
3. Average of 5 basalts, Island section, W.A. (Fig. 1).
4. Average of 11 basalts, Lefroy Peninsula, W.A. (Fig. 1).
5. Average of 6 basalts, Mt. Monger, W.A.
6. Average of 3 basalts, Eulaminna, W.A. (Williams, 1969).

† Total Fe as Fe_2O_3.

7. Uralitized basalt, Scotia, W.A.
8. Average of 5 basalts, Kookynie, W.A.
9. Average of 123 basalts from the Kalgoorlie–Norseman area, W.A.
10. Average of 84 pre-metamorphic dolerites and gabbros from the Kalgoorlie–Norseman area, W.A.
11. Average tholeiitic basalt (major elements from Manson, 1967, and trace elements from Prinz, 1967).
12. Average oceanic tholeiite (Engel *et al.*, 1965).
13. Average oceanic tholeiite (Manson, 1967).
14. Average 'central' basalt (Nockolds, 1954).
15. Average continental basalt (major elements from Manson, 1967, and trace elements from Condie *et al.*, 1969).
16. Average Canadian Shield basalt (major elements from Wilson *et al.*, 1965, and trace elements from Baragar and Goodwin, 1969).
17. Average Archaean tholeiite from the Bulawayan Group, Rhodesia (Viljoen and Viljoen, 1969).
18. Archaean tholeiite, Guiana Shield (Cannon, 1964).
19. Archaean tholeiite, Barberton Mountain Land South Africa (Viljoen and Viljoen, 1969).

which may exceed 9 km in thickness in the case of the Norseman Succession. Differentiation indicators such as Cr, Ni, Sr, Y, and Zr show little or no variation through such sequences, showing the sources to be remarkably constant in composition. Prolonged periods of basic volcanism have in some areas been climaxed by a period of acid activity. The acid volcanics and acid porphyry dykes may be attributed to partial crustal melting as suggested by a trace element study of the Norseman porphyries (Larking, 1969) and discussed by O'Beirne (1968).

TABLE 2

Element distribution parameters for total sample population

	123 basalts				*84 dolerites and gabbros*			
	x	m	s	C	x	m	s	C
SiO_2*	51·4	51·0	1·7	3·3%	50·8	50·9	2·0	3·9
Al_2O_3	14·8	15·0	1·0	6·7	14·5	14·7	1·3	8·9
FeO†	10·4	10·3	1·7	16·3	11·7	11·5	1·9	16·2
MgO	6·7	6·5	0·9	13·4	6·9	7·0	1·5	21·7
CaO	10·7	10·7	0·9	8·4	9·9	10·1	1·7	17·1
Na_2O	2·7	2·7	0·4	14·8	2·7	2·7	0·7	25·9
K_2O	0·18	0·17	0·06	33·3	0·25	0·23	0·19	76·0
TiO_2	0·92	0·93	0·19	20·6	1·16	1·01	0·45	38·7
Co*	59	57	6	10·6%	57	59	6	10·5%
Cr	395	410	105	26·5	314	275	82	26·1
Cu	98	89	44	44·8	111	120	35	31·5
Li	5	5	1	22·0	6	7	2	29·5
Ni	161	159	26	16·2	145	134	29	20·0
Rb	9	4	10	110·0	9	13	8	88·8
Sr	105	96	23	22·3	91	96	19	20·8
V	320	311	38	11·8	307	295	39	12·7
Y	22	21	3	15·0	22	23	3·5	15·9
Zn	112	117	19	17·4	107	112	12·5	11·6
Zr	60	67	10	16·6	54	56	10	18·5

mean (x), median (m), standard deviation (s), and relative standard deviation (C).
* Oxides in weight %, trace elements in ppm.
† total iron as FeO.

AFFINITIES

Volcanic belts in the Eastern Goldfields region are remarkably similar to Archaean volcanic systems in Canada and South Africa. The upper portion of the Archaean Barberton Mountain Land sequence in South Africa consists of a thick sequence of tholeiitic basalts overlain by acid volcanics (Viljoen & Viljoen, 1969). Canadian Archaean volcanic belts are similar with the exception that andesites are apparently more abundant. It is felt, however, that the proportion of intermediate material has in some instances been greatly overemphasized, and that the assessment of these belts in terms of calc-alkaline differentiation sequences rather than tholeiitic piles (Wilson *et al.*, 1965; Baragar, 1966, 1968) may not be warranted. The use of colour index as the sole criterion for distinguishing andesites from basalts is questionable. Even when a normative definition has been applied (Baragar and Goodwin, 1969), the inclusion of carbonated samples and numerous analyses of fragmental volcanics may affect rock type abundances. Wilson (1965) indicates that most andesites in some Canadian belts are fragmental. However, fragmental rocks are certainly more amenable to diagenetic and metamorphic alteration than

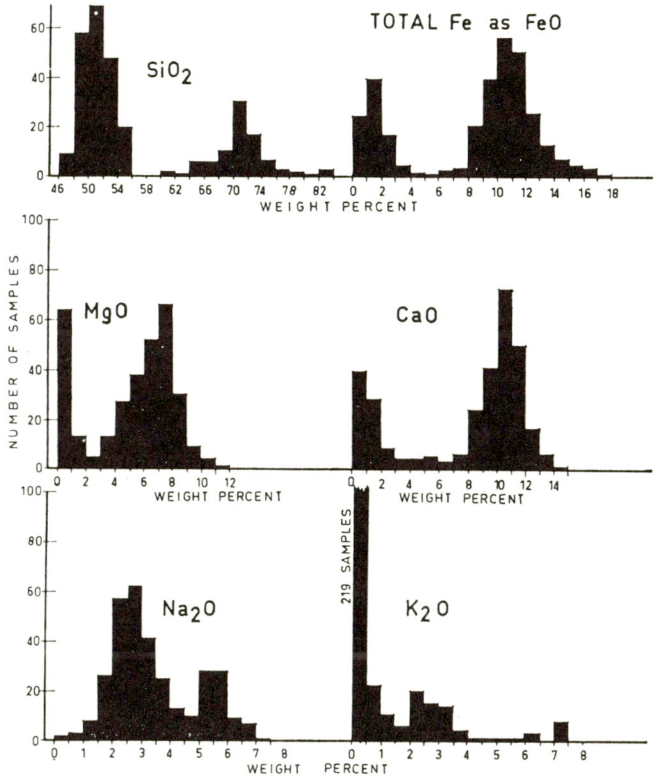

FIG. 3. Distribution of 6 major oxides in all volcanic and related rocks from the Eastern Goldfields region.

the massive and pillow basalt flows with which they are associated, and it is doubtful whether their chemistry truly represents the parent material. Table 4 shows the wide variation in the best-preserved basic fragmentals from volcanic belts in the Eastern Goldfields region.

As shown in Table 1, basalts from Archaean shield areas in Canada, Western Australia, and South Africa are similar in chemical composition, showing tholeiitic affinities with low contents of K, Sr, and Rb, and low Fe_2O_3/FeO ratios. However, comparison of the relative abundances of the major rock types in volcanic belts in the three areas cannot be made until uniform criteria for distinguishing rock classes have been applied.

TABLE 3. *Chemical variations in basaltic pillows*

	Yilmia Hill	*Widgie-mooltha*	*Coolgardie*		*Coolgardie*		*Norseman*	
	skin	*skin*	*margin*	*core*	*margin*	*core*	*margin*	*core*
Sample number	55015	63268	63298	63297	63295	63296	63458A	63458
SiO_2*	49·9	47·53	53·19	54·71	50·74	54·03	53·06	53·43
Al_2O_3	11·3	11·89	15·87	16·18	15·79	15·66	14·23	14·86
Fe_2O_3	1·1	2·94	1·26	1·59	1·41	1·44	0·83	1·03
FeO	11·0	11·45	7·11	6·80	8·32	7·32	9·49	8·72
MgO	11·0	9·74	7·42	5·32	7·82	6·74	6·03	5·86
CaO	13·2	12·09	10·17	10·03	11·87	10·55	10·87	10·76
Na_2O	1·3	1·43	3·87	3·22	2·92	3·12	2·66	2·98
K_2O	0·4	0·18	0·11	0·07	0·09	0·09	0·13	0·15
H_2O (+)	0·6	0·99	0·70	0·65	0·88	0·77	0·84	0·97
H_2O (−)	0·1	0·07	0·14	0·02	0·20	0·13	0·04	0·03
CO_2	—	0·00	0·00	0·00	0·10	0·00	0·95	0·00
TiO_2	0·5	1·38	0·73	0·76	0·72	0·70	0·92	0·88
P_2O_5	—	0·25	0·11	0·12	0·11	0·13	0·15	0·13
MnO	—	0·23	0·20	0·17	0·21	0·23	0·29	0·21
Total	100·4	100·17	100·88	99·64	100·09	100·91	100·49	100·01
Co*		77	50	55	60	54	56	50
Cr		529	350	365	440	475	270	285
Cu		62	21	35	53	53	63	90
Ni		149	146	165	152	140	195	162
Rb		9	3	5	3	<3	4	3
Sr		60	101	106	88	94	69	99
V		352	307	288	356	288	353	270
Y		27	17	17	14	17	21	22
Zn		151	85	80	94	86	154	72
Zr		73	59	48	53	53	54	56

* oxides in weight per cent, trace elements in ppm.

TABLE 4. *Tuffs, pyroclastics, and agglomerates from pillow lavas in the Eastern Goldfields region*

	Tuffs				*Pyroclastics–agglomerates*			
Sample number	63258	63349	63322	63353	63348	63358	63305	63351
SiO_2	51·66	49·51	47·65	46·94	52·09	55·82	49·61	50·53
Al_2O_3	11·63	11·26	15·66	14·24	15·86	15·28	15·80	14·87
Fe_2O_3	1·28	2·38	2·01	1·67	0·23	1·11	2·16	1·03
FeO	17·87	17·01	10·54	11·09	13·36	10·36	13·31	11·59
MgO	4·76	5·75	7·91	9·46	4·36	5·35	5·68	7·77
CaO	7·87	8·17	10·39	11·05	3·68	3·86	5·30	8·02
Na_2O	1·34	0·64	0·58	1·43	4·67	2·58	1·42	3·00
K_2O	0·30	0·25	2·09	0·77	0·18	1·31	1·08	0·14
H_2O (+)	1·31	1·58	1·75	1·06	1·49	1·53	1·93	1·07
H_2O (−)	0·34	0·03	0·02	0·11	0·06	0·06	0·47	0·08
CO_2	0·00	0·00	0·00	1·55	0·00	0·00	0·50	0·00
TiO_2	0·59	1·38	1·16	0·64	2·17	1·52	1·54	0·68
P_2O_5	0·16	0·20	0·26	0·14	0·26	0·19	0·23	0·13
MnO	0·50	1·15	0·25	0·32	0·52	0·31	0·48	0·31
S	—	0·70	—	—	0·45	—	—	—
Total	99·69	100·01	100·27	100·47	99·38	99·28	99·51	99·22

REFERENCES

BARAGAR, W. R. A., 1966. Geochemistry of the Yellowknife volcanic rocks. *Can. J. Earth Sci.* **3,** 9–30.

—— 1968. Major-element geochemistry of the Noranda volcanic belt, Quebec–Ontario. Ibid. **5,** 773–90.

—— & GOODWIN, A. M., 1969. Andesites and Archaean volcanism of the Canadian Shield. *Bull. Oregon Dep. Min. Ind.* **65,** 121–42.

CANNON, R. T., 1964. The geology of the Bartica Assemblage. *Bull. Geol. Surv. British Guiana,* **35,** 26–7.

CONDIE, K. C., BARSKY, C. K., & MUELLER, P. A., 1969. Geochemistry of Precambrian diabase dikes from Wyoming. *Geochim. Cosmochim. Acta,* **33,** 1371–89.

—— & POTTS, M. J., 1969. Calc alkaline volcanism and the thickness of the early Precambrian crust in North America. *Can. J. Earth Sci.* **6,** 1179–84.

DANIELS, J. L., & HORWITZ, R. C., 1969. Precambrian tectonic units of Western Australia. *Geol. Surv. W. Aust. Report,* 37–8.

ENGEL, A. E. J., ENGEL, C. G., & HAVENS, R. G., 1965. Chemical characteristics of oceanic basalts and the upper mantle. *Bull. Geol. Soc. Am.* **76,** 719–34.

FLANAGAN, F. J., 1969. U.S. Geological Survey standards–II. First compilation of data for the new U.S.G.S. rocks. *Geochim. Cosmochim. Acta,* **33,** 81–120.

FLEISCHER, M., 1969. U.S. Geological Survey standards–I. Additional data on rocks G-1 and W-1, 1965–1967. Ibid. **33,** 65–79.

GLIKSON, A. Y., 1970. Geosynclinal evolution and geochemical affinities of early Precambrian systems. *Tectonophysics,* **9,** 397–433.

HALL, H. I. E. & BEKKER, C., 1965. Gold deposits of Norseman. In McAndrew, J. (ed.) *Geology of Australian Ore Deposits,* pp. 101–7. 8th Commonwealth Min. Met. Congr., Melbourne.

LARKING, A., 1969. Acid rocks and their relationship to the gold mineralization at Norseman, Western Australia. B.Sc. (Hons.) thesis, Adelaide University.

MCCALL, G. J. H., 1969. The Archaean succession to the west of Lake Lefroy. *J. Roy. Soc. W. Aust.* **52,** 119–28.

—— & DOEPEL, J. J. G., 1969. Stratiform basic bodies within the Yilmia ophiolites, Western Australia. *Proc. Aust. Inst. Min. Met.* **231,** 67–75.

MANSON, V., 1967. Geochemistry of basaltic rocks: major elements. In Hess, H. H., and Poldervaart, A. (eds.), *Basalts* pp. 215–69. New York: Interscience.

MIYASHIRO, A., 1958. Regional metamorphism of the Gosaisyo–Takanuki district in the Central Abukuma Plateau. *J. Fac. Sci. Univ. Tokyo,* **11,** 219–72.

—— 1961. Evolution of metamorphic belts. *J. Petrology,* **2,** 277–311.

NOCKOLDS, S. R., 1954. Average composition of some igneous rocks. *Bull. Geol. Soc. Am.* **65,** 1007–32.

NORRISH, K., & HUTTON, J. T., 1969. An accurate X-ray spectrographic method for the analysis of a wide range of geological samples. *Geochim. Cosmochim. Acta,* **33,** 431–53.

O'BEIRNE, W. R., 1968. The acid porphyries and porphyroid rocks of the Kalgoorlie area. Ph.D. thesis, University of Western Australia.

PRINZ, M., 1967. Geochemistry of basaltic rocks: trace elements. In Hess, H. H., & Poldervaart, A. (eds.), *Basalts,* pp. 271–325. New York: Interscience.

RILEY, J. P., 1958. Simultaneous determination of water and carbon dioxide in rocks and minerals. *Anal. Chem.* **83,** 42–9.

THAYER, T. P., 1967. Chemical and structural relations of ultramafic and feldspathic rocks in alpine intrusive complexes. In Wyllie, P. J. (ed.) *Ultramafic and Related Rocks* pp. 222–39. New York: John Wiley and Sons.

VILJOEN, M. J., & VILJOEN, R. P., 1969. A reappraisal of granite–greenstone terrains of shield areas based on the Barberton model. A paper read before the South African Upper Mantle Symposium, Pretoria.

VILJOEN, R. P., & VILJOEN, M. J., 1969. The geological and geochemical significance of the upper formations of the Onverwacht Group. In *The Upper Mantle Project,* Geol. Soc. S. Africa Spec. Publ. **2,** 113–51.

WILLIAMS, I. R., 1970. Explanatory notes on the Kurnalpi 1:250 000 geological sheet, W.A. *Geol. Sur. W. Aust. Record* 1970/1.

WILLIAMS, R. E., 1969. An environmental study of base metal sulphide mineralization, Eulaminna, Western Australia. B.Sc. (Hons.) thesis, Adelaide University.

WILSON, H. D. B., ANDREW, P., MOXHAM, R. L., & RAMLAL, K., 1965. Archaean volcanism in the Canadian Shield. *Can. J. Earth Sci.* **2,** 161–75.

WOODALL, R., 1965. Structure of the Kalgoorlie Goldfield. In McAndrew, J. (ed.), *Geology of Australian Ore Deposits,* pp. 71–9. 8th Commonwealth Min. Met. Congr., Melbourne.

16B: ARCHAEAN MAFIC AND ULTRAMAFIC ROCK ASSOCIATIONS IN THE EASTERN GOLDFIELDS REGION, WESTERN AUSTRALIA

J. A. Hallberg
Division of Mineralogy, CSIRO, Floreat Park, Western Australia 6014

D. A. C. Williams
Geology Department, University of Western Australia, Nedlands, Western Australia, 6009

ABSTRACT

Recent studies in Archaean volcanic belts of the Eastern Goldfields region of Western Australia have led to the development of a general picture of Archaean volcanic activity in this area. Three volcanic associations are recognized: a felsic volcanic–porphyry–clastic and volcanogenic sediment association, a pillow basalt–dolerite association and an ultramafic association.

The ultramafic association consists of high-Mg basalts and zoned ultramafic lenses, both with textures indicative of high level intrusion or extrusion, and numerous small layered sills. The association commonly forms the lower parts of individual volcanic sequences, and is overlain by basalt. In some areas felsic volcanic rocks terminate this sequence. It is believed that the mafic and ultramafic associations represent two distinct magma sources. Elements of similarity between these stratiform belts and more recent oceanic crust are discussed.

SUMMARY

In Paper 16B, Hallberg and Williams define three volcanic–sedimentary associations:

1. Felsic volcanic–porphyry–clastic and volcanogenic sedimentary.
2. Basalt–dolerite.
3. High-magnesia basalt–ultramafic lens–layered sill.

A number of attempts have been made to subdivide the rocks of the Kalgoorlie greenstone belt into meaningful associations (Papers 15 and 34). None is entirely satisfactory, but this threefold scheme is realistic. Minor points that can be made are that the term dolerite is not universally used (the common American synonym is diabase); basalt–gabbro would be preferable, and indeed most coarser tholeiitic rocks are of gabbroic texture. The layered sills are not restricted to the third association (the Golden Mile dolerite is tholeiitic), and only those derived from a high-magnesia basalt stem (Mission Sill, Yilmia Sill) (Papers 15 and 16C) fall into class 3.

The first class may well have to be divided into sodic-alkaline and calc-alkaline acid eruptives and their sedimentary equivalents, for Hallberg (personal communica-

 The summary was prepared by the volume editor.

tion, 1973) has lately recognized a very few restricted calc-alkaline centers of Noranda type, at Marda north of Southern Cross in the Yilgarn Greenstone Belt and to the north of Kalgoorlie in the belt under review in this paper. A calc-alkaline, basalt–andesite–dacite–rhyodacite–rhyolite volcanogenic sediment association has major significance in Canada, but minor significance in the eastern part of the Yilgarn Block in Australia. It should perhaps be added to the above scheme as association 4.

The lack of high-magnesia basalts at Norseman has been commented upon above. They are, however, represented in the Mt. Thirsty Group at the top of the Norseman succession, both the Mt. Thirsty and Mission sills being derived from high-magnesia stems.

The authors conclude by drawing an analogy between Archean greenstone belts and present oceans. The marked disparity between these Archean sequences and the ophiolites, including the lack of spilitization, may, they believe, be reconciled with attribution of both types of assemblage to eruption in sites of plate divergence, if the assumption is made that the Archean plates underwent unusually rapid divergence.

16C: GEOCHEMICAL CHARACTERISTICS OF SOME ARCHAEAN GREENSTONE SUITES OF THE YILGARN STRUCTURAL PROVINCE, AUSTRALIA

G. J. H. McCall

Department of Geology, University of Western Australia, Nedlands, W.A. (Australia)

ABSTRACT

Geochemical results representing 225 full major element analyses, mostly by XRF methods, are evaluated. This study is complementary to papers presented by the author and colleagues at the Archaean Symposium of the Geological Society of Australia in Perth in May 1971. Only certain new analyses are given in tabulation, but the variation diagrams presented are based on many additional analyses published in works referred to in the text. The emphasis of this study is a consideration of the relationship of the pre-metamorphic Archaean sills to the volcanics of the same greenstone belts: to consider the original nature of magmas differentiated after emplacement: and to more closely define the pre-metamorphic intrusives.

A contrast is seen between magnesia-rich basic and ultrabasic eruptives, represented amongst both sills and volcanics, and the abyssal, K-poor tholeiite suites of the monotonous pillow lava sequences studied in detail by Hallberg (1970, 1972). A similar contrast is evident, though less emphatically, among the post-metamorphic dykes. Residually iron and alkali enriched suites are evident among the pre-metamorphic and post-metamorphic eruptives. The possibility of some degree of gradation between magnesia-rich and tholeiite suites cannot be discounted: they are intimately intermingled in space and time, and amongst intrusive and volcanic sequences. Alkali/silica plots after Macdonald and Katsura show that, whether or not the magnesia-rich rocks are regarded as tholeiitic or *komatiitic* (following Anhausser, Viljoen and Viljoen), no alkali igneous rocks are represented. Spilites are of no more than extremely rare occurrence. FMA plots indicate that among the pre-metamorphic volcanics and intrusives, there are eruptives, suites and differentiated bodies that resulted from accession of a number of contrasting differentiated melts or magmas from deep in the crust, possibly to suffer further differentiation in situ after emplacement. That is the parental stem to both differentiated and undifferentiated sills, and to volcanics, was a product, in many cases, of an earlier deep-seated differentiation process. A diagram based on solidification indices illustrates this, and shows that the same applies to post-metamorphic dyke intrusions.

These results are discussed with reference to the *komatiites* of the Barberton Mountain Land, South Africa. The resemblance between *komatiite* analyses and olivine tholeiites of Nockolds is stressed, and it is suggested that the highly magnesian greenstone eruptive suites of South Africa and Western Australia might be regarded as an extension of the abyssal tholeiite association, rather than as a discrete compartment in petrology. The anti-uniformitarian, one-burst connotation afforded to these suites is discussed.

SUMMARY

McCall has utilized a profusion of available chemical analyses (including many from Hallberg's program mentioned above) to consider the relationship of the intrusive bodies in the greenstone belts to the volcanics. The essential facts stated in this paper were presented with Paper 15 at the Archean Rock Symposium of the Geological Society of Australia at Perth in 1970, and the use of FMA diagrams and Kuno's solidification indexes showed that many of the large layered sills came from high-magnesia basalt stems, not tholeiitic stems as had been supposed. Particularly interesting is the evidence for multiple intrusion of a progressively differentiating tholeiitic magma stem in the case of the Golden Mile dolerite, the end pulse being ferrodolerite, and the evidence for multiple injection of a repetitive nature in the case of the Jimberlana Dike, which is now known to be a complex product of flowage differentiation and gravity–convection layering, multiple injection, and minor late development of dikes in a differentiating magma invading a fracture zone more than 200 km long and forming a body of carrot-like, bell-jar cross section. The two most significant facts established in this analytical review are (1) that there were two loci of differentiation, one deep in the crust and the other in situ (postemplacement), and that differentiation in depth, producing high-magnesia basalt suites, tholeiites, ferrobasalts, and alkali-rich basalts, was of very similar type of the gravitative–convective in situ differentiation, and (2) that high-magnesia basalts were erupted over a span of at least 1000 million years of Archean history, and are even represented in the poststabilization Jimberlana Dike of earliest Proterozoic age (2420 million years).

The data given by McCall seem to be entirely consistent with the suggestion of Hallberg and Williams that high-magnesia basalts are products of crystal fractionation in depth. The coincidence of the differentiation plot on the FMA diagram for individual layered sills (overall average chemistry) and their components and individual volcanic suites is difficult to interpret in any other way.

III: SEDIMENTOLOGY OF ARCHEAN ROCKS

Russian authorities consider that there is evidence that the earliest sediments possessed an unusual, highly aluminous character, which is reflected in the abundant kyanite and sillimanite of their high-grade metamorphic products (Papers 27, 28, 29, 30, and 32). Ferriferous and dolomitic characteristics are also considered to have been prominent. According to these authorities, the quartzites of the earliest Archean should be considered as largely chemically derived. Derivation of chemical sediments from weathered surfaces is suggested, but opinions differ as to the relative importance of basic rock or granitic rock weathered surfaces as sources. Khain (Paper 30) concludes that granitic terrains were the dominant source; Pavlovskiy (Paper 22) takes the opposing view. Chemical sedimentation may well have been dominant over clastic in the earliest Archean, but clastic sediments appeared in force in the Katarchean greenstone belt systems as far back as 3400 million years.

Studies of the sedimentary associations associated with amphibolite enclaves in the earliest Greenland gneisses have led Windley (unpublished, personal comment) to conclude that they are essentially pelagic and akin to those of present-day oceans. It seems, however, that, although the very earliest sediments may have been entirely chemogenic and deposited in quiet conditions, there is no reason to invoke *deep* oceans; these chemogenic sediments may have been deposited in very quiet, epicontinental seas characterized by very low relief contrasts of land mass and sea floor and under a regime of slight tectonic movements. Even so, some of the oldest rocks of the Archean are quartzites, sandstones, shales, and slates. These would seem to require granitoid source areas. Rounded zircons and monazites are typically present in very ancient Archean sediments, and these certainly cannot be reconciled with a source terrain entirely composed of basic volcanics. Even if such a source existed, the assumption that it was of necessity the primordial sialic crust of the earth is one that is too quickly made (Semenenko, Paper 28).

Studies of Archean sediments of the Southern Hemisphere and Canada, where low grades of metamorphism predominate, have revealed that both chemical and clastic sedimentation alternated; littoral rudites and flyschoid graywackes are abundantly represented alongside banded ironstones in sequences of greenstone belts from 3400 milion years ago upward. There is much evidence favoring a little-weathered source terrain for the clastics in the case of the Fig Tree Group, which has an immature character (Paper 17A); but clastic sediments described from the

Superior Province, Canada, by Donaldson and Jackson (1968) are remarkably mature, and may even be the product of more than one cycle of erosion and deposition.

This volume is not designed to include specialist treatments, but it has been considered appropriate to select one major, definitive paper dealing with the earliest known clastic sediments. The sediments of the Fig Tree Group, described by Condie et al. in Paper 17A, are analogous to those of the Shamvaian in Rhodesia (Paper 7) and the Gorge Creek–Mosquito Creek sedimentary sequences in the Pilbara of Western Australia (Paper 14). The Moodies Group that caps the Fig Tree Group is a peculiarly thick conglomeratic sequence. It is not known whether it has any close analogues in the Shamvaian and in the Pilbara sequence, as we lack detailed sedimentological studies of those sequences.

Condie et al. find that the source of the Fig Tree graywackes was chert, volcanics, granite, metamorphic rocks, and ultramafics, comprising a diverse terrain. Some detritus also stemmed from contemporary volcanism. Such a combination of provenances is the consistent finding of detailed studies of Archean clastic sediments in South Africa, Canada, and Australia. Both foreland and local volcanic sources of contemporaneous age contributed to varying degrees. Mild weathering and rapid erosion of a not too distant source area is indicated for the Fig Tree Group. Dolomitic material was derived from a diagenetic pore fluid. The igneous rock source for the lowest sediments seems to have been strontium depleted. Nickel abundance in graywackes and shales suggests contribution from ultramafic source rocks. The progressive increase in granitic components in the graywackes suggests progressive unroofing of an older granite terrain, and that the greenstones which now occupy the brachysynclinal Barberton trough may have once been more extensive (compare, however, Paper 9). It is concluded that there must have been at least one earlier granite-forming event in Southern Africa than that which affected the Swaziland sequence after it was deposited.

Archean sediments may be remarkably well preserved; it is incorrect to assume that they must, of necessity, be deformed and metamorphically obliterated because of their great age. Indeed, some late Archean sediments show unmatched preservation of sedimentary structures on the mesoscopic scale, although it is true that microtextures are commonly obliterated. Papers that illustrate this, and that could well have been included had space permitted, are those by Donaldson and Jackson (1968), describing late Archean graywackes from North Spirit Lake, Ontario, McCall et al. (1970), describing a thick polymict conglomerate–graywacke littoral sequence from Merougil Creek, Lake Lefroy, Western Australia, Glikson (1971), describing the progression from argillites through to conglomerates in the Coolgardie area to the north of Lake Lefroy, and Dunbar and McCall (1970), describing a unique flysch turbidite–quartz magnetite chemogenic sediment assemblage at Mount Belches to the east of Lake Lefroy.

All these sediments occur *within* Archean greenstone sequences, the clastics more profusely toward the top, and they are quite distinct from platform sediments, although there is some resemblance to them. They are caught up in intense folding, unlike the gently disposed platform sediments. Both they and the volcanics and intrusives within them do, however, display characteristics that suggest a transition from protogeosynclinal to platform environments of deposition.

References

Donaldson, J. A., and Jackson, G. D. (1968). Archaean sedimentary rocks in the North Spirit Lake, North Western Ontario. *Can. J. Earth Sci., 2,* 622–647.

Dunbar, G. J., and McCall, G. J. H. (1971). Archaean turbidites and banded ironstones of the Mount Belches area, Western Australia. *Sed. Geol., 5,* 93–133.

Glikson, A. Y. (1971). Archaean geosynclinal sedimentation near Kalgoorlie, Western Australia. *Spec. Publ. 3, Geol. Soc. Aust.,* 443–460.

McCall, G. J. H., Braybrooke, J. G., and Middleton, D. D. (1970). The Merougil sub-area, *J. Roy. Soc. Western Aust. 53,* 9–32.

PAPERS 17A AND 17B: COMMENTARY

Linking the themes of sedimentology and Archean life forms (Parts III and V) is the special summary prepared for this volume by Trendall, an authority who has traveled all over the globe comparing iron formation developments. This paper also has a close relationship with Paper 2 by Cloud.

17A: PETROLOGY AND GEOCHEMISTRY OF EARLY PRECAMBRIAN GRAYWACKES FROM THE FIG TREE GROUP, SOUTH AFRICA

Kent C. Condie and John E. Macke

Department of Earth Science, Washington University, St. Louis, Missouri

Thomas O. Reimer

Department of Geology, University of Witwatersrand, Johannesburg, South Africa

ABSTRACT

Graywackes of the Fig Tree Group in South Africa were derived from a diverse source area composed principally of chert, volcanics, granitic-metamorphic rocks, and some ultramafics. Chert and quarts pseudomorphs after shard-like fragments and well-formed plagioclase and K-feldspar crystals in some graywackes indicate that some detritus was derived from contemporary volcanism. Immature textures indicate mild source-area weathering and rapid erosion, deposition, and burial. An unusually large amount of dolomite in the graywackes was probably deposited by carbonate-rich pore fluids during diagenesis. A relative Sr depletion (compared to K, Rb, Ca, and Ba) in graywackes of the lower Fig Tree Group (Sheba Formation) is interpreted in terms of an abundance of Sr-depleted igneous rocks in their source area. Large amounts of Ni in all Fig Tree graywackes and shales appear to have been derived from ultramafic source rocks.

A progressive stratigraphic increase in granitic components in the graywackes suggests progressive unroofing of a granitic metamorphic terrane which was initially covered by a thick sequence of Onverwacht or Onverwacht-like volcanic rocks. This source appears to have been located in central Swaziland. The relatively abundant granitic detritus in the graywackes indirectly records one or more pre-Fig Tree (3.4–4.0 b.y.) granite-forming events in southern Africa.

INTRODUCTION

Clastic sedimentary rocks from the Fig Tree Group in the Barberton Mountain Land in South Africa are of particular scientific interest in that they represent some of the earliest known, well-preserved sediments. Recent Rb-Sr age studies indicate that the rocks in this region are > 3.0 b.y. old (Allsopp and others, 1968). Volcanic rocks beneath the Fig Tree Group have U-Pb ages of about 3.4 b.y. (Van Niekerk, 1967) suggesting an age of approximately 3.35 b.y. for the base of the Fig Tree Group. It is the purpose of the present investigation to describe the mineralogy and composition of graywackes from the Fig Tree Group in an attempt to understand better the nature, composition, and origin of the Earth's early continental crust.

Investigations by the Geological Survey of South Africa and recently by the Economic Geology Research Unit at the University of Witwatersrand have been instrumental in our understanding of the geologic history of the early Precambrian rocks in the Barberton Mountain Land (Visser, 1956; Engel, 1968; Anhaeusser and others, 1968; Anhaeusser and others, 1969). This area is composed of rocks of the Swaziland System, consisting of three lithologically different groups, which are, in ascending order: the Onverwacht Group, a sequence of siliceous to ultramafic volcanic rocks (> 15 km thick—base not exposed); the Fig Tree Group, a sequence of graywackes, shales, tuffs, cherts (~2 km thick); and the Moodies Group, a sequence of quartzites, shales, sandstones, and conglomerates (~3.6 km thick). Although the sequence as a whole is conformable, local unconformities exist between the Fig Tree and Moodies Groups, and to a minor extent, between the Fig Tree and Onverwacht Groups. These rocks are intruded by granites which are approximately 3.0 b.y.

in age (Allsopp and others, 1968). The progression of rocks from the Onverwacht Group to the Moodies Group probably represents the evolution of a thin, warm, relatively unstable crust to a thick, cool, relatively stable crust (Engel, 1968; Anhaeusser and others, 1968).

The Fig Tree Group was recently subdivided into three formations in the western part of the Mountain Land (the Stolzburg syncline region) where the only continuous section is preserved (Reimer, 1967); these formations can also be traced eastward into the Ulundi syncline area (Figs. 1 and 2). They are in ascending order: the Sheba Formation, consisting of graywackes with subordinate shales (750–1200 m); the Belvue Road Formation, a sequence of siltstones, shales, and cherts with subordinate graywackes to the east (Ulundi syncline), and a horizon of trachytic tuff (600 m) to the west (Stolzburg syncline); and finally the Schoongezicht Formation, consisting of about 350 m of graded trachytic tuffs and 200 m of coarse volcanic agglomerates.

In this investigation, a total of 24 samples from the Fig Tree Group were analyzed chemically and petrographically: 15 graywackes and 2 shales from the Sheba Formation, and 7 graywackes from the Belvue Road Formation. Ten additional samples, 5 each from the Sheba and Belvue Road Formations were analyzed only petrographically. Sample locations are shown in Figure 1, and stratigraphic locations, if known, are shown in Figure 2. With exception of a faint foliation in a few of the samples, effects of regional metamorphism are absent.

ANALYTICAL METHODS

Modal analyses were made with a polarizing microscope using a Brinkman automatic point-counter; a minimum of 1000 points were counted on each thin section. Thin sections were etched and stained to facilitate in the identification of K-feldspar, and opaque minerals were identified under reflected light. X-ray diffraction scans of whole rock and separated mineral powders were made to verify mineral identifications.

Major elements (except Mg and Na) were determined by X-ray fluorescence. Rock powders were fused with lithium tetraborate and pelletized with boric acid in a manner similar to that reported by Rose and others (1963). Standard rock samples W-1, T-1, G-2, GA, GR, GH, and Syenite-1 were used as standards for major elements, and G-2 was used for trace elements. Trace elements (Rb, Sr, Zr, Ni, and Ba) were determined directly on pelletized rock powders in the manner described by Condie (1967a).

Mg was determined by atomic absorption spectrometry using the 2852Å absorption line of Mg. Na, Mn, and Sc were determined by nondestructive neutron activation analysis in a manner similar to that described by Stueber and Goles (1967) using Na^{24} (1.37 MeV_γ), Mn^{56} (0.84 MeV_γ), and Sc^{46} (0.89 MeV_γ) gamma-ray peaks, a 2 × 2 NaI (Tl) crystal, and a 400-channel analyzer. G-1 was used as a standard. Estimated analytical error for all elements is $\leq$5 percent.

MINERALOGY

Modal compositions of graywackes from the Fig Tree Group are summarized in Table 1. Texturally, the graywackes are poorly sorted, composed of sand- and silt-sized clasts in a fine-grained ($\leq$0.03 mm) matrix. Their poor sorting probably is caused chiefly by diagenetic destruction of labile clastic components, such as calcic plagioclase and mafic minerals and rock fragments. Such a mechanism, which was invoked by Cummins (1962) to explain graywacke textures, is supported by modal analyses of more than 50 graywackes from the Fig Tree Group. Consistently, samples with large amounts of matrix have small amounts of labile constituents. Also, the fact that the minerals in the matrices embay and penetrate obvious clastic grains in varying degrees (Figs. 3 and 6) indicates a diagenetic origin for these constituents.

All textural gradations between quartz and chert occur in the graywackes; the two extreme textures, however, are most common. Collectively, chert and quartz compose 30 to 50 percent of the Fig Tree graywackes; up to half of this occurs in the matrices and is of probable diagenetic origin. Clastic fragments of quartz and chert vary in size from < 0.1 mm to 5 mm and for the most part are very angular in shape (Figs. 3 and 6). Quartz grains usually exhibit slight or no undulatory extinction; some have well-preserved crystal faces and were probably derived from porphyritic siliceous volcanic rocks in the source area. Some of the smaller quartz and chert fragments are dagger, prismatic, and triangular shaped (Figs. 3 and 6). Such fragments, which exhibit shapes reminiscent of glass shards, appear to be pseudomorphs of shards formed by devitrification, recrystal-

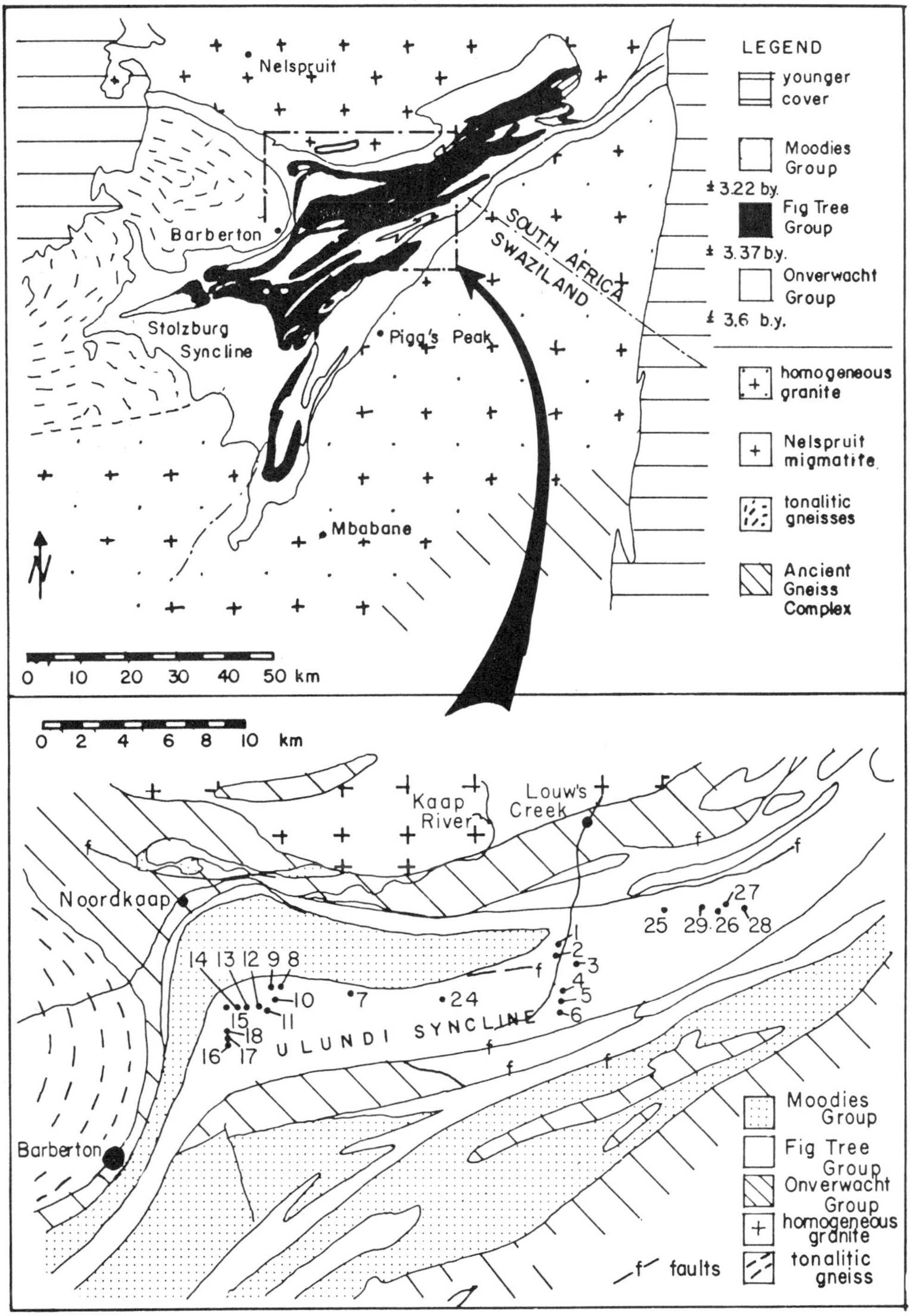

Figure 1. Geologic map of the Barberton Mountain Land, eastern Transvaal, South Africa (top) and geologic map of the area near Barberton showing graywacke sample localities (bottom).

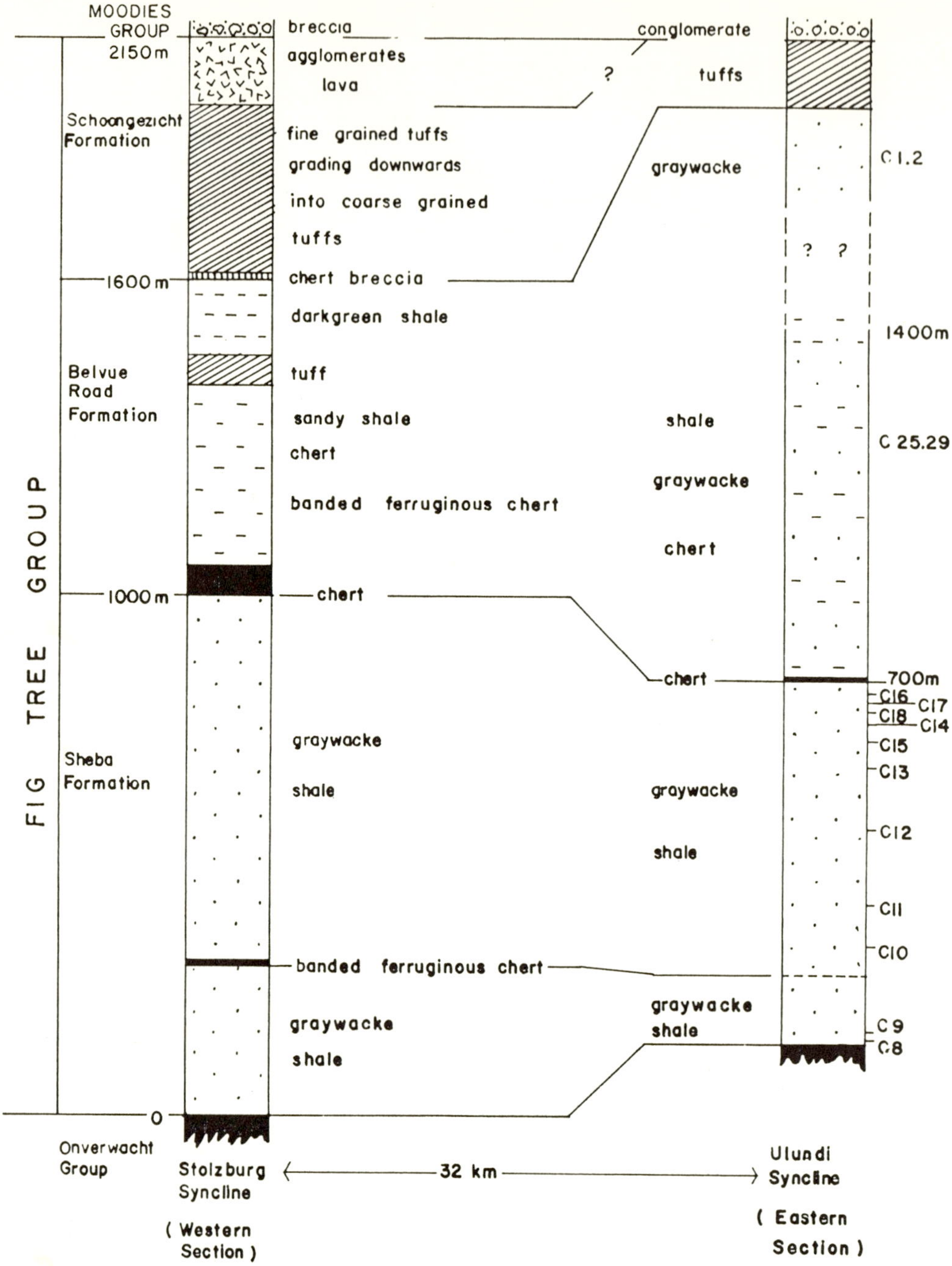

Figure 2. Stratigraphic section of the Fig Tree Group in the Stolzburg and Ulundi synclines; known graywacke sample levels are shown in the Ulundi syncline section.

lization, and perhaps replacement of volcanic glass and associated pyroclastic minerals during diagenesis.

The chert is composed of a microcrystalline mosaic of intergrown quartz with an average grain size of 0.005 mm to 0.02 mm. It commonly contains finely disseminated graphite and pyrite inclusions (Pflug, 1967). Occasional "cherty" fragments exhibit well-developed snowflake structure, a structure which appears

TABLE 1. MODAL COMPOSITION OF GRAYWACKES FROM THE FIG TREE GROUP*

	Sheba Formation		Belvue Road Formation	
	Range	Mean	Range	Mean
Quartz	15–30	24	24–35	30
Chert	13–19	15	3–7	5
Microcline	tr-12	6	tr-25	15
Plagioclase (An_{10}-An_{25})	1–3	2	2–7	4
Mica	0.2–2	1	1–3	2
Chlorite and serpentine	0.1–4	1	1–8	3
Dolomite	2–14	8	tr-2	1
Opaque†	tr-3	2	0.5–3	2
Rock fragment (excl. chert)	1–4	3	3–6	5
Matrix ($\lesssim$ 0.03mm)	30–50	38	25–40	33

tr = < 0.1 percent.

* Ranges and mean values based on 13 samples from the Sheba Formation and 8 from the Belvue Road Formation.

† Chiefly pyrite and limonite

to develop in highly welded tuffs during devitrification (Anderson, 1969). Because true chert fragments and possible devitrified welded tuff fragments are virtually indistinguishable under the microscope, they are combined in the mode (Table 1). Much of the cherty matrix material in these rocks may have been composed originally of glassy volcanic detritus. The occurrence of bedded chert in the Fig Tree Group (Fig. 2) indicates that primary chert deposition occurred during Fig Tree time and that the chert in the graywacke matrices may have formed penecontemporaneously with graywacke deposition. It is possible that silica, derived from contemporary submarine volcanism or from desilication of deeper rocks during dehydration, partly replaced original volcanic detritus in the graywackes preserving, in part, the original grain shapes (such as shards). The abundance of angular chert (and welded-tuff) fragments in the graywackes suggests that nearby chert-rich and welded-tuff-rich source terranes also existed in Fig Tree times.

Microline is usually the dominant feldspar in the Fig Tree graywackes. It occurs as clastic grains up to about 0.5 mm long and often contains perthitic or micrographic intergrowths (Fig. 4). Some microcline is part of obvious granitic rock fragments which also contain quartz, and plagioclase, and mica. Micrographic intergrowths and textural gradations between individual mcirocline grains and granitic rock fragments indicate a granitic source area for at least part of the Fig Tree detritus. If, as is likely the case, most of the quartz sand grains were also derived from granitic source areas, a comparatively large part of the Fig Tree source area must have been composed of granitic or quartzo-feldspathic metamorphic rocks, or both. An increase in the microcline content in graywackes from about 0.1 to 12 percent from the base to the top of the Sheba Formation (samples 8 to 16, Fig. 2) suggests that granitic source terranes became progressively more abundant as erosion continued in the source area. Small euhedral microcline crystals ($\lesssim$0.1 mm long), probably sanidine originally, appear to have been derived from either (or both) contemporary explosive volcanism or siliceous porphyritic volcanic rocks in the source area.

Plagioclase occurs as small, angular to prismatic grains (< 0.1–0.5 mm long) and ranges in composition from An_{10} to An_{35}. Plagioclase of more calcic composition, however, occurs in mafic volcanic rock fragments. Although some of the plagioclase in the Fig Tree graywackes

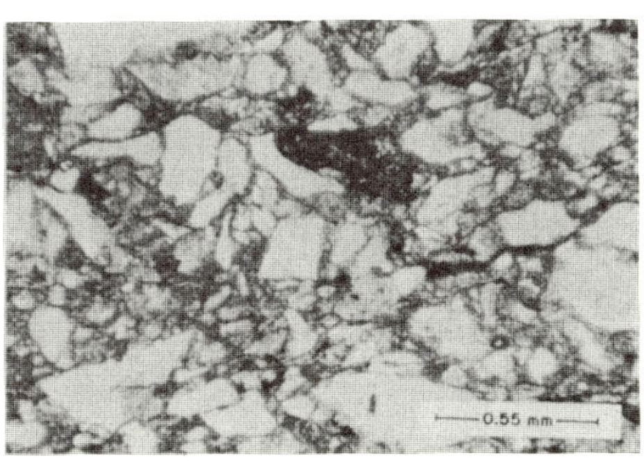

Figure 3. Photomicrograph of a typical Fig Tree graywacke showing the extreme angularity of the clasts. Sample C-12; uncrossed nicols.

Figure 4. Photomicrograph of a feldspar grain showing micrographic intergrowths of quartz. Sample C-29; uncrossed nicols.

was derived from granitic source areas as evidenced by the presence of plagioclase-bearing granitic rock fragments, most of it appears to have been derived from volcanic source areas or from contemporary volcanic activity.

Chlorite, and to a lesser extent, mica (chiefly muscovite) occur as common matrix constituents, as well as occasional larger crystals. These minerals commonly embay and penetrate clastic grains indicating a diagenetic origin for them. Chlorite also occurs as pseudomorphs after biotite. Chlorite, chrysotile, and biotite grains of questionable detrital origin are found also in some of the graywackes. The Fig Tree shale samples contain a considerably larger amount of chlorite and mica than do the graywackes.

The significant amount of dolomite (up to 14 percent) in the matrices of the Fig Tree graywackes is of particular interest because carbonates of any kind are rare in early Precambrian terranes. The dolomite in the Fig Tree graywackes is irregularly distributed in the matrices and often occurs as well-formed rhombs (Fig. 5). The fact that it often penetrates and embays clastic grains indicates a diagenetic origin; a minor amount of questionably detrital, carbonaceous dolomite was also observed. Possible sources for the carbon in the diagenetic dolomite are as follows: (1) clastic carbonate fragments which recrystallized during diagenesis; (2) oxidation of organic material originally deposited in the graywackes; or (3) carbonate-rich pore fluids trapped in the original sediments. The extreme rarity of carbonates in early Precambrian terranes does not favor the first possibility. Although probable organic remains have been described from cherts in both the Fig Tree and Onverwacht Groups (Pflug, 1967; Schopf and Barghoorn, 1967; Engel and others, 1968), it is unlikely that significant amounts of organic material existed during the early Precambrian and, hence, unlikely that the carbon in the dolomite came from organic remains in the graywackes. For these resaons, the third alternative is favored by the authors. The Ca and Mg may have been derived from either trapped pore fluids or from the diagenetic breakdown of calcic plagioclase and mafic silicates (such as pyroxenes and olivine).

Pyrite and limonite are important opaque minerals in the graywackes. Pyrite commonly has well-developed crystal faces and appears to replace all other minerals (Fig. 6). Such a texture suggests a late diagenetic origin for this mineral. Limonite is a relatively abundant diagenetic matrix constituent, especially in the two shale samples. Ilmenite (with leucoxene) composes up to one percent of some of the graywackes; it occurs as irregular, sievelike patches of probable diagenetic origin in the graywacke matrices. Traces of chromite of probable detrital origin and chalcopyrite also occur in some of the graywackes. Detrital zircon, apatite, and tourmaline are also common trace constituents.

Next to chert, the most abundant rock fragments in the Fig Tree graywackes are siliceous volcanic and schist fragments. Rock fragments other than chert range from < 1 to > 3 cm in size and compose up to 6 percent of the rocks. The relative abundances of the major rock fragments (excluding chert) as determined by modal analysis are illustrated for the Sheba Formation as a function of stratigraphic height in Figure 10.

Fragments of siliceous volcanic rock are especially important in the lower part of the Sheba Formation. As it is difficult to distinguish chert from these fragments, only those which

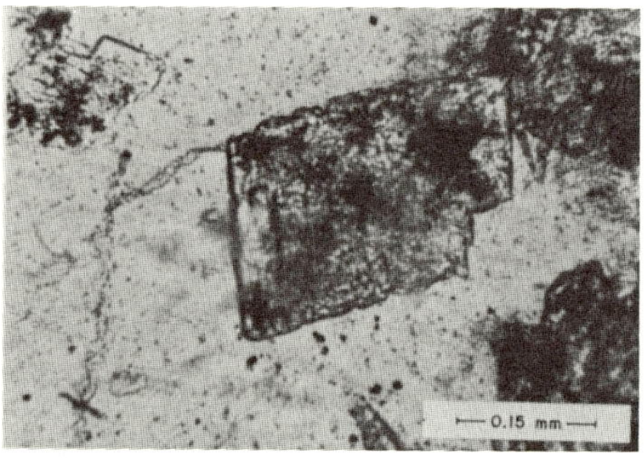

Figure 5. Photomicrograph of a dolomite rhomb surrounded by chert (white) and other graywacke matrix constituents. Sample C-17; uncrossed nicols.

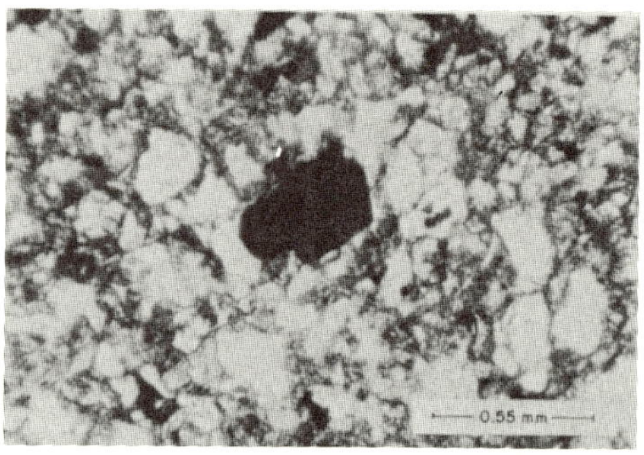

Figure 6. Photomicrograph of a pyrite crystal (black) in a graywacke matrix. Sample C-17; uncrossed nicols.

could be identified confidently as volcanic were plotted in Figure 10; the remainder are grouped with chert; therefore, this category should be considered as composed of minimum values only. Some of these fragments contain phenocrysts of sodic plagioclase (An_{20}-An_{30}) and less commonly, of quartz or K-feldspar. As discussed later, it is probable that most of the siliceous volcanic fragments were derived from siliceous pyroclastic terranes in Onverwacht-like source areas; some, however, were undoubtingly derived from contemporary volcanic activity as evidenced by the presence of probable pseudomorphs of shards.

Mafic volcanic fragments are composed of small plagioclase laths (chiefly An_{30}-An_{40}) embedded in a microcrystalline matrix of chlorite, sericite, dolomite, quartz, and limonite. The matrix material is of ovbious diagenetic origin, and the plagioclase laths may have been originally more calcic in composition. These fragments were probably derived from basalt and andesite terranes in Onverwacht-like source areas.

Metamorphic and granitic rock fragments become increasingly important as a function of stratigraphic height in the Sheba Formation (Fig. 10). The metamorphic fragments are chiefly quartz-sericite and quartz-chlorite schists and fine- to medium-grained quartzites. Because it is difficult to distinguish fine-grained quartzites from chert, questionable fragments were grouped with chert. Hence, the quartzites referred to in Figure 10 are chiefly the medium-grained quartzites. Due to the coarse grain size of granites and gneisses, fragments of these rocks are uncommon in the graywackes. However, as previously mentioned, the existence of all gradations between individual microcline, quartz, and plagioclase clasts and granite-gneiss fragments indicates that much of the graywacke detritus was derived from granite-quartzofeldspathic gneiss source terranes. Shaly rock fragments are abundant in some of the graywackes. However, they are not included in Figure 10 because it is not possible to distinguish extraformational from intraformational shale fragments.

The variety of mineral and rock fragments found in the Fig Tree graywackes suggests a diverse source area with at least some material derived from contemporary volcanic activity. The source area must have contained varying amounts of chert, volcanic rocks, and granitic-metamorphic rocks. The extreme angularity of most clasts and the relative abundance of clastic feldspar suggest that most source materials were not extensively weathered nor carried great distances prior to their deposition. Work still in progress suggests that much of the material was derived from central Swaziland, from 50 to 100 km to the southeast (Fig. 1). Because of the mineralogical and rock-fragment composition of the Fig Tree graywackes and shales, an unusually high initial Sr^{87}/Sr^{86} ratio (0.712) reported for Fig Tree shales by Allsopp and others (1968) is probably most accurately interpreted in terms of a significant amount of older source material in these rocks (granitic detritus) with a high Rb/Sr ratio.

MAJOR ELEMENT GEOCHEMISTRY

Major element analyses of the Fig Tree graywackes and shales are compiled in the Appendix and compared with other graywackes and average igneous rocks in Table 2. The major element variation within the graywackes of the Sheba and Belvue Road Formations is approximately the same as reported by Condie (1967a) for a suite of graywackes from Wyoming of about the same age. A greater variation in CaO is, however, observed in the Fig Tree graywackes and is probably related to the variable amounts of diagenetic dolomite in these rocks.

The oxide totals of those samples for which MgO analyses are available (*see* Appendix) range from 93 to 96 percent. In terms of the amount of chlorite, limonite, and dolomite in these rocks, the remaining 4 to 7 percent can easily be accounted for by H_2O and CO_2.

The similarity in composition of most graywackes to Ca-rich granite (granodiorite-quartz diorite) has been previously discussed (Condie, 1967a and 1967b). The Fig Tree graywackes, however, when compared to most other graywackes and to Ca-rich granite (Table 2) are enriched in Mg and, in the case of the graywackes of the Sheba Formation, are depleted in Al, Na, and K. In the Sheba graywackes, such differences are due to relatively large amounts of chlorite and dolomite and small amounts of total feldspar and mica as compared to most graywackes. The graywackes of the Belvue Road Formation have more typical concentrations of feldspar and mica and, hence, also of Al, Na, and K. The extremely high Fe and variable Ca and Na contents of the Fig Tree shales are related to the high limonite and variable dolomite and feldspar contents of these rocks.

The deficiency in Al in most of the Fig Tree graywackes, especially from the Sheba Formation, is particularly striking (Table 2). If this

Table 2. Major Elements in Graywackes and High-Ca Granite
(oxides in wt. percent)

	Precambrian Graywackes					Paleozoic Graywackes			Mesozoic Graywackes			High-Ca Granite
	1	2	3	4	5	6	7	8	9	10	11	12
SiO_2	66.2	59.8	65.5	64.4	64.7	69.7	66.7	70.6	71.1	69.7	67.5	67.2
TiO_2	0.52	0.55	0.53	0.62	0.57	0.5	0.54	0.64	0.50	0.6	0.5	0.5
Al_2O_3	10.2	12.9	12.5	15.5	13.4	14.3	..	12.6	13.9	14.3	13.5	15.5
Fe_2O_3*	7.01	6.56	6.94	6.54	6.27	4.6	6.12	4.97	3.0	3.8	4.5	4.2
MgO	4.50	4.44	3.30	3.12	3.23	1.8	..	1.51	1.3	1.2	2.2	1.6
CaO	1.97	3.18	4.69	2.22	3.04	1.3	3.09	1.61	1.8	1.9	2.4	3.5
Na_2O	1.80	2.83	2.41	3.74	2.99	3.1	1.90	2.76	3.7	3.5	3.6	3.8
K_2O	1.58	2.23	1.57	2.44	2.02	1.4	1.75	2.19	2.3	2.4	1.7	3.0
Al_2O_3/Na_2O	5.7	4.6	5.2	4.2	4.5	4.6	..	4.6	3.8	4.1	3.8	4.1

* Total Fe as Fe_2O_3
.. Dash signifies no determination

1. Average Precambrian graywacke of the Sheba Formation (this work). 2. Average Precambrian graywacke of the Belvue Road Formation (this work). 3. Average of 7 Southern Rhodesia graywackes compiled in Visser (1956, p. 63). 4. Average Wyoming Precambrian graywacke, (Condie 1967a). 5. Precambrian graywacke (Pettijohn, 1963). 6. Average Harz Mountain graywacke, Paleozoic (Mattiat, 1960). 7. Composite of German graywacke, Paleozoic (data *of* K. H. Wedepohl *in* Pettijohn, 1963). 8. Average Rensselaer graywacke, Paleozoic (Ondrick and Griffiths, 1969). 9. Composite lower Mesozoic graywacke, New Zealand (Reed, 1957). 10. Average lower Mesozoic graywacke, New Zealand (Reed, 1957). 11. Average Franciscan graywacke, Jurassic (Bailey and others, 1964). 12. Turekian and Wedepohl (1961).

deficiency were due to intense source-area weathering leaving Al behind, Fe also should be left behind; Fe, however, is enriched in the Fig Tree graywackes compared with Ca-rich granites and with post-Precambrian graywackes. Also, the fact that the graywackes are texturally immature, indicates that the original graywacke detritus was not highly weathered nor transported far from its source. This being the case, the rather high quartz + chert/feldspar + rock fragment and Al_2O_3/Na_2O ratios (3.5 and 5.7, respectively) in the Sheba graywackes indicate that their *apparent* mineralogical and chemical maturity is controlled chiefly by the composition of source materials and diagenetic processes rather than by the amount of weathering and erosion.

The low Na and K contents of the graywackes from the Sheba Formation are clearly evident in Figure 7. Also shown on the figure is the compositional range of the Precambrian Wyoming graywackes reported by Condie (1967a); these graywackes have Na_2O and K_2O concentrations typical of many graywackes. The graywackes of the Belvue Road Formation occupy an intermediate position between the Sheba and Wyoming graywackes. Analyses of graywackes of similar age to the Fig Tree Group from Southern Rhodesia (Hall, 1938) are also plotted on the figure and like the Fig Tree samples, show two groupings, one deficient in Na_2O and K_2O and the other with a more "normal" composition. Mineralogical studies of the Fig Tree and Wyoming graywackes (as well as of other graywackes) indicate that the increase in Na_2O and K_2O, shown in Figure 7, is controlled by a corresponding increase in granite-gneiss detritus at the expense of vol-

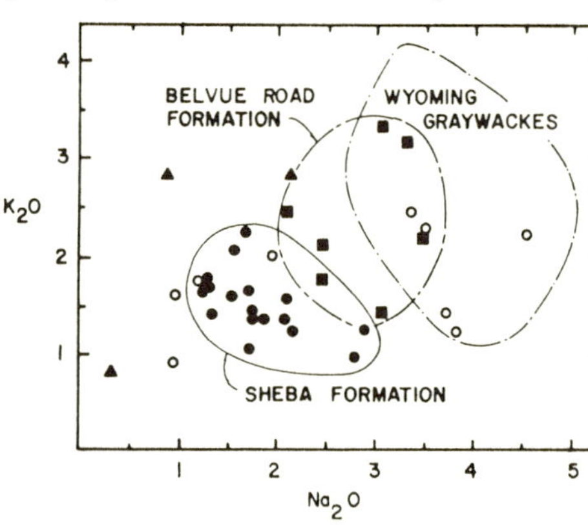

Figure 7. K_2O *versus* Na_2O plot of the Fig Tree graywackes and shales compared to Precambrian graywackes from Southern Rhodesia (Hall, 1938) and Wyoming (Condie, 1967a). Symbols: ●, Sheba Formation graywackes; ▲, Sheba Formation shales; ■, Belvue Road Formation graywackes; ○, Southern Rhodesia graywackes.

canic, and, in particular, mafic volcanic detritus.

TRACE ELEMENT GEOCHEMISTRY

Graywacke trace element variation within both the Sheba and Belvue Road Formations (*see* Appendix) is similar to that of the early Precambrian graywackes from Wyoming (Condie, 1967a). In comparison to some other graywackes of Precambrian and Paleozoic age, however, those of the Sheba Formation are depleted in Zr and Sr and greatly enriched in Ni (Table 3). Compared with high-Ca granites, these samples are lower in Rb and Sr and greatly enriched in Ni; there is also a relative depletion in Sr compared to K, Rb, Ca, and Ba (Table 4). The graywackes of the Belvue Road Formation have trace-element concentrations and ratios more nearly like the other graywackes in Table 3 and like high-Ca granites; they are, however, somewhat enriched in Ba and, like the graywackes of the Sheba Formation, greatly enriched in Ni. The absolute depletion in Rb, K, and Sr in the Sheba Graywackes (Fig. 8, Tables 2 to 4) relative to some other graywackes and to Ca-rich granites is perhaps most easily explained as a "dilution effect" caused by a relative abundance of nonalkali and nonalkaline earth-bearing minerals (that is, quartz, chert, chlorite, limonite, and probably dolomite) in these rocks.

When plotted on a Rb/Sr *versus* Rb plot (Fig. 8), the Sheba graywackes fall off the normal "igneous-rock trend." Unlike the Belvue Road, Wyoming, and Tyee graywackes, no mixture of common igneous rocks can account for these graywackes. Possible causes for this peculiar Sr depletion are as follows: (1) preferential loss of Sr during weathering and erosion; (2) preferential loss of Sr to pore waters during diagenesis; and (3) the presence of Sr-depleted igneous rocks in the source areas. The fact that K and Rb appear to be removed more rapidly or at about the same rate as Sr during weathering and erosion (Butler, 1953; Dasch, 1969), together with the observed textural immaturity of the Fig Tree graywackes, indicates that weathering and erosion are unlikely causes for the relative Sr depletion. The possibility that Sr was preferentially lost in an aqueous phase during diagenesis cannot be eliminated with the present data. However, if this were the case, it is difficult to explain why graywackes of the Belvue Road Formation escaped such losses. For these reasons, the relative Sr depletion in the Sheba graywackes is thought to be most readily accounted for by corresponding Sr depletion in the graywacke source materials. Condie and others (1969) discuss the origin of Sr-depleted basalts. It is possible that basaltic magmas fractionated along Sr-depletion trends and yielded a series of Sr-depleted volcanic (and perhaps granitic) rocks which later composed the source areas of the Sheba graywackes. Recent geochemical studies of R. P. and M. J. Viljoen (1969, personal commun.) indicate that

TABLE 3. TRACE ELEMENTS IN GRAYWACKES
(in ppm)

	1	2	3	4	5	6	7
Ba	319	626	..*	290	..	380	750
Mn	771	727	529	750	626	600	750
Ni	290	160	91	40	108	43	8
Rb	54	90	88	51	..	..	..
Sr	98	354	424	120	..	260	190
Zr	134	184	196	300	..	400	413
Sc	13	13	..	..	..	8	26
K/Sr	136	55	48	121	..	..	96
Rb/Sr	0.64	0.27	0.23	0.43	..	..	..
K/Rb	216	204	230	285	..	..	..
Ca/Sr	151	66	37	184	..	..	61
Sr/Ba	0.33	0.56	..	0.4	..	0.7	0.3

* Dash signifies no determination.

1. Average Precambrian graywacke from the Sheba Formation (this work). 2. Average Precambrian graywacke from the Belvue Road Formation (this work). 3. Average Wyoming Precambrian graywacke (Condie, 1967a). 4. Composite of German graywacke (data *of* K. H. Wedepohl *in* Pettijohn, 1963). 5. Average Canadian Precambrian graywacke (MacPherson, 1958). 6. Normanskill graywacke, New York (Weber and Middleton, 1961). 7. Rensselaer graywacke (Ondrick and Griffiths, 1967; Weber and Middleton, 1961).

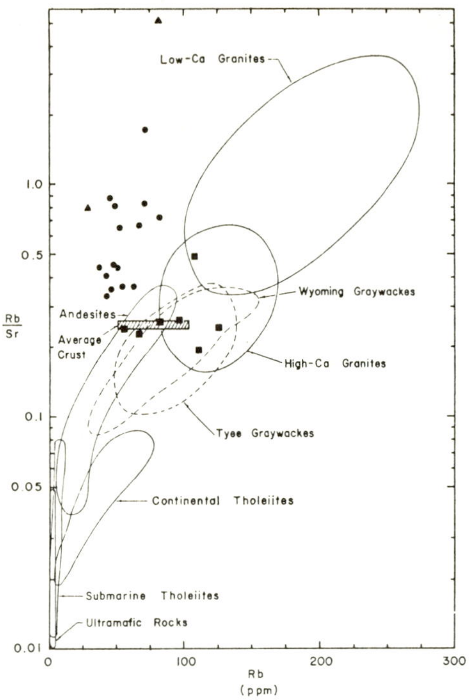

Figure 8. **Rb/Sr** ***versus*** **Rb plot of Fig Tree graywackes and shales compared to the major igneous-rock groups and to Wyoming (Condie, 1967a) and Tyee (Peterman and others, 1967) graywackes. Symbols same as Figure 7.**

some basalts in the lower Onverwacht Group contain only 15 to 50 ppm Sr. As such basalts compose a large part of the Onverwacht section, the lateral equivalents of these rocks may have composed the source areas of the Sheba graywackes. In terms of the data of Green and Ringwood (1967), the relative Sr depletion in the Sheba graywackes appears to record indirectly a major period of shallow ($\lesssim$ 15 km) magma fractionation accompanied by extensive wall-rock reaction early in the Earth's history.

The large enrichment of Ni in the Sheba graywackes (and in both shale samples analyzed) is especially apparent when plotted against the concentration of an alkali or alkaline-earth element such as Rb (Fig. 9). This enrichment may be explained by either of the following: (1) intense weathering of a relatively large amount of mafic volcanic rock in the source area resulting in a three- to fivefold increase of Ni in the detritus or, (2) mild weathering followed by rapid erosion and burial of detritus derived from ultramafic rocks in the source areas. The fact that intense weathering of source materials is unlikely in terms of the textural immaturity of the graywackes renders the first explanation unlikely and, therefore, the second explanation is favored. The existence of an ultramafic component in the source area would also assist in explaining the large amounts of Fe and Mg in the Fig Tree graywackes. A relatively large amount of Ni and Cr in Fig Tree shales has been interpreted recently in terms of an ultramafic source area for these rocks (Danchin, 1967). The concentration of other elements as well as the mineralogical composition of both the graywackes and shales of the Sheba Formation indicate, however, that ultramafic rocks composed, at most, only a relatively small part (~10 percent) of the original graywacke-shale source material. The graywackes of the Belvue Road Formation appear to have had much less contribution from ultramafic rocks as evidenced by their low Ni contents (Fig. 9).

SECULAR COMPOSITIONAL TRENDS IN THE FIG TREE GRAYWACKES

Figure 10 shows stratigraphic variations in the content of rock fragments and major and trace elements in graywackes of the Sheba Formation. Corresponding stratigraphic positions are given in Figure 2. The data show, with increasing stratigraphic level, a decrease in volcanic rock fragments, Ti, Na, and the Na/K ratio, a slight decrease in Zr, an increase in granitic and metamorphic rock fragments, and

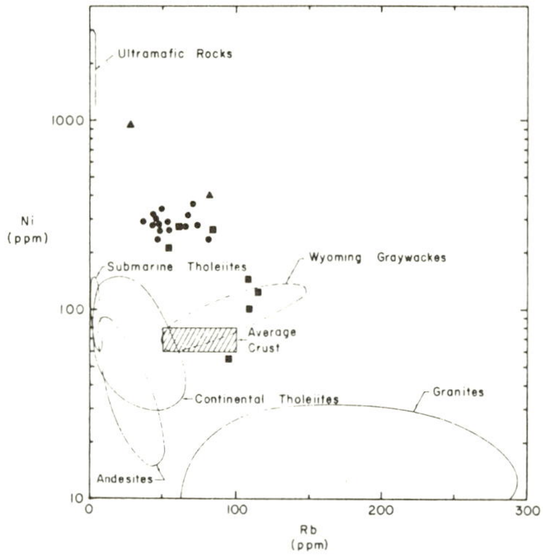

Figure 9. **Ni** ***versus*** **Rb plot of Fig Tree graywackes and shales compared to igneous-rock groups and to Wyoming graywackes (Condie, 1967a). Symbols same as Figure 7.**

TABLE 4. AVERAGE VALUES OF TRACE ELEMENTS IN IGNEOUS ROCKS AND FIG TREE GRAYWACKES (in ppm)

	Ultramafic Rocks	Submarine Tholeiites	Continental Tholeiites	Andesite	High-Ca Granite	Low-Ca Granite	Fig Tree Graywackes	
	1	2	3	4	5	6	7	8
Ba	0.4	12	352	270	420	840	319	626
Mn	1040	1230	1380	1200	540	390	771	727
Ni	1500	102	120	18	15	5	290	160
Rb	1	2	36	31	110	170	54	90
Sr	20	125	428	385	440	100	98	354
Zr	45	96	224	110	140	175	134	184
Sc	16	45	36	30	14	7	13	13
Rb/Sr	0.05	0.02	0.08	0.08	0.25	1.70	0.64	0.27
K/Rb	200	700	183	430	229	246	216	204
Sr/Ba	50	10	1.2	1.4	1.0	0.12	0.33	0.56
K/Sr	10	11.2	154	34.6	57	420	136	55
Ca/Sr	1250	663	164	130	44	59	151	66

1. Goles (1967). 2 and 3. Compiled from many sources. 4. Taylor and others (1969). 5 and 6. Turekian and Wedepohl (1961). 7. Average Precambrian graywacke from the Sheba Formation (this work). 8. Average Precambrian graywacke from the Belvue Road Formation (this work).

a slight increase in Ca and Sr. Near the top of the Sheba Formation (samples 14–16) a marked increase in K, Ba, and Rb also occurs. Such trends suggest an over-all increase, with a marked increase near the top of this formation, of a granitic-metamorphic component in the source material at the expense of a volcanic component. The increase in K-feldspar over this same interval (as previously mentioned) also records this change in source material. It is also interesting to note that the rock-fragment distribution shown in Figure 10 suggests cyclic variations in the source area. The amount of granitic detritus continues to increase upward into the Belvue Road Formation and the lower part of the overlying Moodies Group. Granitic pebbles compose up to 2.5 percent of conglomerates in the lower Moodies Group (Reimer, 1967). Higher in the Moodies Group, however, the feldspar content of the sandstones decreases; such a decrease appears to reflect the recycling of lower Moodies Group sediments. This progressive increase in sialic detritus appears to represent the progressive unroofing of a granitic-metamorphic terrane southeast of the Barberton Mountain Land which was initially covered by a thick assemblage of volcanic rocks of probable Onverwacht affinities.

In order to evaluate more fully the possibility that much of the volcanic detritus in the Sheba Formation was derived from lateral equivalents of the underlying Onverwacht Group, it is of interest to compare the distribution of volcanic rock fragments in the graywackes of the Sheba Formation with the known distribution of volcanic rocks in the Onverwacht Group. The Onverwacht Group is composed of a thick sequence of chiefly mafic volcanics with a thick section of siliceous volcanics in the upper part and two thinner siliceous sections, one near the top and one near the base. This distribution is illustrated in Figure 11 where the left curve represents a moving average of the ratio of siliceous to mafic volcanics. An inverted stratigraphic section of the Sheba Formation on the right side of the figure shows this same ratio for volcanic rock fragments in the Sheba graywackes; points on the curve correspond to sample sites. The two maxima observed on each curve and a third possible maximum at the base of the Sheba Formation may correlate as suggested by the dashed lines. Such correlations, if correct, would suggest that much of the volcanic detritus in the Sheba graywackes was derived from Onverwacht or Onverwacht-like volcanics.

INFERENCES REGARDING THE NATURE AND ORIGIN OF THE EARLY CRUST IN SOUTH AFRICA

The results of this investigation indicate the early crust that served as a source area for the Fig Tree graywackes in South Africa was composed of a diversity of igneous rock types. As evidenced by the Onverwacht Group, great quantities of volcanic rock were extruded, probably onto the mantle > 3.4 b.y. ago;

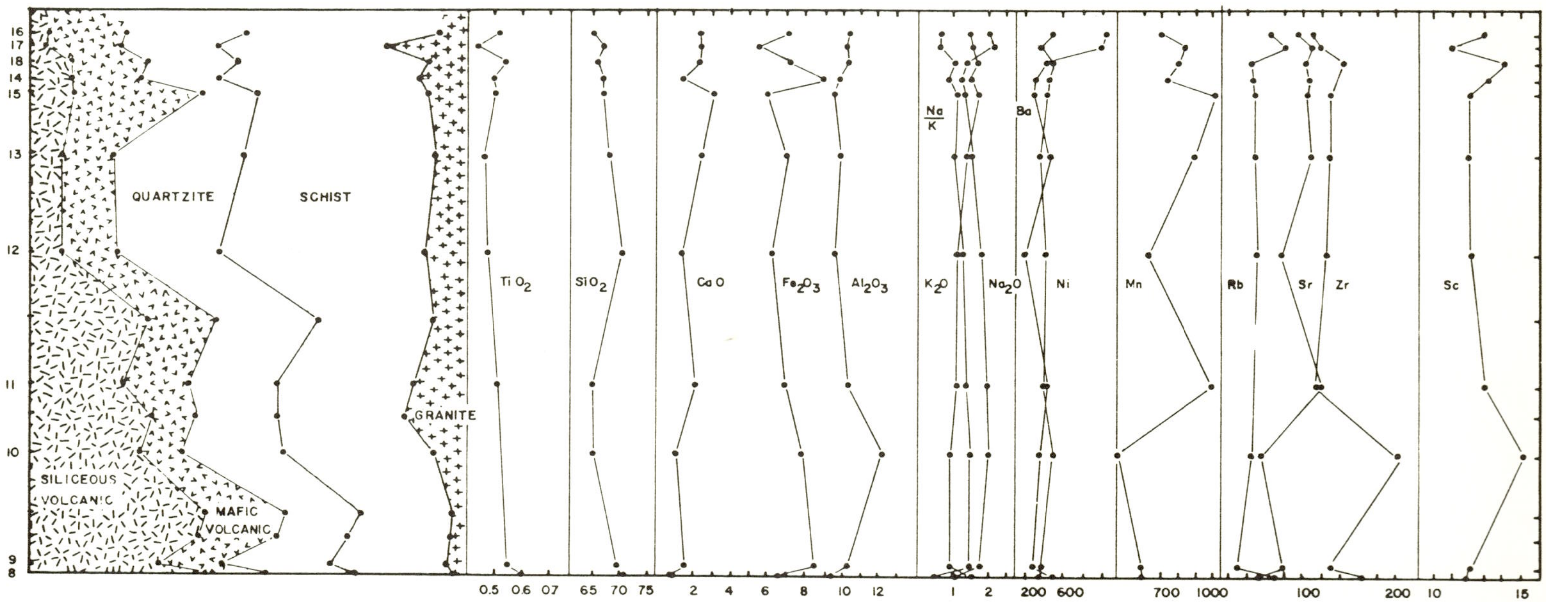

Figure 10. Secular stratigraphic variation in the content of rock fragments and in major and trace elements in graywackes from the Sheba Formation in the Ulundi syncline. Sample numbers correspond to those in Figure 2; thickness of the section is about 700 m.

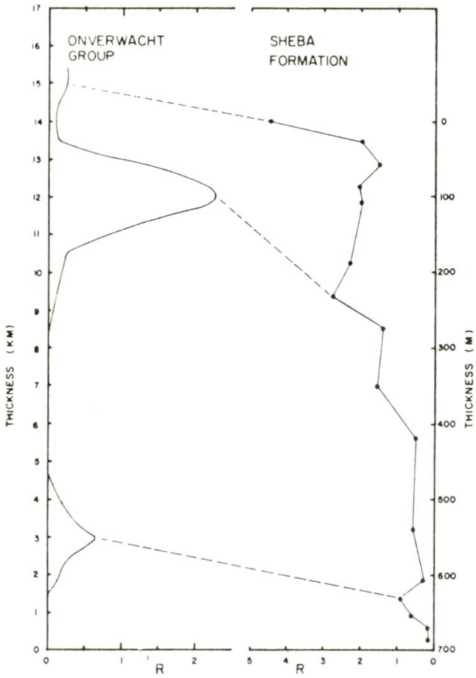

Figure 11. Plot of the ratio of siliceous to mafic volcanic component (R) in the Onverwacht Group and in volcanic rock fragments of the Sheba graywackes as a function stratigraphic height. (Sheba section is inverted; points in this inverted section correspond to sample levels.)

intermittent volcanic activity continued into Fig Tree and Moodies times. It is possible that much of the silica now in the form of chert in the Fig Tree and Onverwacht sections was derived from submarine volcanic emanations or desilication of deeper rocks. The fact that granitic detritus is abundant in the Fig Tree Group, yet granitic rocks of pre-Fig Tree age have not been found associated with the underlying Onverwacht Group, indicates that the Onverwacht section or its lateral equivalents were not the sole source of Fig Tree sediments. The granitic-metamorphic source rocks for the Fig Tree sediments may be preserved in the Ancient Gneiss complex described by Hunter (1968) in central Swaziland. Recent dates of about 3.34 b.y. have been reported from this complex (R. D. Davies, 1969, personal commun.). The increasing amount of sialic component in the Fig Tree Group with stratigraphic level indicates that significant parts of the early (proto-) crust in the area had been engulfed with granitic rocks prior to and perhaps during Fig Tree time. Although a major period of granite formation is recognized at about 3.0 b.y. in South Africa (Allsopp and others, 1968), this earlier period of granite formation may be recorded only indirectly in the sediments derived from its erosion. Van Niekerk (1967) reports an age of 4.1 b.y. for a granite pebble from the base of the Moodies Group; if confirmed, such a date would indicate that the sediments of the Fig Tree and Moodies Groups record, in part, perhaps the earliest granite-forming event in the Earth's history.

The great period of granite formation at about 3 b.y. is commonly associated with rapid crustal thickening (Engel, 1968, Anhaeusser and others, 1969) in this area. The fact, however, that still one or more earlier periods of granite formation occurred in this region suggests that significant crustal thickening may have occurred prior to Fig Tree time. Also, recent interpretations of calc-alkaline volcanic rocks in early Precambrian greenstone belts in the Canadian Shield that are similar to the Swaziland System in the Barberton Mountain Land (Condie and Potts, 1969) indicate that early greenstone belts may have reached thicknesses approximating present-day crustal thicknesses (30–40 km) prior to their engulfment with granite. These two factors suggest that the pre-Fig Tree crust may have been significantly thicker than the minimum thickness of 16 km exposed in the Onverwacht Group. Because of extensive granitization and metamorphism of this early crust during later orogenies, its lateral extent remains unknown; remnants of it, however, may be preserved in the Ancient Gneiss complex in central Swaziland.

The problem of the partial melting of the Earth prior to 3.0 b.y. ago, and especially prior to 3.5 b.y. ago, to account for this early magmatism has been discussed by Condie (1968). It is possible that local concentrations of volatiles, principally water, in the mantle beneath southern Africa depressed silicate melting points, thus allowing significant partial melting of the mantle. It may be that such local "water pockets" in the mantle defined the sites of early crustal formation.

CONCLUSIONS

(1) The mineralogy and composition of graywackes from the Fig Tree Group indicate a source area of diverse composition; chert, volcanic, and granitic-metamorphic source materials are most prevalent.

(2) The presence of chert and quartz shard-like pseudomorphs and of well-formed plagioclase and K-feldspar crystals in some of the graywackes suggests some detritus was derived

from contemporary volcanic activity.

(3) Immature graywacke textures indicate that weathering of source materials was not pronounced and that the original graywacke detritus was not transported great distances ($\lesssim$100 km) prior to deposition and burial.

(4) Dolomite, which is an important phase in many of the Fig Tree graywackes, appears to have been deposited by carbonate-saturated pore fluids during diagenesis.

(5) A relative depletion of Sr in the Sheba graywackes, as compared to K, Ba, Ca, and Rb, is interpreted in terms of an abundance of Sr-depleted igneous rocks in the source areas. The fact that the Belvue Road graywackes do not show such a depletion would, in turn, indicate that their source areas contained more normal Sr-bearing igneous rocks.

(6) A large concentration of Ni in the Sheba graywackes is interpreted as indicating a significant, although not major, ultramafic component in the source area of these rocks.

(7) Progressive compositional changes in graywackes of the lower Fig Tree Group continuing into the lower Moodies Group suggest progressive unroofing of a granitic-metamorphic terrane probably located southeast of the Barberton Mountain Land.

(8) The relatively abundant granitic detritus in the graywackes of the Fig Tree Group indirectly records one or more pre-Fig Tree granite-forming events, one event perhaps as old as 4 b.y.

ACKNOWLEDGMENTS

The authors are especially appreciative of the cooperation of Dr. O. R. van Eeden, of the Geological Survey of South Africa and to M. J. and R. P. Viljoen of Witwatersrand University for their assistance in collecting samples. Written correspondence with the Viljoens also has been of great help in clarifying geologic problems in the Barberton Mountain Land. The help of the staff at the nuclear reactor at the University of Missouri at Rolla is acknowledged as is the assistance with analytical work of C. K. Barsky, M. J. Potts, J. Davis, and S. Sun. The research at Washington University was sponsored by National Aeronautics and Space Agency research grant NGR 26-008-006 and the computer time by National Science Foundation grant G-22296. Field and laboratory investigations in South Africa were sponsored by a bursary awarded jointly by the Deutscher Akademischer Austauschdienst, Bad Godesberg, Germany, and the Department of Cultural Affairs, Pretoria, South Africa.

APPENDIX

MAJOR AND TRACE ELEMENT CONCENTRATIONS IN GRAYWACKES FROM THE FIG TREE GROUP. (OXIDES IN WT. PERCENT, TRACE ELEMENTS IN PPM)

Sheba Formation

Sample	Shales									
No.	C-3	C-5	V-1*	V-2*	K-1*	C-4	C-6	C-7	C-8	C-9
SiO_2	56.0	63.7	54.11	63.32	63.88	65.1	66.0	63.8	72.0	70.6
TiO_2	0.46	0.72	1.00	0.59	0.59	0.50	0.45	0.56	0.60	0.54
Al_2O_3	10.3	12.3	17.54	10.79	10.97	9.48	11.5	10.5	9.34	10.3
Fe_2O_3†	14.4	11.0	8.88	6.96	6.67	7.88	6.26	7.74	6.82	8.55
CaO	2.17	0.13	0.22	1.95	1.70	3.19	2.05	1.51	0.80	1.67
K_2O	0.72	1.70	2.72	1.44	1.63	1.35	0.90	1.61	1.54	0.97
Na_2O	0.28	0.81	2.14	2.09	1.27	1.77	2.80	1.30	1.22	1.71
MgO	..	..	6.97	5.67	5.50	3.35	..	..	..	..
Mn	1260	290	500	..	..	1090	800	600	610	610
Rb	28	83	..	..	..	53	46	70	73	36
Sr	85	16	..	..	..	142	129	81	41	80
Zr	34	112	..	..	..	113	140	150	166	123
Ni	958	400	..	..	..	264	232	362	280	291
Ba	516	316	..	..	..	408	156	286	331	243
Sc	28	21	..	..	..	13	10	14	13	13
Rb/Sr	0.80	5.19	..	..	..	0.37	0.36	0.86	1.78	0.45
K/Rb	212	170	..	..	..	211	163	191	175	225
K/Sr	170	883	..	..	..	79	58	166	312	102
Sr/Ba	0.06	0.05	..	..	..	0.34	0.82	0 28	0.12	0.33
Ca/Sr	444	58	..	..	..	161	114	133	140	149

Appendix (*Continued*)

Sample No.	C-10	C-11	C-12	C-13	C-14	C-15	C-16	C-17B	C-18	C-25	x̄‡
SiO_2	66.0	66.2	70.6	68.0	67.0	66.8	65.2	67.0	66.0	57.2	66.2
TiO_2	0.69	0.51	0.47	0.46	0.49	0.49	0.51	0.43	0.53	0.42	0.52
Al_2O_3	12.5	10.3	9.97	10.1	10.0	9.61	10.7	10.3	10.3	9.48	10.20
Fe_2O_3†	8.23	7.02	6.40	7.05	9.14	6.07	7.21	5.42	6.87	4.95	7.01
CaO	1.34	2.29	1.51	2.47	1.47	3.00	2.21	2.43	2.39	1.60	1.97
K_2O	1.13	1.26	1.26	1.55	1.32	1.35	1.98	2.14	1.45	1.18	1.58
Na_2O	2.13	2.06	1.84	1.50	1.34	1.76	1.55	1.64	1.70	2.90	1.80
MgO	..	..	3.50	..	..	..	..	..	..	..	4.50
Mn	490	1010	670	910	750	1040	710	860	810	600	771
Rb	49	43	52	49	45	47	67	83	43	61	54
Sr	59	126	78	109	50	101	97	115	104	164	98
Zr	206	120	124	127	103	123	109	115	131	157	134
Ni	342	283	287	266	306	280	317	257	318	266	290
Ba	274	285	207	315	224	231	526	502	284	519	319
Sc	15	13	12	12	13	12	13	11	14	11	13
Rb/Sr	0.83	0.34	0.67	0.45	0.90	0.47	0.69	0.72	0.41	0.37	0.64
K/Rb	192	244	202	262	243	238	246	213	280	161	216
K/Sr	160	83	134	118	219	110	170	153	115	60	136
Sr/Ba	0.21	0.44	0.38	0.35	0.22	0.43	0.18	0.23	0.37	0.31	0.33
Ca/Sr	163	130	138	162	210	212	163	151	164	69	151

Belvue Road Formation

Sample No.	C-1	C-2	C-24	C-26	C-27	C-28	C-29	x̄‡
SiO_2	65.8	65.2	59.3	51.9	51.2	60.8	64.4	59.8
TiO_2	0.50	0.53	0.67	0.69	0.66	0.38	0.43	0.55
Al_2O_3	12.4	12.6	14.5	14.3	15.0	10.6	10.7	12.9
Fe_2O_3†	5.65	5.38	7.63	8.73	8.89	4.72	4.89	6.56
CaO	2.07	2.24	2.98	4.49	3.10	4.99	2.36	3.18
K_2O	1.62	2.32	2.03	3.20	3.07	2.00	1.36	2.23
Na_2O	2.44	2.10	3.45	3.04	3.32	2.48	3.01	2.83
MgO	4.95	4.08	4.29	..	..	..	..	4.44
Mn	570	450	740	980	760	970	620	727
Rb	65	106	95	124	109	81	53	90
Sr	283	215	363	507	568	318	222	354
Zr	146	165	206	226	245	153	150	184
Ni	272	142	53	114	101	234	202	160
Ba	571	476	610	1090	701	489	455	626
Sc	12	9	17	17	14	10	9	13
Rb/Sr	0.23	0.49	0.26	0.24	0.19	0.25	0.24	0.27
K/Rb	207	181	177	214	234	205	213	204
K/Sr	48	90	46	52	44	52	50	55
Sr/Ba	0.49	0.45	0.59	0.47	0.81	0.65	0.49	0.56
Ca/Sr	53	75	58	63	39	112	76	66

.. indicates no determination.
* analyses from Visser (1956): V-1 and V-2
Koen (1947): K-1
† total Fe as Fe_2O_3.
‡ x̄ = mean values, excluding shales.

REFERENCES CITED

Allsopp, H. L., Ulrych, T. J., and Nicolysen, L. O., 1968, Dating some significant events in the history of the Swaziland System by the Rb-Sr isochron method: Canadian Jour. Earth Sci., v. 5, p. 605-619.

Anderson, J. E., Jr., 1969, Development of snowflake texture in a welded tuff, Davis Mountains, Texas: Geol. Soc. America Bull., v. 80, p. 2075–2080.

Anhaeusser, C. R., Mason, R., Viljoen, M. J., and Viljoen, R. P., 1969, A reappraisal of some aspects of Precambrian shield geology: Geol. Soc. America Bull., v. 80, p. 2175–2200.

Anhaeusser, C. R., Roering, C., Viljoen, M. J., and Viljoen, R. P., 1968, The Barberton Mountain Land: A model of the elements and evolution of an Archean fold belt: Geol. Soc. South Africa Trans., v. 71 (Annexure), p. 225–254.

Bailey, E. H., Irwin, W. P., and Jones, D. L., 1964, Franciscan and related rocks and their significance in the geology of western California: California Div. Mines and Geology Bull. No. 183, 177 p.

Butler, J. R., 1953, The geochemistry and mineralogy of rock weathering. (1) The Lizard area, Cornwall: Geochim. et Cosmochim. Acta, v. 4, p. 157–178.

Condie, K. C., 1967a, Geochemistry of early Precambrian graywackes from Wyoming: Geochim. et Cosmochim. Acta, v. 31, p. 2135–2149.

—— 1967b, Composition of the ancient North American crust: Science, v. 155, p. 1013–1015.

—— 1968, Possible role of water in the Earth's thermal history: Jour. Geophys. Research, v. 73, p. 5466–5468.

Condie, K. C., Barsky, C. K., and Mueller, P. A., 1969, Geochemistry of Precambrian diabase dikes from Wyoming: Geochim. et Cosmochim. Acta, v. 33, p. 1371–1388.

Condie, K. C., and Potts, M. J., 1969, Calc-alkaline volcanism and the thickness of the early Precambrian crust in North America: Canadian Jour. Earth Sci., v. 6, p. 1179–1184.

Cummins, W. A., 1962, The graywacke problem: Liverpool Manchester Geol. Jour. v. 2, p. 168–179.

Danchin, R. V., 1967, Chromium and nickel in the Fig Tree Shale from South Africa: Science, v. 158, p. 261–262.

Dasch, E. J., 1969, Sr isotopes in weathering profiles, deep-sea sediments, and sedimentary rocks: Geochim. et Cosmochim. Acta, v. 33, p. 1521–1552.

Engel, A. E. J., 1968, The Barberton Mountain Land: Clues to the differentiation of the earth: Geol. Soc. South Africa Trans., v. 71 (Annexure), p. 255–270.

Engel, A. E. J., Nagy, B., Nagy, L. A., Engel, C. G., Kremp, G. O. W., and Drew, C. M., 1968, Alga-like forms in Onverwacht Series, South Africa: Oldest recognized lifelike forms on earth: Science, v. 161, p. 1005–1008.

Goles, G. G., 1967, Trace elements in ultramafic rocks, *in* Wyllie, P. J., *Editor*. Ultramafic and Related Rocks: New York, John Wiley & Sons, Inc., p. 352–362.

Green, D. H., and Ringwood, A. E., 1967, The genesis of basaltic magmas: Contr. Mineralogy and Petrology, v. 15, p. 103–190.

Hall, A. L., 1938, Analyses of rocks, minerals, ores, coal, soils and waters from Southern Africa: South Africa Geol. Survey Mem., no. 32, 875 p.

Hunter, D. R., 1968, The Precambrian terrain in Swaziland with particular reference to the granite rocks: Ph.D. thesis, Witwatersrand Univ., Johannesburg, South Africa.*

Koen, G. W., 1947, Die Geologie van die Shebagebied in die Barbertonse Distrik: M.Sc. thesis, Pretoria Univ., Pretoria, South Africa.*

Mac Pherson, H. G., 1958, A chemical and petrographic study of Precambrian sediments: Geochim. et Cosmochim. Acta, v. 14, p. 73–92.

Mattiat, B., 1960, Beitrag zur Petrographie der Oberharzer Kulmgrauwacke: Contr. Mineralogy and Petrology, v. 7, p. 242–280.

Ondrick, C. W., and Griffiths, J. G., 1969, Frequency distribution of elements in Rensselaer graywacke, Troy, New York: Geol. Soc. America Bull., v. 80, p. 509–518.

Peterman, F. E., Hedge, C. E., Coleman, R. B., and Snavely, P. D., Jr., 1967, Sr^{86}-Sr^{87} ratios in some eugeosynclinal sedimentary rocks and their bearing on the origin of granitic magma in orgenic belts: Earth and Planetary Sci. Letters, v. 2, 433–439.

Pettijohn, F. J., 1963, Chemical composition of sandstones—excluding carbonate and volcanic sands, *in* Data of Geochemistry, 6th ed.: U.S. Geol. Survey Prof. Paper 440S, Chap. 5, 19 p.

Pflug, H. D., 1967, Strukturierte organische Reste aus über 3 Milliarden Jahre alten Gesteinen Sudafrikas: Die Natur wissen Schaften, v. 54, p. 237–241.

Reed, J. J., 1957, Petrology of the lower Mesozoic rocks of the Wellington District: New Zealand Geol. Survey Bull., v. 57, 68 p.

Reimer, T. O., 1967, Die Geologie der Stolzburg Synklinale, Barberton Bergland (Transvaal-Sudafrika): Diplom-Arbeit, Geothe-Univ., Frankfurt, Germany.*

Rose, H. J., Adler, I., and Flannigan, F. J., 1963, X-ray fluorescence of light elements in rocks and minerals: Appl. Spectroscopy, v. 17, p. 81–85.

Schopf, J. W., and Barghoorn, E. S., 1967, Alga-

* available on interlibrary loan

like fossils from early Precambrian of South Africa: Science, v. 156, p. 508–511.

Stueber, A. M., and Goles, G. G., 1967, Abundances of Na, Mn, Cr, Sc, and Co in ultramafic rocks: Geochim. et Cosmochim. Acta, v. 31, p. 75–93.

Taylor, S. R., Capp, A. C., Graham, A. L., and Blake, D. H., 1969, Trace element abundances in andesites, II Saipan, Bougainville, and Fiji: Contr. Mineralogy and Petrology, v. 23, p. 1–26.

Turekian, K. K., and Wedepohl, K. H., 1961, Distribution of the elements in some major units of the Earth's crust: Geol. Soc. America Bull., v. 72, p. 175–192.

Van Niekerk, C. B., 1967, The suitability of extrusive acid volcanic rocks for U-Pb radiometric dating: Ph.D. thesis, Capetown Univ., Capetown, South Africa.

Visser, D. L. J., *Compiler*, 1956, The geology of the Barberton area: South Africa Geol. Survey Spec. Pub., no. 15, 251 p.

Weber, J. N., and Middleton, G. V., 1961, Geochemistry of the turbidites of the Normanskill and Charny Formations: Geochim. et Cosmochim. Acta, v. 22, p. 200–243.

Manuscript Received by The Society December 29, 1969

Revised Manuscript Received March 30, 1970

Present Address (Condie): Department of Geosciences, New Mexico Institute of Mining and Technology, Socorro, New Mexico 87801

17B: BANDED IRON FORMATION IN THE ARCHEAN

A. F. Trendall

Banded iron formation (BIF) is a rock type distinguished by a number of remarkable characteristics; not the least remarkable of these is that, although it is familiar to almost every geologist who has worked in Precambrian terrain, BIF has received comparatively little close attention, as a result of which most aspects of its origin and significance are either controversial or frankly speculative. No other sedimentary rock differs so radically from any material currently being laid down in an observable environment, not only in its chemical and mineral composition, but also in the striking contrast between its extreme homogeneity along the banding and its heterogeneity across it; there is no present sediment with such regular and persistent alternations of two such contrasting constituents.

To appreciate the problems posed by the Archean iron formations, it is useful to refer also to those of the Proterozoic, first, because the varying abundance of BIFs in time provides an important clue to their interpretation and, second, because the post-Archean BIFs have been most closely studied, and thus throw comparative light on the earlier ones. However, before giving a descriptive summary of either, the following two paragraphs provide, successively, a brief guide to published literature and some notes on nomenclature.

The first detailed studies of iron formation were those by the U.S. Geological Survey on the Lake Superior iron ranges. The results form an impressive sequence of monographs, which established these ranges as an informal but widely accepted "type area" for the rock (Irving and Van Hise, 1892; Van Hise and Bayley, 1897; Leith, 1903; Bayley, 1904; Van Hise and Leith, 1911). This acceptance was reinforced by the later work of Gruner (1922, 1946), White (1954), and Gundersen and Schwartz (1962) on the Mesabi Range, and by that of Goodwin (1956, 1960) and Moorhouse (1960) on the Gunflint, its continuation into Canada; by the work of Aldrich (1929) on the Gogebic Range; and especially by the work of James (1951, 1954, 1955, 1958) in Michigan. By 1960, knowledge of the Lake Superior iron formations and of those elsewhere in the North American Precambrian (Gross, 1965) overshadowed work on other continents, although accounts had been published of selected iron formations of Australia (Miles, 1953, 1955), India (Dunn, 1941; Spencer and Percival, 1952), Brazil (Dorr, 1969; Guild, 1957), and South Africa (Hall, 1918, 1930). Although a good deal of work had been carried out in Russia, the fact that only a negligible amount was available in translation prevented any important interaction with the English-language literature. The discovery of the rich Hamersley iron ore province of Western Australia in 1960 was followed by close stratigraphic and petrographic study of the iron formations of the Hamersley Basin (Trendall and Blockley, 1970). This work established (Trendall, 1968) that, in spite of their basic compositional resemblance to the Lake Superior iron formations, those of the Hamersley Basin differed in many significant ways, and showed by inference that every iron formation merited close study in its own right, since similarity with those of Lake Superior could not be assumed. During the last

This article was especially prepared by the author for this volume.

few years two important symposia on iron formation have been held: in Kiev in August 1970, and in Duluth in November 1972. The collected papers of both these symposia [respectively, UNESCO, 1973, and *Economic Geology, 68*(7)] jointly form a reference point in iron-formation literature that will stand for many years, since they include contributions from and concerning all continents.

A variety of names have been applied to iron formation, many of them with mainly local significance; Brandt et al. (1972) have reviewed the nomenclature. The main point to appreciate is that such terms as jaspilite, taconite, itabirite, hematite, quartzite, and iron chert are not applied systematicaly in the literature to rocks known to have either different genesis or different petrography; rather, they are virtual synonyms that have arisen through insufficient communication. The term BIF is here used to include banded sedimentary rocks with 15 percent or more iron, in which the iron minerals are interbanded with chert or carbonate. Caution should be used before classifying pyritic sediments as "sulfide facies iron-formation," with the implication that they are cogenetic with cherty rocks. It should be borne in mind that the facies classification of James (1954) was initially conceptual, and that lateral stratigraphic intergradation of pyritic rocks and cherty iron formation is, at best, a rare and untypical phenomenon.

The attempt of O'Rourke (1961) to demonstrate that a long-accepted restriction of BIF to the Precambrian was invalid resulted in a clear indication, by the paucity of Phanerozoic analogues available after search, that BIF is indeed a characteristic Precambrian rock. Although objective quantitative data are insufficient for proof, it is the view of most present workers that BIF occurs commonly throughout the Archean but is not a major constituent of most stratified sequences; by contrast, its relative abundance increased enormously about 2000–1800 million years ago, and thereafter fell abruptly to virtual absence during the later Precambrian and Phanerozoic.

Archean BIFs most typically consist of red, white, brown, yellow, gray, or clear chert bands alternating with black, dark red, or dark bluish bands consisting largely of magnetite or hematite, generally with some silica. Individual bands range in thickness from less than 1 mm to about 15 mm, although there are no systematic measurements published. The median thickness of both band types is probably about 5 mm, and there are usually about equal thicknesses of both types. The total iron content is normally between 20 and 30 percent, and there is a real paucity of rocks intermediate between bedded charts with a very small iron content and BIF proper. The usual stratigraphic thickness of continuous Archean BIF units is between 10 and 50 m, although a few exceed 100 m. Continuous strike lengths of up to 75 km occur, but most fall far short of this. Although the banding is clearly very consistent laterally on the single-outcrop scale, detailed studies of the lateral extent of the individual bands have not been made. Like those of the Proterozoic, Archean BIFs occur with a wide variety of other rocks; although a consistent association with volcanics has been suggested, the abundance of volcanic rocks in Archean sequences generally makes it hard for any nonvolcanic rock not to be so associated!

The following broad contrasts between Archean BIFs and those laid down during the terminal burst of BIF abundance—for convenience, referred to as Proterozoic below—appear to be valid:

1. Archean BIFs are relatively less abundant, and individual BIF units are both stratigraphically thinner and less laterally extensive.

2. Archean BIFs are more restricted in textural variety; they tend to be thinly and evenly banded, whereas the Proterozoic BIFs vary from the evaporite-like Hamersley Basin rocks to the clearly high energy facies of those of Labrador.

3. In composition, Archean BIFs tend to be restricted to the "oxide facies" of James (1954), and to be essentially bimineralic (either quartz-magnetite or quartz-hematite), whereas the Proterozoic BIFs normally have carbon dioxide, sodium, potassium, calcium, and magnesium as important minor constituents.

The distribution in time of BIFs—from presence, to abundance, to absence—lacked any persuasive explanation before Cloud (1968, 1972) elegantly integrated it with a hypothesis for the evolution of photosynthesizing organisms. In summary, the first photosynthesizers had to have a means of immediate oxygen disposal to avoid being poisoned (burnt) by their own oxygen; it was disposed of by oxidation of the abundant ferrous iron in the ambient water, with resultant precipitation of the ferric iron. The process was slow and limited during the Archean because the organisms could only flourish within very narrow environmental limits imposed by the availability of ferrous iron. But as photosynthesizers overcame this handicap by evolving their own oxygen-mediating enzymes, their chemical dependence on their environment enormously decreased; in the resultant burst of continuing oxygen production, the oceans were abruptly flushed of their remaining iron.

By combining this suggestion with the quite low basinal iron concentrations required for deposition of the Hamersley Basin BIFs suggested by Trendall and Blockley (1970), and also with the proposal of Holland (1973) that the oceans themselves could have provided this order of concentration without a special local volcanic source of iron, it can be proposed that BIFs formed during the Precambrian as a biochemical precipitate whenever a suitable depositional environment arose, and that the abundance and facies of BIF were controlled by biochemical evolution, in a primary and secondary way, respectively. The "suitability" of an environment was broadly similar to that for Phanerozoic evaporite accumulation: long-term basinal stability accompanied by plumbing able to replenish "used" water, and a virtual absence of incoming detritus. Only during the final Proterozoic burst was BIF sometimes laid down in environments other than ideal, to give the distinctive high-energy facies best exemplified by Dimroth's (1973) descriptions of the Quebec iron formations.

In summary, the existence of Archean BIFs signifies (1) in broad terms the recurrent establishment of colonies of photosynthesizing organisms in aqueous environments where ferrous iron was available, and (2) in sedimentological terms a fairly shallow but very stable environment with free circulation where these organisms could respond to sunlight free from the inhibiting effects of a blanket of incoming terrigenous debris.

References

Aldrich, H. R. (1929). The geology of the Gogebic Iron Range of Wisconsin. *Wisconsin Geol. Nat. Hist. Survey Bull. 71,* 279p.

Bayley, W. S. (1904). The Menominee iron-bearing district of Michigan. *U.S. Geol. Survey Monogr.* 46, 313p.

Brandt, R. T., Dorr, J. V. N., Gross, G. A., Gruss, H., and Semenenko, N. P. (1972). Problems of nomenclature for banded ferruginous-cherty sedimentary rocks and their metamorphic equivalents. *Econ. Geol. 67:* 682–684.

Cloud, P. (1968). Atmospheric and hydrospheric evolution on the primitive earth. *Science: 160,* 729–736.

____. (1972). A working model of the primitive earth. *Am. J. Sci.: 272,* 537–548.

Dimroth, E. (1973). Superior-type iron formations in Canada, with special reference to localities in Quebec. *Econ. Geol.: 68,* 000–000.

Dorr, J. V. N. (1969). Physiographic, stratigraphic and structural development of the Quadrilatero Ferrifero, Minas Gerais, Brazil. *U.S. Geol. Survey Prof. Paper 641-A,* 110p.

Dunn, J. A. (1941). The origin of banded hematite ores in India. *Econ. Geol.: 36,* 355–370.

Goodwin, A. M. (1956). Facies relations in the Gunflint iron formation. *Econ. Geol.: 51,* 565–595.

____. (1960). Gunflint Iron Formation of the Whitefish Lake Area, District of Thunder Bay. *Ontario Dept. Mines Ann. Rept., 69,* 41–63.

Gross, G. A. (1965). Geology of iron deposits in Canada—general geology and evaluation of iron deposits. *Geol. Survey Canada, Rept. 22,* Vol. 1, 181p.

Gruner, J. W. (1922). Organic matter and the origin of the Biwabik iron-bearing formation of the Mesabi Range. *Econ. Geol.: 17,* 407–460.

____. (1946). The mineralogy and geology of the taconites and iron ores of the Mesabi Range, Minnesota. St. Paul, Minnesota, Iron Range Resources and Rehabilitation Commission and Minnesota Geological Survey, 127p.

Guild, P. W. (1957). Geology and mineral resources of the Congonhas District, Minas, Gerais, Brazil. *U.S. Geol. Survey Prof. Paper 290,* 90p.

Gundersen, J. N., and Schwartz, G. M. (1962). The geology of the metamorphosed Biwabik Iron-Formation, Eastern Mesabi District, Minnesota. *Minnesota Geol. Survey Bull. 43,* 139p.

Hall, A. L. (1918). Asbestos in the Union of South Africa. *Union of South Africa Geol. Survey Mem. 12,* 152p.

____. (1930). Asbestos in the Union of South Africa, 2nd ed. *Union of South Africa Geol. Survey Mem. 12,* 324p.

Holland, H. D. (1973). The oceans: a source of iron in iron-formations. *Econ. Geol.: 68.*

Irving, R. D., and Van Hise, C. R. (1892). The Penokee iron-bearing series of Michigan and Wisconsin. *U.S. Geol. Survey Monogr. 19,* 534p.

James, H. L. (1951). Iron formation and associated rocks in the Iron River district, Michigan. *Geol. Soc. Am. Bull.: 62,* 251–266.

____. (1954). Sedimentary facies of iron-formation. *Econ. Geol.: 49,* 235–291.

____. (1955). Zones of regional metamorphism in the Precambrian of northern Michigan. *Geol. Soc. Am. Bull., 66,* 1455–1488.

____. (1958). Stratigraphy of Pre-Keeweenawan rocks in parts of northern Michigan. *U.S. Geol. Survey Prof. Paper 314-C,* 44p.

Leith, C. K. (1903). The mesabi iron-bearing district of Minnesota. *U.S. Geol. Survey Monogr. 43,* 316p.

Miles, K. R. (1953). Banded iron formations in Western Australia. *In* Geology of

Australian ore deposits, Empire Mining Metall. Congr., 5th, Australia, 1953, 159–171.

____. (1955). The geology and iron ore resources of the Middleback Range area. Adelaide, *Geol. Survey South Aust. Bull. 33,* 247p.

Moorhouse, W. W. (1960). Gunflint Iron Range in the vicinity of Port Arthur. *Ontario Dept. Mines Ann. Rept., 69,* p. 1–40.

O'Rourke, J. E. (1961). Paleozoic banded iron-formations. *Econ. Geol., 56,* 331–361.

Spencer, E., and Percival, F. G. (1952). The structure and origin of the banded hematite jaspers of Singhbhum, India. *Econ. Geol.: 47,* 365–383.

Trendall, A. F. (1968). Three great basins of Precambrian banded iron formation deposition: a systematic comparison. *Geol. Soc. Am. Bull.: 79,* 1527–1544.

____, and Blockley, J. G. (1970). The iron formations of the Precambrian Hamersley Group, Western Australia, with special reference to the associated crocidolite. *Geol. Survey Western Aust. Bull. 119,* 366p.

UNESCO. (1973). Genesis of Precambrian iron and manganese deposits. Proc. Kiev Symp., 1970 (*Earth Sciences, 9*).

Van Hise, C. R., and Bayley, W. S. (1897). The Marquette iron-bearing district of Michigan. *U.S. Geol. Survey Monogr. 28,* 608p.

____, and Leith, C. K. (1911). The geology of the Lake Superior region. *U.S. Geol. Survey Monogr. 52,* 641p.

White, D. A. (1954). The stratigraphy and structure of the Mesabi range, Minnesota. *Minnesota Geol. Survey Bull. 38,* 92p.

IV: THE HIGH-GRADE TERRAINS

Part IV includes some of the most important papers in this volume. The greenstone belts are not difficult to unravel provided the geologist is willing to map exhaustively and study the rocks themselves in detail in the laboratory. The high-grade terrains are less easy to unravel, for much of the original nature of the rocks and the original structure has been obliterated. Yet it is the unraveling of the problem of the anomalously vast granite–gneiss expanses separating the greenstone belts that may lead to a real understanding of the nature of the primordial crust and the early history of continent–ocean differentiation.

The first four papers deal with the North Atlantic craton, which, despite its fragmentation by later events, is perhaps the most intensively studied and the best understood of such areas. In Paper 22, Pavlovskiy summarizes three of the four very important Archean cratons in Russia, all high-grade terrains (although that of the Ukraine includes low-grade metamorphics). A note on the fourth, also of high-grade character, the Anabar shield in Siberia, is added in commentary, together with a discussion of the Archean *nuclear structural patterns.* Wynne-Edwards, in Paper 23, revolutionizes our ideas about the incorporation and survival of Archean rocks in Proterozoic mobile belts, and shows that gray gneisses may derive from very old magmatic granites and be virtually indestructable, being reworked again and again and emerging much as before. Papers 25 and 26 by Wilson and Hepworth cover the enigmatic question of the origins of the granulite facies metamorphics of Archean high-grade terrains.

Three topics of major importance are considered in Part IV: the nature and sequence of metamorphism and its distribution in space, the significance of the nuclear terrains of the permobile stage in the earth's development, and the nature of the primordial crust of the earth.

The Russian and North Atlantic Archean provinces (Papers 18–22 and 38) are of dominant high metamorphic grade; there is certainly a difference in this respect to the Archean terrains of Canada, South Africa, and Australia. Metamorphic grade is no parameter of age, and some of the oldest Katarchean rocks are little metamorphosed (Papers 5 and 14).

Greenschist facies characterizes greenstone belts, the change to amphibolite facies and above being typically abrupt at the contact with bordering expanses of granite–gneiss terrain. Granulite metamorphism can commonly be recognized patchily developed within this terrain (Papers 18–20 and 24–26); amphibolite metamorphism is commonly retrogressively superimposed. Granulite facies is by no means restricted to the Archean, although widely developed, but there is some doubt whether it is any more markedly developed here than in younger rocks; it is certainly characteristic of the Proterozoic in its welt-like linear belts that transect the older cratons (Paper 23). There appears to be a close analogy between this linear ultrametamorphism and the less linear, more patchy ultrametamorphic developments of Archean high-grade terrains. Both characteristically display paleosomes and nebulites, agmatites and pegmatites. A roadside cutting through the Grenville or Kapuskasing 950 million year old mobile belts of the Canadian Proterozoic appears in no way different to one through Katarchean high-grade terrain of the Wheat Belt or Darling ranges in Western Australia.

A good case has been made for the depletion of the underlying granulite zone by migration of volatile elements into the overlying hood of granitic rocks (Papers 6 and 21). The narrow and ill-defined zones of high-grade metamorphism at the margins of greenstone belts reflect contact metamorphism of late synorogenic type, related to invasion by the nearby granite. Local hot spots of metamorphism characterize greenstone tracts, but it is erroneous to suppose that such tracts are of consistent high grade just because of such recognitions.

Isograds of metamorphism may run oblique to margins of greenstone tracts. Where narrow paleosomes occur in high-grade granite–gneiss terrain, migmatitic selvages may sheath them (Paper 19), and they may show the high metamorphic grade of the host rocks throughout.

Granulite metamorphism may well have occurred in some cases during the low-grade metamorphism of the adjacent greenstone tracts; the granite–gneiss areas are essentially uplifed protocratonic areas, reworked one or more times by successive metamorphisms, and they may be expected to include material pushed up from their high-grade core relative to the neighboring greenstone troughs of subsidence. However, it is clear that metamorphic grade is a factor of metamorphic depth, not purely depth of burial, for sequences of up to 30 km of greenstones show little metamorphic enhancement with stratigraphic depth. We should envisage metamorphism as moving in decreasing isograd zones outward and upward from the center of the protocraton, an uplift zone of nonlinearity or weak linearity, and, like rift zones and oceanic ridges, the site of high heat flow; these nucleii and the later Proterozoic linear mobile belts (Paper 23) are the Archean forerunners of the modern rift zones and oceanic ridges. The granites change character upward and outward from the Protocratonic nucleus, too, and in the contact zone with the greenstone tracts the characteristic granite expression is in the form of larger or smaller bodies of essentially invasive, new granite, of quite high level character, and contemporaneous with at least the last stages of metamorphism and folding of the belt. Despite this character, they may be remobilized and reworked material from the protocontinental nucleus, derived from a deeper metamorphic zone, or they may be new, mantle-derived accessions of acid material.

Not all granulite metamorphism is contemporary with the metamorphism of the greenstone belts, for evidence from the North Atlantic Craton and the Aldan Region (Papers 20, 22, and 38) clearly indicates a superimposed granulite metamorphism upon a lower-grade original metamorphism, the superimposition taking place hundreds of millions of years afterward. In this case, a later mobile belt metamorphism is superimposed on an original metamorphism. The small, thin protocratonic crustal nucleii of the Katarchean were much more liable to such reworking than the thicker, later cratonic stabilizations.

The nuclear terrains, characterized by brachysynclines fringing granite–gneiss domes (Papers 22, 29, and 32) are supposed to characterize an early, permobile stage in Earth history (see the discussion of plate tectonics in Part VII). It is true that very ancient Archean terrain does show these features (that of the Katarchean Pilbara of Western Australia has lately been spectacularly displayed in ERTS satellite photography), and such patterns of contrasting granitic domes and fringing troughs moulded by their uprise are in all probability related to the early degassing of the earth. The volatile, mobile alkali- and silica-rich fraction from the mantle migrated to certain point foci, their situation controlled by intersections of a very early, global fracture network comparable with the lunar grid system (a network either cosmically or endogenously produced, in the latter case possibly by initial global shrinkage; see Part VIII).

Such patterns do persist with less emphasis into the late Archean; they are quite noticeable in the Abitibi belt of Canada and very pronounced in late cratonic stabilizations of West Africa and Guiana. They are even recognized in the Paleozoic of the Erzegebirge in Germany (Paper 28). It seems that permobility has been wrongly regarded as something purely Katarchean. It is, of course, the vertical regime of magmatic uprise and new crust generation of the modern rift zones and oceanic ridges, although the products were quite different in this early phase of earth history.

Permobility is an expression of migration of volatile-charged magmas or volatiles alone into fracture-controlled intersections on major lineaments, themselves the sites of upwarping. In the early Archean these effects occurred at a great number of centers spread evenly throughout the global surface; but as cratons enlarged, extensive global lineaments controlled such effects, and the same processes were restricted to localized expressions along such lineaments. The transition from extreme permobility to the localized, linear permobility of the Proterozoic and later systems was gradual, and analogues of granitic permobility, such as characterizes the Katarchean, are seen in quite young, linear tracts in Peru and the Himalayas (Ashgirei, 1972; Pitcher, 1972). The changeover may have been controlled by a number of factors, among which may be cited the growth of large cratonic areas and consequent large-scale crustal inhomogeneities, or a change from an extensive pattern of shrinkage-induced lineaments to a pattern of a few immense, global lineaments (the latter being either cosmically or endogenously produced).

It is of interest to note that Russian authorities suggest (Paper 22) that the early Proterozoic was characterized by abundant development of diapiric domes. The Vredefort and Sudbury structures (the latter, despite its complexity, essentially a dome occupied by granites in the first stage of its development) are of the same early Proterozoic age. It is popular to attribute these unique structures to extraterrestrial agencies, but this is by no means an established fact. In the case of Sudbury, the deformation into an elongated funnel form of an original circular structure required by such a hypothesis is at variance with the present attitude of the Onaping formation, with the lack of deformation of the older Huronian, and with sedimentological evidence from the Whitewater Series. We should not, perhaps, entirely discount the alternative possibility that these unusual structures are features of permobility, extreme expressions of global degassing involving release of volatile-charged magma or gas concentrations at major lineament intersections. It may well be that their unusual high-pressure shock features are due to endogenous chemical explosions of immense magnitude, possibly involving hydrogen.

The third major topic, the nature of the primordial crust, is more suitable treated in summary form in Part VIII.

References

Ashgirei, G. D. (1972). The problem of the origin of the continental earth crust. *Proc. 24th Int. Geol. Congr., 3,* 3–9.

Pitcher, W. S. (1972). The coastal batholith of Peru—some structural aspects. *Proc. 24th Int. Geol. Congr., 2,* 156–163.

PAPERS 18 THROUGH 21: COMMENTARY

The Scottish Highlands is a classic area for geological research, having been the birthplace of many important structural and metamorphic concepts. The oldest part, the *Lewisian,* contains Archean rocks—the *Scourian.* The terrain is not easy to unravel because of overprinting by a later, middle Proterozoic metamorphism associated with a mobile belt—the *Laxfordian.*

Scourian rocks occur as relics, mostly of banded gneiss of migmatite aspect associated with knots and lenticles of darker gneiss and some unquestionable metasediments metamorphosed to amphibolite and granulite facies assemblages. Among the rocks of late Scourian age, only at Loch Maree are there tracts that call to mind the extensive sedimentary belts of Canada, Australia, and South Africa. Mica schists, semipelites, marbles, and quartz-magnetite schists, together with graphitic and calc-magnesian schists, structurally overlie the early Scourian and are separated from the older rocks by unconformity. Thick basic sheets within the Loch Maree Group may be lavas, and the assemblage may represent a greenstone belt of late Scourian age. The interleaving of this sequence with the gneisses, apparently as a result of tectonic plications, is interesting.

The possibility that a Pre-Scourian assemblage older than 2800 million years is represented in the Lewisian is discussed. Metasedimentary relics in the Hebridean islands of Benbecula and South Uist could form part of such an older assemblage.

This paper is a result of conclusions reached by treating the Scottish Archean largely in isolation. The Scottish example is simply an eastern segment of a once-continuous North Atlantic craton. In Papers 19, 20, 21, and 38, the fragmented components of the North Atlantic craton are treated together in an approach that has led to a fuller understanding of the origins of the individual component fragments.

In Paper 19, the authors distinguish *high-level terrains,* characterized by abundant greenstone belts that reflect the deposition of lavas and sediments in narrow, elongate basins sited over linear fractures that penetrated to the mantle, and *deeper-level terrains,* which consist of granulite-facies gneiss, aluminous mica schists, and pyrobolites together with their retrogressed equivalents. The greenstone belts may have been located along the margins of thin elongate plates of a thin, early Archean crust, which, in the greenstone troughs, was depressed so as to cause anatexis and generation of granite magmas; the latter rose preferentially along basement–cover interfaces so that the greenstone belts are bordered by the granite plutons which deformed them into arcuate and irregular patterns. The structural style of the deeper-level terrains is thus one of domes, basins, and eye folds.

Locally, the root zones of greenstone belts are found as strips of greenschists bordered by unconformable strips of migmatite.

One argument postulates the origin of greenstone belts on a thin sialic crust on sunken segments; these could be in troughs between separating sialic blocks of primordial crust or possibly of more ancient protocontinental nucleus origin. Alternatively, they may be viewed simply as sequences formed in downwarps of a primordial oceanic, ultramafic–mafic crust. It seems likely that the truth may lie in a combination of the first two. These authors assume the reality of a primordial sialic crust (cf. Papers 6, 36, and 37), downwarped or downfaulted beneath the thick greenstone accumulations, and subjected to anatexis and remobilization to provide the granite sheath to the greenstone belts, which, significantly, invariably yields younger ages to radiometric analysis. One of the most important observations of these authors is that there is such a sheath, or a migmatite selvage to smaller greenstone wisps in deeper-level granite–gneiss terrains. This sheath corrodes and obscures the basal contact of the greenstone belt, which is rarely detectable.

Geophysical methods may eventually provide evidence concerning the presence or absence of sialic crust beneath large greenstone accumulations. Doubt has been thrown on the concept of an underlying crust following the recognition of immense thicknesses of greenstone in the order of 30 km thick in the greenstone belt south of Kalgoorlie (Gemuts and Theron, 1973). Furthermore, there is no evidence that the volcanic piles of modern island arcs or abyssal tholeiites of the ocean floor are laid down on sialic foundations. However, it is certain that Archean relationships could not have been exactly the same as those of the present day, so one cannot use the direct analogy as a concrete objection to the concept of Windley and Bridgwater. Nevertheless, the opening up of the Atlantic rift between continental blocks does suggest the existence of smaller-scale rifts between more numerous and smaller protocontinental blocks that may have suffered limited opening to form the greenstone troughs of the Archean. The concept of high- and deeper-level terrains presented by Windley and Bridgwater is illustrated by excellent diagrams.

Depth should not perhaps be seen as depth of burial, but rather as metamorphic (geothermal) depth. The granite–gneiss terrain has been subsequently uplifted and eroded to expose rocks from a greater metamorphic depth, but high-grade rocks may well have occurred closer to the surface in the granite–gneiss areas than in the neighboring greenstone belts, the contrast being due to differences in thermal environment. This could explain how one can see a 30-km section of greenstones without any significant grade variation. Thermal conditions may have differed considerably between the positive granite–gneiss areas of uplift and the negative greenstone troughs of subsidence, heat flow having localized in the former, as in the modern rift zones to which these zones and the Proterozoic moble belts such as the Grenville Belt of Canada may be analogous.

The remobilization sheath may have been derived from the older sialic substratum beneath the greenstones, whatever its origin. Equally, it may come from the root of the geosyncline itself (the old, outmoded tectogene concept), or it may have come from lateral migration of the low-melting elements from the mantle beneath the greenstone trough. It may have come from the mantle beneath the uplifted intervening area that lay between the greenstone belts. Thus, there are at least four possibilities.

There is the interesting observation that ultrabasic pods line up and map stratigraphically in the greenstone remnants within the high-grade terrain, and thus could well be of surface volcanic origin. This fact could have relevance to nickel search and to obtaining a fuller understanding of the thin remnants of greenstones or other supracrustals in granite–gneiss terrains.

The authors seem to see the domical granite plutons of the Archean nuclear terrains as separating greenstone belts that are each separate individual depositories; in other words, each granite dome margin is an obscured unconformity. This picture probably requires modifying; there are granites *and* granites in the Katarchean, and some of the granites may be obscure remnants of older, basement granite–gneiss rocks and have hidden unconformities at their margins, whereas others may be new magmatic invasions of dome form intruding the greenstones.

The North Atlantic province of the Archean is a craton that dates, as a fully stabilized cratonic entity, from about 2500 million years ago; it was later fragmented by the opening of the North Atlantic. Remnants in Labrador, east and west Greenland, and northwest Scotland can be reassembled to form a triangular cratonic block measuring 600 km along each side. The craton is enclosed by early Proterozoic mobile belts that include reworked Archean material. Dominated as it is by gneisses of granulite and amphibolite metamorphic facies, in contrast to the Superior Province of Canada (Papers 8 and 33), where such high-grade metamorphics are in juxtaposition with large tracts of greenschist facies assemblages in greenstone belts, this craton had a prolonged active history between 3800 and 2600 million years ago before being fully stabilized into a craton. In the succeeding 1000 million years there was renewed activity along its borders, but the cratonic nucleus itself remained essentially stable, not being similarly affected.

Ancient gneisses, the Amtsoq gneisses (McGregor, 1973), dated at 3787 ± 85 million years by the Rb/Sr method, (Oxford Isotope Geology Laboratory and McGregor, 1971), include basic and ultramafic enclaves; they were metamorphosed shortly after the invasion of a regional basic dike swarm, the *Ameralik Dikes,* and then overlain by supracrustals for which an age of deposition of 3700–3600 million years can be assumed, by virtue of the date of 3600 million years obtained from layered anorthosites that intrude them. Intrusive gneisses invaded the entire sequence slightly later. It is questionable whether correlations of events should be extended throughout the entire craton, but widespread granite invasion at about 3000 million years ago can be justifiably extrapolated in this manner.

A granulite–high amphibolite metamorphism affected a wide area between 2900 and 2800 million years ago, a scatter of dates up to 2500 million years indicating possible slow cooling or late and localized granitic injection. After this event there was only localized structural and metamorphic disturbance. Some low-grade supracrustals form scattered relics that yield ages as low as 2560 million years. Calc-alkaline igneous bodies, together with carbonatite complexes, invaded the existing assemblage of rocks between 2800 and 2000 million years ago, and prolific swarms of basic and ultramafic dikes were emplaced between 2600 and 1900 million years ago.

The *Amtsoq gneisses,* the oldest dated crustal rocks so far identified on the earth (but see also Paper 21), include both basic and ultramafic material, but no proven supracrustals, although inclusions of iron silicate bearing gneiss are reported and are likely to be of supracrustal origin. The suggestion has been advanced that these gneisses may be primordial crust. The Ameralik Dikes do not invade some of the included areas of supracrustals (*Malene supracrustals*), and these are believed to be younger cover.

There is an interlayering between these Malene supracrustals and the interfolded older gneisses and their transecting dikes that may be explained by tectonic interleaving of foundation and cover; or possibly it may be due to slices of oceanic and continental crust without systematic age relationships. Inherent to this second explanation is the suggestion that not all of the basic–ultramafic assemblages not transected by the dikes are necessarily part of the younger cover; nevertheless, if they are older or equivalent-aged material, they must have been mechanically interposed and derived from a quite different geotectonic setting some distance away.

The Malene supracrustals include amphibolites, mica schists, quartzites, and marbles, and have a predominant basic component. They form narrow units averaging only 1 km in width. Pillow lavas are locally preserved, and there are also pods of ultramafic rocks that may similarly preserve primary textures and that appear to belong largely to the supracrustal sequence. Although the metasediments could in most cases be referred to chemogenic processes, the unusually wide Isua Belt (2–3 km) includes a basal quartzite–graywacke sequence beneath garnetiferous schists and banded ironstones overlain by basic volcanic greenschists. This belt is of controversial age because of the low-grade greenschist to amphibolite metamorphism. The K/Ar dating has yielded values over the remarkable range of 3500–1800 million years! This sequence appears to be a shelf rather than a deep-water or oceanic sequence, but there is a resemblance between these supracrustals and those of the Superior, Slave, Yilgarn, Rhodesian, South African, and Pilbara Archean cratons. No high-magnesia basalts (i.e., komatiite equivalents; Papers 6, 16a, and 16b) have been recognized within the North Atlantic craton, and there are no major sedimentary units comparable with the Fig Tree and Moodies groups (Papers 5 and 26), Shamvaian (Paper 7), and Gorge Creek–Mosquito Creek (Paper 14) groups of Africa and Australia. The supracrustal belts are invariably narrow and show strong deformation. No gold quartz association has been recognized, in contrast to the greenstone belts of the Canadian, African, and Australian cratons.

These differences may be due to the fact that the belts of the North Atlantic Craton are products, at least in most cases, of an older period of the Earth's early development in the Archean, a period in which the protocontinents and proto-oceans were just becoming differentiated, but had only a very limited degree of development. We must assume that the crust was thin and simply could not support thick supracrustal sequences such as characterize the late Archean.

The analogy is drawn between the supracrustals of the North Atlantic Craton and narrow belts within the ancient gneissic protocontinental nucleii of Swaziland and Rhodesia. The Johannesburg Dome and Western Yilgarn (Wheat Belt–Darling Range) of Australia might also have been mentioned. The oldest greenstone belts may be as old as 3500–3600 million years in the North Atlantic Craton. The youngest supracrustals are the *Taro gneisses* of East Greenland, dated at 2600 million years, and the Loch Maree supracrustals of Scotland (Paper 18) of similar age. Chromititic anorthosites are widespread and range up to 2 km thick in layered bodies. Such bodies are not characteristic of the low-grade greenstone sequences of the Archean.

The continental crust of the Godthaab area of Greenland is older than any so far established on the globe (although there are reports from Russian sources of similarly ancient or even older terrain from Enderby Land, Antarctica). There is, however, no reason to suppose it is primordial. The authors suggest that, if we hope to find evidence concerning the nature of the crust prior to the formation of this protocontinental nucleus, we must closely examine the enclaves in such very ancient gneisses, which could yet provide clues to the nature of the crust prior to the first continent–ocean differentiation.

There is only about 15 percent of supracrustal rock in the entire North Atlantic Craton. The authors discuss the immense tracts of silicic gneiss in these Archean terrains, and conclude that, despite the paucity of features indicating their origin, they are more likely to be derived from reworked granitic terrain than from a volcanic crust, as Bowes et al. (1971) have suggested, or from transformation of geosynclinal sediments, according to a once popular concept now somewhat out of favor.

Paper 21 is an appropriate follow-up to the Paper 20, presenting as it does geochemical comparisons between segments of the fragmented North Atlantic Craton. It is based on analyses of 700 samples from Scotland and Greenland carried out for 20 elements.

References

Bowes, D. R., Barooah, B. C., and Khoury, S. G. (1971). The original nature of the Archaean rocks of north west Scotland. *Spec. Publ. 3, Geol. Soc. Aust.,* 77–92.

Gemuts, I., and Theron, A. (1973). Stratigraphy and mineralisation of a portion of the Archaean between Norseman and Coolgardie, Western Australia. *Abstract Vol., Sec. 3, ANZAAS, 45th Congr., Perth,* p. 88.

McGregor, V. R. (1973). The early Precambrian gneisses of the Godthaab district, West Greenland. *Phil. Trans. Roy. Soc. Lond., A, 273,* 343–358.

Oxford Isotope Geology Laboratory and McGregor, V. R. (1971). Isotopic dating of very early amphibolite facies gneisses of the Godthaab district, West Greenland. *Earth Planetary Sci. Letters, 12,* 245–259.

18: SCOURIAN–LAXFORDIAN RELATIONSHIPS IN THE LEWISIAN OF NORTHWEST SCOTLAND

J. Sutton and Janet Watson

Department of Geology, Imperial College, London, England

ABSTRACT

The two principal metamorphic complexes which make up the Lewisian of northwest Scotland are the Scourian complex (2600 - 2200 m.y.) and the Laxfordian complex (1650-1450 m.y.). Phenomena related to the two cycles were originally distinguished on the basis of their relations with a swarm of post-Scourian, pre-Laxfordian basic dykes and by evidence of the superposition of successive metamorphic assemblages and tectonic patterns. Confirmation by isotopic dating followed. Both Scourian and Laxfordian complexes are predominantly gneissose and include rocks of granulite and amphibolite facies. Most of the Laxfordian complex consists of regenerated Scourian material. Supracrustal rocks which show no indications of Scourian metamorphism and which may be of post-Scourian age are of restricted occurrence and their relationships with the regenerated Scourian are obscure. Belts of older supracrustal rocks have been identified and there are some indications that the Scourian complex itself incorporates both a previously crystalline basement and a metamorphosed cover-series.

Remarkable features of the Laxfordian regeneration are, firstly, the survival of small relict masses, sometimes only a few kilometres across, which retain Scourian structural patterns and textures, and, secondly, the rapid variations of Laxfordian tectonic styles. The paper discusses these features, as well as the limitations of the criteria employed in the separation of the two complexes and some possible geochemical differences between the rock assemblages characteristic of Scourian and Laxfordian complexes.

INTRODUCTION

The crystalline basement which underlies late Precambrian and Phanerozoic supracrustal sequences along the northwest seaboard of Scotland has long been known by the general name of Lewisian Gneiss. It was realized when the region was mapped by the Geological Survey of Great Britain, that this crystalline complex incorporated materials of many ages and had been subjected to profound structural and metamorphic modifications. The memoir by Teall, Clough, *et. al.*, provides one of the earliest well-documented accounts of the effects of reworking in such a complex (Peach, *et al.*, 1907, see

TABLE I: PRINCIPAL EVENTS IN THE EVOLUTION OF THE LEWISIAN OF NORTHWEST SCOTLAND

Sedimentation and basic igneous activity		Orogenic activity	Age m.y.
		LAXFORDIAN CYCLE:	
	products constitute the Laxfordian complex	(Late Laxfordian: deformation, (metamorphism, migmatization, (pegmatite-injection))))	1650 - 1450
		(Early Laxfordian: deformation, (metamorphism))	
?Supracrustal formations (Loch Maree Series) Basic and ultrabasic dykes (including "Scourie dykes")			2190
		SCOURIAN CYCLE:	
	products constitute the Scourian complex	(Inverian: deformation, (metamorphism (	2200
		(Pegmatite-injection (	2300
		(Early Scourian, deformation, (metamorphism, migmatization	2600
Pre-Scourian sediments, basic and ultrabasic igneous rocks			
		PRE-SCOURIAN CYCLE:	
		? formation of early gneiss complex.	

Chapters 3, 5, and 9-12). This evidence of reworking has formed the basis for a chronological division in terms of successive orogenic cycles which was proposed by Sutton and Watson (1951) and elaborated by Dearnley (1962) and Evans (1965). The time-spans of the principal metamorphic events have been established in outline by radiometric dating (Giletti, Moorbath, and Lambert, 1961; Evans, 1965).

The sequences of events so determined are summarised in Table I. It is seen that two polyphase tectonic and metamorphic events are recognized, the earlier Scourian events belonging to the Superior chelogenic cycle and the later Laxfordian events to the Svecofennide chelogenic cycle (Sutton, 1963). A third, pre-Scourian cycle may eventually prove distinguishable.

The crystalline complexes of the Lewisian exhibit in a small, well-exposed and accessible terrain, evidence of the repeated tectonic and metamorphic regeneration of basement material. The scale on which these phenomena are displayed is relatively small and much of the area has been studied in considerable detail. Our principal purpose is to examine certain aspects of the Laxfordian regeneration of Scourian complexes to facilitate comparisons with similar phenomena developed in the great shield areas of other continents.

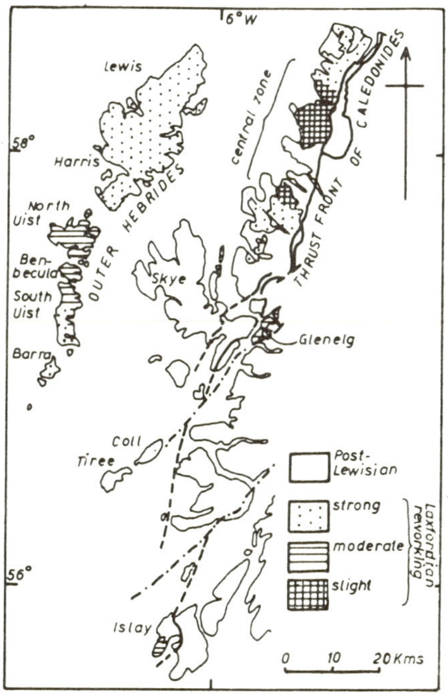

Figure 1: Sketch-map showing variations in the intensity of Laxfordian regeneration.

CRITERIA USED IN THE SEPARATION OF SCOURIAN AND LAXFORDIAN PHENOMENA

The imprint of Laxfordian metamorphism and/or deformation is seen over most of the Lewisian, and for the purposes of a regional classification almost the whole outcrop would probably have to be assigned to a Laxfordian province. Nevertheless, unmodified or partially-modified Scourian rocks and structural patterns are preserved in several massifs which have been mapped using a variety of field criteria (Figure 1). The largest of these massifs, occupying the central part of the Lewisian outcrop of the Scottish mainland, is some 65 kms in width. In many places within it Scourian metamorphic assemblages, fabrics, and structures are almost unaffected by later modifications. Other relict masses are considerably smaller and usually show some effects of Laxfordian deformation and recrystallisation. At the borders of the Scourian massifs, effects of Laxfordian metamorphism and deformation usually increase rapidly across a narrow transition zone, but areas of little-modified rocks survive locally in these border zones and even in the thoroughly reworked complexes beyond. As a whole, however, the reworked complexes exhibit Laxfordian structural patterns, textures, and mineral assemblages.

In general terms, the little-modified Scourian complexes and the Laxfordian complexes derived from them by regeneration have much in common. Both consist predominantly of banded gneisses, often of migmatitic aspect. The darker bands of both are basic, ultrabasic, or intermediate in composition and the lighter bands, which often enclose knots and lenticles of darker material, are predominantly quartzo-feldspathic. Both Scourian and Laxfordian complexes contain occasional narrow belts of undoubted metasediments. The grade of metamorphism is invariably high and both complexes include rocks of granulite and amphibolite facies.

In the southern part of the mainland outcrop, the Laxfordian complex includes several areas of altered supracrustal rocks which form distinct belts. These belts, consisting of a variety of metasediments interbanded with thick sheets of basic igneous rock, are predominantly schistose rather than gneissose and are only locally migmatized. They constitute the Loch Maree Series which, as will be seen later, may represent remnants of a late- or post-Scourian supracrustal formation.

In view of the general similarities between the regenerated Laxfordian and its Scourian parent rocks, it may be useful to examine the means which have been employed to separate the effects of Scourian and Laxfordian processes. The main criteria used are listed below.

(i) A swarm of basic and ultrabasic dykes intrudes the Scourian gneisses and cuts across Scourian structures. Members of this swarm of dykes (which may be called the Scourie dykes, from an important type) were deformed, metamorphosed, and migmatized during Laxfordian episodes and hence their degree of modfication has been used as a guide to the intensity of Laxfordian reworking. The Scourian and Laxfordian complexes were originally defined by reference to their relationships with the dyke swarm (Sutton and Watson, 1951, p. 295). More recent work has revealed complications in these relationships which will be discussed later.

(ii) Successive sets of structures and successive textures and mineral assemblages provide widespread evidence of repeated deformation and metamorphism in the Lewisian. Since both the Scourian and the Laxfordian events were polyphase (Table I), the superposition of structures or metamorphisms does not in itself demonstrate Laxfordian reworking; but where younger sets affect time markers such as the Scourie dykes, their place in the time-sequence may be established. The superposition of Laxfordian structures is especially clearly demonstrated at the margins of certain Scourian massifs (e.g., Sutton and Watson, 1951, 1962).

(iii) Although the Scourian and Laxfordian gneisses are broadly similar, there are certain characteristics which over some small areas have provided means of distinction, though their validity on a regional scale is more questionable. As a series of very rough generalizations, it appears that the rocks of the Scourian complex are predominantly greyish, brownish, or white and have rather uniform coarse granoblastic textures. They are usually poor in potash feldspar and lack abundant veins of pegmatite or granite. The regenerated Laxfordian gneisses are often streaked with pink or white and show conspicuous variations in grain-size and texture. They are often rich in potash-feldspar and, in migmatitic complexes, are penetrated by abundant veins of pegmatite or granite gneiss. The tectonic pattern of the Laxfordian complexes is often characterised by conspicuous small-scale folding and by the dominance of a NW-SE axial direction.

(iv) The separation of Scourian and Laxfordian events by reference to the field criteria mentioned above has been confirmed in some instances by radiometric analysis (Giletti, Moorbath, Lambert, 1961; Evans and Tarney, 1964; Evans, 1965). The available age-determinations are represented graphically in Figure 2, and their evidence can be summarised as follows: (a) pegmatites and granites associated with Laxfordian migmatites fall in the range 1650-1410 m.y.; (b) gneisses assigned in the field to Laxfordian complexes and to members of the Loch Maree Series have maximum apparent ages of about 1650 m.y. and range down to 1120 m.y.; (c) members of the Scourie dyke swarm give maximum dates of about 2200 m.y. and show a scatter down to about 1120 m.y.; (d) rocks assigned in the field to unmodified or partially-

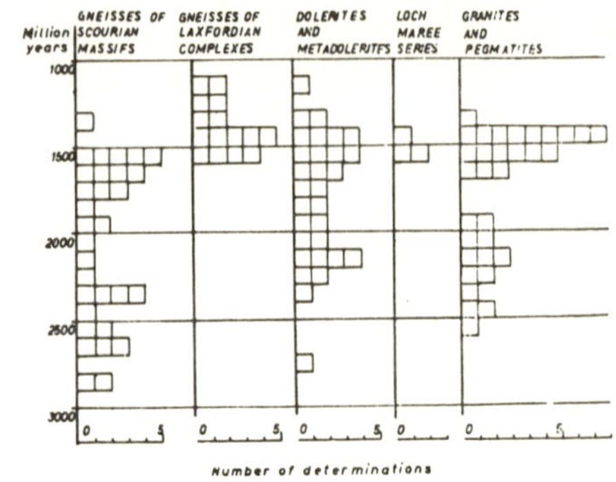

Figure 2: Histograms showing the variations in apparent radiometric age of the main components of the Scourian and Laxfordian complexes.

modified Scourian complexes show a wide scatter ranging from about 2800 m.y. to 1150 m.y. These results clearly reflect the overprinting of Laxfordian events and of some minor late thermal events on both the Scourian gneisses and the post-Scourian dykes. The period of intrusion of certain members of the Scourie dyke swarm is put at about 2190 m.y. by Evans and Tarney (1965). Several Scourian episodes have been distinguished radiometrically by Evans (1965) but the successive episodes of Laxfordian metamorphic history have not so far been separated by radiometric dating.

(v) Although the Loch Maree Series may represent a post-Scourian cover-succession, the relationships of this series with the reworked gneisses are obscure and have contributed relatively little to the elucidation of tectonic and metamorphic histories.

VARIATIONS IN LAXFORDIAN REGENERATION

Variations in tectonic reworking

A remarkable feature of the Lewisian of northwest Scotland is the wide distribution of small massifs in which relicts of Scourian structural and textural patterns are preserved, and in which the overprinting of Laxfordian structures is relatively weak. Such relicts are encountered through a tract at least 100 to probably 250 kms broad, when measured across the Laxfordian tectonic trend (Figure 1). The effects of Laxfordian tectonic and metamorphic regeneration extend well beyond the limits of the relict massifs and there is no indication that the Lewisian area as a whole lies near a northeastern or southwestern Laxfordian orogenic front. It appears, therefore, that even in the interior of the Laxfordian belt, tectonic reworking was patchy.

These variations in intensity of reworking appear to bear some relationship both to large-scale Laxfordian structures and to the form of the Laxfordian migmatite front. On the mainland, for example, the large Scourian massif which forms the central zone is flanked to the northeast by a wide expanse of Laxfordian migmatites which are almost free from Scourian relicts. The migmatite front lies only 2 kms northeast of the limit of thorough tectonic reconstitution. On the southwestern flank of the Scourian massif, as Bailey has pointed out (1951), the situation is different. Here there is a broad transition zone in which migmatites are only sporadically developed and in which several small Scourian relicts are preserved. The front of regional Laxfordian migmatites lies in the vicinity of Loch Torridon, some 30 kms southwest of the limit of tectonic reconstitution. In the Outer Hebrides also, there are considerable areas which have suffered tectonic and metamorphic reconstitution but not migmatisation, and in these areas traces of modified Scourian structures can often be identified.

In the southern part of the mainland outcrop, there are four principal regions in which pre-Laxfordian structural patterns are preserved (Figure 3). In each of these regions, the compositional banding of the gneiss is clearly transected by the Scourie dykes and at numerous localities—for example, at the western end of Loch Maree—quartzo-feldspathic streaks and veins showing migmatitic relationships with the more basic portions of the gneiss are cut by dykes. The dykes themselves retain an obviously intrusive habit and show relatively little evidence of deformation. Their primary sub-ophitic texture is often still recognisable and primary pyroxenes and feldspars are locally preserved. In the broader zones which separate these massifs, evidence of repeated Laxfordian deformation and metamorphism is almost universal. Schistosities and linear structures, with associated sets of minor folds, produce a complex tectonic pattern which involves the gneisses, the Scourie dykes and the Loch Maree Series. The dykes appear as amphibolites or hornblende-schists usually lacking in relicts of igneous minerals or textures. Their primary relationships are extensively modified, their originally intrusive habit is seldom apparent and they are generally concordant with the compositional banding of the adjacent gneisses. The obliteration of the original discordance between dykes and gneisses is a common feature of Laxfordian tectonic reconstitution which was described as long ago as 1907 (Clough, *in* Peach, *et al.*, 1907, pp. 205, 246-247).

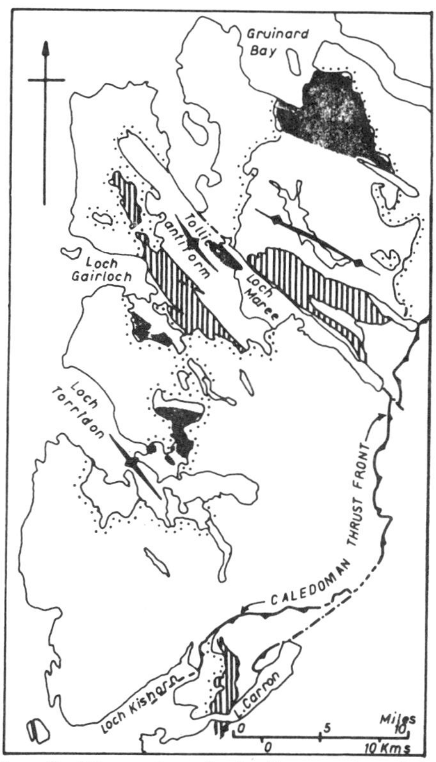

Figure 3: The southern Laxfordian complex of the mainland, showing the principal outcrops attributed to the Loch Maree Series (vertical lines) and the principal massifs in which Scourian structures are preserved (stippled): the bold lines mark the late Laxfordian antiforms referred to in the text.

These variations in the intensity of Laxfordian distortion of earlier structures suggest that there was a differentiation of the crystalline complex into zones of active reworking and massifs which remained more or less passive. Since the zones of reworking invariably exhibit several successive sets of Laxfordian structures (Peach, *et al.*, 1907, chapter 12; Sutton and Watson, 1951, p. 259; Park, 1964), it appears that the contrasts in behaviour were initiated in an early Laxfordian event and maintained through a considerable period.

In the southern part of the mainland, a rather consistent arrangement of structures suggests that the development of active and passive zones was linked with the development of major fold-structures (Figures 3 and 4). The large Scourian massif which forms the central zone of the mainland is flanked at its southwest margin by the broad Carnmore antiform over which members of the Scourie dykes swarm are deflected to produce an open arch. This arch is in turn flanked to the southwest by a complementary synform which carries in its core metasediments and basic rocks of the Loch Maree Series. The narrow relict massif at the west end of Loch Maree is similarly followed southwestward by an antiform involving Scourie dykes (the Tollie antiform) which is in turn followed by a belt of the Loch Maree Series interpreted as a complex synform (Park, 1964). All these structures were identified by the officers of the Geological

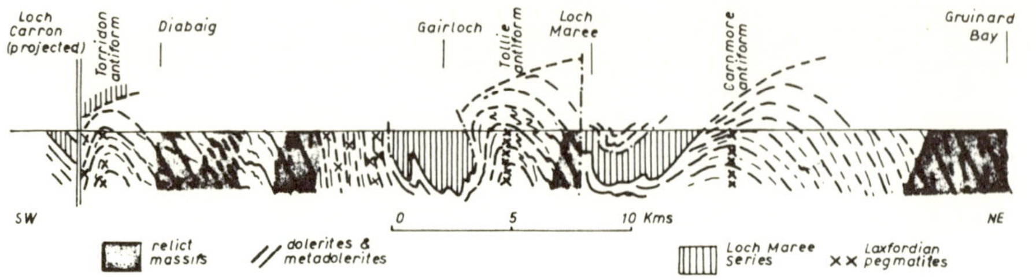

Figure 4: Diagrammatic cross-section of the area shown in Figure 3.

Survey (Peach, *et al.*, 1907, Chapters 12-13). Between Loch Gairloch and Loch Torridon, a number of small regions in which relict Scourian structures survive occur in a zone which as a whole shows moderate evidence of Laxfordian deformation. To the southwest of this zone, Loch Torridon is straddled by an antiform involving Scourie dykes (Sutton and Watson, 1951). There are no autochthonous Lewisian outcrops in the region immediately to the south of this antiform. However, a thrust-wedge of Lewisian in the marginal zone of the Caledonides between Loch Kishorn and Loch Carron includes a belt of basic rocks and mica schists which have been compared with the Loch Maree Series. When the thrust-mass is restored to its probable place of origin, some 15 kms to the ESE (Kennedy, 1955) it is seen that this belt falls roughly in place to the southwest of the projected continuation of the Loch Torridon antiform (Figure 3). There thus appears to be a juxtaposition, three times repeated, of the following units: (i) massif showing relatively light Laxfordian deformation (northeast), (ii) antiform of Scourie dykes, (iii) synform occupied in part by rocks of the Loch Maree Series (southwest).

Because all observers are agreed that the broad antiforms and synforms referred to above are late structures superimposed on earlier Laxfordian folds, thrusts, and fabrics affecting both the Loch Maree Series and the Scourie dykes, the downfolds of the Loch Maree Series cannot be regarded as simple synclines of a cover-series overlying a reworked basement. Nevertheless, the structural relationships summarised in the very schematic section of Figure 4 suggest that the passive blocks which were differentiated at an early stage in the period of Laxfordian deformation behaved, at least towards the end of the period, as anticlinal wedges separating synformal zones in which tectonic and metamorphic reconstitution was more advanced. The sense of displacement along movement-zones which appear to predate the Loch Torridon antiform suggests upward movement at earlier stages (Figure 5). This interpretation would be consistent with the pattern of Laxfordian structures at the northeastern margin of the central zone: here, the fronts of reworking and of migmatization dip beneath the Scourian massif and the arrangement of shear-belts and minor structures suggests that the massif tended to ride up over the reworked migmatites to the northeast (Sutton and Watson, 1962).

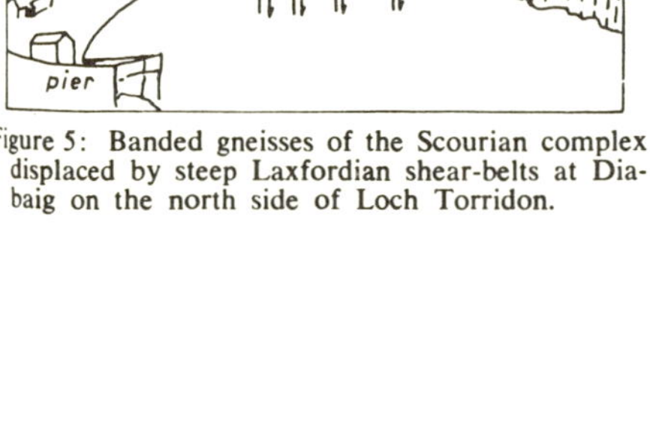

Figure 5: Banded gneisses of the Scourian complex displaced by steep Laxfordian shear-belts at Diabaig on the north side of Loch Torridon.

Variations in Laxfordian metamorphism and migmatization

Throughout most of the area affected by Laxfordian metamorphism, the characteristic mineral assemblages are now of amphibolite facies. In the Outer Hebrides there are indications that these assemblages were preceded by Early Laxfordian assemblages of granulite facies (Dearnley, 1962, pp. 151-153) but neither in the islands nor on the mainland is there any regional development of the greenschist or epidote-amphibolite facies. On the other hand, there is very considerable variation in the completeness of the Laxfordian recrystallisation. In the central zone of the mainland, a variety of charnockitic gneisses and many of the post-Scourian dolerites appear to have remained virtually unaltered. In the smaller relict massifs, pre-Laxfordian minerals are often partially replaced by new minerals, though the pre-existing textures may be mimetically preserved. The lowest degrees of modification are seen as a rule in the cores of basic or ultrabasic Scourian bodies, in the interior of large metadolerites and in certain metasediments all of which appear to lag behind the more reactive acid gneisses (Sutton and Watson, 1951; Dearnley, 1963). Wherever relicts of pre-existing minerals are preserved, it can be seen that they have been replaced directly by minerals of amphibolite or granulite facies without the preliminary development of lower grade minerals.

The uniformly high grades of Laxfordian metamorphism suggest that reconstitution took place in a crustal environment where temperatures and pressures were everywhere fairly high, and that variations in the amount of recrystallization were not related solely to PT gradients. It follows that standard systems of metamorphic zoning related to such gradients are difficult to apply and, in practice, zoning of Laxfordian complexes has been based largely on textural and structural criteria (Sutton and Watson, 1951; Dearnley, 1963).

Several factors other than temperature-pressure variations may have influenced the extent of Laxfordian metamorphism. There is often, both locally and regionally, a very close connection between the amount of recrystallisation and the amount of penetrative deformation. Where movement was concentrated in definite zones, the more passive blocks between them often retain pre-Laxfordian minerals and textures while the rocks of the movement-zones are entirely recrystallised. Such a contrast is illustrated on a large scale by the alternation of small Scourian massifs and broader zones of Laxfordian activity in the southern region already referred to (Figure 3). On a smaller scale, it may be seen within the large Scourian massif of the mainland which is traversed by numerous Laxfordian movement-belts only a few metres in width. At many localities coarse granoblastic Scourian charnockites pass within a metre or so into finer-grained but not cataclastic Laxfordianised hornblende- or biotite-granulites at the margins of these belts. The effect of deformation may be due in part to its role in facilitating the migration of water. The Scourian host-rocks, being largely of granulite facies, are poor in water and this fact may have helped to delay the onset of recrystallisation.

The relationship between metamorphism and migmatization in the Laxfordian is more complex. As has been noted, the fronts of migmatization and metamorphic recrystallization are close to each other along the northeastern margin of the main Scourian massif of the mainland. Zones defined by changes in structural and metamorphic style fringe the migmatites and it seems clear that metamorphism and migmatization were closely connected. In other areas, however, migmatization, with its accompanying metamorphic changes, appears to have been preceded by one or more phases of metamorphism which had little spatial connection with migmatization. In the southern part of the mainland (Figure 3), widespread tectonic and metamorphic reconstitution took place before the formation of the broad antiforms and synforms already discussed. Migmatization and pegmatite-injection, on the other hand, are restricted in extent. Pegmatites are emplaced in the axial-plane region of the Carnmore and Toullie antiforms (Figure 4; Peach, *et al.*, 1907, p. 212); injection and permeation near Gairloch and Loch Torridon are assigned to late phases of Laxfordian activity (Sutton and Watson, 1965; Park, 1964).

In the Outer Hebrides, the effects of Early Laxfordian recrystallization are seen over most of the archipelago and late Laxfordian recrystallization is only slightly less extensive (Dearnley, 1962). Late Laxfordian migmatization, on the other hand, is confined mainly to Harris and South Uist. The migmatite front cuts sharply across the tectonic pattern established during the initial Laxfordian reworking for many tens of kilometres in western Harris (Dearnley, 1962, Pl. 9). These varying geometrical and chronological relationships suggest that the process of metamorphic reconstitution is not always directly linked, at the present levels of erosion, with migmatization.

The megascopic signs of migmatization in Laxfordian complexes are provided by the development of coarse stringers, eyes and layers of quartzofeldspathic material often rich in potash-feldspars. Usually, though not always, these developments within the host-rocks are associated with the appearance of apparently intrusive veins and sheets of granite-gneiss and pegmatite. Although such injected material may form up to about half the volume of the migmatite complex, it always takes the form of small bodies intimately penetrating the gneisses; there are no larger homogeneous granites.

In Laxfordian pegmatites and granite-gneisses, the content of total alkalies may reach 12 per cent and K_2O alone ranges from about 4 to 10 per cent. There are few published analyses of the country-rock gneisses of Laxfordian migmatites, but these are known to contain significant amounts of potash feldspar. Analyses of Scourian acid gneiss and of rocks which have suffered Laxfordian metamorphic reconstitution without megascopic indications of migmatization give values of Na_2O+K_2O of 4-8 per cent, and of K_2O of 0.5 - 3.0 per cent. These figures suggest enrichment in alkalies, especially potassium, during Laxfordian migmatization. No economic mineralization is associated with either Scourian or Laxfordian events in Britain, though Laxfordian pegmatites in some localities carry accessory minerals containing U, Th, Nb, Ce, Zn, and Pb, and bring about scapolitisation in the country rocks.

THE RÔLE OF BASIC IGNEOUS ROCKS

Two generations of basic and ultrabasic rocks have been widely recognised in the Lewisian and their contrasting habit, structures, and metamorphic history have contributed much to the understanding of the processes of reworking. Not unnaturally some complexities and ambiguities have been encountered but these do not seem to affect the main points at issue. The two principal basic assemblages are:-

(i) Basic and ultrabasic rocks occurring as integral parts of the Scourian complex. These rocks, which include most of the "early basic" rocks of Geological Survey authors, are partially or completely metamorphosed and not infrequently act as host-rocks to Scourian migmatites. They occur as pods, lenses, and concordant layers enclosed in and often veined by more acid Scourian gneisses. Radiometric dates range up to about 2800 m.y. (Evans, 1965).

(ii) Basic and ultrabasic rocks intrusive into Scourian gneiss complexes. These rocks, which include the "Scourie dykes" already referred to, are sometimes unaltered, sometimes thoroughly metamorphosed. The majority were intruded as dykes with WNW or NW trends but where their instrusive habit has been modified by Laxfordian deformation they appear as almost concordant bands and lenses often veined by Laxfordian pegmatites and granites. Radiometric dates range up to about 2200 m.y.; a few much older dates are regarded as anomalous (Evans and Tarney, 1964).

In the Scourian massifs, basic rocks of two generations are as a rule immediately distinguishable, since the younger dyke-set intrudes metamorphosed Scourian basic rocks and cuts acid veins and agmatitic structures produced in these rocks by Scourian migmatization. The same fundamental relationship can often be recognized also in regions of Laxfordian tectonic and metamorphic reworking, for although the primary characters of the younger dykes have been extensively modified, it is often possible to see that they transect older basic migmatites. Both generations of basic rocks may act as hosts during Laxfordian migmatization, and the problem of separation becomes increasingly difficult in Laxfordian migmatites.

The separation of older and younger generations of basic and ultrabasic rocks is thus dependent mainly on the evidence of field-relationships. Chemical data are insufficient as yet to provide reliable criteria for separation. The pre-Scourian suite includes peridotitic varieties closely comparable with many igneous cumulates (O'Hara, 1961a) and more abundant basic types some of which are comparable with olivine-gabbros (Bowes, *et al.*, 1964). The Scourie dykes are characteristically tholeiitic (O'Hara, 1961; Dearnley, 1962; Burns, 1964) but hypersthene-dolerites, olivine-dolerites, and picrites are associated with them. Anorthosites are prominent in the Hebrides where some are regarded as post-Scourian (Dearnley, 1962).

The Scourie dykes and associated intrusions which were emplaced after the main phases of Scourian metamorphism and deformation but before the main phases of Laxfordian reactivation have been used as time markers separating the effects of Scourian and Laxfordian processes. Their metamorphic history, however, has been complex. Dykes cutting little-altered Scourian gneisses in the central massif of the mainland sometimes show an incomplete mineralogical reconstitution involving the development of hornblende with or without garnet. O'Hara (1961) has attributed this reconstitution to the effects of emplacement in country rocks which were at moderately high temperatures. Tarney (1963) has shown that older members of the swarm tend to be more thoroughly altered than later members and Evans and Tarney (1964) have obtained radiometric dates for certain members of the swarm which are very close to those assigned to the later Inverian phase of Scourian metamorphism (Table I).

In the Outer Hebrides, many dykes assigned to the post-Scourian swarm carry metamorphic hypersthene - diopside-plagioclase (-hornblende) assemblages which are earlier than the widespread Late Laxfordian amphibolite-facies metamorphism. Dearnley (1962; 1963) attributes these assemblages to an Early Laxfordian phase of granulite-facies metamorphism and suggests that the hornblende-garnet assemblages in dykes of the central zone of the mainland record incipient metamorphism of the same phase. Two possibilities have thus been envisaged—firstly, that the emplacement of the dyke swarm began during the waning stages of the Scourian event, when high regional temperatures promoted recrystallization; and, secondly, that recrystallization took place under different conditions during two successive phases of Laxfordian metamorphism. It is to be noted that gneissose country-rocks of the Outer Hebrides seldom retain granulite-facies assemblages equivalent to those in the Scourie dykes, a fact attributed by Dearnley to the well-established tendency of basic rocks to lag behind more acid types during later Laxfordian recrystallization.

These various observations and suggestions are concerned with phenomena which are becoming familiar in many terrains where basic dykes are emplaced in a crystalline basement which had suffered earlier phases of granulite facies metamorphism. The occurrence of granulite-facies assemblages in basic dykes, some of which are possibly equivalent to the Scourie dykes, has been widely recorded in southern Greenland (where, again similar assemblages are seldom seen in the country rocks), East Africa, and elsewhere. The fact that these features seem to be so commonly produced suggests the operation of a general cause rather than of some peculiarity in the local history of the Lewisian. O'Hara has concluded (1961; 1961a) that the features of the Scourie dykes in the central zone of the mainland resulted from emplacement at considerable depths while regional temperatures were fairly high. While this suggestion would be compatible with the fact that the level of erosion appears to reveal a deep section through the Laxfordian belt, it may be at variance with the fact that some Scourian gneisses have yielded isotopic ages of 2400-2800 m.y., indicating a reduction of temperatures sufficient to set the radiometric clock well before the period of dyke-intrusion. Moreover, the megascopic characters of the dyke-swarm appear to be very much like those of younger swarms emplaced at shallow depths. The consistent orientation, the wide extent of the swarm, which is some 300 kms. in breadth, and the occurrence of a large number of relatively thin intrusions, all indicate emplacement in more or less rigid rocks during a period of regional crustal extension. On the other hand, the anhydrous nature of the crystalline assemblage as a whole may have influenced the initial course of Laxfordian recrystallization in the dykes by promoting the development of pyroxenes.

The simple primary forms and consistent orientations of the Scourie dykes give them an added value as structural markers recording the effects of Laxfordian deformation. Variations in their structural patterns suggest variations in the Laxfordian stress-systems which might be difficult to establish by reference to the reworked Scourian gneisses alone. In the southern part of the mainland and over much of the Outer Hebrides, the dykes are extensively folded in a manner which indicates that at the start of deformation the plane of the intrusions contained a Laxfordian direction of shortening. In parts of the northern Laxfordian complex of the mainland, on the other hand, the dykes are rarely folded, but are disrupted and rotated. The direction of shortening here seems to have been almost perpendicular to the dykes (Sutton and Watson, 1962).

Variations in the style of Laxfordian deformation are reflected in rapid lateral variations in the field-relationships of Scourie dykes which are especially well illustrated in the southern part of the mainland (Figures 3 and 4). Within the relict Scourian massifs where Laxfordian deformation is slight, the dykes retain many primary characters and can be seen to cut banding and migmatitic structures in the gneisses they intrude. At the margins of the massifs, distortion and rotation tend to bring dykes and gneisses into parallelism and strong Laxfordian schistosities, lineations and folds are overprinted on both. Some authors have concluded that the little-deformed dykes of the relict massifs are younger than those in the adjacent reworked areas and on this basis, Bowes has questioned the validity of the post-Scourian dykes as time-markers (Bowes and Ghaly, 1964).

Although the little-altered dykes mentioned by Bowes and his colleagues appear to be pre-Laxfordian and retain their primary characters simply as a result of their tectonic situation, the occurrence of late-Laxfordian or post-Laxfordian intrusions might be expected in view of the abundance of Sanerutian and Gardar minor intrusions in south Greenland. Corresponding intrusions may exist in the Outer Hebrides; an unmodified appinitic body for example, has been encountered to the east of Carloway in Lewis which is possibly of late Laxfordian age.

THE RÔLE OF METASEDIMENTARY ROCKS

The Loch Maree Series of the southern part of the mainland outcrop has already been mentioned. Metasedimentary rocks of this series include mica-schists, rocks of semi-pelitic composition, marbles, quartz - magnetite schists, graphite-schists, and a variety of calcareous and magnesian schists. They are associated with thick basic sheets, probably representing lavas or minor intrusions. The Loch Maree Series occupies a number of belts—at least two representing the cores of synforms—which are flanked by gneisses tectonically and metamorphically reworked during the Laxfordian events (Figures 3 and 4).

Within the outcrops of the Loch Maree Series there are no recorded discordant basic intrusions comparable with the Scourie dykes; migmatites are only locally developed, the bulk of the rocks being schistose or granulitic rather than gneissose; and no evidence that Laxfordian metamorphism had been preceded by earlier high-grade metamorphism or migmatization has been found. The contacts of the belts are sufficiently clearly defined to be easily mapped, though narrow bands of metasediment appear in gneisses and *vice versa* near the contact-zones. No discordance betwen the layering of the gneisses and that of the Loch Maree Series has been observed. Zones of mylonite, ultramylonite, or of granulated rocks are commonly developed near the contact-zones.

The general contrasts between the Loch Maree Series and the Scourian gneisses or their reworked derivatives, have led many authors to conclude that the metasediments were deposited after one or more early gneiss-forming episodes had already taken place. The absence of recognizable equivalents of the Scourie dykes led Gunn (*in*, Peach, *et al*, 1907) to the view that they were deposited after emplacement of the dyke swarm—a view which would be consistent with the idea that they represented part of a cover-succession laid down between the Scourian and Laxfordian tectonic events. The length of the Scourian event revealed by radiometric dating shows, however, that there would also have been time for the deposition of a cover-series between the early Scourian and Inverian episodes (Table I).

The present outcrops of the Loch Maree Series appear to occupy high structural positions above the main body of regenerated Scourian gneisses (Figure 4). These outcrops do not appear to represent primary downfolds of a cover-series since the synforms which they occupy are thought to refold earlier isoclinal folds and thrusts (Peach and Horne, 1907;

Bowes, *et al.*, 1964). Slices of retrograded gneiss appear to be interleaved with the metasedimentary rocks; indeed, one such slice overlies metasedimentary rocks in the core of the Loch Maree synform. An analogy may perhaps be drawn with the relationships of basement and cover in the Caledonides of Scotland. In the Glenelg region (Ramsay, 1958, 1963) Lewisian basement interleaved with strips of Moinian cover as a result of early isoclinal folding and thrusting was subsequently refolded by broader and more open antiforms and synforms. It could be suggested that the southern part of the mainland outcrop of Lewisian passed through a similar tectonic evolution in Laxfordian times. The level of present-day erosion in this part may lie near that of the distorted basement/cover unconformity; elsewhere it appears to lie well below this unconformity.

Assemblages of metasedimentary rocks which appear to have suffered high-grade metamorphism, migmatization and deformation prior to the Laxfordian cycle have been encountered at a few localities in the mainland and much more widely in the Outer Hebrides. In South Harris, such assemblages include pelites, semi-pelites, quartzites, marbles, and a variety of calc-gneisses which show evidence of at least one phase of granulite-facies metamorphism older than the Late Laxfordian period of migmatization (Davidson, 1943; Dearnley, 1963). The range of original lithologies recorded here and at other localities is not unlike that of the Loch Maree Series and the possibility of separating the two groups on lithological criteria therefore seems remote. The metasedimentary rocks of South Harris are intruded by anorthosites, gabbros, tonalites, and peridotites which are assigned by Dearnley (1962; 1963) to the same general period of intrusion as the Scourie dykes but which have been compared by Davidson (1943) with basic and ultrabasic rocks incorporated in the Scourian complex of the mainland. These metasedimentary rocks in Harris appear to pass into gneisses which themselves acted as host-rocks during Laxfordian migmatization, a relationship which seems to be emerging more generally from recent work of Myers, Coward and others in the Outer Hebrides. In the southern part of the mainland outcrop, migmatized metasedimentary rocks including kyanite gneisses have been recorded from localities where Laxfordian migmatization is virtually absent and where Scourie dykes show only a moderate degree of alteration (Clough, in Peach *et al.*, 1907, pp. 237-238; Sutton and Watson, 1951, pp. 245-248). From these various lines of evidence it seems clear that relicts of a pre-Scourian or early Scourian supracrustal series are preserved in addition to the younger supracrustal assemblages of the Loch Maree Series.

Although the distribution of these very early metasedimentary relicts is still incompletely known, there seems to be some evidence that they may be concentrated in certain parts of the Scourian complex, while the intervening gneiss tracts may be almost lacking in such relicts. If this impression is correct, it may eventually be possible to differentiate between a pre-Scourian cover-series and a still older crystalline basement. Such a basement could well exceed 3000 m.y. in age, since the early Scourian metamorphism is dated at over 2600 m.y. (Evans, 1965). Dearnley and Dunning (1968) have concluded, from the relationships of certain deformed pre-Laxfordian pegmatites, that pre-Scourian gneisses are incorporated in the complexes of Benbecula and South Uist. If this possibility is to be further explored, the most hopeful terrain would seem to be the central zone of the mainland where Laxfordian reworking is at a minimum. Pre-Scourian metasedimentary relicts seem to be almost absent from this massif in which the oldest detectable metamorphism is of charnockitic facies. The anhydrous assemblages, the rather low K_2O content and the scarcity of pegmatites suggest a depletion in volatile components which might be compatible with the reworking of a pre-Scourian crystalline complex. Preliminary investigation (Evans, 1965) shows that the Rb content and $^{87}Sr/^{86}Sr$ ratio of representative samples are abnormally low for crustal rocks. Evans concluded that if normal crustal values of Rb have ever been present, Rb must have been removed by profound metamorphism soon after the original differentiation of pre-Scourian sialic material from the mantle. These geochemical peculiarities indicate a line of approach which might lead to the recognition of rocks dating back to some 3000 m.y. along the western border of the European continent.

REFERENCES

Bailey, E. B. (1951)—Scourie dykes and Laxfordian metamorphism; Geol. Mag., vol. 88, pp. 153-165.

Bowes, D. R. and Ghaly, T. S. (1964)—Age relations of Lewisian basic rocks, south of Gairloch, Ross-shire; Geol. Mag., vol. 101, pp. 150-160.

........, Wright, A. E., and Park, R. G. (1964)—Layered intrusive rocks in the Lewisian of the North-west Highlands of Scotland; Quart. J. Geol. Soc. London, vol. 120, pp. 153-184.

Burns, D. J. (1966)—Chemical and mineralogical changes associated with the Laxfordian metamorphism of dolerite dykes in the Scourie-Loch Laxford area, Sutherland, Scotland; Geol. Mag., vol. 103, pp. 19-43.

Davidson, C. F. (1943)—The Archaean rocks of the Rodil district, South Harris, Outer Hebrides; Trans. Roy. Soc. Edin., vol. 61, pp. 71-112.

Dearnley, R. (1962)—An outline of the Lewisian complex of the Outer Hebrides in relation to that of the Scottish mainland; Quart. J. Geol. Soc. London, vol. 118, pp. 143-176.

........(1963)—The Lewisian complex of South Harris; Quart. J. Geol. Soc. London, vol. 119, pp. 243-307.

.... and Dunning, F. W. (1968)—Metamorphosed and deformed pegmatites and basic dykes in the

Lewisian complex of the Outer Hebrides and their geological significance; Quart. J. Geol. Soc. London, vol. 123, pp. 353-378.

Evans, C. R. (1965)—Geochronology of the Lewisian basement near Lochinver, Sutherland; Nature, vol. 207, pp. 54-65.

.... and Tarney, J. (1964)—Isotopic ages of Assynt dykes; Nature, vol. 204, pp. 638-641.

Giletti, B. J., Moorbath, S. and Lambert, R. St. J. (1960)—A geochronological study of the metamorphic complexes of the Scottish Highlands; Quart. J. Geol. Soc. London, vol. 117, pp. 233-264.

Kennedy, W. Q. (1955)—The tectonics of the Morar anticline and the problem of the northwest Caledonian front; Quart. J. Geol. Soc. London, vol. 110, pp. 357-390.

Knorring, R. Von and Dearnley, R. (1960)—The Lewisian pegmatites of South Harris, Outer Hebrides; Min. Mag., vol. 32, pp. 366-378.

O'Hara, M. J. (1961)—Petrology of the Scourie dyke, Sutherland; Min. Mag., vol. 32, pp. 848-865.

........(1961a)—Zoned ultrabasic and basic masses in the early Lewisian complex at Scourie, Sutherland; J. Petrol., vol. 2, pp. 248-276.

Peach, B. N. Horne, J. *et al.* (1907)—The geological structure of the North-west Highlands of Scotland; Mem. Geol. Surv. Scotland.

Ramsay, J. G. (1958)—Moine-Lewisian relations at Glenelg, Inverness-shire; Quart. J. Geol. Soc. London, vol. 113, pp. 487-520.

Sutton, J. (1963)—Long-term cycles in the evolution of the continents; Nature, vol. 198, pp. 731-735.

.... and Watson, J. (1951)—The pre-Torridonian metamorphic history of the Loch Torridon and Scourie areas in the North-West Highlands and its bearing on the chronological classification of the Lewisian; Quart. J. Geol. Soc. London, vol. 106, pp. 241-307.

........(1962)—Further observations on the margin of the Laxfordian complex of the Lewisian near Loch Laxford, Sutherland; Trans. Roy. Soc. Edin., vol. 65, pp. 89-106.

Tarney, J. (1963)—Assynt dykes and their metamorphism; Nature, vol. 199, pp. 672-674.

19: THE EVOLUTION OF ARCHAEAN LOW- AND HIGH-GRADE TERRAINS

Brian F. Windley and David Bridgwater

ABSTRACT

Two main types of Archaean terrain are distinguished which have different rock units, metamorphic grade and structural style.

High-level terrains, characterised by abundant greenstone-schist belts, formed by the deposition of lavas and sediments in narrow elongate basins sited over linear fractures which penetrated to the mantle. The lower parts of the successions consist dominantly of basic volcanics and ultramafic rocks with anorthosites. Ore deposits and associated volcanics resemble those in areas such as Fiji and Cyprus situated on modern mid-oceanic ridges or rift zones. The greenstone belts may likewise be located along the boundaries of small elongate plates characterising a thin early Precambrian crust. The greenstone belts depressed the underlying sialic crust causing anatexis and the generation of granitic magmas which rose preferentially along the basement-cover interfaces so that the greenstone belts are bordered by granitic plutons which deformed them into arcuate irregular shapes.

Deeper-level terrains consist of granulite-facies gneisses, aluminous mica schists and pyribolites or their retrogressed equivalents. The structural style of these deep-seated rocks is characterised by domes, basins or eye folds. Locally the root zones of greenstone belts are found as thin strips of greenschists bordered by conformable belts of migmatites. Basic stratiform complexes typified by chromite-layered calcic meta-anorthositic, basic and ultrabasic layers were intruded as regionally concordant bodies into parts of the early crust and are now found as tectonic lenses and layers in the catazonal gneisses. First reports of calcic chromiferous anorthosites from the lunar highlands allow the suggestion that the early anorthositic evolution of the terrestrial crust may have affinities to that suggested for the surface of the moon.

The higher-level greenstone belt terrains became closed to the loss of radiogenic daughter elements earlier than the catazonal basement rocks. Thus geologically older rocks may give younger radiometric ages.

INTRODUCTION

Present ideas on the character of the early crust have been largely developed as a result of the intensive study of Archaean terrains such as the Superior Province of Canada, Western Australia and Rhodesia-South Africa (Fig. 1), which contain major greenstone belts associated with high-level granite plutons. These areas originally attracted considerable attention because of the important economic mineral deposits in the greenstones; they are now being studied extensively because of the insight that the almost unmetamorphosed supracrustal successions forming the belts can give about the depositional environment in Archaean times.

In contrast, comparatively little attention has been paid to the Archaean terrains which contain less well-preserved greenstone belts and which consist essentially of migmatitic gneisses, the majority of which have passed through high-grade (mostly granulite-facies) metamorphism. The North Atlantic Archaean craton, remnants of which occur from eastern Labrador through Greenland to North West Scotland (Fig. 1) is used in this paper as a model for this second type of Archaean terrain. Published accounts from the Archaean areas of the Baltic, Ukrainian and Aldan Shields (Salop, 1968; Salop & Scheinmann, 1969) and from the high-grade terrains of Madagascar suggest that the features noted from the North Atlantic craton are common to many early Precambrian high-grade terrains.

The object of this paper is to compare the characters of the two types of early Precam-

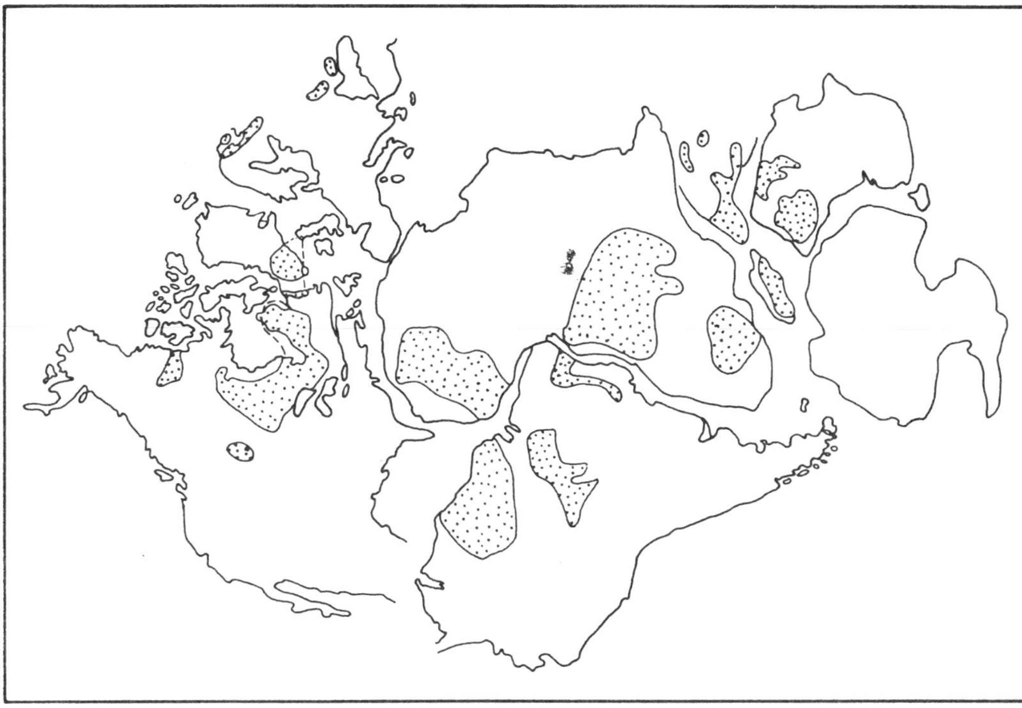

Fig. 1. Archaean cratons (> 2500 m.y.) on a pre-continental drift map of Laurasia and Gondwanaland.

brian geological environments and, by emphasizing some of the features found in the higher-grade terrains, to try and obtain a more balanced picture of the Archaean than is so far available in the literature.

TERRAINS DOMINATED BY LOW-GRADE METAMORPHIC ROCKS

Two major rock units are recognised in the low-grade high-level Archaean cratons: (1) greenstone belts (2) granites.

(1) GREENSTONE BELTS

(a) *General form*

The characteristic features of greenstone belts have been summarised by Anhaeusser *et al.* (1969). Greenstone belts are elongate, well-defined basins infilled with volcanic and sedimentary deposits and separated from each other by what are generally described as 'granites'. The total thickness of supracrustal material within the belts commonly exceeds 10 km (see Table I) and may reach 15 km or more (for example, the Onverwacht Group of the Barberton Mountain Land [Viljoen & Viljoen, 1970]). Individual rock units thin rapidly towards the present margins of the belts; this, combined with an abundance of coarse clastic sediments especially near the margins, suggests that the outcrops now seen give a reasonable impression of the original size of the depositional basins (Table I).

TABLE I

Estimated sizes of some Canadian greenstone belts

Greenstone Belt	*Thickness*	*Length*	*Breadth*	*Reference*
Michipicoten	13 km	90+ km	20 – 30 km	Goodwin & Shklanka (1967)
Yellowknife	13	200+	100	Baragar (1966)
Kirkland Lake	11 }	350 }	150 }	Moorhouse (1965)
Noranda	7 }			Stockwell (1957)
Birch-Uchi	10	40	50 – 65	Bateman (1940)
Red Lake	8	60	25	Gross & Ferguson (1965)
				Grant, Gross & Chinnery (1965)

(b) *Rock sequences*

The supracrustal successions which comprise the greenstone belts show a remarkable similarity, both within single cratons and from craton to craton. It is possible that neighbouring belts were originally connected by thinner successions formed on the intervening platforms (Wood, 1966) and that in a few cases rock units of a particular type from one belt may correspond to similar rocks in a neighbouring belt. However, as a general case, these similarities represent a general evolutionary pattern in the depositional environment in individual belts rather than chronostratigraphic units.

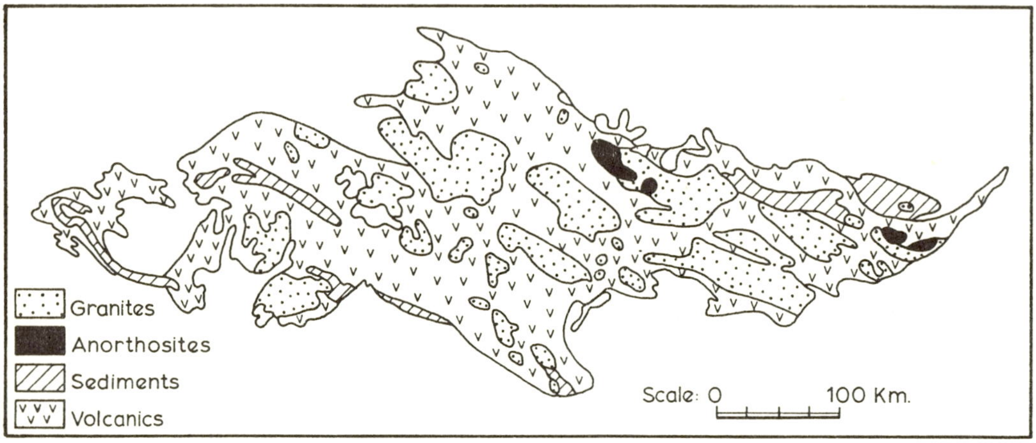

Fig. 2. The Abitibi greenstone belt of the Superior province (after the Tectonic Map of Canada, 1969).

The lower part of the supracrustal piles is dominated by ultramafic and primitive basic igneous rocks with dunites, peridotites, pyroxenites, gabbros and anorthosites. Most of the basic and many of the ultramafic rocks formed as extrusives. There are minor intercalated sedimentary layers including cherts and banded ironstones (Anhaeusser *et al.*, 1969). Layered intrusions with some anorthositic units, described as comparable to the Bushveld complex, may be present in this part of the succession (Fig. 2) (see Chibougamau Lake—Mawdsley & Norman, 1935, Allard, 1959; Bell Lake—Freeman, 1939; Barberton—Viljoen & Viljoen, 1970). The lower units are followed by greenstones with tholeiitic basalts, andesites, dacites and rhyolites, often as pyroclastics, tuffs, breccias and agglomerates. The igneous rocks generally show calc-alkaline trends (Baragar, 1966; Anhaeusser *et al.*, 1969; Condie & Potts, 1969). The volcanic rocks are interbedded with thin limestones, greywackes, quartzites and some banded ironstones and rare carbonaceous shales.

The upper part of the successions is flysch-like and is dominated by locally derived clastic sediment, conglomerates, breccias, grits and some chemical sediments such as cherts. Sedimentary structures frequently suggest a shallow water depositional environment.

(c) *Structural features*

The greenstone belts are thought to be sited over fundamental linear fractures which controlled the rise of basic magmas from the mantle (Anhaeusser *et al.*, 1969) and gave rise to linear fissure eruptions (Goodwin, 1968). Some individual greenstone belts in Western Australia and Canada (Fig. 3) are

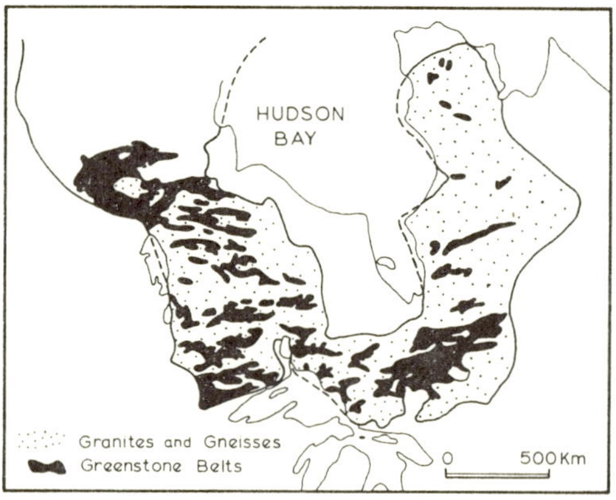

Fig. 3. The greenstone belts of the Canadian Shield (after Goodwin, 1968).

at least 650 km long and the original fractures may have extended for at least 1600 km if it is assumed that smaller belts falling along the same line were controlled by the same crustal feature. The axial regions of parallel belts may be 80-250 km apart.

Internal structures, such as upright folds showing no vergence and vertically stretched pillows (Wood, 1966) and conglomerate pebbles, have been used by Martin (1969) and Anhaeusser *et al.* (1969) to support the idea put forward by Macgregor (1951) that the basins sank under their own weight of dense basic and ultrabasic rocks into a less dense granitic substratum. If the estimate of 12-15 km suggested for the thickness of the early Precambrian crust by Anhaeusser *et al.* (1969) is correct, then these basins must have depressed the general level of the mantle/crust boundary at the time of their maximum infilling.

(d) *Mineralisation*

The greenstone belts are characterised by considerable concentrations of base metals (especially copper-zinc-nickel) and gold and silver, which are now generally regarded as genetically related to the volcanic and ultrabasic rocks.

(e) *Comparison with younger supracrustal sequences*

The rapid deposition of considerable thicknesses of volcanic and sedimentary rocks in subsiding basins invites a comparison with the similar rocks found in more recent geosynclinal sequences. However the general form, size and distribution of the greenstone belts is very different to those of the much larger sedimentary basins of the more modern geosynclines. The chemical properties of the volcanics from lower parts of the greenstone belts and the types of mineralisation found in them resemble those in Fiji (Ibbotson, 1962; Houtz & Phillips, 1963), Cyprus (Hutchinson, 1965) and perhaps the West Greenland Tertiary province (Clarke, 1970) where primitive basic and ultrabasic magmas and associated ore concentrates have been derived from the mantle along deep fracture zones which are often sited at the margins of major crustal plates. In our opinion the general form and distribution of the greenstone belts, the properties of their volcanic and sedimentary sequences, and the type of mineralisation and structures present within the belts, all point to the presence of a thin crust in the Archaean which was more liable to fundamental fracture and less able to bear the weight of an extensive geosynclinal sequence than was the crust in later periods in the Earth's evolution. If the fractures, along which the greenstone belts were situated, are regarded as having extended to the mantle and as marking the boundaries of Archaean crustal plates, then it appears that these were smaller and more elongate than the plates which characterise the present crust. In common with several other authors we would regard the greenstone belts as representing sedimentation and volcanic activity in an environment which differed markedly from that found in modern geosynclines (Brock, 1959; Donaldson & Jackson, 1965; Goodwin, 1968; Salop & Scheinmann, 1969). We would regard the thickness of the crust and the size of the crustal plates as two of the most important factors controlling the type of sediments and volcanic rocks formed. Some of the larger greenstone belts formed after 3000 m.y. ago (such as the Abitibi belt of Canada) which extend for up to 250 km across could be regarded as transitional in type between the smaller depositional basins of the pre-3000 m.y. old crust, such as Barberton, and the much larger Proterozoic and Phanerozoic geosynclines.

(2) GRANITES

At their present relatively high exposure levels the greenstone belts are not usually surrounded by their original basement material. The cover-basement contacts have been obscured by the intrusion of late transgressive homogeneous granitic plutons. Similar plutons intruded the centres of the basins. They are mostly granodiorites to potassic granites. The plutons within the belts are surrounded by narrow thermal aureoles in which the greenschists are upgraded to amphibolites facies. There is a similar increase in metamorphic grade towards the bordering granitic rocks. The plutons have deflected the otherwise linear belts into arcuate shapes and thus the structural style of these high-level cratons is characterised by an irregular curvilinear strike of the supracrustal belts. Due to the vertical rise of these granitic nodes into and between the supracrustals, the latter are in places preserved as triangular synclines between granitic domes (Macgregor, 1951) or as mantles around such domes similar to those in Finland (Eskola, 1949).

The granitic rocks surrounding the belts frequently pass laterally into migmatitic gneisses. These have received comparatively little attention in regional accounts of the

Archaean, partly no doubt due to the lack of immediate economic potential, and partly perhaps to the difficulty of dealing with gneissose terrains in poorly exposed areas. Some writers, for example Goodwin (1968) and Viljoen & Viljoen (1970), have described these gneisses as essentially younger than the greenstone belts they surround, while others, for example Hunter (1957) and Martin (1969), regard them at least in part as older basement.

TERRAINS DOMINATED BY HIGH-GRADE METAMORPHIC ROCKS

The main rock units, metamorphic characters and structural style of the high-grade metamorphic terrains differ markedly from those of the greenstone belts and their immediate surroundings.

The main rock units may be described under five headings:

(1) Rare belts of low- to medium-grade supracrustals which have never been at a high metamorphic grade. These are mostly meta-volcanics (greenschists) with well-preserved but deformed original structures.

(2) Belts of supracrustals (mostly amphibolites with ultrabasics, semi-pelites and pelites with conspicuous sillimanite, cordierite, mica assemblages) which have been affected by the same high-grade metamorphism as the surrounding gneisses. Primary structures are rarely preserved.

(3) Quartzo-feldspathic granitic gneisses, migmatites and agmatites. These are the predominant rock types.

(4) High-grade layered basic complexes with prominent anorthositic units and isolated anorthosite and gabbroic anorthosite masses in the gneisses.

(5) Late intrusive granites and related rocks.

(1) *Low to medium-grade supracrustal belts*

The high-grade gneisses contain occasional conformable layers of low- to medium-grade supracrustal rocks which rarely exceed 2-3 km in width and which consist largely of basic volcanic rocks with subsidiary ultrabasic pods and metasedimentary units; see for example the Tartoq Group (Fig. 4) (Higgins & Bondesen, 1966; Higgins, 1968) and the Ravns Storø supracrustal belt (Windley *et al.*, 1966). The centres of these units commonly preserve original structures such as pillows and volcanic bombs, and show greenschist facies metamorphic assemblages with chlorite, epidote, plagioclase, quartz, calcite and hornblende/actinolite. The margins commonly show a higher grade metamorphism (amphibolite facies) and are much more heavily deformed. The layers frequently become tectonically thinned along their strike and the central zone of less deformed, low-grade metamorphic rocks is lost. The ultrabasic rocks within these schist belts are commonly tectonic pods consisting of dunites, harzburgites or their metamorphic equivalents (serpentinites, chlorite, tremolite, anthophyllite rocks). Sediments include conglomerates with stretched pebbles, quartzitic schists and calcareous schists and other comparable belts contain banded ironstones.

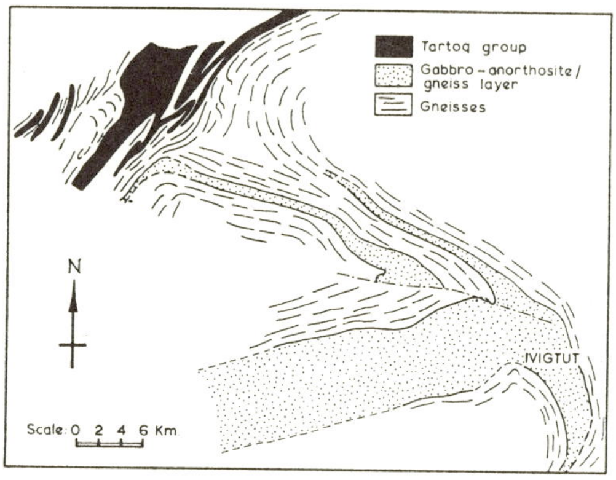

Fig. 4. The root of a greenstone belt (the Tartoq Group) in high-grade basement gneisses, Ivigtut region, SW Greenland (modified after Windley *et al.*, 1966).

Typically the rocks display few minor folds, but show a consistent, penetrative, steep to vertical mineral lineation and a coincident lineation defined by stretched pillows, conglomerate pebbles and agglomerate clasts. They appear to have undergone an extensive vertical stretching similar to that described from the major greenstone belts. Although unconformities are rare (Nautiyal, 1965; Stowe, 1968), there may be a discontinuity between the regional strike of the basement rocks and that of the greenschist layers (Fig. 4) (Henriksen, 1969).

Between the supracrustal sequences and the adjacent gneisses there is generally a transitional border zone which is the site of intense gneissic veining and migmatisation. The greenschists are thus separated from the basement gneisses by conformable belts, up to about one km in thickness, of migmatitic gneisses and pegmatites within which the basement gneisses have been flattened, giving rise to refoliated gneisses in a series of shear belts, and remobilised by partial melting. Migmatising material has encroached upon the borders of the supracrustals with the result that there may be a marginal zone of foliated granitic material containing abundant inclusions of greenschists (Higgins, 1968). In our opinion these belts of relatively low-grade supracrustals surrounded by higher grade gneisses may represent the tectonically thinned root zones of greenstone belts of the type described earlier, while the surrounding gneisses are the remains of an older basement. The main differences between the thin layers of conformable greenstones in the high-grade terrains and the major greenstone belts of the Canadian, South African and Australian Shields are in size, intensity of deformation, and the fact that instead of high-level intrusive granites along their contacts, these smaller greenstone belts are bordered and disrupted by migmatites. These differences are all explicable in terms of the erosion levels now exposed.

(2) *High-grade supracrustal belts*

Belts of recognisable supracrustal rocks which have passed through the same general high-grade metamorphism as the surrounding gneisses are a common minor constituent of nearly all high-grade Archaean terrains. The individual sequences preserved rarely exceed 1-2 km in width (unless repeated by folding or strike faulting) and many occurrences are limited to slices of supracrustal rock a few metres wide intercalated with granitic gneisses. The original thickness of these successions is not known since the top and bottom are nearly always lost and there commonly has been considerable tectonic thinning. However the successions show a remarkable similarity from area to area both within one craton and from widely separated areas and while there is no proof that the present outcrops represent the true thickness of the original supracrustal piles, the general impression given is one of rather thin successions characterised by rapid alteration of lithological units.

Four main lithological types are recognised: —basic amphibolite and pyroxene amphibolite units of basaltic composition; ultrabasic pods and sheets; aluminous metasedimentary units with conspicuous cordierite, sillimanite, garnet, biotite; graphite schists and gneisses with minor impure marbles and quartzites containing garnet and other Fe/Mg-bearing minerals; and variable thicknesses of hypersthene biotite gneisses of granodioritic or tonalitic composition. Sequences of this general type have been noted from Labrador (Bridgwater, 1970), West Greenland (Berthelsen & Noe-Nygaard, 1965), East Greenland (Bridgwater & Gormsen, 1968, 1969) and North West Scotland (Coward *et al.*, 1969) and are recorded from many areas in the Archaean of the Baltic, Ukrainian and Aldan Shields (Salop & Scheinmann, 1969). Descriptions by Wilson (1958) suggest that sequences of this type are also present in the Archaean granulites of Western Australia. In some areas particular units such as the iron/magnesium-bearing quartzites (Salop & Scheinmann, 1969) or the marbles may reach a considerable thickness, but in general the sequences are characterised by a rapid alternation of thin lithological units. Original depositional features are rarely preserved.

The close association of aluminous material with basic amphibolite and the occasional preservation of original volcanic features suggests that the original supracrustal sequences consisted largely of pelitic sediments and basic lavas. The origin of the ultrabasic and the concordant granitic sheets within the successions is more controversial. The ultrabasic rocks generally occur as tectonic pods or minor intrusions. Where they are found as sheets they may be discordant and presumably represent intrusives. However they commonly occupy characteristic positions within the local successions and may be associated with distinctive units in the surrounding metasediments which suggests that many were extrusive. The concordant granitic layers can be variously interpreted as strongly recrystallised acid volcanic or greywacke units, slices of an older basement, or younger concordant sheets of granite intruded into the successions. Thin layers of sillimanite or calc-silicate schists, which commonly persist for considerable distances within these granitic units, suggest that at least some are metasedimentary or metavolcanic in origin. Conformable layers of foliated amphibolite up to 2-3 km wide are found within the gneisses, apparently unassociated with metasedimentary rocks. Rare deformed pillows and bomb-like structures may be preserved locally and a supracrustal origin for many of these layers seems likely. The contacts of the layers with the adjacent gneisses may be sharp and surprisingly unaffected by migmatisation.

The mineral assemblages found in the pelitic rocks generally point to a high-temperature, medium- to low-pressure environment with a predominance of cordierite and sillimanite and a general scarcity of kyanite in these rocks unless they have been involved in younger orogenesis.

The amphibolitic units in the granulite facies sequences consist of orthopyroxene, clinopyroxene, garnet, brown hornblende, plagioclase and quartz (the pyribolites of Berthelsen, 1960; Windley, 1969*a*). These are commonly downgraded to clinopyroxene, green hornblende, plagioclase, quartz, biotite assemblages. The presence of mica-schist and amphibolite layers in gneisses is usually taken to indicate that the bulk of the Archaean gneissic terrains was formed by the metamorphism, homogenisation and gneissification of original supracrustal sequences. Using this interpretation the pelitic and amphibolitic units are regarded as preferentially preserved units from originally much more extensive supracrustal piles in which the primary lithological variation has been entirely homogenised and obliterated. However, in well-exposed ground this interpretation can be shown in places to be invalid. Detailed mapping by McGregor (1969) has demonstrated that a 'typical' area of West Greenland Archaean basement, consisting of alternating conformable lithostratographic units of high-grade supracrustal rocks and grey gneisses, can be separated into three chronological divisions: an older group of granitic gneisses cut by strongly deformed basic (amphibolite) dykes, a group of supracrustal rocks, and a younger group of granitic gneisses which intruded both the older gneisses and migmatised the supracrustal rocks. Similarly Bridgwater & Gormsen (1969) noted that the agmatitic gneisses which form the major part of the Archaean of South Greenland differ markedly from the gneisses which are found close to the well-defined high-grade layers of recognisable supracrustal origin, and interpret this change as representing an original basement-cover relationship. While these relationships are only clear in well-exposed ground we suggest that they may be more common than has been generally recognised. At least it would seem unnecessary to assume that Archaean terrains composed of gneisses with supracrustal relics were formed as the result of a single cycle of sedimentation, metamorphism and introduction of granites.

If it is accepted that the present distribution and lithological character of the high-grade schist belts reflects at least partially their original size and composition, then a clear resemblance can be seen between them and the major greenstone belts described earlier, although the high-grade rocks have been more affected by metamorphism, migmatisation and tectonic intercalation with the surrounding gneisses. In our opinion this resemblance is not fortuitous. We suggest in common with such workers as McGregor (1969) and Hunter (1957) that the high-grade Archaean terrains are typically structurally complex, containing metasupracrustal units, older basement gneisses and younger granitic units. In any one broad area of the crust we would also suggest that the high-grade schist belts generally represent older sequences than those preserved in the major greenschist belts from the same area. While it seems possible that different parts of the crust stabilised at different times and no strict chronological correlations are feasible, it would appear reasonable to suggest that the high-grade Archaean terrains are composed for the larger part of material older than that composing the major greenstone belts. Furthermore on present evidence we can see no evidence to suggest that the basement units within the high grade terrains should represent a primeval crust formed before the first sediments. We would prefer to suggest that they are also structurally complex and to regard the isolated fragments of metasedimentary origin within them as representing even older cover sequences now almost completely disrupted by subsequent mobility and tectonism.

(3) *Quartzo-feldspathic gneisses, migmatites and agmatites*

Quartzo-feldspathic gneisses with a general granitic to granodioritic composition are the major rock unit in the high-grade Archaean terrains. They vary considerably in lithology and internal structure, ranging from essentially homogeneous granites and tonalites to enormous areas of migmatitic gneisses and agmatites. Almost all the rocks in this group show an impressed foliation which has commonly destroyed much of the original character of the rocks and which thereby causes considerable difficulty in the interpretation of their origin. Three types of quartzo-feldspathic gneiss are of particular interest: homogeneous granodioritic gneisses, banded migmatites and agmatites.

The homogeneous quartzo-feldspathic gneisses commonly form the main country rock surrounding the rock units described in this chapter. They are rather monotonous rocks with very little variation in composition or structure. Where they are preserved as granulite-facies rocks these gneisses typically have a brown, reddish or greenish weathered surface in the field which frequently makes them appear more basic than they really are. In granulite-facies areas the gneisses consist essentially of quartz, plagioclase, hypersthene and a little biotite and generally lack potash feldspar. The mafic minerals are generally subordinate to the quartz and feldspar. In amphibolite-facies terrains the gneisses are white to grey and contain biotite, hornblende, quartz, plagioclase and commonly potash feldspar.

Most of the Archaean rocks in Greenland and eastern Labrador appear to be polymetamorphic. The general impression on a regional scale is one of retrogression from early granulite to late amphibolite facies. This retrogression is frequently accompanied by a change in structural style. Locally however there is evidence to suggest that the granulites themselves are polymetamorphic and in some cases appear to have passed through an even earlier amphibolite facies. It appears that many areas passed back and forth between amphibolite and granulite facies several times and that the dominant mineral facies now preserved in the rocks depends on local factors rather than any fundamental differences between high- and medium-grade areas. Basic and ultrabasic rocks frequently lag behind the surrounding quartzo-feldspathic gneisses during changes in metamorphic facies. As a general rule they more commonly preserve high-grade mineral assemblages than the adjacent gneisses. The change from high-grade to medium- or low-grade is frequently accompanied by a marked change in colour. Hypersthene-bearing gneisses generally contain brown feldspar and appear more mafic in hand specimen than they really are. The stable amphibolite facies gneisses are generally grey or white. In several areas partly retrogressed quartzo-feldspathic rocks show a peculiar purple cast (Berthelsen, 1960), relict inclusions of gneiss with brown feldspar and hypersthene giving some indication of the origin of these rocks.

The origin of homogeneous granodioritic or enderbitic gneisses is uncertain. In the Godthåb region, McGregor (1969) has suggested that their homogeneity is an original feature and that these rocks are essentially igneous in character, representing early granites. In Labrador (see Bridgwater, 1970) the writers have interpreted the concordant layers of granitic gneiss intercalated with recognisable metasediments as probably acid volcanic in origin. Similar explanations have been put forward for the origin of the homogeneous granodioritic gneisses in the Baltic, Ukrainian and Aldan Shields (Salop, 1968).

Banded gneisses with alternating units of different mineral composition are common. Many of these can be seen to be the product of the metamorphism of original supracrustal rocks. However V. R. McGregor (*pers. comm.*) has shown that the early gneisses with a well-defined layering from the Godthåb area in West Greenland were developed by the deformation and metamorphic differentiation of an originally homogeneous granodioritic rock and thus the presence of layering in the gneisses cannot be taken as evidence for original lithological variations in the rocks.

Some Archaean gneiss terrains consist almost exclusively of agmatites containing innumerable fragments of basic and ultrabasic (generally hornblendic) rocks in a leucocratic granodioritic matrix. The total extent of these agmatites is enormous on the southeast coast of Greenland where they have been mapped for some 300-400 km and they are thought to form the major constituent of the reworked Archaean rocks for a comparable distance to the north (Bridgwater & Gormsen, 1968, 1969). The fragments vary considerably in composition; identifiable rocks include some derived from high-grade supracrustal sequences similar to those described above, inclusions derived from layered anorthositic and ultrabasic igneous complexes, and the broken-up remains of migmatites and even earlier agmatites. However a high proportion of the fragments cannot be identified as being derived from either the supracrustal sequences or discrete basic igneous bodies (such as those described in the next section) and the ultimate origin of these rocks must remain a matter for some speculation. Among the most interesting fragments are small, well-rounded pods ranging from a few centimetres to tens of metres across, of magnesium-rich olivine-orthopyroxene rocks (harzburgites). These commonly occur in small groups within otherwise moderately inclusion-free granodiorite. They may be zoned with rims of anthophyllite, talc and actinolite sometimes enclosed in phlogopite; however as a rule they show remarkably little reaction with the surrounding gneisses with which they are chemically in complete disequilibrium. Other fragments of unknown origin include large amounts of mafic hornblende-biotite-rich material, which could be the remains of some early basic igneous rock broken up and partly digested by the host rock. There is, however, also the possibility that they represent a basic segregation from the granodioritic host rocks which form the matrix to the agmatite complex.

(4) *High-grade layered basic complexes with prominent anorthosites*

Calcic anorthositic rocks form a characteris-

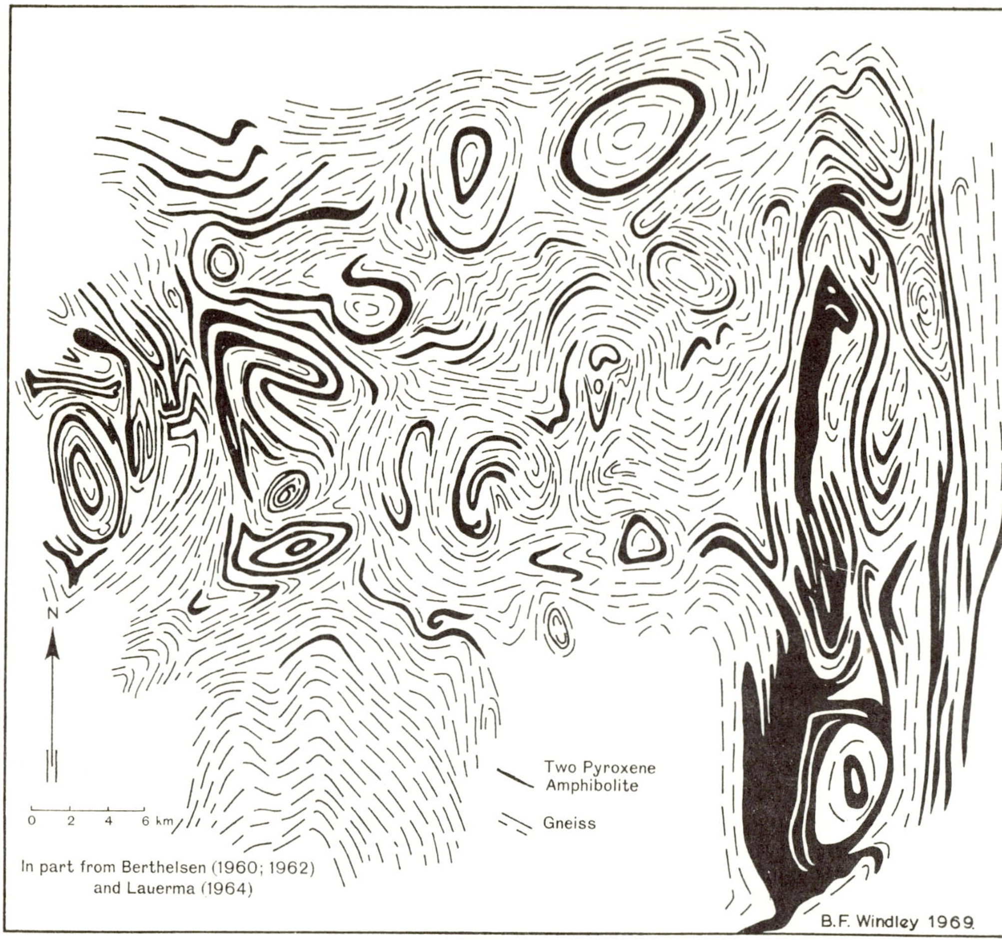

Fig. 5. The catazonal structural style of the high-grade basement terrain of the Fiskefjord region, W. Greenland.

Whatever the ultimate origin of these agmatitic rocks two general points seem to be of importance; firstly, their enormous lateral extent suggesting that there were large areas of mobile rock forming under catazonal conditions during the Archaean; secondly, their composite and varied nature suggesting that this mobility was present for a considerable time. In our opinion agmatites of this type are the kind of rock which could be expected to form in an early thin unstable crust onto which supracrustal rocks were deposited in small basins to be broken up and recycled as small changes in local heat flow and stress occurred.

tic component of several deep-seated Archaean regions. Notable examples occur in the North Atlantic craton, particularly in West Greenland where they occur throughout a 600 km coastal strip (Windley, 1969*b*). They also occur in East Greenland (Bridgwater & Gormsen, 1968, 1969), possibly the Ness area of the Outer Hebrides, North West Scotland (Watson, 1969), Sittampundi, Madras, India (Subramaniam, 1956), Kondapalli, India (Leelanandam, 1967), Sakeny, Madagascar (Boulanger, 1959) and Messina-Soutpansberg, Rhodesia-South Africa (Sohnge, 1946; van Zyl, 1950).

The rocks occur as conformable layers,

usually not more than two km thick and a few tens of kilometres long, and as smaller lenses and pods in high-grade, commonly granulite facies, gneisses or their retrogressed equivalents. They have been recrystallised under high-grade metamorphic conditions and folded and thinned during deformation.

There are pure anorthosites (> 90% plagioclase), gabbroic anorthosites (80-90% plagioclase) and anorthositic gabbros (70-80% plagioclase) which may be intercalated with layers of amphibolite, dunite, peridotite and pyroxenite; all these are internally well layered on a mesoscopic scale. The anorthositic rocks are sometimes chromite-layered, individual chromitites being usually up to a few metres thick; in the Fiskenaesset complex of West Greenland, the chromitites attain a thickness of 20 m (Ghisler & Windley, 1967). A characteristic of most of these early anorthosites is their calcic plagioclase which typically ranges from about An_{80} to An_{100}. Where the rocks have been retrogressed or migmatised lower An values are found.

The layered nature of the anorthosite complexes and in particular the stratified nature of the chromitites together with rare grading and slump structures of apparent igneous origin suggests that these rocks represent large cumulitic stratiform igneous intrusions. It is possible that the anorthositic components have become partially separated from complementary gabbroic and peridotitic cumulates at the base of the crust (Stueber & Murthy, 1966). In East Greenland many of the anorthosites are found close to marked changes in the surrounding country rocks—for example along boundaries separating the homogeneous gneisses from high-grade supracrustal schists and amphibolites. This suggests that originally the anorthositic complexes were emplaced as flat-lying sheets along cover-basement interfaces which probably formed the most marked structural breaks in otherwise homogeneous gneissic crustal material. At an early stage in crustal development it seems likely that the layered anorthosites may have formed at the surface.

These early Precambrian anorthosites, which are thought to have formed prior to 3000 m.y. ago, are chemically similar to the calcic, chromiferous anorthosites reported from the moon (Wood *et al.*, 1970). These chemical affinities, together with the age of the terrestial rocks, allows the suggestion that both planetary bodies went through parallel stages of anorthositic development early in their evolution (Windley, 1970).

(5) *Late intrusive granites*

Toward the end of their plutonic activity, the consolidating Archaean cratons were intruded by a variety of igneous rocks with some resemblances to the post-orogenic suite of Phanerozoic fold belts. They range considerably in character from anatectic granites with gradational migmatitic contacts against the surrounding rocks (McGregor, 1966), to cross-cutting plutons. In areas in which the gneisses had preserved their early granulite facies mineralogy, the late intrusions tend to be hypersthene-bearing complexes ranging from norite to charnockite—for example, the Skjoldungen complex of southeast Greenland (Bridgwater & Gormsen, 1969). In other areas, where the host gneisses had amphibolite facies grade, the late intrusives tend to be diorites, appinites and biotite granites. However, in contrast to the calc-alkaline suites formed in the areas affected by Proterozoic orogenesis, the late Archaean post-orogenic suites do not show the widespread development of massive andesine-labradorite anorthosites of Adirondack type. In our opinion the comparative rarity of these bodies in the early Precambrian may be due to the suggested absence of a thick crust in the period before 2500 m.y. ago.

The structural style of Archaean basement regions is different from that in most later mobile belts. Regional isotropy, but anisotropy at outcrop scale, are the important characteristics.

On a regional scale, especially in areas of mediocre exposure, these largely gneissose terrains are essentially homogeneous in that they lack a dominant structural trend. Dome, basin or eye fold structures such as those described by Berthelsen (1960) may be developed over a large area (Fig. 5). In some cases these can be shown to have formed by the superposition of one fold phase on another in the manner ilustrated by Ramsay (1962) during reworking of earlier basement, but mostly the impression given is one of plastic movement of material to form complex flow folds of the type suggested by Carey (1953) and Wynne-Edwards (1963). Both explanations suggest a high degree of mobility, perhaps indicating that the crust was less rigid during its early history than in later times.

On account of this regional structural isotropy or homogeneity the separation of Archaean cratons into major sub-units (or structural regimes) affected by various types of orogenic activity is less easy than in younger Precambrian or Phanerozoic mobile belts.

When changes of structural character occur on a regional scale they are commonly the result of a later deformation which has impressed an apparently simple pattern on earlier, more complex structures (Windley, 1969*a*).

In well-exposed ground the rocks may vary rapidly in structural style and metamorphic grade and a single outcrop may show many of the variations seen over much of the craton. Evidence of several reworked cover-basements and several episodes of metamorphism, migmatisation and deformation may occur in quite small areas. The history of these catazonal levels is one of instability expressed by intense fragmentation and interslicing of rock units to produce regional homogeneity.

INTERMEDIATE ENVIRONMENTS

The present exposure level of some cratons probably represents an intermediate crustal environment between the two extremes just described. The Rhodesia craton was thought by Macgregor (1951) to comprise largely high-level greenstone-schist belts encircling granitic batholiths. However recent work by the Rhodesian Geological Survey (Bliss, 1969) has established that it has a more complex structure, consisting of three types of granites and gneisses formed at different times in different tectonic environments (Fig. 6) as follows:

(1) An early basement consists of gneisses that migmatise conformable schist belts.

(2) Intermediate stage, massive to highly foliated granites, probably derived by local melting of the basement, have both intrusive and gradational contacts with surrounding rocks.

(3) Late, massive, post-orogenic granites intrude adjacent schists and gneisses with sharp transgressive contacts.

According to Stowe (1968) the oldest relics in the basement gneisses are granulite-facies rocks which locally are overlain unconformably by low-grade schist belts. Only scattered remnants of the granulites occur as they have been highly altered by the extensive greenstone belt/late granite activity.

Further south in the Barberton Mountain Land in the Kaapvaal craton the earliest gneisses and migmatites form a series of discrete plutons or domes which are intrusive into the Onverwacht Group, the lower part of the Barberton greenstone belt (Viljoen & Viljoen, 1970). Thus although this is a model of an Archaean greenstone granite belt (Anhaeusser *et al.*, 1969), the Barberton Mountain Land must be seen as an essentially high-level terrain with no representatives of unmodified high-grade basement rocks.

Regions such as the Superior and Great Slave provinces of Canada, and Western Aus-

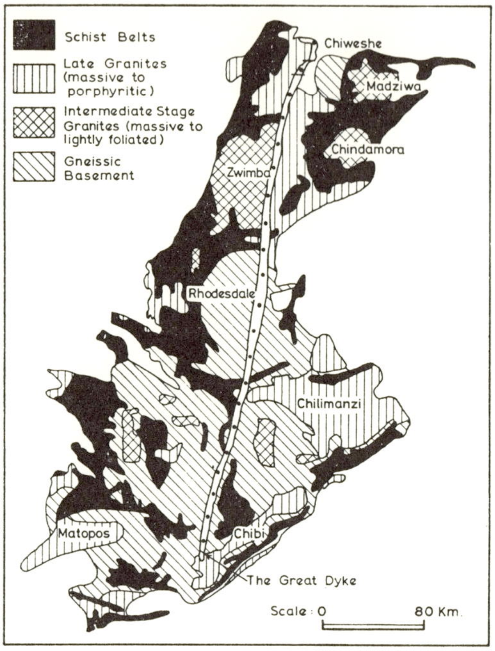

Fig. 6. The structure of the Rhodesian craton showing three generations of granitic/gneissic rocks between the schist belts (after Bliss, 1969).

tralia, are generally considered to consist largely of greenstone belts and 'granites', but it is probable that some basement gneisses exist even though in a modified form. They might be expected to occur ideally in areas farthest from the greenstone belts and beyond the screen of bordering granitic plutons. We suggest that intermediate relationships such as those in the Rhodesia and Kaapvaal cratons might well occur in these regions. This leads to the question: are there any high-level regions consisting entirely of greenstone belts and granitic plutons?

DISCUSSION

The varied character of Archaean cratons suggests that they formed at different crustal levels. Although the high and deep levels described are extreme stages of development, we consider that some cratons may approximate to them while others fall between them in intermediate environments.

There is no evidence that the Archaean developed in 'fold belts', as supracrustal deposition and plutonic activity appears to have taken place over the whole crust (Dearnley, 1966). There were no geosynclines, but rather abundant elongate narrow basins of sedimentation while at a plutonic level there was just as widespread high-grade activity. This type of Archaean mobility continued from earliest geological time to the period 2500-2000 m.y. which was one of relative stability when major basic dyke swarms were injected.

In the earliest Archaean the basins were probably very small and easily broken up by reactivation of the unstable basement giving rise to abundant agmatites. With time the basins become larger and were better preserved on a gradually stabilising sialic substratum. We consider, with Salop (1968), that it is inadvisable to erect a division around 3000 m.y. separating Katarchaean from Archaean (Dearnley & Dunning, 1968; Sutton, 1967), as greenstone belts were apparently forming from as early as 3400 m.y. in Barberton Mountain Land to as late as 2700-2900 m.y. in Canada.

The type of contact a greenstone-schist belt has with its country rocks depends on the degree of reworking and regeneration of the basement in the form of reactivated and intrusive granites. All transitions may be found from pre-greenstone high-grade basement to post-greenstone granites (Fig. 7). At deep levels the root zones of the belts may be preserved in linear deformation belts of reworked basement as thin strips of cover rocks bordered by conformable belts of migmatites; at intermediate levels the basins are separated by domes of remobilised modified basement gneisses; at higher levels the broader belts of greenstones are bordered by intrusive granites.

The relatively young isotopic ages obtained from high-grade Archaean terrains are not a reflection of their geological age but probably indicate that these deep-level rocks did not become closed systems until their uplift following that of the major greenstone belts. The ages obtained can be regarded as a function of depth as much as time in an analogous way to the relationships described by Harper (1967) from younger fold belts. If this is correct, it is the high-grade cratons which offer the possibility of getting back to the oldest events in the continental crust.

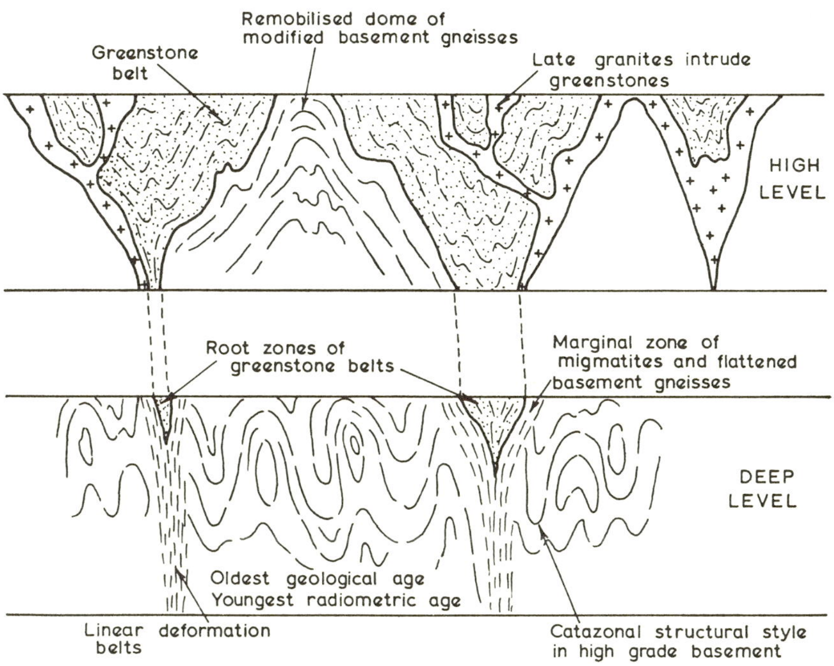

Fig. 7. Diagrammatic section through the deep- and high-level terrains of the Archaean.

ACKNOWLEDGMENTS

We wish to thank K. Ellitsgaard Rasmussen, Director of the Geological Survey of Greenland, for permision to publish this paper. We would also like to thank V. R. McGregor and A. K. Higgins for critical reading of the manuscript and helpful discussion. The responsibility for ideas expressed is the writers' alone. One of us (B.F.W.) acknowledges a British grant from the Natural Environment Research Council.

REFERENCES

ALLARD, G. O., 1959: The Dore lake complex: a metamorphosed layered complex (Chibougamau district, Quebec). *Va J. Sci., 10,* pp. 292-293.

ANHAEUSSER, C. R., MASON, R., VILJOEN, M. J., & VILJOEN, R. P., 1969: A reappraisal of some aspects of Precambrian shield geology. *Bull. geol. Soc. Am., 80,* pp. 2175-2200.

BARAGAR, W. R. A., 1966: Geochemistry of the Yellowknife volcanic rocks. *Can. J. Earth Sci., 3,* pp. 9-30.

BERTHELSEN, A., 1960: Structural studies in the Precambrian of West Greenland II. Geology of Tovqussap nunâ. *Bull. Grønlands geol. Unders., 31* (also *Meddr Grønland, 123,* (2)).

———, & NOE-NYGAARD, A., 1965: The Precambrian of Greenland; *in* Rankama, K. (Ed.), *The Precambrian, Vol. 2,* pp. 113-262.

BLISS, N. W., 1969: Thermal convection in the Archaean crust? *Nature, Lond. 222,* pp. 972-973.

BOULANGER, J., 1959: Les anorthosites de Madagascar. *Annls geol. Madagascar, Fasc. 26,* pp. 1-71.

BRIDGWATER, D., 1970: Observations on the Precambrian rocks of Scandinavia and Labrador and their implications for the interpretation of the Precambrian of Greenland. *Rapp. Grønlands geol. Unders., 28,* pp. 43-47.

———, & GORMSEN, K., 1968: Precambrian rocks of the Angmagssalik area, East Greenland, *Rapp. Grønlands geol. Unders., 15,* pp. 61-71.

———, & GORMSEN, K., 1969: Geological reconnaissance of the Precambrian rocks of southeast Greenland. *Rapp. Grønlands geol. Unders., 19,* pp. 43-50.

BROCK, B. B., 1959: On orogenic evolution, with special reference to southern Africa. *Trans. geol. Soc. S. Afr., 62,* pp. 326-365.

CAREY, S. W., 1953: The rheid concept in geotectonics. *J. geol. Soc. Aust., 1,* pp. 67-117.

CLARKE, D. B., 1970: Tertiary basalts of Baffin Bay: possible primary magma from the mantle. *Contr. Miner. Pet. 25,* pp. 203-224.

CONDIE, K. C., & POTTS, M. J., 1969: Calc-alkaline volcanism and the thickness of the early Precambrian crust in N. America. *Can. J. Earth Sci., 6,* pp. 1179-1184.

COWARD, M. P., FRANCIS, P. W., GRAHAM, R. H., MYERS, J. S., & WATSON, J., 1969: Remnants of an early meta-sedimentary assemblage in the Lewisian complex of the Outer Hebrides. *Proc. Geol. Ass., 80,* pp. 387-408.

DEARNLEY, R., 1966: Orogenic fold belts and a hypothesis of earth evolution; *in* Ahrens, L. H., Press, F., Runcorn, S. K., & Urey, H. C. (Eds), *Physics and Chemistry of the Earth, 7,* pp. 1-114. Pergamon, Oxford.

———, & DUNNING, F. W., 1968: Metamorphosed and deformed pegmatites and basic dykes in the Lewisian complex of the Outer Hebrides and their geological significance. *Quart. J geol. Soc. Lond., 123,* pp. 353-378.

DONALDSON, J. A., & JACKSON, G. D., 1965: Archaean sedimentary rocks of North Spirit Lake area, northwestern Ontario. *Can. J. Earth. Sci., 2,* pp. 622-647.

ESKOLA, P., 1949: The problems of mantled gneiss domes. *Quart. J geol. Soc. Lond., 104,* pp. 461-476.

FREEMAN, B. C., 1939: The Bell River complex, northwestern Quebec. *J. Geol., 47,* pp. 27-46.

GHISLER, M., & WINDLEY, B. F., 1967: The chromite deposits of the Fiskenaesset region, West Greenland. *Grønlands geol. Unders. Rapp. 12.*

GOODWIN, A. M., 1968: Evolution of the Canadian Shield. *Proc. geol. Ass. Can., 19,* pp. 1-14.

HARPER, C. T., 1967: The geological interpretation of potassium-argon ages of metamorphic rocks from the Scottish Caledonides. *Scott. J. Geol., 3,* pp. 46-66.

HENRIKSEN, N., 1969: Boundary relations between Precambrian fold belts in the Ivigtut area, southwest Greenland. *Spec. Pap. Geol. Ass. Can., 5,* pp. 143-154.

HIGGINS, A. K., 1968: The Tartoq group on Nuna qaqortoq and in the Iterdlak area, south-west Grønland. *Rapp. Grønlands geol. Unders., 17.*

———, & BONDESEN, E., 1966: Supracrustals of pre-Ketilidian age (the Tartoq Group) and their relationships with Ketilidian supracrustals in the Ivigtut region, south-west Greenland. *Rapp. Grønlands geol. Unders., 8.*

HOUTZ, R. E., & PHILLIPS, K. A., 1963: Interim report on the economic geology of Fiji. *Econ. Rep. geol. Surv. Fiji, 1.*

HUNTER, D. R., 1957: The geology, petrology and classification of the Swaziland granites and gneisses. *Trans. geol. Soc. S. Afr., 60.*

HUTCHINSON, R. W., 1965: Genesis of Canadian massive sulphides reconsidered by comparison to Cyprus deposits. *Can. Min. metall. Bull., 68,* pp. 286-300.

IBBOTSON, P., 1962: The geology of the Tavua area, Viti Levu. *Bull geol. Surv. Fiji, 8.*

LEELANANDAM, C., 1967: Occurrence of anorthosites from the charnockitic area of Kondapalli, Andhra Pradesh. *Bull geol. Soc. India, 4,* pp. 5-7.

MARTIN, H., 1969: Problems of age relations and structure in some metamorphic belts of southern Africa. *Spec. Pap. Geol. Ass. Can., 5,* pp. 17-26.

MAWDSLEY, J. B., & NORMAN, G. W. H., 1935: Vhibougamau Lake map-area, Quebec. *Mem. Can. geol. Surv., 185.*

MACGREGOR, A. M., 1951: Some milestones in the Precambrian of southern Rhodesia. *Proc. geol. Soc. S. Afr., 54,* pp. 27-71.

McGREGOR, V. R., 1966: Migmatization and deformation in Ameralik, Godthåb Commune, West Greenland, and their effect on a swarm of basic dykes. *Rapp. Grønlands geol. Unders., 11,* pp. 27-28.

———, 1969: Early Precambrian geology of the Godthåb area. *Rapp. Grønlands geol. Unders., 19,* pp. 28-30.

NAUTIYAL, S. P., 1965: Precambrian of Mysore plateau. *Indian Sci. Congr. 53rd, Calcutta, Sect. Geol. Geograph.*, pp. 1-14.

RAMSAY, J. G., 1962: Interference patterns produced by the superposition of folds of similar type. *J. Geol., 70,* pp. 466-481.

SALOP, L. I., 1968: Precambrian of the U.S.S.R. *Rep. 23rd Sess. int. geol. Congr. Czech. Sect. 14,* pp. 61-74.

———, & SCHEINMANN, YU M., 1969: Tectonic history and structures of platforms and Shields. *Tectonophys., 7,* pp. 565-597.

SÖHNGE, P. G., 1946: The geology of the Messina copper mines and surrounding country. *Mem. geol. Surv. Un. S. Afr., 40.*

STOCKWELL, C. H., 1957: Geology and economic minerals of Canada. *Geol. Surv. Can. Econ. Geol. Ser., No. 1.*

STOWE, C. W., 1968: Intersecting fold trends in the Rhodesian Basement Complex south and west of Selukwe. *Trans. geol. Surv. S. Afr., 71, Annex.*, pp. 53-78.

STUEBER, A. M., & MURTHY, V. R., 1966: Strontium isotope and alkali element abundances in ultramafic rocks. *Geochim. cosmochim. Acta, 30,* pp. 1243-1259.

SUBRAMANIAM, A. P., 1956: Mineralogy and petrology of the Sittampundi complex, Salem district, Madras State, India. *Bull. geol. Soc. Amer., 67,* pp. 317-390.

SUTTON, J., 1967: The extension of the geological record in the Pre-Cambrian. *Proc. Geol. Ass., 78,* pp. 493-534.

VILJOEN, M. J., & VILJOEN, R. P., 1970: Archaean volcanicity and continental evolution in the Barberton region, Transvaal; *in* Clifford, T. N., & Gass, I. G. (Eds), *African Magnetism and Tectonics,* pp. 27-49. Hafner, Darien.

WATSON, J., 1969: The Precambrian gneiss complex of Ness, Lewis, in relation to the effects of Laxfordian regeneration. *Scott. J. Geol., 5,* pp. 269-285.

WILSON, A. F., 1958: Advances in the knowledge of the structure and petrology of the Precambrian rocks of south-western Australia. *J. Proc. R. Soc. West. Aust., 41,* pp. 57-83.

WINDLEY, B. F., 1969*a*: Evolution of the early Precambrian basement complex of southern West Greenland. *Spec. Pap. geol. Ass. Can., No. 5,* pp. 155-161.

———, 1969*b*: Anorthosites of southern West Greenland. *Mem. Am. Ass. Petrol. Geol., 12,* pp. 899-915.

———, 1970: Anorthosites in the early crust of the earth and on the moon. *Nature, Lond., 226,* No. 5243, pp. 333-335.

———, HENRIKSEN, N., HIGGINS, A. K., BONDESEN, E., & JENSEN, S. B., 1966: Some border relations between supracrustal and infracrustal rocks in south-west Greenland. *Rapp. Grønlands geol. Unders., 9.*

WOOD, D. S., 1966: The Rhodesian basement. *10th Annu. Rep. Sci. Results, Univ. Leeds.*

WOOD, J. A., DICKEY, J. S. JR., MARVIN, U. B., & POWELL, B. N., 1970: Lunar anorthosites. *Science, 167,* pp. 602-604.

WYNNE-EDWARDS, H. R., 1963: Flow folding. *Am. J. Sci., 261,* pp. 793-814.

ZYL, J. S. VAN, 1950: Aspects of the geology of the N. Soutpansberg area. *Annls Stellenbosch Univ., 26.*

Brian F. Windley,
Department of Geology,
The University,
Leicester,
England.

David Bridgwater,
Geological Survey of Greenland,
Copehagen,
Denmark.

20: THE ARCHAEAN CRATON OF THE NORTH ATLANTIC REGION

D. Bridgwater
Geological Survey of Greenland, Copenhagen, Denmark
Janet Watson
Imperial College, London, England
B. F. Windley
Leicester University, Leicester, England

CONTENTS

The North Atlantic Archaean craton, exposed in parts of Greenland, Labrador and northwest Scotland, is a high-grade gneiss terrain which contrasts with Archaean granite–greenstone-belt terrains such as those of southern Africa. The tonalitic to granitic banded or agmatitic gneisses which occupy most of the craton are considered to be derived largely from granitic bodies emplaced within the crust. Early granitic gneisses of this type in Godthåbsfiord are at least 3800 Ma in age and it is suggested that a granitic basement of similar age extended over much of the craton. Most of this early basement was reworked and interleaved with metamorphosed supracrustal rocks, with layered anorthositic complexes and with abundant tonalitic gneisses derived from younger intrusions. Identifiable metavolcanics and metasediments, forming narrow belts in the gneisses, occupy less than 20% of the craton; they include highly-metamorphosed basic, ultrabasic and intermediate-acid volcanic rocks with associated intrusions and predominantly chemical sedimentary rocks. Clastic sediments are preserved in the lower part of the Isua supracrustal belt where they are overlain by banded ironstones and metavolcanics.

All these rocks suffered profound deformation and metamorphism which destroyed their primary relationships and culminated in the development of fold interference patterns without linear grain and in granulite or amphibolite-facies metamorphism ending at about 2800 Ma. Tectonic and metamorphic episodes over the next thousand million years were more localized and served to differentiate the Archaean craton from border-zones of early Proterozoic mobility.

1. Regional setting

The opening of the North Atlantic Ocean in Phanerozoic times fragmented and dispersed portions of an important Archaean province which had been defined as a cratonic geological entity some 2500 Ma ago. The object of this paper is to consider as an integral unit the remnants of this ancient craton which are now located along the Labrador coast of Canada, in East and West Greenland and in northwest Scotland. When regrouped (figure 1) they are seen to constitute a roughly triangular massif with sides 600 km in length which is enclosed by mobile belts of early Proterozoic age. Major parts of the mobile belts surrounding the old craton (for example the Baffin island 'Hudsonian', the Nagssugtoqidian of Greenland and the Laxfordian of Scotland) are composed of reworked Archaean material suggesting that the features now

[*Editor's Note:* Figures 2 and 3 have been omitted owing to limitations of space.]

preserved in the old block may have been much more extensive than they are at present (Bridgwater & Windley 1972).

The central Archaean craton is dominated by gneisses of amphibolite or granulite facies and in many respects contrasts markedly with provinces of similar age, such as the nearby Superior province of Canada, which are dominated by an association of massive granitic rocks with weakly-metamorphosed supracrustal greenstone belts. In spite of the very high grades of metamorphism attained, the early history of the North Atlantic province can be established in some detail and in west Greenland this history can be taken back to 3800 Ma. Tectonic, igneous and metamorphic activity continued over almost the entire province and its border-regions

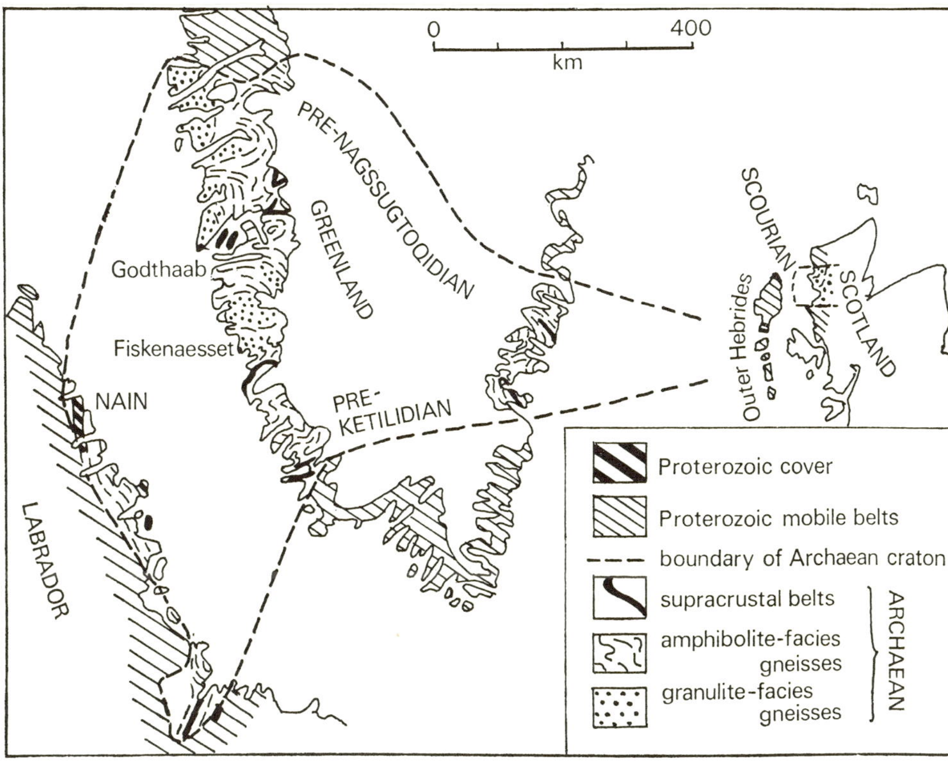

FIGURE 1. Outline map of the Archaean craton of the North Atlantic.

down to about 2600 Ma, when large areas became effectively stabilized. Activities were continued or renewed for about the next 1000 Ma in the border-zones which behaved in a different manner from the central Archaean craton over this time-span.

Our principal objects in this paper are to summarize the information concerning the early history of the craton, to provide an account of its geological components and of the tectonic and metamorphic styles displayed to serve as a basis for comparison with Archaean provinces of granite–greenstone-belt type, and to examine briefly the evidence relating to the stage at which the craton became differentiated from its border-zones.

2. Outline of geological history

The history of repeated deformation, metamorphism, igneous intrusion and accumulation of supracrustal groups recorded in the craton spanned a period of nearly 1500 Ma. The sequences of events established by field geologists from the relationships of the principal components, and some fixed points dated by isotopic methods are summarized in table 1. The early rock-groups and structures are best known in the Godthåbsfjord area of west Greenland which escaped the full effects of high-grade metamorphism at a later date (see McGregor, this volume). An early basement of deformed granitic rocks in this area (table 2) has yielded a Rb/Sr isochron age of 3980 ± 170 Ma and an Pb/Pb whole rock age of 3760 Ma (O.I.G.L. & McGregor 1971 – amended to 3750 Ma).

Metamorphosed supracrustal rocks are shown to be younger than these very ancient gneisses and are in turn older than a variety of gneisses of intrusive origin. 100 km to the south the Fiskenaesset meta-igneous complex with prominent anorthosite layers is seen to be intrusive into amphibolitic horizons believed to represent supracrustal rocks similar to those of the Godthåbsfiord region. If one accepts the tentative model age of 3600 Ma obtained from these rocks as the date of their intrusion (Evenson, Murthy & Windley, in prep.) then the supracrustals were probably deposited at some time between 3600 and 3790 Ma.

Broadly similar sequences of early events can be made out for many widely separated parts of the ancient craton (tables 1 and 2) and there is no reason to believe that Godthåbsfjord was in any way unique (see McGregor, this volume). Whether or not the various lithological units such as supracrustal belts and anorthosites can be correlated even over short distances is a matter of conjecture which can only be resolved by a combination of detailed field mapping and isotope studies. At present it seems reasonable to regard some of the younger events, for example the formation of large amounts of granitic material around 3000 Ma, as events which occurred throughout the craton and to leave correlation of older events such as the deposition of widely separated supracrustal groups more open. A valuable time marker recognized in every part of the massif is provided by a 'blanket' metamorphism of granulite or high amphibolite grade affecting all the rocks mentioned above. Dates for the late stages of this regional event fall about 2900 to 2800 Ma with a scatter down to about 2500 Ma which may reflect slow cooling and the local injection of granitic rocks. We regard this 2800 Ma high-grade metamorphic event as representing a turning point in the history of the Archaean massif after which structural and metamorphic effects became localized in specific areas. Scattered relics of low-grade supracrustal rocks apparently deposited on a high-grade basement but yielding radiometric ages in excess of 2500 Ma were incorporated in the craton during this late stage and numerous calc-alkaline intrusions and some carbonatites were emplaced in the general period 2000 to 2800 Ma ago (table 1). Between 2600 and 1900 Ma ago the craton was intruded by numerous swarms of basic and ultramafic dykes represented throughout east Labrador, west and east Greenland and the Precambrian of northwest Scotland. Extensive faulting occurred within the craton at this time. Some of the ancient faults have been reactivated by Cainozoic movements.

TABLE 1. MAJOR GEOLOGICAL EVENTS IN THE NORTH ATLANTIC CRATON

eastern Labrador	Ma		Greenland	Ma	northwest Scotland	Ma
Regional swarms of dolerite dykes. Faulting in craton. Possible late Archaean–early Proterozoic orogenic activity in area east of Labrador trough, west of coastal Archaean block	> 1800 < 2500	12	Regional swarms of dolerite and ultrabasic dykes. Faulting, unstable conditions in the areas to north and south of craton	2000–2600	Dolerite and ultrabasic dykes. Local metamorphism and deformation	2000–2400 2200–2500
—		11	Local calc-alkaline and carbonatitic activity in southeast Greenland	> 2000 < 2600	Pegmatites tonalitic granites and minor calc-alkaline intrusions	2400–2600 2600
Emplacement of granites, widespread pegmatites retrogression of granulite facies rocks. Extensive break up of older structures in southern part of Archaean coastal strip	< 2500	10	Emplacement of granites, widespread pegmatite swarms, infolding and greenschist facies metamorphism of late supracrustal sequences, retrogression of granulite facies rocks	> 2500 < 2800		
—		9	Deposition of supracrustal sequences (mainly basic and acid volcanic sequences) ?Retrogression	> 2500 < 2800	Deposition of supracrustal sequence (Loch Maree series). Local retrogressive metamorphism.	
Cordierite granulite facies metamorphism preserved in northern part of block		8	Culmination of cordierite granulite or amphibolite facies metamorphism, emplacement of andesine-anorthosite hypersthene-monzonite complexes. Widespread tectonic activity forming interference patterns	2800–2900	Granulite or amphibolite grade metamorphism	2900
Intrusion of granitic-tonalitic suite to form sheets in high grade gneisses		7	Emplacement of major calc-alkaline suites probably over extended time interval. Widespread mobilization of earlier rocks with major phase of formation of migmatite complexes. Intrusions of Nûk gneiss suite Godthåbsfjord	> 2800–? < 3300 3100	Tonalitic intrusions? Anorthosites (Hebrides) and basic and ultrabasic intrusions	
—		6	Tectonic interleaving of all earlier rocks (probably by a variety of processes and in more than one event)			
—	not dated	5	Intrusion of layered igneous complexes characterized by presence of chromite-bearing calcic anorthosites	?3600		
Deposition of supracrustal rocks, basic and ultrabasic volcanics, aluminous pelites, thin quartzites and impure calc-silicate horizons. Possibly some acid volcanics		4	Deposition of supracrustal sequences (largely basic ultrabasic and intermediate volcanic rocks in southern part of craton, possibly higher proportion of pelitic material further north)		Deposition of supracrustal sequences – mainly sediments with ? volcanic components including ultrabasic rocks	
—		3	Basic dykes (Ameralik dykes)			
Presumed older basement now forming one of the components in the gneiss complex		2	Deformation and metamorphism (3–1: demonstrable in Godthåbsfjord district. Inferred elsewhere)	3800	Presumed early basement	
—		1	Emplacement of granitic rocks with inclusions of basic material			

Table 2. Events recorded at three localities in Greenland

	(A) *Godthåbsfjord and immediate surroundings (modified from McGregor* 1972)	(B) *Fiskenaesset and northern Frederikshåb district (after Dawes* 1970 *and recent mapping by the Geological Survey of Greenland)*	(C) *Southeast Greenland (from Bridgwater & Gormsen* 1968, 1969; *Andrews* et al. 1971 *and Andrews* et al. *in prep.)*
12	Shearing with local loss of radiogenic argon *ca.* 1800 Ma	Local shearing causing sporadic loss of radiogenic argon *ca.* 1800 Ma	Local intrusions of carbonatite, appinite and calc-alkaline plutons 2000–2600 Ma
11	Dolerite dykes, some slightly metamorphosed > 1800 < 2600 Ma	Dolerite dykes (3 or more generations) K/Ar determination on ENE dyke 2428 Ma (Bridgwater 1971)	Dolerite dykes
10	Regional closure of K/Ar and Rb/Sr mineral systems *ca.* 2500 Ma regional pegmatite swarms, intrusion of granites	Regional uplift and closure of K/Ar systems *ca.* 2500 Ma. Some problematical earlier K/Ar ages	Regional uplift ages 2300–2700 Ma. Widespread slight alteration
9	Intense local deformation	Retrogression, probably major period of granite emplacement particularly in the south of the area	Retrogression of granulite facies rocks leaving small relics. Possibly some of granites under (6) below belong to this suite
8	High grade metamorphism. Granulite facies in areas to north *ca.* 2800 Ma metamorphism apparently continued after major folding	High-grade metamorphism. Granulite facies in northwest, amphibolite facies to the south *ca.* 2800 Ma	High-grade metamorphism. Emplacement of norite anorthosite and hypersthene granite suites at Skjoldungen and Angmagssalik probably during major period of folding.
7	Complex folding	Complex folding	Regional formation of interference patterns
6	Intrusion of syntectonic calc-alkaline suites (rocks now forming the Nûk gneisses). Rb/Sr whole rock system set at 3050 Ma	Granitic activity, migmatization of anorthosites and supracrustal units injection of granite sheets	Major emplacement of granite–granodiorite–diorite suite, varying between syntectonic to late tectonic. Locally affected by granulite facies metamorphism (?9, above). Anorthosites severly broken up
5	Emplacement of anorthosites	Emplacement of anorthosites and associated igneous bodies	Emplacement of anorthosite sheets and associated more basic rocks. ? Some ultrabasic bodies
4	Thrusting to form alternating units of supracrustal rocks and gneisses emplacement of ultrabasic lenses	—	Thrusting to form concordant layers of gneiss and supracrustal rock (particularly well seen in relics of Archaean rock near Angmagssalik
3	Deposition of basic lavas and sediments (Malene supracrustals)	Deposition of supracrustal units (mainly basic lavas) well preserved on Ravns Storø and on the nunataks. Correlation uncertain between different groups	Deposition of supracrustal units – mainly basic to the south, more aluminous to the north. Some ultrabasic rocks in sequences
2	Injection of basic dykes	Basic dykes pre-dating supracrustal units on the nunataks – mainly synkinematic types proven. Possibly some normal dolerites now metamorphosed. Granitic gneisses with inclusions of basic and ultrabasic material interpreted as earlier supracrustal rocks	Basic dykes – particularly in northern part of the area
1	Deformation of original granitic suite to form Amîtsoq gneisses Rb/Sr whole rocks isochron at 3750 Ma. Inclusions of small basic and ultrabasic masses		Granitic rocks forming early gneisses. Contain migmatitic inclusions of basic and ultrabasic material interpreted as older supracrustal rocks

The correlation between the oldest parts of the above sequences of events is clearly difficult and relies on lithological units such as supracrustal layers which are probably not unique to any one time. The columns are placed side by side to give an impression of the complexity of the Archaean and to suggest that no one single model can be used for the whole block

3. Early crustal components (> 3000 Ma)

The oldest components of the province – indeed the oldest dated rocks of the Earth's crust – are the Amîtsoq gneisses of the Godthåb district which escaped extreme deformation and granulite-facies metamorphism in later times (table 2). They are mainly banded gneisses of granitic to granodioritic composition which sometimes carry megacrysts of potash feldspar. McGregor considers (O.I.G.L. & McGregor 1971) that the Amîtsoq gneisses were formed mainly by recrystallization and flattening of an originally more uniform granitic suite. The Amîtsoq gneisses contain enclaves of basic and ultrabasic material now represented by inclusions of amphibolite, hornblendite, dioritic gneiss, micaceous gneiss and quartzose gneiss carrying either magnetite or iron silicates. No inclusions of rocks of proven supracrustal origin occur in these gneisses and in their published account the O.I.G.L. & McGregor consider that the Amîtsoq gneisses may represent primordial granitic crust.

The stratigraphical position of the Amîtsoq gneisses relative to other units in the Godthåb district was established by McGregor from field evidence concerning the distribution of a swarm of basic dykes (the Ameralik dykes, see below) which cut, and are deformed with, Amîtsoq gneisses but do not enter younger supracrustal groups or intrusive gneisses. The radiometric age of the Amîtsoq gneisses is shown by isotopic studies at Oxford University referred to above which indicate that they date back to at least 3800 Ma ago.

According to V. R. McGregor (personal communication and this volume) the type Amîtsoq gneisses occur as part of the gneiss complex over an area of at least 130 × 25 km. Outside this area in which the presence of the Ameralik dykes clarified field relations, the presence of rocks of equivalent age is less easy to demonstrate, partly because of lack of detailed mapping, partly because of the intensity of later metamorphic effects and, perhaps most importantly, because of the lack of amphibolite dykes in sufficiently large numbers to allow the methods used by McGregor in Godthåbsfjord to be applied. Local sequences of events in the Archaean gneisses throughout the craton show patterns similar to that described by McGregor, however, and in our opinion rocks similar in age to the Amîtsoq gneisses make up an appreciable part of the whole block. In Labrador the amphibolite-facies Hopedale gneisses on Ford island (to the east of Nain, 56° 30′ N) show a history of metamorphism and deformation before the injection of amphibolitic basic dykes which were themselves later metamorphosed. In the Fiskenaesset area south of Godthåb (table 2) rare amphibolite dykes are found in gneisses regarded by one of us (B. F. W.) as a basement into which the meta-igneous complex was intruded at Fiskenaesset (see below). However, amphibolite dykes are seen to cut supracrustal rocks and the anorthosite complex (J. Myers, personal communication 1972).

In the Dalagers Nunatak region southeast of Fiskenaesset, Dawes (1970), has recorded a sequence of events (table 2) which is at least as complex as that recorded from Godthåbsfiord and which could well prove as interesting geochronologically. The oldest rocks in that area are acid and basic gneisses which were deformed, cut by a variety of basic dykes, overlain by a supracrustal succession consisting dominantly of basic volcanics, metamorphosed under amphibolite facies conditions and intruded by granites in the period before 2500 Ma ago. Dawes tentatively correlates the major supracrustal sequences with the Ravns Storø schists of the Fiskenaesset district now thought to be older than the 2800 Ma-old granulite facies metamorphism in the areas to the north (Kalsbeek 1972). This would place his older gneisses in the same relative position in the regional sequence of events as the Amîtsoq gneisses of

Godthåbsfjord. Dawes interprets the basic and ultrabasic material in his oldest group of rocks as of possible volcanic origin, an interpretation followed by many of the geologists working in the Frederikshåb area to the south (see, for example, Kalsbeek & Leake 1970; Andrews, in press).

In southeast Greenland thin amphibolite dykes are found in gneiss units which appear to have passed through a more complex history than the major supracrustal belts in the same area. This relationship (which is particularly clearly seen in the remnants of Archaean rocks preserved in the Angmagssalik area) has been interpreted in terms of a series of basement slices thrust together with younger cover rocks (Bridgwater & Gormsen 1969). Many of the gneisses interpreted as basement in southeast Greenland are agmatitic and contain numerous inclusions of amphibolite, dioritic material and ultrabasic (peridotite and olivine-orthopyroxene) rocks. In contrast to the interpretation favoured by O.I.G.L. & McGregor for the primeval origin of the Amîtsoq granitic gneisses, Windley & Bridgwater (1971) regard this arrangement as possible evidence of still earlier events involving the break-up of supracrustal material by intrusive granitic material broadly equivalent in age to the Amîtsoq gneisses.

In the Outer Hebrides of western Scotland, and possibly also in the Scottish mainland, a similar distinction between banded gneisses, largely of tonalitic or granodioritic composition and units representing an ancient cover-series can be drawn (Coward *et al.* 1969, see also Dearnley & Dunning 1968). While it remains to be seen whether any of these possible basement assemblages will turn out to be as old as the Amîtsoq gneisses, it does seem reasonable to infer that remnants of a gneiss complex pre-dating both the earliest important supracrustal groups and the earliest anorthosites are distributed across almost the full width of the Archaean massif.

These remnants, some or all of which are at least 3600 Ma in age, are broadly granitic and, although the majority are tonalitic or granodioritic, it is of interest to note that potash feldspar is conspicuous in the Amîtsoq gneisses which locally verge on a truly granitic composition. These gneisses (O.I.G.L. & McGregor 1971) bear no resemblance to lunar materials and equally do not resemble oceanic crustal materials. The extent of granitic gneisses in the North Atlantic Archaean suggests the existence of considerable areas of continental crust before the accumulation of the earliest dated greenstone-belts. The survival of folds and foliations predating the Ameralik dykes, and the preservation at some localities of the primary structural relationships of these dykes show that regeneration of basement complexes at least as old as 3800 Ma could be achieved without loss of coherence, that is without wholesale remelting.

The Ameralik dykes of Godthåbsfjord represent the oldest well-documented dyke-swarm yet recognized. The swarm, which now has a general north-northwest trend, contains hundreds of amphibolized dykes up to a few metres in width. Primary discordances, apophyses, chilled edges and anorthositic inclusions are well preserved where later deformation has been slight; but many dykes have been rotated into parallelism with the gneiss foliation, folded, boudinaged or migmatized by granitic veins extending from the gneiss. Little or no evidence remains as to their primary textures and minerals.

Intercalated conformably with the Amîtsoq gneisses of Godthåbsfjord are belts of supracrustal rocks up to 1 km in breadth (the Malene supracrustals) which include amphibolites, mica-schists and less common marbles and quartzites. As these supracrustal units contain no amphibolite dykes McGregor suggested in his earlier papers that they represent a cover deposited on an earlier gneiss basement and subsequently interleaved tectonically with reworked gneisses and dykes. In his most recent publication a second hypothesis is offered as an alternative in which the relationship is explained in terms of the interleaving of continental and oceanic crust with no proven age relationships to one another.

The Malene supracrustals are representatives of an assemblage which recurs in units of much the same scale, character and range of variation at many points throughout the craton. These units are seldom much more than 1 km broad and are generally, though not always, sufficiently clearly defined to be mapped out from the adjacent gneisses. Almost all contain a high proportion of basic material (amphibolites, hornblende-gneisses, basic granulites) which appears to be of igneous origin. Some of these basic rocks were undoubtedly volcanic as they exhibit pillow-structures. Geochemical studies of amphibolites from the Frederikshåb area of West Greenland led Kalsbeek & Leake (1970) to an interpretation in terms of an assemblage of basic lavas and tuffs. Pods of dunite, harzburgite and serpentinite (sometimes with layers of chromite) commonly occur in the basic units or in the gneisses immediately adjacent to amphibolite layers. We interpret most of these as associates of the original supracrustal sequence, although some have clearly been moved tectonically as pips during later deformation. The ultrabasic rocks frequently preserve more of their primary textures and mineralogy than other rocks in the sequence, presumably because their physical properties allowed them to respond to deformation by sliding along their contacts rather than by recrystallization.

Many of the supracrustal sequences contain garnet–amphibole–quartz–feldspar rocks of intermediate composition some of which include fragments of granitic material and which are thought to have been derived from andesitic or dacitic tuffs and agglomerates. A variety of feldspathic gneisses which lack distinctive features may be capable of a similar interpretation. Demonstrable metasediments form a comparatively small part of the successions. Siliceous layers found in conjunction with ultrabasic and basic rocks have been interpreted as derived from cherts; aluminous units commonly consisting of biotite, sillimanite or kyanite, garnet, quartz and feldspar with minor amounts of graphite and sulphide have generally been interpreted as derived from pelitic sediments (Coward *et al.* 1969) though a volcanic origin is not impossible. Calc-silicate layers probably represent impure limestones or oozes. Major quartzitic units are apparently absent from the typical supracrustal successions – thin quartz-rich layers often with magnetite, sillimanite, garnet or iron-rich silicates are widespread as minor components of the successions. No primary clastic structures have been noted from the majority of the supposed metasedimentary rocks which can be interpreted as chemical precipitates or volcanic tuffs rather than detrital sediments. A major exception to this is found in the Isua belt of supracrustal rocks from Godthåbsfjord (Keto 1969) which differs from the majority of successions preserved in that the lower part of the succession consists of quartzites and metagreywackes. These are overlain by garnet-chlorite schists and banded ironstones including both oxides and carbonates. The top of the succession consists of a greenschist series interpreted as derived from basic volcanics. The total succession is up to 2 to 3 km thick – that is considerably more than the majority of the supracrustal successions. It is metamorphosed under upper greenschist to amphibolite facies conditions. The age of the Isua successions is controversial. They, and the rocks surrounding them, have yielded K/Ar ages varying from 1800 to 3500 Ma. Their comparatively low metamorphic grade has been used to suggest that they are younger than other supracrustal successions but this criterion is no longer regarded as reliable (see, for example, the revised position of the Ravns Storø succession). If the typical supracrustal successions preserved in the Archaean block are regarded as having been formed under oceanic or at least deep water conditions then the Isua supracrustals appear to be more in the nature of shelf deposits.

The supracrustal assemblages referred to above have something in common with those which characterize the greenstone belts of the Superior province, the Yilgarn block of western Australia and the Rhodesian and Kaapvaal cratons. The range of lithological types is broadly equivalent and the dominance of basic material is common to most assemblages. So far, however, rocks with the distinctive chemistry of the komatiites (Anhaeusser *et al.* 1969) or major units comparable with the clastic Moodies or Shamvaian sediments of the southern African supracrustal belts have not been recognized. The supracrustal belts of the North Atlantic province are narrow (figure 1) and form only a small proportion of the gneiss complex. Their rocks are strongly deformed (often lacking any primary structures), highly metamorphosed and at present have not shown the gold–quartz mineralization of the major greenstone belts. As a working hypothesis Windley & Bridgwater (1971) suggested that many of these differences can be explained by regarding the North Atlantic craton as revealing a deeper crustal section than that exposed in Archaean provinces of the granite-greenstone type. A second possibility – not necessarily in opposition to the first – is that we are dealing in the North Atlantic craton with a slightly earlier phase of crustal evolution (either in time or in stage of development). Major greenstone belts of the Barberton–Superior province type may characterize a phase when the crust had differentiated into areas of continental and oceanic type and was sufficiently stable to bear thick successions of basic and ultrabasic rock. The thinner supracrustal belts of the high-grade terrains could have formed at a stage in crustal development when proto-continents and proto-oceanic crust had only separated on a small scale. It is possible that the smaller remnants of highly-metamorphosed supracrustal rocks recorded from the granitic terrains bordering the major greenstone belts – for example those recorded from Swaziland (Hunter 1970) and Rhodesia (Stowe 1968) which are considered by these authors to be older than the neighbouring major belts – may provide appropriate material for comparison with the amphibolite belts of the North Atlantic craton.

The make-up of the North Atlantic craton confirms the evidence from other provinces that groups resembling greenstone-belts in lithology and relationships continued to form over a very long period. The oldest belts in the craton may be as old as 3500 to 3600 Ma (Evenson, Murthy & Windley 1972). The youngest – the Tartoq belt of West Greenland (Higgins 1968) and the Loch Maree Series of northwest Scotland – are thought to post-date the 2900 Ma metamorphic event and are probably not older than about 2600 Ma (table 1).

Anorthositic rocks are widespread throughout the massif and in its reworked border-zones; the only region where Archaean anorthosites have not yet been found is Labrador. In west Greenland, metamorphosed and folded anorthositic layers with a strike-length of at least 500 km (Windley 1969) are remarkably continuous. In East Greenland (Bridgwater & Gormsen 1969) and in the Outer Hebrides, anorthosites occur mainly in pods up to a few kilometres in length associated with basic and ultrabasic bodies which together probably represent the disrupted remnants of stratified igneous complexes. Although the primary igneous stratigraphy of these remnants has been greatly modified, the resemblance of individual fragments to recognizable units within the better-preserved bodies is close enough to suggest that the suites developed under similar conditions. Individual anorthositic gabbro pods now less than 1 m in diameter may retain the original igneous textures remarkably well (figure 3, plate 7).

In the Fiskenaesset region of West Greenland, calcic anorthosites form part of a metamorphosed layered igneous complex up to 2 km thick and 200 km along the strike. The complex is bordered conformably along much of its outcrop by layers of amphibolites (interpreted as metavolcanics) associated with minor marbles and mica-schists which are flanked in turn by gneisses carrying deformed rare remnants of the amphibolitic dykes which were mentioned above. Many of the smaller anorthositic bodies both in Greenland and in Scotland are similarly bordered by metamorphosed supracrustal assemblages and it is inferred that the igneous complexes to which they belong were emplaced along zones of tectonic weakness such as the interface between slices of original basement gneiss and younger cover rocks. Indeed in the absence of widespread equivalents to the Ameralik dykes the characteristic occurrence of meta-anorthosites in zones separating gneisses of unknown origin from supracrustal rocks is one of the most useful indications of the possible presence of an older pre-supracrustal gneissose basement in many parts of the craton.

In the Fiskenaesset complex, magnetite-bearing ultramafic rocks (dunites, peridotites, pyroxenites) are succeeded towards the original top by a lower layered series of mafic rocks, by mafic gabbros, leucogabbros with marked cumulate textures, and by anorthosites, chromitites and garnet-anorthosites. This layered structure represents a gravity-differentiated sequence, the fractionation trend of which is well preserved despite subsequent metamorphism (Windley, Herd & Bowden, in the press).

The next important additions to the crust were tonalitic and dioritic rock-masses emplaced within pre-existing units as pods, sheets or more irregular intrusions. In the Godthåb area, the deformed and flattened derivatives of these bodies make the Nûk gneisses which are interleaved conformably with the older Amîtsoq gneisses and Malene supracrustals. The type Nûk gneisses have yielded a Rb/Sr isochron age of 3084 Ma (see McGregor, this volume). McGregor estimates that over 50% of the rocks exposed in the area mapped in detail around Godthåb is made up of Nûk gneiss. Elsewhere in the craton the amount of gneiss formed from rocks intruded after the formation of the early supracrustal belts and anorthosites is known with less certainty. Over large areas of southeast Greenland many of the gneisses show intrusive contacts against the local supracrustal units and it has been estimated that up to 80% of the gneiss around Tingmiarmiut may have been derived from rocks which were intruded between the formation of the local supracrustal belts and the period of high-grade metamorphism around 2900 Ma. Probably more than one event is represented by these rocks since they range from syn- or pretectonic sheets of gneissified diorite and granodiorite which only locally preserve their intrusive contacts to major cross-cutting bodies which break up the structures preserved in the supracrustal rocks and the earlier post-supracrustal intrusives. The youngest phases, which include large bodies of norite, andesine-anorthosite and mangeritic-charnockitic members, appear to have been emplaced as late- or post-tectonic intrusions around the peak of high amphibolite and granulite facies metamorphism dated at 2800 to 2900 Ma ago. Some of the intrusive rocks are massive, free of foreign inclusions, show no sign of local derivation and are thought to represent a major addition to the crust at the level exposed. Other more nebulitic bodies and many of the agmatitic gneisses of the east coast of Greenland probably represent remobilized basement, partly remobilized rocks from the supracrustal successions and perhaps early members of the post-supracrustal intrusive suite remobilized by later metamorphism. So far we have seen little evidence which suggests to us that major parts of Archaean gneisses were formed by *in situ* recrystallization of supracrustal rock of appropriate composition such as acid volcanic or greywackes and we regard the post-supracrustal suite as essentially intrusive in origin.

Reviewing the lithological constitution of the whole massif a few general points may be worth commenting on. Perhaps the most important is the evidence that continental crust was in existence 3800 Ma or more ago and is older than any proven supracrustal sequences. The very old ages obtained from the Amîtsoq gneisses as well as the survival of primary intrusive relationships in the Ameralik dykes shows that although the basement underwent repeated reworking this did not involve wholesale remelting of the crust throughout the craton. In this respect, the gneissose craton differs markedly from the areas immediately surrounding the typical greenstone belts in southern Africa, Canada and Australia where the majority of granitic rocks show clear intrusive relationships with the low-grade greenstone belts and where there is seldom any isotopic evidence for the presence of an older basement. It gives support to workers such as Hunter (1970) and Stowe (1968) who have suggested that older basements underlie the well-preserved greenstone belts in southern Africa, and that the contact relationships have been modified by later granite injection and thrusting. However, in our opinion the fact that sialic material was already present 3800 Ma or more ago does not necessarily indicate the presence of a primordial granitic crust. It seems rather unlikely that rocks formed some 700 Ma after the formation of the Earth should be regarded as primordial.

We regard rocks such as those of the lower parts of the Barberton mountainland volcanics as models for crust of oceanic type formed some 1000 Ma after the formation of the Earth, and perhaps approaching primordial basalts in chemical character, and the Amîtsoq gneisses as models for early (but not necessarily primitive) granitic crust of continental type. It would perhaps be reasonable to assume that the oldest rocks likely to be preserved would be granitic, since in unstable environments the denser rocks might be rapidly returned to the mantle and recycled, but this does not mean any granitic rocks we see now are primordial. For a type of crust which could predate differentiation into oceanic and continental régimes perhaps we should look to the agmatitic gneisses which are common in both the North Atlantic craton and the granitic gneisses surrounding the major greenstone belts. We would not expect these to be primordial – nor necessarily to contain older crustal elements than are seen in the more clearly-defined granitic and greenstone terrains; they may, however, give an indication of what the earliest crust was like before it stabilized sufficiently to allow the differentiation of major continental and oceanic areas.

A second point of interest concerns the relatively small contribution, in terms of area of outcrop, made to the massif by rocks of proven supracrustal origin. Much of the complex, as noted above, consists of gneisses which we consider were largely derived from granitic rocks emplaced within the crust, while rocks of obvious supracrustal derivation form a relatively small part of the craton. Clearly the proportion of the various components may be expected to vary from area to area, but taking the areas mapped in detail (for example the Outer Hebrides, the Frederikshåb and Fiskenaesset areas of southwest Greenland and the Godthåbsfjord area), we would suggest that an average of 10 to 15 % of the craton is made up of demonstrable supracrustal material (table 3). Areas where the proportion of obvious supracrustal material is lower than 10 % are generally those where massive granitic activity at a late stage in the local development of the craton has both diluted the proportion of supracrustal rocks present and partially destroyed the evidence of their origin.

When the recognizable supracrustal groups and the intrusive bodies in the high-grade complex have been accounted for, large areas of banded gneisses which appear to have lost almost all diagnostic primary structures remain. In the Lewisian complex of northwest Scotland, for example, such undistinctive gneisses occupy over 70 % of the outcrop; most are of tonalitic or granodioritic composition, but mafic and ultramafic pods and granitic layers are widespread (see, for example, Watson 1965). Sheraton (1970) and Bowes, Barooah & Khoury (1971) have favoured a volcanic origin, on account of geochemical considerations and of the characteristic banding; such a suggestion implies the dominance of andesitic-dacitic volcanism over basic volcanism on a regional scale. An alternative suggestion favoured by some workers in both Scotland and Greenland (e.g. Kalsbeek 1970; Andrews 1972) is that the gneisses are derived from the recrystallization *in situ* of sedimentary rocks of appropriate composition.

TABLE 3. COMPONENTS OF SOME ARCHAEAN COMPLEXES

	southwest Greenland						
	Godthåbsfiord area	Fiskenaesset area	Frederikshåb area	Fiskefiord area	northwest Scotland	Rhodesian craton	Canadian Archaean craton
percentage area of	(1)	(2)	(3)	(4)	(5)	(6)	(7)
(*a*) supracrustal sequences < 2900 Ma	—	—	5	—	2	—	30–40
(*b*) supracrustal sequences > 3000 Ma with associated basic-ultramafic intrusives	11	11	11	19	7–12	18	} 60–70 [(*b*)–(*d*)]
(*c*) gneisses demonstrably older than (*b*)	39 } 89 [(*c*)+(*d*)]						
(*d*) gneisses and granites later than (*b*) and > 2400 Ma	50	85	83	81	85	82	
(*e*) anorthositic and associated complexes	—	4	—	—	2	—	—
map area (km^2)	200	3000	1350	1800	4400	190 000	2×10^6

(1) From map, McGregor, this volume.
(2) From information supplied by F. Kalsbeek (see Kalsbeek *et al.* 1972). Figures for (*b*) refer to amphibolite layers in gneiss.
(3) From map compiled by J. Escher (1971).
(4) From a sketch-map published by Windley & Bridgwater (1971). Figures for (*b*) refer to amphibolite and pyribolite layers in gneiss.
(5) From maps of Institute of Geological Sciences, Coward *et al.* (1969) and others.
(6) From Phaup (1971).
(7) From Stockwell (1968).

As we have stated above, our own views are strongly against accepting a sedimentary or acid volcanic origin for granitic gneisses of this kind without direct evidence. Our preferences arise partly from the fact that many Archaean gneisses formerly accepted as supracrustal have been shown to have originated either as a basement older than undoubted supracrustal associates or as later intrusive granitic bodies. In younger mobile belts such as the Caledonides or the Alps, where the level of erosion might favour the preservation of rocks of supracrustal origin, surprisingly small proportions of the gneisses present appear to have been formed from the geosynclinal successions deposited in the period before the main tectonism and metamorphism; recent remapping and geochronological studies, again, suggest that the proportions of regenerated basement are higher than was formerly thought. Geochemical evidence is difficult to interpret for several reasons; rocks of volcanic origin cannot be distinguished with confidence from rocks of intrusive origin by geochemical criteria; supracrustal rocks converted to gneisses in a single cycle cannot easily be distinguished from reworked supracrustal rocks in a polycyclic basement; and the geochemical effects of the 'blanket' high-grade metamorphism affecting the gneisses still remain to be established. Field evidence is also usually inconclusive, since authentic primary structures are virtually absent. The banding of the gneisses and the interlayering of gneisses with undoubted supracrustal rocks are largely the results of repeated deformation (p. 506). The transitional contacts which have been described from areas where small supracrustal remnants are enclosed in gneisses (see e.g. Coward *et al.* 1969) undoubtedly indicate local transformation of supracrustal parent-rocks to banded gneisses but the scale of the phenomenon does not, in our view, justify the conclusion that such a transformation has taken place on a regional scale; there is an analogy with the transitional contacts of xenoliths and aureole-rocks associated with certain granite plutons, which a few decades ago were cited as evidence that the plutons were products of large-scale granitization but which are now generally accepted as marginal phenomena.

With these considerations in mind, we consider it is now more realistic to suggest a granitic origin for the majority of granitic gneisses of unknown parentage than to invoke a direct derivation from volcanic or sedimentary materials. The easiest way to produce a granitic or tonalitic gneiss is to deform and recrystallize a granite, a tonalite or an earlier gneiss and in our opinion the demonstration in this volume by McGregor that banded gneisses of the type so prevalent in high-grade Archaean complexes have been produced in this way represents a major contribution to basement geology.

4. Structural and metamorphic history

Deformation and metamorphism recurring over a time-span of more than a 1000 Ma have affected the Archaean province and the resulting structures are complex. It may be helpful to group the episodes concerned rather loosely into four deformational-metamorphic 'events', using the relationships of supracrustal belts and igneous bodies as markers:

(1) Events older than the emplacement of the Ameralik dykes in the early basement.

(2) Events associated with the interleaving of basement and cover.

(3) Events which occurred during the regional event of *ca.* 2900 Ma.

(4) During the initial stages of stabilization.

(1) The effects of the earliest episodes can only be distinguished where little-deformed Ameralik dykes cut folds and foliations in the Amîtsoq gneisses and their equivalents. McGregor considers that the banded Amîtsoq gneisses cut by the dykes were formed by metamorphic differentiation from originally more homogeneous granitic rocks and recognizes augen gneisses formed by flattening of rocks containing potash-feldspar megacrysts. The modifications envisaged by McGregor suggest deformation in a high-grade metamorphic environment at least 3600 Ma ago.

(2) Throughout the Archaean massif, as in other high-grade Archaean provinces, the contacts between adjacent rock-units are characteristically conformable (figure 2, plate 7). It is clear that this parellelism of lithological boundaries on all scales does not reflect a primary stratigraphical arrangement but is the result of tectonic and metamorphic modifications imposed on assemblages of very different ages. In the Godthåbsfjord area units of the early basement of Amîtsoq gneisses, distinguished by McGregor by their relationships with Ameralik dykes, are conformably interleaved with belts of Malene supracrustals and with units of Nûk gneisses

derived from still younger intrusive bodies. No angular discordances are seen at the borders of the Amîtsoq gneisses although the Ameralik dykes invading these gneisses do retain discordant contacts at a few localities where effects of post-dyke deformation are exceptionally slight.

Concordant belts of metamorphosed supracrustal rocks broadly similar to the Malene supracrustals in thickness, metamorphic state and lithological range are, as noted above, distributed through most of the Archaean terrains in Labrador, Greenland and Scotland, in association with gneisses considered to include both younger intrusive rocks and an older basement. The conclusion follows that cover/basement units were interleaved tectonically in such a way that their internal structures were brought into parallelism with their contacts both on a regional and on a local scale. Evidence as to whether the interleaving was achieved by downfolding of the cover and subsequent flattening of the synclines so formed, to produce parallel-sided belts, or by the trapping of cover slices on dislocations between moving slabs of basement is conflicting and both processes may have occurred. Symmetrical arrangements of horizons within the supracrustal belts which would favour a synclinal interpretation have only rarely been recognized (for example in the Sagleq area of Labrador where the supracrustal units form an isoclinally repeated succession). The Ravns Størø supracrustal belts and the Grydefjeldet belt of respectively southwest and southeast Greenland show possible bilateral symmetry, but its significance remains to be established. The layered rocks of the Fiskenaesset meta-igneous complex, which lies mainly within a supracrustal belt, are commonly arranged in isoclinal synclines which could have been formed simply by flattening of an initially basin-shaped complex.

Table 4. Trace element abundances in Archaean gneisses of northwest Scotland

element	no. of samples	average		crustal average of Taylor (1964)	
uranium†	24	0.24	parts/10^6	2.7	parts/10^6
rubidium‡	254	11.00	parts/10^6	9.0	parts/10^6
thorium‡	254	1.00	parts/10^6	9.6	parts/10^6
yttrium‡	254	9.00	parts/10^6	33.0	parts/10^6
K/Rb‡	254	763.00		230.0	
K/Rb§	—	1500.00		—	

† From Moorbath, Welke & Gale (1969).
‡ From Sheraton (1970).
§ From Holland (1965).

The samples studied include charnockitic Scourian gneisses affected by the 2900 Ma event and a variety of amphibolized gneisses thought to have been derived from similar parent rocks.

The degree of flattening seen in the supracrustal belts and in the meta-igneous complexes where primary structures remain is less than that which might be expected to accompany the very large reduction in width of the sequences measureable in the field, and it appears that simple shear may have played a major part in the early deformation of the Archaean block. Many of the gneisses contain vertical zones of highly sheared rocks which vary from a few metres to a kilometer or more in width. Some at least of these zones of high deformation appear to have developed early in the history of the block since the deformed rocks can be mapped as lithological units around later structures.

Evidence for interleaving basement and cover by the thrusting together of a series of cover/basement slices is presented by McGregor (this volume) and thrusting certainly played a large part in the early intercalation of Archaean metasupracrustal rocks and older basement in the Angmagssalik area (Bridgwater & Gormsen 1969).

Whatever the mechanism, it is clear that the interleaving of basement and cover took place at an early stage, since the supracrustal belts behave as parallel-sided layers in the structural 'succession' which defines the interference patterns discussed below. Furthermore, the intrusion of the post-supracrustal igneous suites (including perhaps in some areas the anorthosite complexes) appears to have been partly controlled by the tectonic intercalation of supracrustal rocks with older basement. There seems also little doubt that the conformable layering produced at this early stage has been accentuated by penetrative deformation which was broadly homogeneous when considered on a regional scale. The obliteration of unconformities on a regional scale is matched by the obliteration of discordant intrusive contacts, as between Ameralik dykes or Nûk gneisses and their country rocks and by the almost complete loss of primary structures (apart from a few examples of very strongly elongated pillow-structures) from the supracrustal rocks. Such changes must clearly have taken place while the rocks were ductile, that is, in a metamorphic environment.

(3) The last stages of deformation whose effects are regionally displayed are those which were broadly contemporaneous with the 'blanket' high-grade metamorphism which terminated at some time after 2900 Ma and before about 2700 Ma. Over much of the central Archaean massif in Greenland the most widespread structural patterns and the dominant metamorphic assemblages date from these episodes and similar patterns and assemblages are preserved as remnants far out in the reworked marginal tracts – for example, within the Laxfordian belt of northwest Scotland and within the Nagssugtoqidian in west Greenland.

In West Greenland, where the structures developed during and before the 2900 Ma event are preserved over a tract several hundred kilometres in length, the tectonic pattern is characterized by fold interference structures on a scale of a few kilometres to several tens of kilometres and is remarkable for the absence of a linear 'grain' (see, for example, Berthelsen 1960). Dome and basin forms and refolded folds are outlined by steeply dipping foliations, the relative scarcity of gentle dips suggesting that the domes and basins have peaked rather than broad closures. Where structures of comparable age survive in Scotland there is a similar absence of linear grain but the dips are predominantly low, suggesting a broader and flatter type of interference-pattern. Small-scale interference-patterns visible in single outcrops sometimes, though not very often, mimic the structures displayed on a regional scale. Such minor structures are displayed even by chromitite layers in anorthosites, a phenomenon that suggests a remarkable ductility at the time of deformation.

The distinctive tectonic pattern developed in the North Atlantic craton at or soon after 2900 Ma resembles that seen in several other provinces dominated by high-grade gneisses, notably in the Limpopo belt of southern Africa (Mason 1972, this volume, p. 463). If the high-grade craton presents a deep section beneath an original 'roof' of granitic domes and greenstone-belts, the pattern might be indirectly related to the gravitational rise of the granitic domes, since such a movement could create a constrictional stress with a vertical extension direction consistent with the forms of the steep-sided interference structures.

The characteristic metamorphism shown by the rocks displaying these complex patterns is of high amphibolite or granulite facies, the rocks being predominantly gneissose, coarse grained and often lacking in strong mineral fabrics. Assemblages of cordierite granulite facies are widely distributed, occurring in Labrador and West and East Greenland. In the central part of the outcrop on the Scottish mainland, the gneisses have charnockitic affinities. Relict granulite-facies assemblages partly replaced by assemblages of amphibolite facies are perhaps even more widespread and suggest that a large part of the craton may have approached granulite grade at the height of the metamorphism.

Some original variations can be recognized, however; there is no evidence that the region of Godthåbsfjord, where the rocks are now of amphibolite facies, has suffered retrogression and, as has been seen, this region provides a 'window' through which early events in the history of the complex can be recognized. The southern part of the Hopedale gneisses in Labrador, the southern Fiskenaesset area and the Frederikshåb area in southwest Greenland give no evidence of having passed through granulite facies conditions (Kalsbeek 1970, personal communication 1972). In southeast Greenland granulite facies rocks are only locally developed and where they are found show clear evidence of developing from earlier amphibolite-facies rocks (Bridgwater & Gormsen 1969; Andrews *et al.* 1971) either as local hot spots or as large-scale marginal phenomena in areas surrounding the late Archaean leuco-norite-charnockite complexes of Skjoldungen and Angmagssalik. The granulite facies metamorphism appears in southeast Greenland to have taken place essentially statically without the disruption of the structures developed under amphibolite facies conditions, a conclusion also reached by V. R. McGregor (this volume) for the development of granulite facies conditions in the Nordland area to the north of Godthåb. Similar 'windows' which escaped the most profound metamorphism and remained in amphibolite facies may exist in the southern part of the Archaean outcrop on the Scottish mainland and the northern part of the Outer Hebrides, though in both regions the picture has been complicated by early Proterozoic reworking. In the Glenelg region of the Scottish mainland, assemblages including eclogites and kyanite-garnet gneisses occurring in the basement of the Caledonides are probably of similar age but have not, so far, been securely dated: kyanite which occurs, but without true eclogites, in the Outer Hebrides and east Greenland may have formed during Proterozoic deformation.

The 'blanket' metamorphism dated at 2900 or 2850 Ma in West Greenland (Evenson *et al.* in press and O.I.G.L. & B. F Windley, in preparation, respectively), 2800 Ma in southeast Greenland (Gulson & Krogh 1972) and at 2900 Ma or somewhat later in Scotland (Moorbath, Welke & Gale 1969; Pidgeon & Bowes 1972), was imposed on all rock types mentioned in the previous section, blotting out much of the geological and isotopic evidence relating to the earlier history. Since it affects supracrustal rocks which must initially have been at low temperatures as well as older gneisses containing dyke swarms one can infer that the metamorphism reflected an actual rise in crustal temperature in the section of the crust now available for study. This would suggest that the period of high-grade metamorphism had a definite beginning as well as an end at the crustal level now exposed and that it was preceded by a period of lower temperatures. Whether these lower temperatures coincided with a period of crustal stability or whether the crust remained essentially mobile throughout the period before 2900 Ma remains to be established, as does the real significance and duration of the high regional temperatures attained during the 'blanket' metamorphism.

Since Ramberg (1951) suggested that a massive change in the bulk composition of the lower parts of the crust might accompany granulite facies metamorphism there have been various attempts to relate the relative abundances of mobile elements to the metamorphic history of the rocks in the craton. The granulitic Scourian gneisses of the Scottish mainland were shown by Holland (1965), Sheraton (1970) and Moorbath *et al.* (1969) to be low not only in potassium (a feature they share with many gneisses in other parts of the craton) but also in uranium, thorium, yttrium and rubidium, a character ascribed by some authors to depletion during metamorphism in which the more mobile elements moved upwards (cf. table 4). However, interpretation of the available results is far from simple since most studies have been carried out on complexes made up of different proportions of rocks of very different ages and possibly very different pre-metamorphic characters. For example, the ancient Amîtsoq gneisses of West Greenland have lead isotope ratios indicating early U/Pb ratios lower than those of any known terrestrial rock (O.I.G.L. & McGregor 1971), but relatively high rubidium and potassium contents. On the other hand, the much younger Nûk gneisses in the same area have low potassium contents like many granitic rocks formed in other parts of the crust before 3000 Ma ago. Regional geochemical studies on terrains containing an unknown mixture of these two gneiss groups involved in later metamorphism could easily be interpreted incorrectly and it seems that much more information is required about the primary composition of identifiable rock units within the complex before the effects of regional metamorphism can be interpreted. The fact that Rb/Sr and whole-rock lead studies give reasonable isochrons from gneisses in various parts of the complex shows there must have been some migration of material around 2700 to 2900 Ma ago and it would be absurd to assume that such a migration and homogenization was restricted to the elements used in age-determination studies. However, so far, individual studies on the bulk chemistry of rock groups within the complex have shown surprisingly little change in composition which could be attributed to the massive migration of material. The primary differentiation trends of the Fiskenaesset meta-igneous complex do not appear to have been strongly affected by the periods of heavy deformation and metamorphism through which the body has passed (Windley *et al.* in press). Ultrabasic lenses throughout the complex show very little alteration except for a marginal rim of biotite or phlogopite suggesting that water and potassium penetrated into these bodies for a few centimetres only. Preliminary rock-staining studies on granulite facies gneisses from Sukkertoppen in West Greenland and the Tingmiarmiut area of southeast Greenland have shown the high-grade rocks to contain large amounts of potassium feldspar, while the much more comprehensive studies of Kalsbeek, Ghisler & Thomson (1972) have shown no statistical differences in the proportions of potassium feldspar and plagioclase in sands derived from granulite-facies and amphibolite facies terrains in the Fiskenaesset district. If migration of major elements has occurred then it does not appear to have been on the scale envisaged by earlier writers.

5. Stages of stabilization

Until the ending of the 2900 Ma metamorphic event, there appear to have been no very consistent differences between the central craton of the North Atlantic region and the enveloping zones which later became parts of the early Proterozoic mobile belt (figure 1) either in terms of rock-types, or in terms of tectonic and metamorphic styles. Small massifs and remnants of Archaean gneisses which escaped wholesale Proterozoic reworking occur well within the Nagssugtoqidian and Laxfordian belts while reworked Archaean rocks form the greater part of these belts. Allowing for the effects of reworking, it can be inferred that soon after 2900 Ma a complex of gneisses of granulite or amphibolite facies incorporating narrow supracrustal belts and anorthosites not only existed at depth throughout the present Archaean craton, but also extended at least some hundreds of kilometres into the border zones: no limit can yet be set to the original extent of this complex.

The geological events which took place over the next thousand million years, 2700 to

1700 Ma, served collectively to differentiate the central massif from its surroundings: whereas the tectonic and metamorphic patterns already established were modified only to limited extents in the massif, they underwent extensive reworking in the border regions where several major episodes of deformation and metamorphism took place. Most of these episodes were Proterozoic, affecting dyke-swarms emplaced at about 2600 to 1900 Ma as well as earlier rocks, but some appear to date from the closing stages of the Archaean and in both Greenland and Scotland several workers have suggested that the areas which later marked the border-zone between the Archaean block and the mobile belts to the north and south, had been areas which either remained unstable from about 2600 Ma onwards or which were reactivated at a very early stage in the Proterozoic (see Bridgwater, Escher & Watterson, this volume, p. 513). In Scotland a period of metamorphism and deformation giving rise to a number of steep movement-zones in the Archaean gneisses has been called the Inverian episode and has been dated at around 2200 to 2400 Ma (Evans 1965).

The sedimentary and igneous rocks added to the crust after the 2900 Ma event fall into two principal categories, those of the first resembling earlier contributions in composition and relationships, those of the second being distinctive in both respects. In the first category, two assemblages are important. A few groups of supracrustal rocks and associated intrusives, dominated by basic material but including acid volcanics and pelitic and calcareous metasediments, appear to have accumulated after the 2900 Ma event. These groups, which show metamorphism of low grades but which are strongly deformed and interleaved with their basement, include the Tartoq belt of southwest Greenland (Higgins 1968) and the Loch Maree series of northwest Scotland (table 1). Small intrusions of tonalites, diorites and granites with minor carbonatites and appinites bearing some resemblance to the more widespread tonatitic suite emplaced before the main episode of high-grade metamorphism are common in southeast Greenland and occur more locally in northwest Scotland; examples of this suite have yielded ages of 2600 Ma in the southern Outer Hebrides (Francis, Moorbath & Welke 1971) and 2200 Ma in southeast Greenland (Bridgwater 1970). Local developments of granulite-facies metamorphism are seen in the vicinity of some of these intrusions.

New rocks in the second category are represented by swarms of basic dykes, mostly tholeiitic (Dearnley 1962; Chadwick 1969), but including a variety of other basic and ultrabasic (picritic) types (Berthelsen & Bridgwater 1960). These swarms were developed on a regional scale – the Scourie dyke swarm of northwest Scotland, for example, is not less than 250 km across the strike direction – and although only a few examples have so far been accurately dated it seems clear that very large numbers of dykes were emplaced in one or more phases after about 2600 and before about 2000 Ma. This period in which mantle material was tapped and steep crustal fractures opened not only in the central massif but also in many parts of the marginal belts was marked by a style of geological activity conspicuously different from that of earlier periods. Dykes emplaced in the craton tend to have simpler forms than those of the bordering zones some of which (e.g. the Kangamiut dykes in West Greenland) appear to have been emplaced during phases of movement (Windley 1970). Some dykes show partial or complete recrystallization to assemblages of amphibolite or even granulite facies which have been variously attributed to emplacement in hot country rocks (by, for example, O'Hara 1961; Moorbath & Park 1972; Tarney 1972) and to subsequent reheating (Dearnley 1962).

Basic dykes emplaced near the margin of the craton were partly amphibolized and deformed by Proterozoic activity.

Faulting affected the craton during its last stages of stabilization. A network of several generations of faults traverses the crystalline basement displacing many of the basic dykes. Some areas have undergone more complex faulting than others. The Fiskefjord area of West Greenland has a prominent set of northeast trending faults with displacements up to 1.5 km (Berthelsen & Bridgwater 1960) and there is an important northeast trending fault at least 200 km long in West Greenland with a lateral displacement of up to 2 km (Windley, 1972).

Some faults movements took place under sufficiently deep-seated conditions to be responsible for recrystallization of the wall rocks. The Fiskefjord faults (Berthelsen & Bridgwater 1960) and some faults in Amitsuarssuqasuaq fiord in West Greenland have caused amphibolization of dolerite dykes. In one example, 30 cm of amphibolite was produced by a fault with a displacement of 10 m (Windley, 1972). Larsen (1971) has recorded a K/Ar age of 1865 Ma from one such amphibolized dyke. This may indicate that movement on the deep-seated faults cutting the Archaean basement took place at approximately the time at which regional deformation was affecting the Proterozoic mobile belts to the north and south.

Bridgwater and Windley wish to thank the Director of the Geological Survey of Greenland for permission to publish relevant information used in this paper and for constructive conversations with their colleagues especially V. R. McGregor, J. Myers and F. Kalsbeek in the Geological Survey of Greenland. Bridgwater would like to acknowledge the receipt of a grant from the Danish Natural Science Research Committee enabling him to travel to southern Africa for comparative studies of the Archaean rocks in that area.

References (Bridgwater *et al.*)

Andrews, J. 1972 Stratigraphic, structural and metamorphic features of Archaean (pre-Ketilidian) rocks in the Frederikshåb area in *The early Precambrian of Scotland and related rocks of Greenland.* University of Keele (in the Press).

Andrews, J., Bridgwater, D. B., Gulson, B. & Watterson J. 1971 Reconnaissance mapping of south-east Greenland between 62° 30′ N and 60° 30′ N. *Rapp. Grønlands geol. Unders.* **35**, 32–37.

Berthelsen, A. 1960 Structural studies in the Precambrian of western Greenland Pt 2, Geology of Torqussap Nuna *Meddr Grønland* **123**, 1–223.

Berthelsen, A. & Bridgwater, D. 1960 On the field occurrence and petrography of some basic dykes of supposed Precambrian age from the southern Sukkertoppen district, Western Greenland. *Meddr Grønland* **123**, 1–42.

Bowes, D. R., Barooah, B. C. & Khoury, S. G. 1971 Original nature of Archaean rocks of northwest Scotland. *Spec. Publ. geol. Soc. Austr.* **3**, 77–92.

Bridgwater, D. 1970 A compilation of K/Ar age determinations on rocks from Greenland carried out in 1969. *Rapp. Grønlands Geol. Unders.* **28**, 47–55.

Bridgwater, D. 1971 Routine K/Ar determinations on rocks from Greenland carried out for G.G.U. in 1970. *Rapp. Grønlands Geol. Unders.* **35**, 52–60.

Bridgwater, D. & Gormsen, K. 1968 Precambrian rocks of the Angmagssalik area, East Greenland. *Rapp. Grønlands Geol. Unders.* **15**, 61.

Bridgwater, D. & Gormsen, K. 1969 Geological reconnaissance of the Precambrian rocks of south-east Greenland. *Rapp. Grønlands Geol. Unders.* **19**, 43–49.

Bridgwater, D. & Windley, B. F. 1972 Anorthosites, post-orogenic granites, acid volcanic rocks and crustal development in the North Atlantic shield during the mid-Proterozoic *'Granite 71' Symposium, Geol. Soc. S. Afr.* (in the press).

Chadwick, B. 1969 Patterns of fracture and dyke intrusion near Frederikshåb, south-west Greenland. *Tectonophysics* **8**, 247–264.

Coward, M. P., Francis, P. W., Graham, R. H., Myers, J. S. & Watson, J. 1969 Remnants of an early metasedimentary assemblage in the Lewisian complex of the Outer Hebrides. *Proc. geol. Ass.* **80**, 387–408.

Dawes, P. R. 1970 Bedrock geology of the nunataks and semi-nunataks in the Frederikshåbs Isblink area of southern West Greenland. *Rapp. Grønlands geol. Unders.* **29**, 1–60.

Dearnley, R. 1962 An outline of the Lewisian complex of the Outer Hebrides in relation to that of the Scottish mainland. *Q. Jl geol. Soc. Lond.* **118**, 143–176.

Dearnley, R. & Dunning, F. W. 1968 Metamorphosed and deformed pegmatitic and basic dykes in the Lewisian complex of the Outer Hebrides and their geological significance. *Q. Jl. geol. Soc. Lond.* **123**, 353–378.

Escher, J. C. 1971 The geology of Akuliaq. Doc es Sci. Thesis, Lausanne University.

Evans, C. R. 1965 Geochronology of the Lewisian basement near Lochinver, Sutherland. *Nature, Lond.* **207**, 54–56.

Evenson, M. M., Murthy, V. R. & Windley, B. F. In prep. Rb–Sr dating of Fiskenaesset complex, West Greenland.

Francis, P. W., Moorbath, S. & Welke, H. J. 1971 Isotopic age data from Scourian intrusive rocks on the Isle of Barra, Outer Hebrides, north-west Scotland. *Geol. Mag.* **108**, 13–22.

Gulson, B. L. & Krogh, T. E. 1972 U/Pb zircon studies on the age and origin of post-tectonic intrusions from South Greenland. *Rapp. Grønlands geol. Unders.* **45**, 48–53.

Higgins, A. K. 1968 The Tartoq Group on Nunaquqortoq in the Itlerdlak area, south-west Greenland. *Rapp. Grønlands geol. Unders.* **17**, 1–17.

Holland, J. G. 1965 Geochemical studies in the Lewisian. Thesis for D. Phil. Oxford.

Hunter, D. R. 1970 The ancient gneiss complex of Swaziland. *Trans. geol. Soc. S. Afr.* **73**, 107–150.

Kalsbeek, F. 1970 The petrography and origin of gneisses, amphibolites and migmatites in the Qasigialik area, south-west Greenland. *Bull. Grønlands geol. Unders.* **83**, 1–70.

Kalsbeek, F. 1972 Field work in the Fiskenaesset area, southern west Greenland. *Rapp. Grønlands Geol. Unders.* **45**, 22–24.

Kalsbeek, F. & Leake, B. E. 1970 The chemistry and origin of some basement amphibolites between Ivigtut and Frederikshaab, south-west Greenland. *Bull. Grønlands Geol. Unders.* **90**, 1–36.

Kalsbeek, M., Ghisler & Thomsen, B. 1972 The composition of sands from the Fiskenaesset region, south-west Greenland. *Bull, Grønlands geol. Unders* (in prep.).

Keto, L. 1969 Järnmalmfyndigheten vid Isua, Väst Grönland. *Abst. IX Nord. Geol. Vintermøde Lyngby*, 1970.

Larsen, O. 1971 Reconnaissance K/Ar dating of samples from West Greenland between Søndre Stromfjord and Frederikshaab Isblink. *Rapp. Grønlands geol. Unders* **35**, 44–48.

Mason, R. 1973 The Limpopo mobile belt – southern Africa. *Phil. Trans R. Soc. Lond.* A **273**, 463–485 (this volume).

McGregor, V. R. 1973 The early Precambrian gneisses of the Godthåb district, West Greenland. *Phil. Trans. R. Soc. Lond.* A **273**, 343–358 (this volume).

Moorbath, S., Welke, H. & Gale, N. H. 1969 The significance of lead isotope studies in ancient, high-grade metamorphic basement complexes as exemplified by the Lewisian rocks of north-west Scotland. *Earth Pl. Sci. Lett.* **6**, 245–256.

Moorbath, S. & Park, R. G. 1972 The Lewisian chronology of the southern region of the Scottish mainland. *Scott. Jl Geol.* **8**, 51–74.

O'Hara, M. J. 1961 Petrology of the Scourie dyke, Sutherland. *Min. Mag.* **32**, 848–865.

Oxford Isotopic Geology Laboratory & McGregor, V. R. 1971 Isotopic dating of very early Precambrian amphibolite facies gneisses from the Godthaab district, West Greenland. *Earth Pl. Sci. Lett.* **12**, 246–259.

Phaup, A. E. 1971 Provisional map of the granitic rocks of the Rhodesian craton. Excursion guide-book 'Granite 71' Symposium, *Geol. Soc. S. Afr.*

Pidgeon, R. & Bowes, D. R. 1972 Zircon U–Pb ages of granulites from the central region of the Lewisian, north-west Scotland. *Geol. Mag.* **109**, 247–58.

Ramberg, H. 1951 Remarks on the average chemical composition of granulite facies and amphibolite or epidote–amphibolite facies gneisses from the Godthaab district, West Greenland. *Medd. Dansk geol. Foren.* **12**, 27–34.

Sheraton, J. W. 1970 The origin of the Lewisian gneisses of north-west Scotland, with particular reference to the Drumbeg area, Sutherland. *Earth Pl. Sci. Lett.* **8**, 301–310.

Stockwell, C. H. 1968 Geochronology of stratified rocks of the Canadian shield. *Can. J. Earth Sci.* **5**, 693–698.

Stowe, C. W. 1968 Intersecting fold trends in the Rhodesian basement south and east of Selukwe. *Geol. Soc. S. Afr.* annexure to **71**, 53–78.

Tarney, J. 1972 The Scourie dyke suit in Assynt in *The early Precambrian of Scotland and related rocks of Greenland.* University of Keele (in the press).

Taylor, S. R. 1964 Abundance of chemical elements in the continental crust. *Geochim. et cosmochim. Acta* **28**, 1273–1285.

Watson, J. 1965 Lewisian, in *The geology of Scotland* (ed. Craig) Oliver & Boyd.

Windley, B. F. 1969 Anorthosites of southern West Greenland. *Mem. Am. Ass. Petrol. Geol.* **12**, 899–915.

Windley, B. F. 1970 Primary quartz ferro-dolerite/garnet amphibolite dykes in the Sukkertoppen region of West Greenland. In *Mechanism of igneous intrusion* (ed. G. Newall and K. Rast) **6**, 79–92. (Also *Geol. J.*, sp. issue no. 2.)

Windley, B. F. 1972 Regional geology of early Precambrian high-grade metamorphic rocks in West Greenland. Pt. 1. Kangnaitsoq to Ameralik fjord. *Rapp. Grønlands geol. Unders*, **46**, 1–46.

Windley, B. F. & Bridgwater, D. 1971 The evolution of Archaean low- and high-grade terrains. *Spec. Publ. Geol. Soc. Aust.* **3**, 33–46.

Windley, B. F., Herd, R. K. & Bowden, A. A. The Fiskenaesset complex, West Greenland, Pt. 1. A preliminary study of the stratigraphy, petrology and whole-rock chemistry. *Bull. Grønlands geol. Unders.* (in the press).

21: A GEOCHEMICAL COMPARISON OF MAJOR ARCHAEAN GNEISS UNITS FROM NORTHWEST SCOTLAND AND EAST GREENLAND

J. Tarney, A. C. Skinner, and J. W. Sheraton

ABSTRACT

A detailed geochemical study has been made of some of the major gneiss groups of the Lewisian of the Scottish mainland, and comparable reconnaissance studies have been made of Archaean gneiss units of equivalent age in parts of the Outer Hebrides and East Greenland. The study incorporates the results of analyses for 24 major and trace elements in over 700 samples. Major differences occur between some of these groups, particularly with regard to K, Rb, Th and U, these elements being strongly depleted in gneisses with charnockitic affinities, giving high K/Rb ratios and low Rb/Sr ratios. These differences probably go back at least 2600 m.y., and the geochemical characteristics have in large part survived subsequent tectonic activity and retrogression. A model for the early Archaean crust is indicated in which the lower crust is depleted in the above elements, probably assisted by mantle degassing. Metamorphism at granulite facies with net potash loss gives rise to rocks of charnockitic character in the lower crust; the alkalies and other elements expelled contribute to metasomatism and migmatisation at higher crustal levels.

SUMMARY

The main fact that emerges is that the charnockitic gneisses show depletion in the potassium, rubidium, thorium, and uranium, with consequent high K/Rb and low Rb/Sr ratios. This difference may be referred to some time prior to 2600 million years ago. The lower part of the Archean crust of the continental areas is believed to have suffered depletion in these elements, assisted by mantle degassing. The alkalies and other elements moved to higher crustal levels where they contributed to metasomatism and migmatization. The sources of the residual charnockitic granulite facies rocks are thought to have been mainly of orthogneiss character, gneisses derived from igneous rocks in the first instance.

This conclusion is in line with widely accepted modern concepts that view granulite metamorphic rocks of charnockite character as being impoverished residuals left behind by upward migration of these elements. Windley (1969) did not favor this wholesale impoverishment of charnockitic terrains, because the differences in content of the key elements between anorthosite layers and their hosts were difficult to reconcile with an upward streaming of dispelled mobile elements through the entire rock mass. Later (see Paper 38), however, Windley accepted this model as a possible answer to the unquestionable potash depletion in granulites, although he suggested as an alternative explanation that the potash-deficient charnockitic granulites may have been derived from isochemical metamorphism of a preexisting potash-deficient tonalite suite, such as characterizes very ancient, large-scale, postgeosynclinal granitic invasions (Paper 5).

Reprinted from *Rept. Sec. 1,* 24th Int. Geol. Congr., Ottawa, 1972, p. 162. The summary was prepared by the volume editor.

Reference

Windley, B. F. (1960). The evolution of the Precambrian basement of south west Greenland. *Spec. Paper 5, Geol. Assoc. Canada,* 155–161.

PAPER 22: COMMENTARY

Pavlovskiy notes that Frolova, in her pioneer work on the Aldan region (of the eastern USSR), adduced that the gneisses had a sedimentary origin, and that the source for the sedimentary material was the weathering and transformation of basic rocks, a conclusion that, in 1964, found support from the distinguished sedimentologist, Ronov. The sediments are supposed to have been chemogenic quartzites and high-alumina rocks (now converted to kyanite-, corundum-, sillimanite-, and cordierite-bearing metamorphics); such sediments were possibly derived from lateritized basalt profiles. This implication concerning early *weathering* and *lateritization* is of very general interest.

The granites of the Aldan region were emplaced after sedimentation during the course of deformation and metamorphism. Migmatites give ages from 2600–3000 million years. The oldest Archean intrusions include anorthosites within the Iyengra Series of schists, and such intrusions were much later granitized and in part converted to charnockites; but the later granitization and granulite metamorphism were quite unrelated to the emplacement of the anorthosites, and took place between 2400 and 2300 million years ago. The oldest Iyengra metasediments and associated basic schists are thought, on geochronologic grounds, to be late Archean, although they could be Katarchean. They form a sequence up to 9 km thick. Very great ages have been obtained on the isolated Kurultino-Gonamskiy complex (K/A dating yields 4520 million years on hypersthene, 3400 million years on amphibole, and 2590 million years on biotite), but the relation of this complex to the Iyengra Series remains uncertain, and the age dating is inconclusive because of this spread of mineral-based numbers. Carbonate rocks are found in subordinate members of the Iyengra Series. There are two superimposed sequences, the Timpton Series of aluminous gneisses, schists, and subordinate basic rocks, and the Dzheltula Series of aluminous schists and gneisses and carbonate rocks. The upper part of the Aldan succession appears to show characteristics approaching those of platform sedimentation, in contrast to the geosynclinal character of the lower part. The most striking feature of the Aldan region is the pattern of large and small granite–gneiss domes, and this is the area that has provided the type example of "nuclear" terrains (Salop, 1972).

In the Ukraine, a very ancient nuclear terrain has also been recognized, much of it by drilling. This terrain includes components ranging from 3600–2600 million years old. There is an absence of delineation in this terrain of sharply defined boundaries between rigid blocks and mobile zones. The greenstone assemblages formed large isolated basins including basic lavas, clastic sediments, and ferruginous quartzites (banded iron formation), together with sills and stocks of basic and ultrabasic rocks. The Konsko-Verkhovtsevskaya Series, an effusive sedimentary sequence, was later subjected to deformation, metamorphism, and granitization with locally variable intensities. Round and irregular domes of granite and migmatite enclosing high-grade paleosomes of the Konsko-Verkhovtsevskaya Series are separated by distinctive brachysynclines preserving the same series in a relatively unmetamorphosed state. The oldest granites are plagioclasic, sodic granites. This nuclear basement was disrupted in the early Proterozoic, and the Krivoi-Rog and Orekhovo-Pavlograd series, consisting of basic effusives, terriginous deposits, and ferruginous quartzites (banded iron formation) were superimposed, straddling the craton along sutural depressions. No evidence of "aboriginal" (primordial) granites has yet been found in the Ukraine or in the Aldan region.

Pavlovskiy goes on to consider the Baltic Shield. Here the Kola Series consists of a sequence of gneisses and schists 8 km thick. Hypersthene gneisses, interbedded aluminous gneisses, and schists are overlain by garnetiferous gneisses that are thought to have once been graywackes, and these in turn by biotite gneisses. Amphibolites are interlayered with the gneisses and there is evidence of polymetamorphism, a granulite metamorphism being overprinted by a later emphibolite metamorphism. Pavlovskiy regards this as *not* being a geosynclinal series, but in saying this seems to be implying that it is nuclear, rather than linear; he does not appear to imply a platform origin, although he may be implying a shelf (epicontinental) origin.

The Kola Series yields radiometric dates around 3600 million years, and is undoubtedly one of the oldest Katarchean sequences represented on the preserved surface of the Earth. Yet it is described as overlying transgressively another even older sequence, forming the Monche, Chuma, and Volchiye Tundra series, a gabbro–norite–labradorite massif. This sequence of basites may well be 4000 million years old or more, according to Pavlovskiy, but geochronologic data appear to be lacking. The suggestion is made that this complex is upper mantle material, a product of primordial zonal fusion. This complex is separated from the overlying Kola Series by a layer of eluvial, coarse, rounded to angular conglomerates, solely composed of material derived from the complex and quite devoid of granitic phenoclasts. Granites, the oldest of which are oligoclasic sodic granites, are invasive into the Kola Series. They are synkinematic and in part domical in pattern. This granite invasion appears to have commenced about 3500 million years ago (K/Ar and Rb/Sr dating).

The Monchegorsk layered intrusives, pyroxenites, peridotites, and norites, with which nickel deposits are associated, are quite distinct from the Monche Tundra massif (see above). This complex is intruded along a gently sloping fault zone, and is subsequent to the Kola folding; it is dated at 3250 million years ($\mp$150). It must be regarded as a protoplatformal "trappean" manifestation, and as such is one of the earliest such major intrusive manifestations recorded. The Kola series and associated rocks were straddled by the Saamides (the Polmos and Poros effusive–sedimentary sequences), which were deposited in long narrow sutural depressions in the cratonic surface. Granitization, localized in these areas 2800–2600 million years ago, provides an upper limit for this platform sedimentation, which must have been broadly contemporary with that of the Witwatersrand in South Africa. Once again, Pavlovskiy finds no evidence of an "aboriginal" granite and concludes that there is no need to hypothesize such a granite.

The Baltic Shield was overprinted by systems of geosynclinal character, the Belomorian, in the Proterozoic about 2100 million years ago.

The remainder of Pavlovskiy's paper is dialectic in character. The claim is made that in these very ancient terrains we *do* see Hutton's elusive "beginning," and arguments are advanced against Donn et al. (1965), who believe that the Katarchean is a cycle superimposed on an earlier silicic, primordial granite crust.

The published works of Pavlovskiy and his associates provide an admirably systematic presentation of the facts as they see them; this summary is the best on the Archean of Russia known to the editor. True, it omits the Anabar Shield, of which Rabken (1968) provides us with a concise summary.

The *Anabar Shield* is the nucleus of Suess "Angaraland" in central Siberia. It is a small triangular shield, a window in an extensive platform of upper Proterozoic and Lower Paleozoic rocks, of 250-km side length; it is made up of the following units:

3. Khapchan Series: garnet hypersthene gneiss, pyroxene gneiss, marbles, calciphyres.
2. Upper Anabar Series: meso- to leucocratic gneisses, pyroxene–plagioclase crystalline schists.
1. Daldyn Series: meso- to melanocratic pyroxene gneisses and pyroxene–plagioclase crystalline schists, quartzites, pyroxene-garnet and cordierite, sillimanite or sapphirine-bearing bands of aluminous gneiss in the order of 10 m thick.

The structure is a steep isoclinal fold system. The assemblages are regarded as geosynclinal, volcanic–sedimentary, and intruded by basic sheets. Regional metamorphism extended from 2700–2000 million years ago (K/Ar dating on pyroxenes and amphiboles), and remetamorphism occurred between 2000 and 1900 million years ago.

22: EARLY STAGES IN DEVELOPMENT OF THE EARTH'S CRUST*

Ye. V. Pavlovskiy

ABSTRACT

The question of the time and manner of the formation of the earth's crust is one that has for a long time interested not only geologists, but also natural scientists, philosophers and a wide circle of social scientists.

The yoke imposed on the human mind by the clergy for centuries held back attempts to penetrate into this endlessly fascinating but for a long time mysterious area of knowledge. In 1788 James Hutton published his conclusion that geological observations give no indications of the existence of any signs at all of the initial process. At the end of the 18th century this conclusion was a heresy, a challenge to the Bible and to religious faith in general. Hutton's sin was that he denied the divine act of creation of the earth and managed to do without the assistance of an omnipotent Creator. Hutton's views, which played an important role in the development of science, were undoubtedly progressive for his age, although in our own day the propaganda for these views made by some scientists, for example A. Holmes, creates the impression of great misunderstanding. These views essentially resemble those of the geological agnostics, who attempted to base their position on the general extremely strong metamorphism of the oldest formations and the high plasticity of the metamorphic complexes, as a result of which all traces of the primary composition of the rocks, their structures and textures, stratigraphy and tectonics and the remains of organisms in them have all utterly disappeared. The fertile soil of such ideas, which have long been contradicted by concrete observations on the various continents, still continues to support such concepts as, for example, orthodox actualism, the belief in the existence of a primary granitic magma and the cyclical nature of geologic processes. Now let us look at the facts.

The Aldan Shield

Twenty years ago, N. V. Frolova began to publish a number of papers (1950, 1951, 1955, 1961, 1962) in which she showed that the thick series of Archean gneisses of the Aldan shield is primarily of sedimentary origin, the banding of the gneisses for the most part coinciding with the primary bedding. In this way great opportunities were opened up for detailed mapping, for studying the tectonics of the Archean rocks and for exploration for deposits of phlogopite that are confined to particular levels in the Archean section. By now these conclusions, which revealed the potentialities of the stratigraphic method, have been widely incorporated into the practical conduct of investigations of old metamorphic series not only of the Aldan shield, but also of other similar structures, and are now considered self-evident. Frolova synthesized a vast amount of material characterizing the chemical composition of all the Archean para-rocks of the Aldan region and arrived at the conclusion that the source material for the formation of the

*Translated from *Ranniye stadii razvitiya zemnoy kory, Izvestiya AN SSSR, ser. geol.*, 1970, no. 5, p. 23–29. The author is with the Geological Institute of the USSR Academy of Sciences, Moscow.

ancient sedimentary series were basic rocks of the type of basalts. In other words, she demonstrated the absence of any "primordial" granitic basement within the area of the Aldan shield. This conclusion was first received skeptically by many, but subsequent researches, particularly those of A. B. Ronov, confirmed it. This is what Ronov wrote in 1964: "The idea that the Archean rocks of the Aldan shield were formed by the weathering and transformation of the oldest basic volcanic rocks was first expressed fourteen years ago by N. V. Frolova. Although she had at her disposal very few data on the chemical composition of the most important types of rocks making up the Archean section of the Aldan region, and based her conclusions almost exclusively on semiquantitative mineralogical calculations, she in general correctly pointed out the path to the solution of this problem" (Ronov, 1964, p. 740).

Frolova considered the distinctive sedimentation processes that occurred in the Archean, and showed that the origin of all the main types of sedimentary rocks belonging to the Archean of the Aldan shield could be explained by the theory of sedimentary differentiation proposed by L. V. Pustovalov (1940). Chemogenic deposits whose ultimate source were the products of lateritic weathering of the basaltoids in the substrate accumulated in a vast basin. In this way she was able to explain the distinctive composition of the entire thick Archean sedimentary series, particularly the very distinctive association of chemogenic quartzites with high-alumina rocks (kyanite-corundum rocks and sillimanite and cordierite schists and gneisses) that was first noted by D. S. Korzhinskiy (1939).

In addition, Frolova showed (1950, 1953, 1961, 1962) that in the geologic history of the Aldan shield the granites appeared *after* the accumulation of the thick sedimentary series, in the course of its deformation, metamorphism and ultrametamorphism.

SUMMARY

Returning to Pavlovskiy's main thesis, his key points are as follows:

1. There is no evidence from any Archean craton in the USSR to suggest the existence of a primordial granite crust, or indeed any need to hypothesize such a crust. In stating this he argues from the factual geological evidence, adopts a pragmatic approach, and does not involve himself in theoretical reasons for postulating the existence of such a crust or introduce any modern concepts of reworking of sialic crust (Papers 23, 23a, 20, and 38).

2. In the three Archean cratons of the USSR, sedimentary and volcanic sequences were later invaded by granite or granitized. This is the characteristic Archean sequence of events, applicable to individual major volcanic–sedimentary, granite–gneiss–migmatite cycles.

3. The interpretation of Frolova and Ronov of the early chemogenic sediments as being derived from the weathering of preexisting basaltic terrain is supported.

In evaluating the last conclusion, which is controversial (Papers 30A and 30B), the high-grade character of the greater part of the Archean in the USSR must be borne in mind. We know for certain that there are very ancient Katarchean clastic sediments that could only have been derived from preexisting silicic terrains in South Africa and Australia—evidence of granite and acid volcanic phenoclasts in rudites—and thus all Katarchean sediments cannot be attributed to sources in crusts on basic rocks. The latter correlation is only applicable to the chemogenic sediments, including the remarkable high-alumina sediments; these may be associates of the clastics, alternating with them in some Katarchean sequences, although they were possibly the only sediments laid down in the very shallow, *early* Katarchean seas. In any case, the hypothesis of derivation from basic crusts does not seem to take fully into account the predominantly submarine nature of Archean sediments, other than the earliest platform sediments; we must pose the question, "Are extensive lateritic crusts likely to be developed on basalts in a submarine environment"? The hypothesis also seems to ignore modern acceptance of the reality of an Archean anoxygenic protoatmosphere (see K. A. Kvenholden, *Geochemistry and the Origin of Life,* and H. Lepp, *Geochemistry of Iron,* both in this Benchmark series). Could such profiles develop under such an atmosphere?

The question of chemogenic sedimentation and the origin of the high-alumina metasediments of the early Archean remains one of the most interesting fields for future research in the Archean. In this connection, it seems appropriate to mention a paper by Sidorenko et al. (1971), even though the rocks referred to as Archean are not of the Kola Series, but occur overlying the Belomorian (Lower Proterozoic). The field mapping of these rocks of a Proterozoic mobile belt discloses a succession through decreasing metamorphism from granulite facies assemblages to unaltered sediments of continental type (a most convincing technique, but only possible where stratigraphy runs oblique to metamorphic boundaries). In this case, a continental red bed sequence seems to have been caught up in profound metamorphism, and the evidence could well be extrapolated to similar Archean granulite terrains, which could include primitive continental sediments and shelf–epicontinental sedimentary sequences laid down under very quiet tectonic conditions and under a paleogeographic regime of slight topographic contrasts. There is no reason to exclude such sedimentary environments from the small, Katarchean protocratons. Such sequences may well have been uplifted with the intervening granitic areas between the Archean greenstone belts, so that all that is preserved of them is their deep, metamorphosed infolded roots.

The importance of *nuclear terrains* (granite domes) in the Katarchean is again emphasized by Pavlovskiy (Paper 29). Salop (1972) has lately provided a detailed account of such a terrain in the Aldan Shield, and similar terrains have been described by MacGregor (1951), Mikhaylov (1969), Choubert (1960, 1965), Goodwin and Ridler (Paper 8), and Ingram (Paper 14). There appears to be an early granite dome element in the Archean of the eastern Yilgarn Block in Western Australia "swamped" by later granite filling of immense distensive zones (Gower and Bunting, 1972). Nuclear terrains are developed in the late extensions of Archean-type cratonic surfaces into the Proterozoic in West Africa, Guyana, and Arizona, and have even been reported from the Paleozoic (Paper 28). They are by no means peculiar to the Katarchean, but have their extreme expression in these ancient systems.

Salop (1972) has concluded that there are two types of domical structure:

1. *Folded ovals:* tectonic forms of high order; groupings of folds in the form of closed rings; oval or irregular systems ranging up to 800 km in diameter or long-axis dimension.

2. *Domes:* simple forms, with layers in their limbs dipping periclinally or complicated by only small folds; diameter or long-axis dimensions of only a few kilometers (up to a few tens of kilometers).

Folded ovals appear to be protocontinental nucleii wrapped around by younger mobile belts. Gneiss domes are the domes of granite that make up the "gregarious granite terrain" (MacGregor, 1951). They are essentially smaller-scale, syntectonic intrusive patterns, components of folded ovals that commonly display diapiric granite intrusive cores with conformable gneiss envelopes. Pavlovskiy suggests that such gneiss dome terrains may be regenerative over a long span of time. There is also evidence for this in Rhodesia.

References

Choubert, B. (1960). Le probléme des structures tectoniques superimposées en Guiane française. *Bull. Soc. Geol. France, Ser. 7, II,* 855–862.

Choubert, G. (1965). Etat actuel de nos connaissances sur la géologie de la Guyane française. *Bull. Soc. Geol. France, Ser. 7, VII,* 129–135.

Donn, W. L., Donn, B. D., and Wilbur, C. V. (1965). On the early history of the earth. *Geol. Soc. Am. Bull., 76.*

Gower, C. F., and Bunting, J. A. Explanatory notes on the Lake Johnston 1:250,000 sheet, Western Australia. *Records Geol. Survey Western Aust.,* 1972, 12, 28p.

MacGregor, A. M. (1951). Some milestones in the Precambrian of Southern Rhodesia. *Proc. Geol. Soc. South Africa, 54,* xxvii–lxxi.

Mikhaylov, B. M. (1969). Geology and commercial minerals of the Liberian Shield. *Izv. Nedra, Moscow,* 179p.

Rabken, M. I. (1968). On the depth and conition of regional metamorphism in the Anabar Shield. *Proc. 23rd Int. Geol. Congr.,* 61–74.

Salop, B. I. (1972). Two types of Precambrian structures: gneiss folded ovals and gneiss domes. *Int. Geol. Rev., 14,* 1209–1228.

Sidorenko, A. V., Lunyeva, O. I., and Nemova, T. V. (1971). The sedimentary origin of granulites of the Kola Peninsula. *Dokl. Akad. Nauk SSSR, 198,* 201–204.

PAPER 23: COMMENTARY

The Grenville structural province of Canada is a linear mobile belt superimposed on Archean and Lower Proterozoic (Aphebian) Huronian rocks. It is shown to consist of a number of elements:

1. Pretectonic basement complex.
2. Pretectonic metasedimentary Grenville granulites.
3. Pretectonic intrusives.
4. Syntectonic intrusives.
5. Posttectonic diabase dikes.

It may be divided as follows:

1. Quartzo-felspathic amphibolite gneisses of Archean age, covering half the area and concentrated toward the north.

2. Quartzites, marbles, and layered paragneisses of Aphebian age, concentrated toward the southwest.

3. Anorthosites, gabbros, mangerites, quartz monzonites, and relic green-rock migmatites of Paleohelikian age, concentrated toward the southeast.

The northeast trending flowage folds of the Grenville are superimposed on east–west trending Kenoran and Hudsonian structures in the Basement Complex. The tectonic overprinting of this polyphase terrain goes back to the 2500 million year old Kenoran (end-Archean) orogeny.

Paper 23 is of paramount importance, defining as it does a type example of a linear mobile belt superimposed on Archean rocks and early Proterozoic platform sediments of a cratonic block. There are many similar superimposed mobile belts in the Canadian Shield, and both the Limpopo Belt of South Africa and the Fraser Belt of Australia provide other analogues. These mobile belts are of interest not only from the point of view of study of subsequent modifications of Archean cratons, but also because in their component elements they closely resemble the intervening granite–gneiss–migmatite (with or without graulite) tracts that separate the Archean greenstone belts and form the major part of any Archean terrain. Roadside cuttings through Grenville rocks east of Sudbury, Ontario, or through the Kapuskasing mobile belt rocks west of Timmins, Ontario, show a remarkable likeness to cuttings through the Archean of the Wheat Belt and Darling ranges in the Yilgarn Block of Western Australia. These intervening granitic areas in Archean terrain only seem to differ from the mobile belts in being markedly less linear. They appear, like the Grenville, to be essentially reworked tracts of preexisting sedimentary, volcanic, and igneous basement. Likewise, the Fraser Range and Limpopo belts include both modified Archean rocks and later, Proterozoic rocks caught up in high-grade, linear metamorphic complexes analogous to that of the Grenville and Kapuskasing. It is by reference to the work of Wynne-Edwards and others working on similar lines that we shall come to fully understand the true nature of the nonlinear Archean granite–gneiss tracts separating the greenstone belts.

23: TECTONIC OVERPRINTING IN THE GRENVILLE PROVINCE, SOUTHWESTERN QUEBEC

H. R. Wynne-Edwards

Department of Geological Sciences, Queen's University, Kingston, Ontario, Canada

ABSTRACT

Systematic mapping of a large area in the central part of the Grenville province has provided structural and textural criteria by which the plutonic and high grade metamorphic rocks can be divided into five tectonic categories determined by their role during the Grenville orogeny. From oldest to youngest these are: (I) a pretectonic basement complex, (II) a pretectonic metasedimentary Grenville Group, (III) a pretectonic, plutonic, intrusive group, (IV) a syntectonic, plutonic, intrusive group, and (V) a post-tectonic family of diabase dykes. The first three categories were deformed and recrystallized during the Grenville orogeny and are respectively represented by (I) quartzofeldspathic and amphibolite gneisses covering half the area and concentrated to the north, (II) quartzite, marble, and layered paragneiss concentrated to the southwest, and (III) anorthosite, gabbro, mangerite, quartz monzonite, and related green-rock migmatitic complexes concentrated to the southeast. These rocks are inferred to be of Archaean, Aphebian, and Palaeohelikian ages respectively.

The structure results from northeast-trending flow folds of the Grenville orogeny superimposed on east-trending structures in the basement complex to the northwest, and on north-trending structures in the rocks to the southeast. These older trends appear to be the results of the Kenoran and Hudsonian orogenies in that order. There is thus a record in this part of the Grenville province of continued tectonic overprinting extending back 2500 m.y. to the Archaean. The Kenoran (about -2500 m.y.), the Hudsonian (about -1750 m.y.), the Elsonian (about -1400 m.y.) and the Grenville (about -950 m.y.) orogenies are each represented, and each contributed 'new' rocks to the total and modified all rocks already present by folding, metamorphism, or both.

INTRODUCTION

With the Grenville Project of the summer of 1964, the Geological Survey of Canada undertook the reconnaissance mapping of the geology of Mont Laurier and Kempt Lake map-areas, comprising 13,000 square miles of the Grenville province of southwestern Quebec. These two map-areas extend from Ste-Agathe westward to south of Maniwaki,, and northward to embrace the headwaters of the Ottawa, Gatineau, Lièvre, Rouge, and Diable Rivers, as far as Casey and Parent on the northern route of the Canadian National Railway (Figure 1). It provides a section through the central part of the Grenville province from within 30 miles of the Grenville front east of Val d'Or to within 20 miles of the Palaeozoic rocks north of Montreal. The systematic and uniform coverage of so large an area affords an unusually comprehensive view of the geological problems of the Grenville province, and has supplied information not obtainable by any other method.

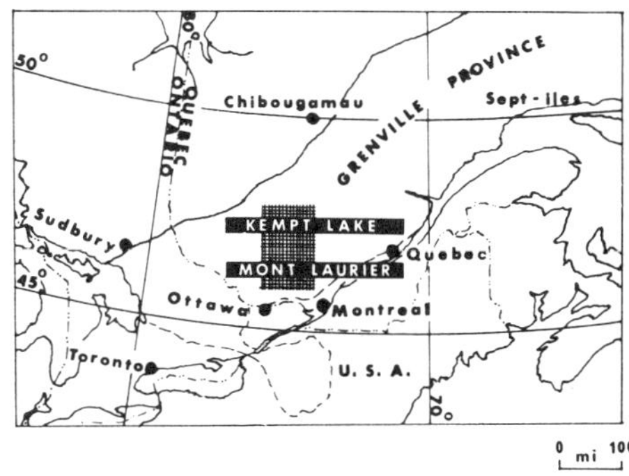

Figure 1: Location map, Mont Laurier and Kempt Lake areas, Quebec.

A preliminary map and a description of the methods used and the results of this survey have been published (Wynne-Edwards, *et. al.*, 1966). These results are here interpreted and extended to show that this part of the Grenville province has successively been subject to the events of the Kenoran (about -2500 m.y.), Hudsonian (about -1750 m.y.), Elsonian (about -1400 m.y.), and Grenville (about -950 m.y.) orogenies, each of these making a new contribution to the sum total of rock types represented and in some fashion modifying those there already. The rocks and structures belonging to each of these events are still distinguishable, although the last overwhelming imprint of the Grenville orogeny produced a dominant northeast-trending structural pattern and wholesale recrystallization to upper amphibolite or granulite facies. Thus, far from being a new contribution to the continental crust about 950 m.y. ago, as has been suggested by proponents of the hypothesis of continental growth, this part of the Grenville province has a history extending back over 2500 m.y. About half the exposed bedrock appears to be part of an Archaean basement complex on which the Grenville Group of metasedimentary rocks was deposited, and the remainder consists of a heterogeneous assemblage of rocks of different ages, almost all of which are much older than the Grenville orogeny, and show signs of having been involved in earlier metamorphic and tectonic events.

[*Editor's Note:* Figures 3 through 6 and much of the section and structure have been omitted owing to limitations of space.]

PREVIOUS WORK

Some parts of Mont Laurier map-area were covered in reconnaissance surveys by Ells (1902) and Adams (1897) that included the results of traverses by Logan and his co-workers in the latter part of the 19th Century. Systematic one-mile mapping was begun in the area by the Quebec Bureau of Mines in 1933, and in the ensuing 30 years maps of roughly two-thirds of Mont Laurier map-area were completed (Figure 2). Most of these are accompanied by brief descriptive reports. The contributors have been Osborne (1935, 1936), Aubert de la Rue (1948, 1953, 1956a, b), Pollock (1956 a, b, 1959, 1960), Katz (1964, 1965), Klugman (1960), and Duesing (1962). In Kempt Lake map-area, the only previous systematic mapping was near Parent by Faessler (1937).

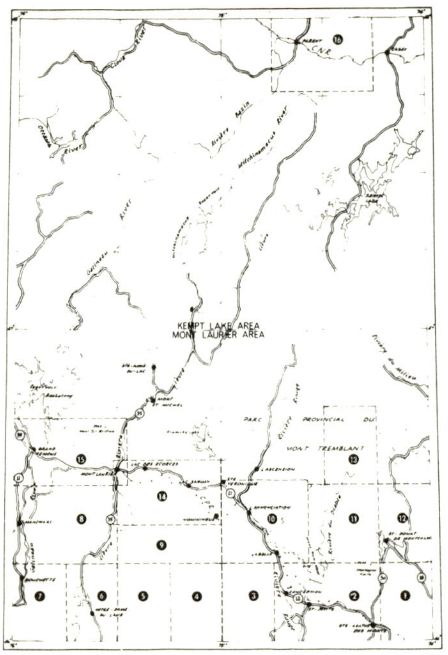

Figure 2: Index map of Mont Laurier and Kempt Lake areas showing one-mile maps of the Quebec Department of Natural Resources.

That the Grenville province contained rocks of different ages, and not simply representatives of the long-famous Grenville Series was clear to Wilson when he first defined it (1925), but the full import of this has only recently been realized. As late as the Royal Society of Canada's symposium on *The Grenville Problem* in 1956 the consensus appears to have been that there was no visible basement for the Grenville Series. In a few later papers there has also been a regrettable tendency to confuse the radiometric 'age' of the Grenville province with the true age of its constituent rocks. On the other hand, Osborne and Morin (1962), Walton and de Waard (1963), Stockwell (1964), and Wynne-Edwards (1964) have suggested that much of the exposed Grenville province is in fact composed of basement rocks older than the Grenville Series (or Group). ['Group' is here preferred to the customary 'Series' because the latter now has a time-stratigraphic connotation. The correlation of the Grenville Group metasedimentary rocks is lithological.] The writer (1964) has emphasized that the only really unifying characteristic of the Grenville province is that the last metamorphic recrystallization took place about 950 ± 150 m.y. ago, and the province is otherwise heterogeneous and the product of a protracted and complex history. The Grenville Group, for example, is largely confined to a radius of 135 miles from Ottawa (the "Ottawa Circle") and is not typical of the whole Grenville province, a fact earlier brought out by Engel (1956).

Stockwell (1964, 1965) assigned the Grenville Group of metasedimentary rocks to an Aphebian age, tentatively correlating them with the Kaniapiskau rocks of the Labrador trough at one end and with the Huronian rocks south of Sudbury, Ontario at the other. Both the Kaniapiskau and Huronian rocks have metamorphic minerals and crosscutting intrusions dating from the Hudsonian orogeny (about -1750 m.y.). He has suggested, as "a working hypothesis", that there is a roughly continuous belt of these rocks through the central part of the Grenville province that has been affected by the Hudsonian orogeny. The rocks to the north and west of this central belt he has assigned to the Archaean, as having been involved first in the Kenoran and then in the Grenville orogenies. To the east and south the main bodies of anorthosite are concentrated, and for these he has suggested a correlation with the large unmetamorphosed anorthosite masses north of the Grenville front in central Labrador, which have the age of the Elsonian "orogeny". The results of the investigation reported here fully justify these postulations, and give them clearer definition and precision over the area studied.

LITHOLOGIC AND TECTONIC ROCK CLASSIFICATION

The bedrock of Mont Laurier and Kempt Lake map-areas is typical of the Grenville province, being made up of plutonic and high-grade metamorphic rocks with potassium-argon radiometric ages close to the mean value of - 945 m.y. characteristic of the Grenville orogeny. [The single exception to this found so far is biotite from porphyritic quartz monzonite belonging to tectonic category III (Geol. Surv. Canada K/A No. 1169, 1965), with a radiometric age of 1205 ± 43 m.y., believed to be a relict from the Elsonian orogeny.] Almost all the rocks are foliated, and most are essentially composed of quartz, feldspar, biotite, hornblende, and pyroxene. A simple petrographic classification on the basis of mineral abundance is not adequate to describe them. A granite, for example, may be massive, gneissic, layered, cataclastic, or contain streaky linear elements; it may be homogeneous or heterogeneous; it may have an igneous or a metamorphic fabric, be coarse or fine grained, porphyritic or not, and may in addition exhibit characteristics that identify it with the amphibolite or with the granulite facies. Each of these features reflects some aspect of its history. Thus although 70 per cent of the rocks exposed in Mont Laurier and Kempt Lake map-areas might generally be classed as granitic, it is possible to divide them into a number of distinct tectonic categories that evolved in different ways. Each tectonic category so devised can next be assigned a relative age on the basis of its role during the deformation and metamorphism of the Grenville orogeny.

Five tectonic categories are distinguished in Mont Laurier and Kempt Lake map-areas, each successively older one having had a more prolonged and complex tectonic history. In brief, there is a group of metasedimentary rocks (the Grenville Group, category II) lying on an older complex predominantly composed of granitic gneisses (I), and intruded first by a sequence of meta-igneous plutonic rocks (III), of which anorthosite, mangerite, and quartz monzonite form the bulk, later by syntectonic gabbro, granite, and syenite (IV), and finally by a few post-tectonic diabase dykes (V). All but the last have been involved in the Grenville orogeny, but the history of the older tectonic categories is much longer and more complex than this, there being both structural and petrological evidence for several major earlier events.

There are several different rock types in each of these tectonic categories; rocks of the same composition may appear in more than one, and the metamorphic grade can also vary independently. The rocks of each tectonic unit are therefore identified as members of a coherent group not purely by their composition or mineralogy, but by textural and structural characteristics that they have in common. The lithology of each rock type or map-unit in Mont Laurier and Kempt Lake map-areas has already been described in accordance with this tectonic classification (Wynne-Edwards, *et al.*, 1966) but a condensed Table of Formations is given in Table I.

TECTONIC CATEGORIES

IV and V. Syntectonic and Later Igneous Rocks

It is best to start with the youngest tectonic categories, the nature and origin of which are not in doubt. In this manner the possible types of tectonic unit are gradually eliminated and the nature of the remaining categories becomes clear. The rocks of the youngest tectonic category (V) are fresh, massive, hypabyssal, intrusive diabase dykes that crosscut all other rocks and are undoubtedly post-tectonic in age. The few found in the map-area are shown in Figure 3.

Older than these are massive or gneissic homogeneous plutonic rocks of two types. The first forms discrete plutons with subcircular outlines and range from mica pyroxenite to alkali syenite in composition. They have hypidiomorphic igneous textures and subhedral feldspar laths. The second type forms smaller, locally crosscutting bodies. Some of these are small plugs of gabbro, but most are composed of medium-grained, pink, leucocratic granite, or of a migmatite that is this granite mixed with other gneisses and with the members of the Grenville Group. The granites are crudely foliated owing to the alignment of biotite.

This syntectonic group of plutonic intrusions might be further divided into late and early tectonic categories depending on the nature of their foliation and the degree to which they are discordant, but they all manifestly are not metamorphic rocks, and their potassium-argon ages are those of the Grenville orogeny (Geol. Surv. Canada K/A Nos. 1056, and 1061, 1965; Wynne-Edwards, *et al.*, 1966). They can thus collectively be classed as syntectonic with respect to the Grenville orogeny, in tectonic category IV. Rocks of tectonic category IV form only about 5 per cent of the total map-area (Figure 3). *This small fraction represents the only rocks that may truly be said to be the same age as the Grenville orogeny and to have an actual age equivalent to their potassium-argon radiometric date of about -945 m.y.* ["Actual", "real" or "structural" age are terms applicable to the date of formation of the rock into a coherent mass for the first time. Subsequent metamorphic recrystallization produces a younger date, sometimes erroneously referred to as the "age" of the rock. This misinterpretation can create considerable confusion.] Even these however, may be partly anatectic in origin and thus also have a previous crustal history.

III. Syntectonic Intrusions

The next tectonic category is composed of large, plutonic masses of anorthosite, gabbro, mangerite (hypersthene syenite), and quartz monzonite. Their distribution is shown in Figure 4. These rocks have a coarse to very coarse grain-size, and are in many cases porphyritic, with large, subhedral tablets of feldspar. Their mineralogy naturally varies with their composition, but is generally that typical of the granulite facies (brown weathering, green or grey perthitic orthoclase and antiperthite, brown hornblende, hypersthene, and minor garnet). This mineralogy is shared by rocks in the granulite facies belonging to other categories, but the plutonic masses of tectonic category III have other unifying characteristics. In addition to their coarse grain-size and generally anhydrous mineralogy, there are places, in particular near the centre of the larger masses, where a hypidiomorphic granular or

TABLE I

Table of Formations in Mont Laurier and Kempt Lake Map-areas

Tectonic Category	Tectonic Age Relative to Grenville Orogeny	Map-Unit*	Formation	Lithology
V	Post-tectonic Intrusions	22	DIABASE	Dykes of fresh diabase
IV	Syntectonic Intrusions		PEGMATITE	Minor pegmatite dykes
		21	BIOTITE SYENITE	Coarse-grained biotite-pyroxene syenite, alkali syenite, and alkali gabbro
		20	GRANITE	Medium-grained, pink biotite alaskite
		19	MIGMATITE	Granite mixed with Grenville Group and other gneisses
III	Pretectonic Intrusions	18	QUARTZ MONZONITE	Porphyritic, coarse-grained quartz monzonite and derived augen gneisses
		17	MANGERITE	Coarse-grained, homogeneous hypersthene syenite
		16	ANORTHOSITE	Anorthosite, norite, and derived cataclastic rocks
		15	GABBRO	Gabbro and hornblende metagabbro, probably not all of one age
		14	GREEN-ROCK COMPLEX	Migmatite of map-units 5-18, all in granulite facies and intimately mixed
II	Pretectonic Sedimentary Rocks	13	GRENVILLE GROUP	Undivided metasedimentary rocks of map units 8-12
		12	HORNBLENDE GNEISS	Hornblende-plagioclase gneiss and amphibolite, same as map-unit 2
		11	MARBLE	White or grey granular crystalline limestone
		10	WHITE ROCK	Calcsilicate rocks, white pegmatite, and pyritic-graphitic gneisses associated with marble
		9	QUARTZITE	White or buff, coarse-grained generally pure quartzite
		8	ALUMINOUS GNEISS	Layered quartz-biotite-feldspar paragneiss, typically with garnet and sillimanite
IC	Pretectonic basement complex, where modified by intrusions of category III	7	MAFIC GRANULITE	Dark, two-pyroxene and hornblende granulite, typically fine grained
		6	LEUCOCRATIC GRANULITE	Fine-grained, massive or streaky cataclastic quartzofeldspathic granulite
		5	CATACLASITE	Fine-grained cataclastic granite gneiss
IB	Pretectonic basement complex, in granulite facies	4	PYROXENE HORNBLENDE GNEISS	Dark, two-pyroxene and hornblende gneiss, typically medium grained
		3	CHARNOCKITIC GNEISS	Medium-grained, locally layered leucocratic granite gneiss with perthite, hypersthene, and brown hornblende
IA	Pretectonic basement complex, in amphibolite facies	2	HORNBLENDE GNEISS	Hornblende-plagioclase gneiss and amphibolite, same as map-unit 12; includes garnetiferous and biotitic varieties that appear metasedimentary
		1	GRANITE GNEISS	Well foliated, typically layered, leucocratic biotite - and hornblende - biotite granite gneisses

*Refers to Preliminary Map (Wynne-Edwards, *et al.*, 1966).

subophitic texture is well preserved. This, together with the presence of inclusions of country rocks, and the large homogeneous, and coherent nature of the plutonic masses, leaves little room for doubt that the rocks were once igneous intrusions. The prevailing texture, however, is cataclastic, the rocks being reduced to augen gneisses or mylonites where cataclasis has been extreme, but more commonly having an inequigranular mortar texture made up of ovoid megacrysts of feldspar set in a crushed and recrystallized ground-mass. This texture is well illustrated in many places by the Morin anorthosite mass near Ste-Agathe, and is also familiar from Buddington's (1939) description of the Adirondack anorthosite body some 180 miles to the south.

In addition to these discrete intrusions, the southern and central parts of Mont Laurier map-area contain rocks mapped as "Green-rock complexes". The dominant rock types in these are granulite and mangerite similar to the rocks around the western side of the Morin anorthosite mass, but intimately mixed with a variety of rocks all with the common mineralogical attributes of the granulite facies as cited above. In this way, map-units 5 to 18 (Table I) are still recognizable constituents. These complexes must represent recrystallized and migmatitic rocks formed by the emplacement of the intrusive rocks of tectonic category III into the basement complex and the Grenville Group (categories IC and II), but later modified by further deformation and recrystallization during the Grenville orogeny. The structural relationships of one such complex in the southern part of Mont Laurier map-area have been discussed by Pollock (1962).

A number of masses of meta-gabbro (Figure 4) have also been included in tectonic category III, as, in common with the rocks described above, they have a relict igneous texture on which partial or complete metamorphic recrystallization has been superimposed. Their tectonic rôle relative to the Grenville orogeny must thus be similar, although they may not be directly related to the other intrusive masses.

The anorthosite and accompanying rocks cut across north-trending structures in the Grenville Group, a relationship used by Logan long ago to establish them as his Upper Laurentian System (1863, p. 839). The intrusions also contain numerous inclusions of quartzite and other rock identifiable lithologically as belonging to the Grenville Group. They are therefore younger than these.

All the rocks of tectonic category III have undergone some metamorphic recrystallization in places. The final products of this recrystallization process are rocks with granulite-facies mineralogy, and with a granular metamorphic fabric in place of the relict subophitic and granulated or cataclastic textures that are more common and which represent intermediate stages of the recrystallization process. Locally, however, olive-green hornblende and garnet or biotite have developed at the expense of hypersthene and plagioclase. This recrystallization and cataclasis show that the rocks are older than the rocks of tectonic categories IV and V and that they have been subjected to a period of post-crystalline deformation and metamorphism. On this basis they can be assigned to a pretectonic (or possibly an early tectonic) rôle relative to the Grenville orogeny. They must have been emplaced in the interval before its culmination and after the deposition and consolidation of the Grenville Group.

II. Pretectonic Sedimentary Rocks

Pursuing the history still further back, the next oldest group of rocks is the metasedimentary Grenville Group. The principal rock types are marble, calcsilicates, quartzite, and aluminous biotite paragneiss. For many years these lithologies have defined the Grenville Series, a name first applied to one of the bands of marble near Grenville, Quebec, south of the present area, but gradually extended by Logan and later workers to include all the marble and intercalated metasedimentary rocks in the Grenville province of southeastern Ontario and southern Quebec. These rocks are better termed the Grenville Group than the Grenville Series, to avoid the now current time-connotation of "Series".

The Grenville Group occurs in all parts of Mont Laurier and Kempt Lake map-areas at least as small masses (Figure 5), but is concentrated in the southwest. The various members of it are intercalated and intricately interfolded, so that all of them may be present within a small area of outcrop. They have not generally been separated on the one-mile maps of the Quebec Department of Natural Resources (Figure 2), but on the four-mile scale of the present mapping their broad regional distribution is readily apparent. Most of the quartzite is concentrated in the centre of Mont Laurier map-area between the broken lines of Figure 5, and in this region quartzite is the dominant lithology of the Group. To the west of this zone, marble and calc-silicate rocks are by far the commonest rocks, whereas to the east the main representative of the Grenville Group is biotite paragneiss. If the rocks of the other tectonic categories are ignored, therefore, as in Figure 5, the southern half of the map-area is made up of three north-trending belts composed predominantly of marble, quartzite, and biotite paragneiss from west to east.

The Grenville Group is cut by the igneous rocks of tectonic categories III, IV, and V, and hence is classed as an older unit, tectonic category II. The rocks are gneissic and layered and have been folded and highly metamorphosed by the Grenville orogeny. An age determination on metamorphic biotite from a typical paragneiss gave 928 ± 30 m.y. (Geol. Surv. Canada, K/A No. 1055, 1965). The rocks can thus be classed as metasedimentary and pretectonic relative to the Grenville orogeny.

I. Pretectonic Basement Complex

More than half the rocks of the area have yet to be assigned to a tectonic category (Figure 6). These rocks are strongly foliated gneisses and granulites, usually with well developed gneissic layering, and hence small scale heterogeneity, but the map-units tend to be uniform in appearance and composition over a wide area, so that they are homogeneous on a large scale. Almost all of Kempt Lake map-area, and large parts of Mont Laurier map-area are underlain by this monotonous gneissic terrain. The gneisses and granulites are generally leucocratic and quartzofeldspathic, but enclose numerous lenses and zones of mafic gneisses and granulites. These also form larger masses that can be mapped individually (Figure 6). The leucocratic rocks typically contain less than 10 per cent dark minerals, of which hornblende and biotite predominate, and the mafic varieties contain between 30 and 80 per cent dark minerals, usually hornblende and pyroxene. This simple bimodal division of the gneisses into leucocratic and mafic varieties is their most persistent and unifying characteristic, and is the basis for their being grouped as a single tectonic unit, a pretectonic basement complex, relative to the Grenville orogeny.

Well-layered, garnetiferous, biotite or hornblende-biotite gneisses in the northwest corner of the area are also included in tectonic category I. They are quartzofeldspathic, but their compositional layering suggests a metasedimentary origin, their composition being akin to that of greywacke. There is no lithological basis for correlating them with the Grenville Group, and they are included in the basement complex because, as shown later, they are believed to be Archaean.

The rocks of tectonic category I, although varying in metamorphic grade and grain-size from place to place, all have metamorphic textures, foliation, and lineation that clearly distinguish them from the igneous rocks of categories III, IV, and V. Both a pretectonic-intrusive and a syntectonic role for them in the Grenville orogeny are thus precluded. They equally cannot represent lateral equivalents of the Grenville Group, firstly because their overall homogeneity is more consonant with a plutonic or igneous parentage than a sedimentary one (with the exception of the layered rocks already noted), and secondly because the familiar rock types of the Grenville Group occur throughout the whole region, at least in small masses, showing that the group must once have formed a more or less continuous cover of roughy constant facies. Tectonic category I must thus be older than the Grenville Group, and its nature suggests that it formed a basement complex on which these rocks were deposited. This is borne out by the absence within it of inclusions of the Grenville Group that cannot be attributed to infolding, and by a structural position that can still be shown to be generally below the Grenville Group, in spite of the complexities of fold geometry. The possibility remains that the rocks represent a lithologically monotonous sedimentary sequence below the Grenville Group but not greatly different in age, and certainly the grade of metamorphism dating from the Grenville orogeny is the same in each at any place and sufficient to obliterate any direct evidence of unconformity between the two. It will be shown later, however, that the gneisses of tectonic category I have relict east-west trending structures which are not shared by the Grenville Group, and that the nature and disposition of its lithologic units are similar to that of Archaean rocks in the Superior province about 40 miles along strike to the west. It seems probable, therefore, that the basement complex is much older.

Three separate pairs of leucocratic and mafic gneisses of the basement complex are included in tectonic category I, each with the same composition and degree of homogeneity, but with distinctive appearance, mineralogy and structure. Set IA consists of quartzofeldspathic and mafic gneisses metamorphosed to middle or upper almandine-amphibolite facies. These typically contain quartz, pink or grey microcline, and grey oligoclase, biotite, green hornblende, and locally epidote, diopside, or muscovite. The metamorphic isograds are shown in Figure 4 and the distribution of the basement complex in Figure 6. Set IA forms the basement complex south of granulite facies isograd in Kempt Lake map-area. The gneisses grade northward into the granulite facies rocks of set IB, a progression marked by the appearance of buff-coloured, white-weathering, perthitic alkali feldspar and brown hornblende in the gneisses between the upper and lower metamorphic isograds in the northern part of the area (Figure 4). Set IB itself occupies the northeastern part of the map-area and is composed of gneisses characterized by brown hornblende, hypersthene, aluminous diopside, and some garnet, as well as the brown-weathering, green perthitic feldspars typical of the granulite facies. The third set, set IC, is largely confined to Mont Laurier map-area, to the south and east a further isograd is shown in Figure 4. This isograd marks an abrupt change from middle amphibolite facies rocks, in which muscovite and epidote are stable, to the granulite facies with the typical granulite facies mineralogy described above. The mineralogy and composition of sets IB and IC are thus the same. The two sets are clearly distinguished texturally. Whereas set IB is composed of medium-grained gneisses, set IC is made up of fine-grained, commonly cataclastic granulites in which streaky lineation is strongly developed and is more prominent than gneissic layering. Their texture is evidently the result of mechanical granulation and partial recrystallization of formerly more coarsely crystalline rocks, and it may take the form of mylonitic laminae, as in the cataclasites, or simply that of an equant, fine-grained, granulated mosaic, commonly with lenticular elements and a prominent lineation, in the rocks mapped as granulite.

The well-defined metamorphic isograd separating sets IC and IA also divides the Grenville Group into granulite and amphibolite facies, and can therefore be traced more or less continuously. Although its position is not everywhere known from our reconnaissance, it clearly outlines the main group of intrusions of tectonic category III, as shown in Figure 4. The fine-grained granulites and cataclasites of the basement complex are thus spatially related to the pretectonic intrusions, an association to be examined further below.

In summary, the rocks of the basement complex change their texture and grade of metamorphism from place to place, and have been divided into three sets. In the northern half of the area they occur both in amphibolite facies and in granulite facies (sets IA and IB), evidently as the result of progressively increasing metamorphic intensity from southwest to northeast during the Grenville orogeny. In the southern and eastern parts of the map-area, around the main masses of pretectonic anorthosite, mangerite, and monzonite of tectonic category III, they occur as fine-grained granulites and cataclasites in granulite facies with mylonitic or cataclastic textures (set IC).

METAMORPHISM

The Grenville Group and the basement complex (tectonic categories I and II) were deformed and recrystallized during the Grenville orogeny, and regionally metamorphosed to granulite and amphibolite facies. The anorthosite and related rocks of tectonic category III were similarly affected, but in these the recrystallization was incomplete, and the deformation largely cataclastic.

The isograds already described (Figure 4) apply to all the metamorphic rocks, and the grade of metamorphism is the same in the Grenville Group and the basement complex wherever they are found together. Two main metamorphic 'zones' are represented: (1) in the central part of the area, where the rocks are either in the quartz-almandine-muscovite or the quartz-almandine-potash feldspar subfacies of the amphibolite facies, and (2) in the northeastern and southern parts of the area, where the rocks are in granulite facies.

The first of these metamorphic zones is characterized by the following minerals in combinations appropriate to the bulk composition: quartz, pink or grey microcline and grey oligoclase, biotite, muscovite, epidote, allanite, sillimanite, almandine, green hornblende, cummingtonite, tremolite, scapolite, diopside, forsterite, chondrodite, grossularite, calcite, and the accessories zircon, apatite, tourmaline, magnetite, and pyrite. The associations hornblende-almandine and hornblende-biotite are particularly common. The second metamorphic zone has rocks containing quartz, green perthitic orthoclase and green oligoclase-andesine (antiperthite), biotite, sillimanite, almandine (in one locality with cordierite), brown hornblende, hypersthene, clinopyroxene, and calcite, with accessory zircon, apatite, spinel, tourmaline, magnetite, ilmenite, and pyrite.

In the northern part of the area, there is a transition between these zones (delimited by the isograds of Figure 4 and described earlier), but to the south the change from wet amphibolite to dry granulite facies is abrupt and may be complete in a distance of 500 feet. It takes place across the composite isograd shown in Figure 4. There is no doubt that this composite isograd is spatially, and hence in some manner genetically, related to the intrusions of tectonic category III, as shown in Figure 4. Close to the larger masses the metamorphic rocks attained granulite facies during the Grenville orogeny; away from them they reached only amphibolite facies. This suggests that the composite isograd embraces rocks baked and dehydrated around the contact of these intrusive masses before the general onset of metamorphism, and hence that it represents a boundary not of metamorphic temperature variation but one between hydrous and anhydrous metamorphic rocks that crystallized at different water vapour pressures, although perhaps similar temperatures, during the Grenville orogeny. Within the area of influence of the large intrusions, the rocks of the basement complex are granulated or cataclastic (tectonic category IC), implying that they were relatively rigid or brittle during the Grenville orogeny as compared with the same rocks remote from the intrusions.

As well as the meta-gabbros shown by different ornament in Figure 4 a few isolated masses of quartz monzonite of tectonic category III occur outside the composite isograd (Figure 4). These rocks, distinguished as map-unit 18a on the geological map (Wynne-Edwards, *et al.,* 1966) are more completely recrystallized than the others of their group, and contain more biotite, suggesting that their small size and relative isolation has brought about their hydration by water derived from the enclosing rocks of the Grenville Group during their metamorphism in the Grenville orogeny.

GRANULITES AND GRANULITE FACIES

Repeated reference to quartzofeldspathic rocks within the granulite facies in Mont Laurier and Kempt Lake map-areas has already been made. The common attributes of these rocks, notably the presence of brown-weathering green perthite and an abundance of hypersthene and dark hornblende, are spectacular and readily recognizable features in the field. In the early stages of mapping there was a natural tendency to group them as a co-genetic suite, in the manner suggested by Osborne and Morin (1962), who termed them

TABLE II

Rocks in Granulite Facies, Mont Laurier and Kempt Lake Map-areas

Tectonic Category	Map-Unit*	Formation	Texture	Mode of Origin
IB	3, 4	CHARNOCKITIC GNEISSES	Medium-grained, equigranular	Crystalline basement rocks, flow folded and elevated to granulite facies by intense metamorphism in the northeastern part of the area
IC	5	CATACLASITE	Mylonitic	Crystalline basement rocks, baked by intrusions of category III and hence relatively brittle and dry during Grenville orogeny
	6, 7	GRANULITE	Fine-grained, equigranular, streaky	As map-unit 5, but deformation less intense during Grenville orogeny and cataclasis less pronounced
II	8-13	GRENVILLE GROUP	Layered paragneisses, metamorphic	Metasedimentary rocks baked by intrusions of category III and hence relatively dry during Grenville orogeny
III	14	GREEN-ROCK COMPLEX	Generally metamorphic	Mixed intrusive rocks of tectonic category III and older rocks of the Grenville Group and basement complex (categories II and IC) baked by the intrusions and subsequently plastically deformed and recrystallized during the Grenville orogeny
	16-18	QUARTZ MONZONITE, MANGERITE, AND ANORTHOSITE	Relict igneous or cataclastic, rarely metamorphic	Deep-seated, dry, plutonic intrusions that crystallized under the conditions of the granulite facies and were later cataclastically deformed and partly recrystallized by the Grenville orogeny

*Refers to Preliminary Map (Wynne-Edwards, *et al.*, 1966).

"green rocks", assigned them to a group of "Laurentide plutonites" that include anorthosite, and stated that they occurred southeast of a "greenrock line" and characterized the "Grenville A subprovince". The distinctive appearance and mineralogy of the granulite facies, however, reflects a particular metamorphic environment rather than a mode of origin, and any rock brought into this environment and able to crystallize within it inevitably assumes the mineralogy stable within it. The monoclinic potash feldspar and perthite testify to high temperatures, and the fine string perthites and the scarcity of the common hydrous minerals of metamorphism indicate that water vapour pressures were low.

The mineralogy of the granulite facies thus reflects a particular metamorphic environment; the texture of any rocks with that mineralogy reflects the manner in which that metamorphic environment was reached. Thus there may be igneous rocks, metaigneous rocks, highly metamorphosed rocks, and polymetamorphosed or previously dehydrated rocks, all with the same mineralogy but with radically different textures.

Table II sets forth the various rock-types in granulite facies exposed in Mont Laurier and Kempt Lake map-areas. They are representative of each of these different origins, each type having reached its final condition by a different route.

As suggested earlier (Wynne-Edwards, 1964, 1965), the prevalence of cataclastic granulite facies rocks in the Grenville province, being the result of dehydration prior to the metamorphism of the Grenville orogeny, might indicate that these rocks belonged to an older crystalline basement complex. This has proved to be too simple, as Table II shows, and in fact the best cataclastic textures are exhibited by pretectonic igneous rocks or by basement rocks baked by them, but not by the basement complex elsewhere. Interestingly, the contrast in texture, and hence in mechanical behaviour between the granulite of Mont Tremblant (basement complex IC) and garnet-sillimanite paragneiss of the Grenville Group on the shore of Lac de la Montagne-Tremblant was noted and discussed more than 70 years ago by F. D. Adams (1897).

TABLE III

Sequence of "Events" in Mont Laurier and Kempt Lake Map-areas

Event	Rocks Produced	Tectonic Category	Correlated Orogeny*	Age*
Intrusion	Diabase	V		Hadrynian
Deformation, regional metamorphism, and intrusion	Granite, syenite and gabbro	IV	Grenville (-950 m.y.)	Neo-Helikian
Intrusion, contact metamorphism	Anorthosite, mangerite, and monzonite	III	Elsonian (-1400 m.y.)	Palaeo-Helikian
Folding on north-south axial planes			Hudsonian (-1750 m.y.)	Aphebian
Deposition of Grenville Group	Grenville Group	II		Aphebian
Folding on east-west axial planes, low-grade metamorphism?	Complex composed of volcanic rocks and greywackes in east-west belts, and granitic gneisses and intrusions as now exposed in the Superior province to the west	I	Kenoran (-2500 m.y.)	Archaean

*As defined by Stockwell, 1964.

STRUCTURE

Chronological Sequence

The tectonic history inferred thus far is assembled in Table III. The relative ages of these events is known from the direct geological evidence already presented, but the basis for their inferred absolute age and their correlation with the major orogenies defined by Stockwell (1964, 1965) has not been given. As with the discussion of the tectonic categories earlier, it is best to examine this by taking the events in reverse order of age.

The diabase dykes are clearly younger than the metamorphism and structures of the Grenville orogeny. By analogy with other radiometrically analysed dykes in the Grenville province, they are probably Hadrynian (*ca.* -750 m.y.) in age (Fahrig and Wanless, 1963). The events of the Grenville orogeny are well documented and supported by potassium-argon age determinations close to the mean value of -945 m.y. found for the orogeny by Stockwell (1964). This can also be taken as the actual age of the rocks of tectonic category IV. The intrusions of tectonic category III are older, being metamorphosed and deformed. Biotite from a porphyritic quartz monzonite of this group (Geol. Surv. Canada K/A No. 1169, 1965) gave a radiometric date of -1205 ± 43 m.y., significantly older than the other radiometric measurements in the area. Stockwell (1964, 1965) has already observed that the anorthosite masses and related rocks in Labrador and in the Grenville province form a large but coherent group and may thus have been emplaced roughly contemporaneously. He has assigned this event to the Elsonian orogeny (1220-1520 m.y., mean -1400 m.y.). The age determination above is somewhat less than this postulated range, but this is consistent with Stockwell's hypothesis, the rock having lost some

Tectonic Overprinting

Figure 10: Pattern of tectonic overprinting produced by superimposed events of the Kenoran, Hudsonian, Elsonian and Grenville orogenies in Mont Laurier and Kempt Lake map-areas. Each pattern and its letter symbol signifies one of these events, and the number of symbols is the number of events affecting that rock unit in that place. Thus the rocks of tectonic category I have undergone two events in the northwestern part of the area (GK); three to the northeast (GHK), and all four around the Morin anorthosite to the southeast (GEHK).

argon during the later partial recrystallization and deformation of the Grenville orogeny. The rocks of tectonic category III can thus be tentatively assigned a Palaeohelikian age and attributed to the Elsonian orogeny.

All the igneous and meta-igneous rocks of tectonic categories III, IV, and V were intruded into the Grenville Group, which must then be still older. The north-south folding in the southern part of the map-area also involved the Grenville Group, but appears to be an event distinct from and more widespread than the intrusive episode of the Elsonian. The north-south structures, as first noted by Logan (1863), appear to be interrupted by the anorthosite and related intrusive masses, so that the folding must have taken place earlier than the Elsonian. Tentatively, therefore, it may be ascribed to the Hudsonian orogeny and assigned an Aphebian age. It can then be compared with the deformation in the Labrador trough to the northeast, which produced structures of a similar north-south orientation. If this correlation is correct, the Grenville Group of tectonic category II must be pre-Hudsonian and hence probably Aphebian in age.

The basement complex is older still. The widespread simple combination of acid and basic lithologies, the concentration of the basic rocks in east-trending belts, and their association with layered rocks that may once have been greywackes are all features in common with the geology of the Superior province to the northwest. The east-west Keewatin or Abitibi greenstone belts of the Kirkland Lake-Rouyn-Malartic area (Wilson, 1962) abut against the Grenville front near Val d'Or, about 30 miles east of the northwest corner of Kempt Lake map-area. There seems every basis for the correlation of the rocks of both areas, the granitic and silicic rocks of the Superior province becoming quartzofeldspathic gneisses within the Grenville province, the associated greenstones becoming hornblendic and pyroxenic gneisses, and both being refolded on northeast trending axial planes, and also, in places, on the earlier north-south trend of the Hudsonian orogeny. On this basis the basement complex of tectonic category I has been assigned to the Archaean, and the relict east-west structures within it to the Kenoran orogeny which is the known source of the parallel structures in the Superior province. Indeed, the mafic rocks concentrated along the north edge of the map-area (Figure 6) are evidently the direct extension of the Abitibi greenstone belt.

Not all the area has been affected by all these events. The effects of the Kenoran orogeny are confined to the places where the basement complex is exposed; the northwestern part of the area has evidently escaped the north-south folding of the Hudsonian orogeny (Figure 11); and the effects of the Elsonian orogeny are generally restricted to the southern and eastern parts of the area, around the large intrusions. The Grenville orogeny, representing the main tectonic episode of folding and metamorphism, affected the entire area, as it did the whole Grenville province. The effects of this multiple tectonic overprinting are shown in the sketch map of Figure 10. The rocks of the basement complex are everywhere affected by the Kenoran and Grenville orogenies (K, G) and in all but the northwestern part also by the Hudsonian orogeny (H). Around the main intrusions of tectonic category III they are also modified by the Elsonian "orogeny" (E). The Grenville Group was likewise everywhere affected by the Grenville orogeny (G), by the Hudsonian orogeny (H) in all but its western part, and also in places by the Elsonian intrusions (E). The latter were modified by the Grenville orogeny, having been emplaced during the Elsonian, and are labelled GE. The rocks of each tectonic category have been given patterns and symbols to signify the number of tectonic events that has affected them in any one place: a rock which has been successively modified by the Kenoran, Hudsonian, Elsonian, and Grenville events, for example, being labelled GEHK. Only a small proportion of the rocks was crystallized or recrystallized for the first time in the Grenville orogeny; these are labelled G. The resulting diagram is complex, but it has been summarized in Figure 11, which shows the main areas of dominance of each event.

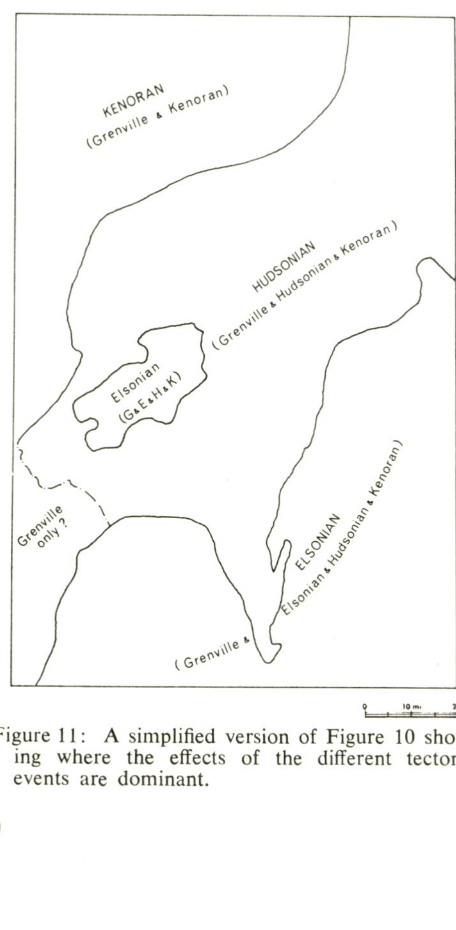

Figure 11: A simplified version of Figure 10 showing where the effects of the different tectonic events are dominant.

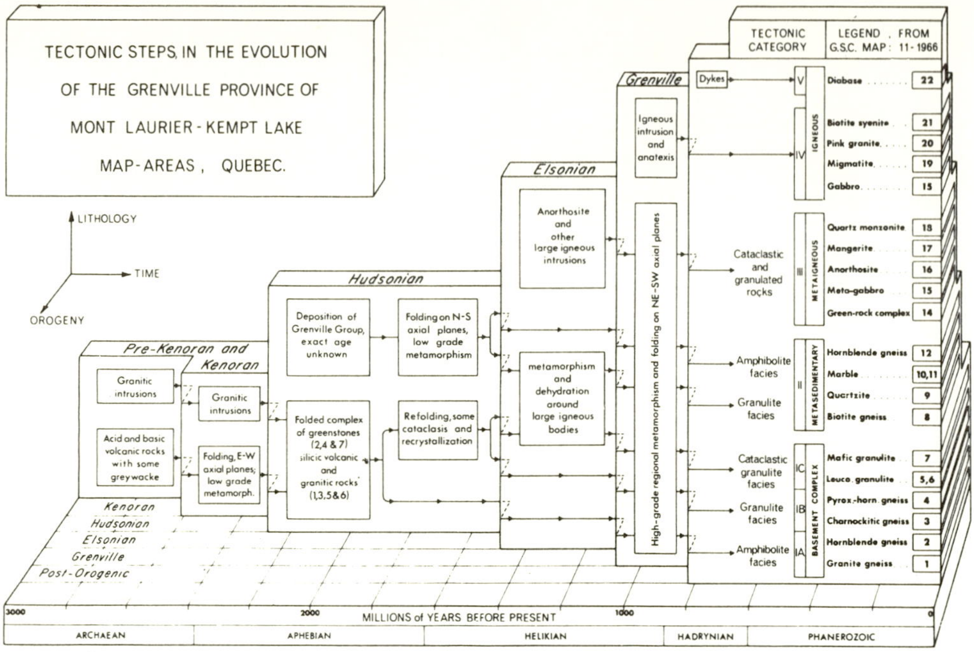

Figure 12: Geological evolution of Mont Laurier and Kempt Lake map-areas showing the progressive growth of the present geology from the Archaean by a series of tectonic steps in which older rocks were reprocessed and new ones were added.

The evolution of the present geology by this sequence of tectonic steps is illustrated in Figure 12, which has time as the horizontal X axis, and the geological legend as the vertical Z axis. Each of the events described can be visualized as a tectonic step, which contributes new rocks to the total present (the 'rise' of the step) and in some manner modifies all other rocks already there (a process represented by the time-interval of the 'tread' of the step). The map legend is thus constructed from a series of steps, like building blocks, added to it at different times, and the history of each block can be traced through all subsequent events. The arrows show the evolutionary paths taken by these blocks in different parts of the map-area from the time of their first appearance. Each orogeny can be thought of as a sort of tectonic 'washing-machine' in which older rocks are cycled, processed, and modified. These tectonic events are not necessarily all orogenies in the sense of mountain-building episodes, but 'orogeny' has been applied, following Stockwell, and for want of a better term, to each of the radiometrically datable tectonic events. The Grenville orogeny and perhaps also the Kenoran are the only ones in which metamorphism and subsequent uplift appear to have taken place, so that they may be the only orogenies *sensu stricto* to have affected the region.

This part of the Grenville province thus contains rocks of Kenoran, Elsonian, and Grenville tectonic episodes. Apart from the Archaean events, these successive tectonic steps appear to have taken place at successively deeper levels so that much of the recorded history may in fact represent the continuous evolution of the Grenville province as an orogenic belt.

Roughly equal proportions of basement rocks and supracrustal metasedimentary rocks are now exposed in the Mont Laurier-Kempt Lake area, so that, statistically at least, the original basement-sediment interface, now much distorted, has been restored to approximately its original crustal level, near sea level. The inferred path of this surface with geological time is traced in Figure 13. Each of the tectonic steps depicted in Figure 12 appears to have been enacted at successively deeper tectonic levels: (1) deposition of the platform sedimentary rocks of the Grenville Group, at or close to sea level; (2) folding which did not influence the present distribution of metamorphic isograds and which, being thus evidently unaccompanied by much metamorphism or dehydration, can be assigned to the epizone; (3) the emplacement of large, relatively anhydrous plutonic intrusions under conditions indicative of the catazone; and, (4) the culminating high-grade regional metamorphism and flow folding of the Grenville orogeny, again of catazonal character.

The complete cycle of the Grenville orogenic belt from initial deposition to the emplacement of diabase dykes under hypabyssal, cratonic con-

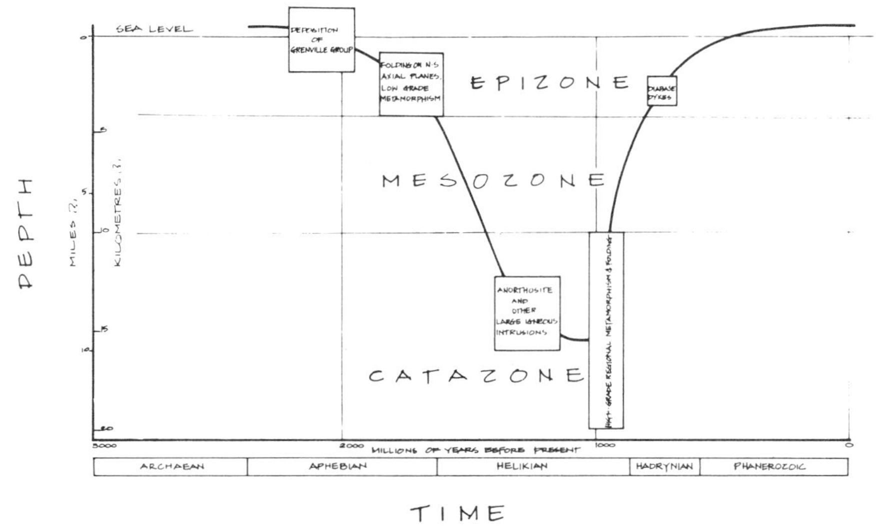

Figure 13: Hypothetical depth-time curve for the basement-supracrustal contact of the Grenville orogenic belt based on the interpretation summarized in Figure 12. In spite of the datable tectonic episodes recorded, the evolution of the belt may have been continuous, with progressive, relatively passive deepening for a period of 1000 m.y. followed by an active orogenic phase, involving uplift and deformation, which occupied less than 300 m.y. before the emplacement of diabase dykes under cratonic conditions.

ditions seems to have required about 1300 million years, with the first 1000 m.y. taken up by burial, and the last 200 or 300 m.y. by active orogenesis and uplift.

The protracted gradual nature of the burial process comes as a surprise if one is accustomed to the tectogene concept of deformation brought about by burial under violent compression, but is further substantiated by the large anorthosite and mangerite bodies, which, strongly in their unmetamorphosed condition north of the Grenville front and locally also within the Grenville province, exhibit primary igneous stratification of the type associated with intrusion under stable conditions. The coarse crystallization of the large anorthosite masses before the main episode of deformation, the existence of primary layering, and the vast extent of their associated contact metamorphic aureoles together indicate stable, deep-seated conditions of emplacement at a time when the country rocks were already hot, but before the main diastrophism. The existence of these intrusions provides considerable independent support for the hypothesis already advanced on other bases (Wynne-Edwards, 1964a, 1965), that the burial of the rocks to the catazone in the Grenville province was a passive process unaccompanied by strong folding, the main folding and deformation being the result of non-uniform viscous flow of the deeply buried rocks towards the surface during isostatic recovery, after they had become sufficiently hot and mobile to yield by recrystallization in response to the accumulated gravitational potential.

The Grenville province is one of the few orogenic belts in which a tectonic cycle has been completed from initial deposition to erosion close to base level. The prevalent high-grade metamorphism testifies to the deep burial of the rocks now exposed, presumably beneath a sedimentary geosynclinal accumulation. The net effect of the whole process of orogenesis appears to have been to restore the original floor of the geosyncline roughly to its ancestral level, the basement being widely exposed and the superincumbent sedimentary load being all but totally removed by erosion. In this way pre-geosynclinal crust is once again exposed, these older rocks being still identifiable, but recrystallized, refolded, in places remelted, and generally updated by the redistribution of radiogenic products.

ACKNOWLEDGEMENTS

The field party of 1964 included ten geologists and ten student assistants and consisted of twenty-five people in all. The enthusiastic co-operation of each one of them is most warmly acknowledged. In particular A. F. Gregory helped enormously with the co-ordination and organization of the project, and shared with the writer the task of mapping with aircraft. E. W. Reinhardt took responsibility for the field party in Kempt-Lake map-area, and P. W. Hay for that in Mont Laurier map-area. The other geologists were D. T. Anderson, V. H. Becker, A. C. Brown, E. H. Gaucher, G. A. Giovanella, and C. M. Nixon. Thanks are also due to S. Duffell of the Crustal Geology Division, Geological

Survey of Canada, who helped in many ways to make the project run smoothly. The compilation of structural, aeromagnetic, and petrological data and the preparation of the geological map on which this paper is based were made possible by the wholehearted efforts of A. F. Gregory (who also critically read the manuscript), C. A. Giovanella, and P. W. Hay. The results are published by permission of the Director, Geological Survey of Canada.

This paper was first prepared while the writer was Visiting Fellow in the Department of Geology and Mineralogy at the University of Aberdeen, Scotland, and he is most grateful for the opportunity so provided.

REFERENCES

Adams, F. D. (1897)—Report on the geology of a portion of the Laurentian area lying to the north of the Island of Montreal; Geol. Surv. Can., Ann. Rept. 1895, pt. J.

Aubert De La Rue, E. (1948)—Nomininque and Sicotte map-areas, Labelle and Gatineau counties; Quebec Dept. Mines, Geol. Rept. 23.

........(1953—Kensington area, Gatineau and Labelle counties; Quebec Dept. Mines, Geol. Rept. 50.

........(1956a)—Trente-et-un-Milles Lake area, Electoral Districts of Papineau, Labelle, Gatineau; Quebec Dept. Mines, Geol. Rept. 67.

........(1956b)—McGill area, Papineau, Labelle, and Gatineau counties; Quebec Dept. Mines, Geol. Rept. 68.

Buddington, A. F. (1939)—Adirondack igneous rocks and their metamorphism; Geol. Soc. Amer., Mem. 7.

Duesing, C. M. (1962)—Preliminary report on Lussier-Tellier area, Joliette and Montcalm counties; Quebec Dept. Nat. Res., Prel. Report 482.

Ells, R. W. (1902)—On the geology of Argenteuil, Ottawa, and part of Pontiac counties, Province of Quebec, and portions of Carleton, Russell and Prescott counties, Province of Ontario; Geol. Surv. Can., Ann. Rept. 1899, pt. J.

Engel, A. E. J. (1956)—Apropos the Grenville; *in* Thompson, J. E., The Grenville Problem, Roy. Soc. Canada., Spec. Publ. 1, pp. 74-96.

Faessler, C. (1937)—Suzer-Letondal area, Laviolette, Saint-Maurice and Abitibi counties; Quebec Bur. Mines, Ann. Rept. 1936, pt. b, pp. 23-36.

Fahrig, W. F. and Wanless, R. K. (1963)—Age and significance of diabase dyke swarms of the Canadian Shield; Nature, vol. 200, No. 4910, p. 934.

Katz M., (1964)—Geology of Cousineau-Rolland area, Montcalm, Terrebonne, and Joliette counties; Quebec Dept. Nat. Res., Prel. Rept. 522.

........(1965)—Lagendre map-area, Joliette and Montcalm counties; Quebec Dept. Nat. Res., Field Work in 1964, p. 12.

Klugman, M. A. (1960)—Doncaster area, Electoral districts of Terrebonne and Montcalm; Quebec Dept. Mines, Geol. Rept. 64.

Logan, W. E. (1863)—Geology of Canada; Geol. Surv. Can., Rept. Prog.

Osborne, F. F. (1935)—Labelle-L'Annonciation map-area; Quebec Bur. Mines, Ann. Rept., 1934, pt. E.

........(1936)—Sainte-Agathe - Saint-Jovite map-area; Quebec Bur. Mines, Ann. Rept., 1935, pt. C.

.... and Morin, M. (1962)—Tectonics of part of the Grenville sub-province in Quebec; *in,* Stevenson, J. S., The Tectonics of the Canadian Shield, Roy. Soc. Can., Spec. Publ. 4, pp. 118-143.

Pollock, D. W. (1956a)—Preliminary report on the Addington-Labelle area, Papineau and Labelle counties; Quebec Dept. Mines, Prel. Rept. 321.

........(1956b)—Preliminary report on Preston-Gagnon area, Electoral Districts of Papineau and Labelle; Quebec Dept. Mines, Prel. Rept. 334.

........(1959)—Preliminary report on Rocheblave area, Electoral Districts of Papineau and Labelle; Quebec Dept. Mines, Prel. Rept. 408.

........(1960)—Preliminary report on Lesage-Rivard area, Electoral District of Labelle; Quebec Dept. Nat. Res., Prel. Rept. 441.

........(1962)—Some petrogenic-structural relationships in an area of Grenville rocks; Trans. Roy. Soc. Can., ser. 3, vol. 56, sec. III, pp. 127-142.

Stockwell, C. H. (1964)—Age determination and geological studies, part II; Geol. Surv. Can., Paper 64-17.

........(1965)—Tectonic map of the Canadian Shield; Geol. Surv. Can., Map 4-1965.

Walton, M. and De Waard, D. (1963)—Orogenic evolution of the Precambrian in the Adirondack Highlands, a new synthesis; Proc. Kon. Ned. Acad. Wetensch., Amsterdam, 66.

Weiss, L. E. (1959)—Geometry of superposed folding; Geol. Soc. Amer., Bull., vol. 70, pp. 91-106.

Wilson, M. E. (1925)—The subprovincial limitations of pre-Cambrian nomenclature in the St. Lawrence basin; J. Geol., vol. 26, p. 325.

Wynne-Edwards, H. R. (1963)—Flow folding; Am. J. Sci., vol. 261, pp. 793-814.

........(1964)—The Grenville province and its tectonic significance; Proc. Geol. Assoc. Can., vol. 15, pt. 2, pp. 53-67.

........(1964a)—Deep zone orogenesis; Proc. XXII Geol. Congress, New Delhi, India, December 1964.

........(1965)—The Grenville province and its tectonic signficance — a reply; Proc. Geol. Assoc. Can., vol. 16, pp. 59-62.

........, Gregory, A. F., Hay, P. W., Giovanella, C. A., and Reinhart, E. W. (1966)—Mont Laurier and Kempt Lake map-areas, Quebec, a preliminary report of the Grenville Project; Geol. Surv. Can., Paper 66-23.

PAPER 24: COMMENTARY

Although some criticism has been voiced by geologists working on the Archean of Western Australia concerning the method used by Arriens to group his data,* Paper 24 presents unique material pertaining to a vast tract of the Archean in Australia, including a Katarchean protocontinental nucleus in the Pilbara (Paper 14), which possibly is part of one of Salop's immense folded ovals (Paper 32).

*On the grounds that the isochron method assumes a common starting point (initial isotopic ratio), and that anomalies could arise if Rb/Sr decay were proportional to the Rb content in the sample.

24: THE ARCHAEAN GEOCHRONOLOGY OF AUSTRALIA

P. A. Arriens

ABSTRACT

Total-rock Rb-Sr isochrons for granites and gneisses in the Yilgarn Block of Western Australia give ages of emplacement which group into three distinct episodes between 3,100–2,900, 2,700–2,550, and 2,300–2,200 million years ago. Most of the granitic rocks exposed throughout the Yilgarn Block have ages between 2,700 and 2,600 m.y. This pattern of near-synchronous granite emplacement and metamorphism affecting the whole length *and* breadth of the Yilgarn Block has little analogy with the tectonism in Palaeozoic or Alpine orogenic belts, where the granitic rocks are confined to narrow zones.

All Archaean rocks dated in the Pilbara Region have ages of 3,100–2,900 m.y.; no age near 2,700 is known. This contrast of geochronology suggests that the Pilbara Archaean nucleus may have been widely separated from the Yilgarn Block during much of Archaean time.

A pegmatite age of 2,700 m.y. has been determined from the East Kimberley Region, and rocks of the Rum Jungle Complex in the Northern Territory have ages between 2,550 and 2,400 m.y. by Rb-Sr and U-Pb methods. These ages are too isolated for making regional comparisons with other Archaean crustal blocks.

SUMMARY

Katarchean ages are obtained from gneisses in the western Wheat Belt in the Yilgarn Block, at Northam, Toodyay, Dale Bridge, and other places. Apparent ages become progressively lower through the craton, eastward, to the Kalgoorlie greenstone belt (Papers 12, 13, 15, 16, 16a, and 16b), where the metamorphic ages of the greenstones are late Archean (2665 million years) (Turek, 1966); the granites are slightly younger, but still late Archean.

The main conclusions of Paper 24, which includes a large number of tabulations, may be summarized as follows:

1. Rb/Sr isotopic data have been synthesized for gneisses, granites, and pegmatite samples collected over a wide area of the Yilgarn Block, all data being total rock measurements on pegmatites. The author notes the lack of any geological map coverage for much of this area, but considers that, despite this, three distinct geochronologic provinces can be distinguished. These are shown in the three line diagrams that are reproduced in facsimile. Three periods of metamorphism are defined at 3100–2900, 2700–2550, and 2300–2200 million years ago. Nearly all the granites and most of the gneisses were emplaced between 2700 and 2600 million years ago, amounting to half the rock content of the 900 by 700 km block. The contrast between the wide area of almost contemporary granitic invasion here and the narrow zones of granitic invasion in later Paleozoic or Alpine orogenic belts is remarked on.

Arriens considers possible sources of systematic error that could explain the systematic variation in ages, but concludes that such an explanation can be virtually discounted. The truth would seem to be that there are relics of an older Katarchean metamorphic terrain incorporated in the late Archean cratonic terrain, relics that still register their separate age despite the proximity of younger granites and gneisses. The last episode of acid magmatism (2300–2200 million years ago) produced only minor bodies in narrow dike form. The fact that these have preserved their identity, although by virtue of their nature they are susceptible to subsequent isotopic disturbances, testifies to the subsequent stability of the block over the 2200 million year period up to the present time.

2. The Pilbara block yields an average Rb/Sr age of 3050 (∓180) million years for the granites that invade the complex greenstone, acid volcanic, and metasedimentary geosynclinal assemblages (Paper 14). The Copper Hills porphyry massif gives an age of 2880 (∓66) million years, but, as de Laeter and Trendall remarked (1970), this young age appeared incompatible with the field relations, which need reexamination. More recently (de Laeter and Blockley, 1972), the existence of a late series of potassic, tin-bearing granites and related swarms of pegmatites has been established; these are dated by the Rb/Sr method at 2670 ∓ 95 million years (Paper 14). These small bodies are believed to have been formed by partial melting of the older granite, which had an initial strontium ratio close to that of primary crustal material.

Arriens clearly demonstrated that the Pilbara is an older Katarchean cratonic stabilization than at least the eastern part of the Yilgarn Block, although the western half may contain relics of the same Katarchean craton (McCall, 1972) or of another, contemporary Katarchean craton. The newly discovered late granites yield ages coincident with those derived from the greater part of the Yilgarn Block. The Pilbara Block must be regarded as a late Katarchean protocontinental stabilization, which shows only minor magmatic granite and basic dike invasions in the late Archean, whereas the western Yilgarn Block must be regarded as a product of prolific granite invasion in the late Archean and metamorphism, which left the Katarchean rocks in the form of small, unmodified remnants, enclaves within this granite terrain.

The western part of the Katarchean craton engulfed in the Yilgarn Block may well have become widely separated from it in much later continental dispersion, and may yet be discovered in the Indian subcontinent.

The gradation of ages through the Wheat Belt, evident in Arriens's results, covers an area of terrain in which there are no distinct structural boundaries delineating the age provinces; such boundaries may well not now exist. Archean tracts flanking greenstone belts may well, in many areas of the world, show this insensible age date progression away from the belts, the closest granites being younger and invasive to the belts and the distal areas of granite containing more and more enclaves derived from the basement to the belts, of greater age than the geosynclinal sequence forming the greenstone belts.

References

de Laeter, J. R., and Blockley, J. G. (1972). Granite ages within the Archaean Pilbara Block of Western Australia. *J. Geol. Soc. Aust., 19,* 363–370.

de Laeter, J. R., and Trendall, A. F. (1970). The age of the Copper Hills Porphyry. *Rept. Geol. Survey Western Aust. 1969,* p. 54–59.

McCall, G. J. H. (1972). The nickel-sulphide bearing ultramafic rocks and their environment in the Archaean of Western Australia. *24th Int. Geol. Congr., 1,* 354–362.

Turek, A. (1966). Rubidium-strontium isotopic studies in the Kalgoorlie-Norseman area. Ph.D. thesis, Australian National University, Canberra.

PAPERS 25 AND 26: COMMENTARY

Wilson remarks that although most of southwestern Australia consists of rocks of Archean age older than 2400 million years, the granulites in their metamorphic definition are restricted to the central and southern Wheat Belt, occurring in patches of a meter to kilometer dimensions of hypersthene granulites surrounded by large expanses of the dominant granite and gneiss rock types, which are characterized by amphibolite metamorphism.

The following are possible origins:

1. Metamorphism of keels composed of postbasement volcanic–sedimentary or intrusive igneous assemblages in synclinal troughs.

2. Recrystallization of the basement itself by depression to hot zones beneath the postbasement troughs, or by wholesale recrystallization of the basement together with the rocks contained in those troughs, which have become fused to the basement.

The original dry or wet character of layered intrusions and volcanic–sedimentary assemblages may have controlled the distribution of dry granulite or wet amphibolite facies assemblages. Surface energy factors related to original coarseness or fineness of grain crystallization could also have exerted a control on distribution. There is evidence of retrograde effects; granulites are later invaded by granites and suffer reversion to lower-grade amphibolite facies assemblages.

Wilson believes that there is some evidence favoring the second mode of origin, in that basement gneisses may locally show superimposed granulite metamorphism. In India, isograds are recognized descending northward from granulite, through amphibolite, to greenschist, transverse to the structural trend of the basement and Dharwars, both of which are affected by the metamorphism. There appears to be a similar northward drop off of metamorphic grade in the Wheat Belt of Western Australia. Such patterns are clearly of erosional origin, being due to planing off of a tilted block of metamorphic rocks parallel to the surface of the earth. It is interesting to note that new evidence from the Jimberlana Dike in Western Australia (Campbell et al., 1969) indicates a downward tilting to the east off the south central Wheat Belt in the manner suggested by Wilson. The probability is that granulite metamorphism and later amphibolite metamorphism were superimposed on an existing terrain, and in some cases a preexisting granite–gneiss terrain, in "blanket" fashion.

Wilson's pioneer studies of the Western Australian Shield led the way to numerous, more-detailed studies of the Yilgarn Block in the last decade. The concept of keels of supracrustals on a primitive granite–gneiss basement complex is difficult to support now in its original form in the light of newer studies (Paper 24). Of the rocks of the shield that we now see exposed, the most abundant are the basic volcanics and intrusions, ultrabasic bodies, acid volcanics, sediments (quartzites, schists, banded ironstones, and clastics), and the vast masses of granites that invaded them after their emplacement in the late Archean. Metamorphism of greenschist grade, with a few hot spots and cold spots, and a rise to amphibolite grade near the contact of the granites characterizes the large greenstone synclines, whereas the neighboring granite expanses display the higher amphibolite to granulite metamorphism. There is as yet little evidence that the granulite metamorphism was not largely, if not entirely, contemporary with the late Archean metamorphism of the greenstones of the Kalgoorlie Belt, which is closely associated with the largely syntectonic invasion of granites 2615 million years ago. The granulite peak metamorphism and the retrogression to amphibolite facies assemblages may have occupied quite a small period of time within a single cycle of events, although the granulite metamorphism may yet prove, in part or whole, to belong to an earlier, Katarchean cycle of events. Although there is little doubt that the western Wheat Belt and Darling Range granitic terrains do contain older relics (Paper 24), there seems to be no detailed analyses published of their lithology or structure, and it is not even known whether they include granulites. The strict distinction between older basement and synclinal keels of supracrustals suggested by Wilson cannot be supported by any geochronologic or field study of detailed areas, and it seems obvious that most of the granitic terrain enclosing the supposed keels is younger and invasive to the rocks of the larger, well-defined greenstone synclines.

The granitic material that envelopes the supposed keels may still, even if apparently younger, incorporate entirely reworked older material. But all that can be said is that the supracrustal synclines that could be keels are enclosed in slightly younger, invasive granite–gneiss terrain, and that this may incorporate such reworked material; if so, it is not recognizable, and there is no distinct boundary evident between keel and basement. The synclinal keel structure of the granulite enclaves does not seem to be well established by detailed mapping, although most of the larger greenstone belts are undoubtedly synclinal in form.

The granulite inclusions, whatever the age of their extreme metamorphism (and more than one age may well be recognized), occur in very small developments closely associated with amphibolite facies developments. There is no evidence of mappable Barrovian zones, the contrasting metamorphic facies being in abrupt contact and intimately intermingled in small masses of rock. The abrupt contiguity in such small masses appears to be more compatible with some control on metamorphic grade exerted by the nature of the rocks themselves, as Wilson has suggested (also Sidorenko and Lunyeva, 1972). The granulite metamorphism may be related to an extensive zone blanketing large areas of rock, but may still be patchily developed because of such controls. In a zone of the required high thermal conditions, still only those rocks of suitable anhydrous character or coarseness of texture may actually develop the mineral associations that characterize the granulite facies.

Hepworth suggests that the cratonic granulites of Uganda, Tanzania, and Kaapvaal craton may form part of the floor upon which the greenstone belts were superimposed, and could be the most primitive crust of which we have knowledge.

References

Campbell, I. H., McCall, G. J. H., and Tyrwhitt, D. S. (1970). The Jimberlana Norite, Western Australia. *Geol. Mag., 107,* 1–12.

Sidorenko, A. V., and Lunyeva, O. I. (1972). Facies of regional metamorphism and primary composition of rocks. *Int. Geol. Rev., 14,* 1297–1306.

25: GRANULITE TERRAINS AND THEIR TECTONIC SETTING AND RELATIONSHIP TO ASSOCIATED METAMORPHIC ROCKS IN AUSTRALIA

A. F. Wilson

ABSTRACT

Although some 'dry' rocks (*e.g.* certain igneous rocks) may give rise to granulites under one metamorphism, many granulite terrains* appear to result from two or more metamorphisms. Tectonic settings are postulated where both the keel of a geosyncline and the depressed basement may be subject to granulite-facies metamorphism. Major gravity anomalies delineate some granulite terrains and suggest their relationship to deep seated geosutures and mantle disturbances. Some granulite terrains merge gradually into the extensive amphibolite-bearing regions of the shields, whereas others are in major fault contact with rocks of much lower grade. These features are illustrated from a survey of the tectonic setting of the Australian granulites. These granulites fall into several age categories: approximate ages are 2,600, 2,000-1,900, 1,700, 1,400-1,300, 1,100-1,000, and 600-700 million years. Several of the Proterozoic granulite terrains are linear belts, and these appear to be welded to Archaean blocks or transgress cratonic regions.

INTRODUCTION

Australia has several distinct regions in which granulites occur. Variations of mineral assemblage, composition of minerals, rock fabric and other features show that the physical and chemical environment controlling formation, preservation and ultimate exposure of these rocks has varied greatly from region to region and within individual regions.

The purpose of this paper, as the first in the symposium on granulites, is to give an outline of the tectonic setting of the Australian granulites. This will involve some discussion not only of the tectonics but also of some fundamental geochemical and general petrological factors.

Australian granulite terrains are shown in Figure 1. Preliminary isotope age data appear to allow a division of the terrains into two major groups; namely, those in which the granulite metamorphism was Archaean, and Proterozoic in age.

1. ARCHAEAN** GRANULITE TERRAINS

Most of southwestern Australia is of Archaean age. The Archaean granulites are restricted, however, to the central and southern portions of the Wheat Belt (shown stippled in Figs 1 and 2).

Throughout the Wheat Belt there are irregular patches of hypersthene-bearing granulites ranging from a few square metres to several square kilometres in area in which the rocks are dominantly of granulite facies (for details, see Fig. 2, and Wilson, 1958, map). The major part of the area, however, is biotite or hornblende granite or gneiss.

It is possible that there are two main modes of origin of the granulites from the Wheat Belt, namely, (i) metamorphism of keels of post-basement troughs of volcanic and sedimentary rocks, or metamorphism of post-basement igneous intrusions; (ii) recrystallisation of the basement itself either by depression to hot zones beneath the post-basement troughs, or by wholesale recrystallisation of the basement together with the contents of the post-basement troughs which have become fused to the basement.

(i) Metamorphism of keels of 'post-basement' troughs

Many of the pyroxene granulites are thought to represent the keels of synclinoria or other relics of folds now enclosed in granitic material. Some of the synclinoria appear to be relics of basins or troughs in which lavas, tuffs, and jaspilites were deposited. Examples have been mapped on the shores of Lake Dumbleyung, near the hospital at Corrigin and at several places in the York-Greenhills and Dangin-Quairading districts (Fig. 2; and Wilson, 1958, map).

* The term 'granulite terrain' is used in the sense of a region in which rocks of granulite facies are found. This usage implies that these may be either dominant in the terrain or occur as relics or patches set within regions of amphibolite or other facies.

** The term Archaean implies an age of at least 2,400 million years.

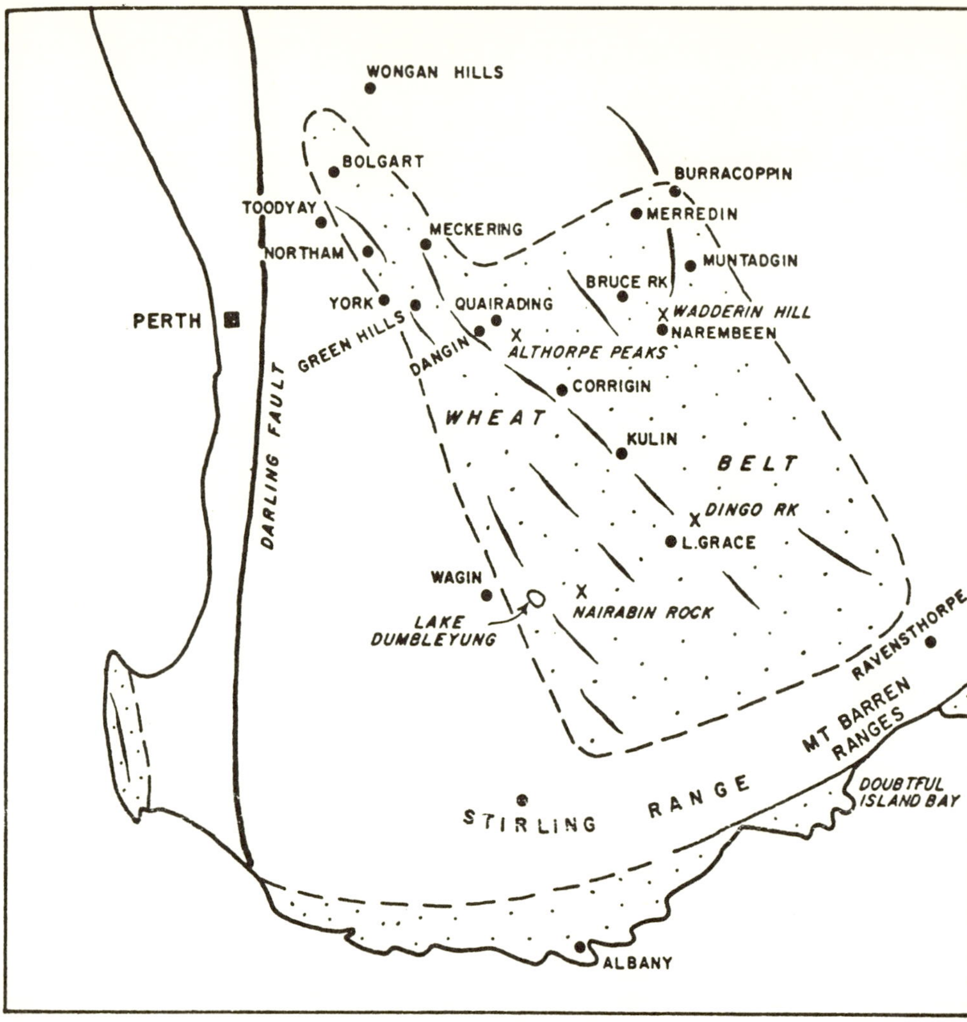

Fig. 2. The Wheat Belt granulite terrain of Western Australia (see Fig. 1) showing the localities mentioned in the text.

Similarly, several masses of basic granulite and ultramafic granulite appear to be recrystallised relics of layered igneous intrusions. Examples have been mapped at Althorpe Peaks (southeast of Quairading), the peninsula on the southern shore of Lake Dumbleyung, east of Nairabin Rock, and south of Corrigin.

In these places it is assumed that during appropriate deep-seated metamorphism original rocks which were largely anhydrous would tend to become granulites, whereas uralitised dolerites or basalts, and other 'wet' rocks would tend to retain much of their 'water' and take on an amphibolitic aspect. In this way we may account for some of the 'wet' zones enclosing (or within patches of) essentially anhydrous pyroxene granulites. Other ways to account for this common phenomenon include consideration of truly retrograde effects, and effect of surface-energy differences related to grainsize differences which in turn may be produced either by original magmatic cooling or by tectonic granulation. Examples of granulites produced in these ways are recognised elsewhere in Australia.

(ii) Recrystallisation of the basement itself

In several places, the basement gneisses themselves appear to have suffered a superimposed granulite facies metamorphism. For example, in the vicinity of Lake Grace and in the Bruce

Rock-Narembeen region many of the acidic rocks of the polymetamorphosed basement are hypersthene-bearing. At Dingo Rock, 16 km eastnortheast of Lake Grace, the acidic rocks of the basement show a contorted gneissose and migmatic character on weathered surfaces and as such cannot easily be distinguished from the normal biotite gneisses of the shield. However, the fresh rock shows no obvious banding, and is greasy dark bluish-grey or greenish-grey which is the typical appearance of the Indian acid charnockites.

It is difficult to decide whether or not the recrystallisation of the basement itself was due to dehydration of the basement during its depression beneath a post-basement geosynclinal trough. The Dingo Rock region, however, is adjacent to a prominent northwesterly-trending belt in which there are many metamorphosed jaspilites, basic lavas and related rocks. This belt appears to continue through Kulin and Corrigin into the Dangin-Greenhills region, and beyond (see Fig. 2, and preliminary reconnaissance map of Wilson (1958)).

More mapping is needed to test the hypothesis that the acidic pyroxenic rocks of Dingo Rock are relics of an early Archaean basement. The task will not be easy, for, during the metamorphism of the geosynclinal trough, it would appear that hornblende or biotite granites intruded not only the rocks of the trough but also the polymetamorphosed basement. Such intrusions and the hydrothermal phenomena associated with them are bound to cause widespread retrograde effects on the basement which had been raised to granulite facies.

Presumed examples of this type of retrograde effect in which pink granitic material cuts or replaces greasy, dark bluish-grey, acidic 'charnockitic' rocks have been noted on a small scale at Dingo Rock and at Wadderin Hill (about 10 km north of Narembeen), and on a large scale at Muntadgin Rock (about 37 km northnortheast of Narembeen).

Another cause of a superimposed granulite imprint on the basement complex may be dependent on a regional increase in temperature without being related to superimposed geosynclinal troughs. This has been noted in India (Pichamuthu, 1965). There the granulites occur in the south near Bangalore, and as one proceeds northwards along the tectonic trend of the gneisses and the rocks of the Dharwar 'System' the grade of metamorphism is found to decrease first towards the amphibolite and then to greenschist facies. This may be explained simply by the hypothesis that in South India we are looking at a deeper portion of the crust which has been exposed by erosion as the 'slab' of Peninsular India has been tilted downwards to the north. In India the isograds are said to cut across both basement and the Dharwar geosynclinal rocks.

In the Archaean of Western Australia a similar pattern seems to be emerging. The granulites of the Dangin region give way along the strike to the northnorthwest to amphibolites and greenschists in the Bolgart region and Wongan Hills. A preliminary delineation of the Archaean granulites shows that few occur north of a line from Burracoppin to Toodyay. This 'isograd' cuts across the regional northwesterly trend, and may indicate a regional tilt northwards and downwards of this portion of southwestern Australia.

Age data and metamorphic grade

The age data for the Wheat Belt imply a widespread metamorphism and granite injection about 2,600 ± 100 m.y. ago (Wilson *et al.*, 1960; Compston & Arriens, 1967). Although most of the data are derived from Rb/Sr techniques, these must be considered to register a minimum age for the rocks that have undergone granulite facies metamorphism, for we are not sure of the mobility of either Rb or Sr during this grade of metamorphism.

High-pressure assemblages characterised by clinopyroxene-garnet or kyanite have not been found. On the other hand, cordierite (with or without garnet) is common in the quartzose gneisses and granulites (*e.g.* near Dangin (Prider, 1945*b*) and in the Dumbleyung district). Sapphirine has been recorded at Dangin (Prider, 1945*a*) and south of Quairading (Wilson, 1962). Forsterite and spinel are not uncommon (*e.g.* near Dangin (Prider, 1945*b*) and in the Dumbleyung district), and fayalite is known from a metagabbro at Burges Find near Burracoppin and from metajaspilites near Dumbleyung and elsewhere.

2. PROTEROZOIC GRANULITE TERRAINS

All Australian granulites other than those of the Wheat Belt appear to yield Proterozoic ages which, however, though largely Rb/Sr ages, should also be considered as minimum ages. It is possible, however, that future techniques may uncover an Archaean history in some of these rocks.

The Proterozoic granulites of Australia show a far greater range in tectonic setting and petrological characteristics than those of Archaean age.

[*Editor's Note:* Material has been omitted at this point.]

CONCLUSIONS ON THE GENERAL TECTONIC SETTING OF AUSTRALIAN GRANULITES

Three main types of granulite terrain emerge from a survey of the granulites of Australia. Similar generalisations appear to be valid for other continents.

These types are those that (1) merge into the extensive amphibolite-bearing regions of the shields, (2) may be of high metamorphic grade and in fault contact with rocks of lower metamorphic grade, and (3) contain large bodies of charnockite or hypersthene granite.

(1) The first type of granulite terrain appears to result from metamorphism of fairly dry rocks to a grade somewhat higher than that normally attained during the first metamorphism of the hydrous rocks of a geosynclinal pile. This can be brought about by metamorphism of the relatively dry igneous or metamorphic basement upon which the geosynclinal rocks have been deposited.

Under these circumstances an unconformity or disconformity may be expected at or near the junction of the granulites and the lower grade metamorphosed younger geosynclinal rocks. Such a structure may be difficult to recognise, however, as planar structure in both basement and geosynclinal rocks may have been pulled into apparent conformity during high-grade metamorphism. Moreover, considerable portions of the lower sections of the geosynclinal rocks may have already been largely dehydrated before the granulite metamorphisms take place. Under such circumstances these lower sections may well be converted to granulites, and may be distinguished from the granulites derived from the basement only by isotope or other geochemical techniques.

This structural setting for granulites may be suspected where large lenticular bodies of granulites are found enclosed in granitic gneisses, some of which (especially near the lenticular granulite bodies) are of granulite facies. The highly metamorphosed keel of a syncline of geosynclinal rocks may be expected near the centre of such a lenticular body.

Tagini (1965) seems to favour a comparable explanation for some of the charnockitic rocks of the Man Complex of the Ivory Coast, West Africa. The ferruginous quartzites (itabirites or banded iron formations) of the Ivory Coast, are thought to mark the unconformity between the Lower Proterozoic metamorphic complex and the Archaean polymetamorphic charnockitic rocks of the Man Complex. Tagini's map (1965, fig. 20, and plate 1) shows the post-Archaean metamorphic rocks and their lenticular keel-like structural attitude. Examples of comparable structures are thought to occur near York in the Wheat Belt, Western Australia (Fig. 2).

In the shield areas a common situation prevails wherein rocks of granulite facies may form only part of a dominantly amphibolite-facies region. In addition to faulting, this may be formed in one or more of the following ways:

(i) A fairly anhydrous igneous rock (*e.g.* acid or basic flow or intrusion) may form part of the geosynclinal rocks. The low water content of the igneous rock may allow it to take on a granulite aspect during prograde metamorphism even though the same thermal flux is able to raise the enclosing relatively hydrous rocks only to amphibolite facies. This is undoubtedly a common cause for the presence of lenses of basic granulite in amphibolite-bearing regions, as in the Arunta Block, the Broken Hill Block, in many places in the Wheat Belt of Western Australia, and elsewhere.

(ii) Large-scale retrograde metamorphism may leave granulite relics by virtue of their structural site or physical and chemical properties.

(iii) A local appearance of granulites may develop in response to a source of heat applied to a restricted portion of the crust. This may be caused by several phenomena, for example, the development of a ridge or cupola on the mantle, or to the near approach of a major basic intrusion or to the concentration of thermal flux at the intersection of two deep-seated faults.

(iv) In a deep-seated fault zone amphibolitic rocks may become granulated, dehydrated and recrystallised, commonly with platy quartz grains and lineated elongate pyroxenes. Many of the granulites of Saxony are of this type, and there are wide linear zones within several of the Australian granulite terrains where this process appears to have taken place. Probable examples are some of the remarkably lineated platy acidic granulites of the York-Greenhills region of the Wheat Belt, the western portion of the Fraser Block and near the Woodroffe Thrust in the Musgrave Block.

(2) The second type of granulite terrain is clearly due to elevation along a fault of a crustal block from depths where granulites are normal rocks. Deep-seated rocks are thus brought into contact with low grade rocks. Most of the granulite terrains of Australia are bounded by major faults (*e.g.* the East Kimberley Block, the Broken Hill Block, and the Fraser Block). The grade of metamorphism of the granulites may be variable from place to place along a single fault zone. Such variations may be due to uneven isothermal surfaces in the crust, or may be due to a scissors-type or other differential movement on the fault.

(3) The third type of granulite terrain contains intrusive pyroxenic plutonic rocks (with or without anorthosites) and is of particular interest in tectonic studies.

The generation of granitic magma from partial melting of crustal rocks is a well-established mechanism. Such magmas may be readily produced from 'wet' rocks such as are normal in a geosynclinal pile. However, considerably higher temperatures are required to generate magma from the 'dry' rocks of a granulite terrain. The fact that large bodies of orthopyroxenic granites have developed in certain parts of the world (*e.g.* Musgrave Block, Grenville Province of Canada, Angola, Scandinavia, etc.) implies not only that very high temperatures were attained, but that adequate pressure was maintained to allow these coarse pyroxene granites to crystallise.

The ferrohypersthene-bearing Ernabella Adamellite in the Musgrave Block was intruded into a synclinal (not anticlinal) position in the granulites (Wilson, 1960*a*). The magma is presumed to have been generated in an environment of low oxygen fugacity, from partial melting of the keel of the syncline, the collapse of which allowed the magma to rise to a higher level in the crust.

Sprigg & Wilson (1960, p. 535) claim, however, that charnockitic granites in the 'Musgrave-Mann granitized metasedimentary belt' intrude fold axial regions, particularly the anticlines. In my experience it is the hornblende-rich granites that tend to be emplaced in anticlines or in cupola apophyses of the charnockitic granites (Wilson, 1960*a*) owing perhaps, to a concentration in these favourable structural sites of such volatiles as were present.

Another possible mode of formation of pyroxenic granites is as follows. Transcurrent movement on a curved fault or on a zone of obliquely intersecting faults is known to produce tensional tectonic domains which favour migration of fluids from deformed zones. Transcurrent movement on major lineaments bordering geosynclinal belts which traverse cratonic regions could localise granite magma in the manner indicated. Deep transcurrent faults in granulite terrains should favour granite formation in lenticular bodies spatially related to the fault zones.

The Mabel Downs Granodiorite, which has charnockitic affinities, is spatially related to the Halls Creek Fault, a major sinistral transcurrent structure in the East Kimberley Block (Fig. 4). A genetic relationship has yet to be proven.

It has been pointed out that the charnockitic granites of the Musgrave Block are commonly related to tight synclinal structures or other folds. However, as the block is so intimately affected by several major deep-seated transcurrent fault zones, the possibility should be investigated in the field that dextral movement on a series of ancient eastnortheasterly-trending shears may have produced many of the tight folds that have localised the granitic intrusions.

Thus, the several granulite terrains of Australia display a wide range of structural and tectonic features. The normal relationship of most of the granulite terrains to major tectonic features such as transcurrent and reverse faults and major gravity anomalies is of great geotectonic significance. Not only in Australia, but also in several other continents there are similar features.

It appears that ancient cratonic nuclei are commonly girdled and cut by geosynclinal belts and mega-shear zones. More and more of these geosutures are being found to have granulites related to them. It is clear that a more thorough understanding of granulite terrains will throw much light on both geochemical and geotectonic principles.

REFERENCES

BINNS, R.A., 1962: Metamorphic pyroxenes from the Broken Hill district, New South Wales. *Mineralog. Mag.*, *33*, pp.320-338.

__________, 1963: Some Observations on Metamorphism at Broken Hill, N.S.W. *Proc. Australas. Inst. Min. Metall.*, *207*, pp.239-261.

__________, 1965: Petrographic Evidence Relating to Movement on the Thackaringa-Pinnacles Fault, Broken Hill District. *Proc. Australas. Inst. Min. Metall.*, *214*, pp.95-101.

BUDDINGTON, A.F., 1952: Chemical petrology of some metamorphosed Adirondack gabbroic, syenitic, and quartz syenitic rocks. *Am. J. Sci.*, Bowen Volume, pp.37-84.

CARRUTHERS, D.S., 1965: An environmental view of Broken Hill ore occurrence. *8th Commonw. Min. Metall. Congr.*, *1*, pp.339-351.

CLARKE, E. de C., PHILLIPPS, H.T., & PRIDER, R.T., 1954: The Pre-Cambrian geology of part of the south coast of Western Australia. *J. Proc. R. Soc. West. Aust.*, *38*, pp.1-64.

COMPSTON, W., & ARRIENS, P.A., 1967: The Precambrian geochronology of Australia. Proc. Confce Geochron. Precambrian stratified rocks, Univ. Alberta, June 1967.

DANIELS, J.L., 1967: Subdivision of the Giles Complex, Central Australia. *Rep. geol. Surv. West. Aust.*, (1966), pp.58-62.

GEMUTS, I., 1965: Metamorphism and igneous activity in the Lamboo Complex, East Kimberley Area, Western Australia. *Rec. Bur. Miner. Resour. Geol. Geophys. Aust.*, 1965/242 [unpublished].

GOODE, A.D.T., & KRIEG, G.W., 1967: The Geology of Ewarara Intrusion, Giles Complex, Central Australia. *J. geol. Soc. Aust.*, *14*, pp.185-194.

GRAY, C.M., : The geology, petrology and geochemistry of the Teizi metanorthosite [in preparation].

GREEN, D.H., & RINGWOOD, A.E., 1967: An experimental investigation of the gabbro to eclogite transformation and its petrological applications. *Geochim. cosmochim. Acta.*, *31*, pp.767-883.

HARMS, J.E., 1965: Geology of the Kimberley Division, Western Australia. *8th commonw. Min. Metall. Congr.*, *1*, pp.66-70.

HILLS, E.S., 1965: Tectonic setting of Australian ore deposits. *8th commonw. Min. Metall. Congr.*, *1*, pp.3-12.
HORWITZ, R.C., DANIELS, J.L., & KRIEWALDT, M.J.B., 1967: Structural layering in the Precambrian of the Musgrave Block, Western Australia. *Rep. geol. Surv. West. Aust.*, 1966, pp.56-58.
HUDSON, D.R., & WILSON, A.F., 1966: A New Occurrence of Sapphirine and Related Anthophyllite from Central Australia. *Geol. Mag.*, *103*, pp.293-298.
__________, WILSON, A.F., & THREADGOLD, I.M., 1967: A new polytype of taaffeite - a rare beryllium mineral from the granulites of central Australia. *Mineralog. Mag.*, *36*, pp.305-310.
JOKLIK, G.F., 1955: The geology and mica-fields of the Harts Range, central Australia. *Bull. Bur. Miner. Resour. Geol. Geophys. Aust.*, *26*.
KAY, J.G., 1958: The high grade metamorphics of Cape Naturaliste, Western Australia. Univ. West. Aust., B.Sc. Hons thesis [unpublished].
__________, 1962: Orientation and Composition of Plagioclase in a Basic Charnockite from Bunker Bay, Western Australia. *J. Proc. R. Soc. West. Aust.*, *45*, pp.111-114.
LOVERING, J.F., & WHITE, A.J.R., 1964: The Significance of Primary Scapolite in Granulitic Inclusions from Deep-seated Pipes. *J. Petrology*, *5*, pp.195-218.
NESBITT, R.W., & TALBOT, J.L., 1966: The Layered Basic and Ultrabasic Intrusives of the Giles Complex, Central Australia. *Contr. Miner. Petrol.*, *13*, pp.1-11.
PICHAMUTHU, C.S., 1965: Regional metamorphism and charnockitization in Mysore State, India. *Indian Mineralogist*, *6*, pp.119-126.
PRIDER, R.T., 1945*a*: Sapphirine from Dangin, Western Australia. *Geol. Mag.*, *82*, pp.49-54.
__________, 1945*b*: Charnockitic and Related Cordierite-Bearing Rocks from Dangin, Western Australia. *Geol. Mag.*, *82*, pp.145-172.
__________, 1958: The Granulites and Associated Rocks of Galena, Western Australia. *F.L. Stillwell Anniv. Vol.*, pp.189-211. Australas. Inst. Min. & Metall.
RICHARDS, J.R., & PIDGEON, R.T., 1963: Some Age Measurements on Micas from Broken Hill, Australia. *J. geol. Soc. Aust.*, *10*, pp.243-260.
ROD, E., 1966: Clues to Ancient Australian Geosutures. *Eclog. geol. Helv.*, *59*, pp.849-883.
SPRIGG, R.C., & WILSON, R.B., 1959: The Musgrave Mountain Belt in South Australia. *Geol. Rdsch.*, *47*, pp.531-542.
TAGINI, B., 1965: Esquisse géotectonique de la Côte d'Ivoire. Société pour le développement minier de la Côte d'Ivoire, 107.
TILLEY, C.E., 1921: The granite gneisses of Southern Eyre Peninsula (South Australia) and their associated amphibolites. *Q. Jl geol. Soc. Lond.*, *77*, pp.75-134.
TUREK, A., & STEPHENSON,N.C.N.,1967: The Radiometric Age of the Albany Granite and the Stirling Range Beds, south-west Australia. *J. geol. Soc. Aust.*, *13*, pp.449-456.
WEBB, A.W., & McDOUGALL, I., 1968: The Geochronology of the Igneous Rocks of Eastern Queensland. *J. geol. Soc. Aust.*, *15*, pp.313-346.
WILSHIRE, H.G., & BINNS, R.A., 1961: Basic and Ultrabasic Xenoliths from Volcanic Rocks of New South Wales. *J. Petrology*, *2*, pp.185-208.
WILSON, A.F., 1947: The charnockitic and associated rocks of north-western South Australia, pt 1: The Musgrave Ranges - an introductory account. *Trans. R. Soc. S. Aust.*, *71*, pp.195-211.
__________, 1952: The charnockite problem in Australia. *Sir Douglas Mawson Anniv. Vol.*, pp.203-224, Univ. Adelaide.
__________, 1957: Some structural notes on charnockitic rocks. *Bull. Mysore Geol. Ass.*,*13*.
__________, 1958: Advances in the Knowledge of the Structure and Petrology of the Precambrian Rocks of South-Western Australia. *J. Proc. R. Soc. West. Aust.*, *41*, pp.57-83.
__________, 1959*a*: Notes on the fabric of some charnockitic rocks from central Australia. *J. Proc. R. Soc. West. Aust.*, *42*, pp.56-64.
__________, 1959*b*: The charnockitic rocks of Australia. *Geol. Rdsch.*, *47*, pp.491-510.
__________, 1960*a*: The charnockitic granites and associated granites of central Australia. *Trans. R. Soc. S. Aust.*, *83*, pp.37-76.
__________, 1960*b*: Co-Existing Pyroxenes: Some Causes of Variation and Anomalies in the optically Derived Compositional Tie-Lines, with Particular Reference to Charnockitic Rocks. *Geol. Mag.*, *97*, pp.1-17.
__________, 1962: A New Occurrence of Sapphirine and the Significance of Sapphirine in the Granulite Terrains of Western Australia. *Am. Miner.*, *47*, pp.207-208, [abstract].
__________, 1964: The petrological features and structural setting of Australian granulites and charnockites. *Int. geol. Congr.*, *22*(13), pp.21-44.
__________, 1966: Metamict allanite from pegmatites cutting basic charnockitic granulites in the Fraser Range, Western Australia. *J. Proc. R. Soc. West. Aust.*, *49*, pp.85-87.
__________, 1969: Problems of exploration for metals in granulite terrains, with particular reference to Australian localities. *This Volume*, pp.375-376.
__________, & HUDSON, D.R., 1967: The discovery of beryllium-bearing sapphirine in the granulites of the Musgrave Ranges (central Australia). *Chem. Geol.*, *2*, pp.209-215.
__________, & MIDDLETON, D.D., 1968: Some petrological features of a spinel-bearing metagabbro in the pyroxene granulites of the Fraser Range, Western Australia. *J. Proc. R. Soc. West. Aust.*, *51*.
WILSON, A.F., COMPSTON, W., JEFFREY, P.M., & RILEY, G.H., 1960: Radioactive Ages from the Precambrian Rocks in Australia. ***J. geol. Soc. Aust.*, *6*, pp.179-196.**

26: CHARNOCKITIC GRANULITES OF SOME AFRICAN CRATONS

J. V. Hepworth

ABSTRACT

Charnockitic granulites of eastern Africa are older than c. 3000 m.y. They form part of the palaeosome of the chelogenic Katarchaean granite, which probably 'overplates' rather than 'underplates' them. Some granulites are associated with greenstone belts, but it is uncertain as to whether they formed their floor.

Some of the granulite suites show original, sedimentary, lithological layering, but no other early structures survive. Most of them are tectonically reworked by later events. Many of them are migmatized and mobilized and are granitoid in habit. Some, however, have survived reworking in intra-cratonic mobile belts without downgrading.

It seems likely that most of the granulites in the Mozambique belt are related to Mozambiquian metamorphism and are not old, cratonic granulites.

SUMMARY

In Uganda, charnockitic granulites are regarded as metamorphic, not magmatic (the classic Indian interpretation). The association of these rocks with alumino-silicate-bearing quartzites suggests a partly sedimentary origin. Conformable basic and acid granulites may represent sediments or volcanics. Palingenesis and anatectic migmatization took place between layers in a subsequent metamorphic stage, and there is evidence of late retrogression to amphibolite facies assemblages. The granulites are claimed to be more than 3000 million years old, but in Uganda the age dating is based on zircons, and Rb/Sr ages indicate a metamorphic age of 2660 million years. There is thus no certainty that the granulite metamorphism is not comparatively late, that is, late Archean.

Hepworth notes the ability of granulites to survive in a readily recognizable form through several superimposed metamorphic events, retaining their metamorphic mineral assemblages.

In Tanzania, there are small relic areas of granulite within granitoid rocks of probable late Archean age (2295–2265 million years). In both Tanzania and Uganda, a granitic cycle followed the granulitic cycle; the latter, in Tanzania, postdated both greenstones and granulites. In Uganda, no greenstones have been recognized.

In Swaziland, charnockites occur within the Ancient Gneiss Complex in a small area measuring 3 km^2. They are devoid of discernible foliation and are intruded by granite. The granulite metamorphism is almost certainly *Katarchean* and is certainly no younger than 2800–2500 million years; it is probably older than 3100 million years. The question has been raised as to whether this is the basement upon which the Onverwacht Group was deposited; is it an older foundation or even a primitive crust?

Hunter (1970) adduced evidence that the Ancient Gneiss Complex in which the granulites occur is at least in part an older foundation. Hepworth has recognized pillow lavas within this complex, and these may either be contemporary with the

Reprinted from *Rept. Sec. 1,* 24th Int. Geol. Congr., Ottawa, 1972, p. 126. The summary was prepared by the volume editor.

Onverwacht Group (engulfed by the gneiss complex) or else part of a much older pre-Onverwacht sequence. The problem, like the similar problem concerning the possible pre-Sebakwian (Papers 7 and 35) remains unresolved.

The granulites of the Mozambique Belt in Tanzania possess a distinctive tectonic style and character, and are taken to represent extreme metamorphic metasedimentary derivatives of the Upper Proterozoic Mozambiquian metamorphic episode. The alternative, that they are slices of basement derived from older cratons and intensively reworked and infolded, cannot, however, be rejected.

Beloussov's equating (1966) of granulites with an impoverished layer beneath the Conrad Discontinuity is not thought to be compatible with evidence that the granitic cycle is superimposed on granulites and retrogrades them. The depth at which granulite metamorphism could have occurred in Archean times is difficult to establish because of geothermal gradient and heat flow enhancement, but McConnell (1967) has noted that the crust must have been thick enough to permit extreme granulite metamorphism. There is evidence of low-pressure phases in granulites, such as cordierite, and Hunter (1970) concludes that the Ancient Gneiss Complex represents low-pressure, andalusite–sillimanite metamorphism. Hepworth finds the contiguity of granulites and greenschist metamorphics in the Swaziland System difficult to reconcile with their contemporaneous metamorphism.

Discussing the emplacement of granite over the Ancient Gneiss Complex, Hepworth invokes overplating rather than the underplating favored by Anhaeusser et al. (Paper 5). The batholithic layer is thought likely to have accumulated above the complex, obliterating the greenstone–foundation contact. Such relationships may be developed widely in Archean terrains and may explain the repeated lack of discrete boundaries between Archean foundations and their unconformable overlays.

References

Beloussov, V. V. (1966). Modern concepts of the structure and development of the earth's crust and upper mantle of continents. *Quart. J. Geol. Soc. London, 122,* 293–314.

Hunter, D. R. (1970). The ancient gneiss complex of Swaziland. *Trans. Geol. Soc. South Africa, 73,* 107–150.

McConnell, R. B. (1967). In discussion of Hepworth (1967), *Quart. J. Geol. Soc. London, 123,* 731–735.

V: EVIDENCE OF LIFE IN THE ARCHEAN

This topic is covered in another volume of the Benchmark Series (*Precambrian Life,* edited by K. A. Kvenvolden). Presentation here is limited to a brief summary.

The stromatolites of the Bulawayan System, first described by A. M. MacGregor (1940), are certainly more than 2600 million years old and are among the most spectacular known occurrences of Archean fossils. Excellent photographic illustrations of these fossils are given by Rutten (1962, pp. 77–83) in a discussion of their significance.

Schopf et al. (1971) confirm their age and consider them to be remnants of associated microorganisms. The most likely age range for these, the earliest known stromatolitic communities, is 3100–2800 million years. Such communities were clearly in existence at the time of deposition of sediments referred to the Katarchean (Paper 7).

Nagy and Nagy (1969) discuss microscopic structures in the Onverwacht Group of the Katarchean Swaziland System (Paper 5) and consider them to be possibly the oldest known fossils on earth. The microscopic structures described show some resemblance to very simple organisms, but their morphology is poor and their size range great. They are cup-shaped, spherical bodies and occur stratigraphically below the Fig Tree Group, from shales and cherts within which fossil algae and flagellates have been described by Pflug et al. (1967). The sediments of the Onverwacht Group are argillaceous cherts and carbonates. The group is now known to be more than 3400 million years old. Cross-bedding and lamination structures indicate subaqueous deposition for the sediments associated with pillow lavas and tuffs of similar deep-water emplacement. Metamorphism is of low grade, but not so low as is commonly supposed. The microscopic structures are not evenly distributed, but occur in pockets within the rocks. The following criteria of biological origin are suggested by Cloud and Licari (1968):

1. Abundance within a narrow size range.
2. Morphological complexity matching that of living organisms.
3. Little resemblance to inorganic structures; features suggestive of reproduction.

These Onverwacht structures, unlike those of the Proterozoic Gunflint, Bitter Springs, and Biwabik formations of North America and Australia do not show these features. The authors noted the danger of later contamination by seepage contaminants of organic origin through permeable rocks. The carbonaceous material was found to be basically aromatic in character, similar to material derived from lignin in later rocks. The origin of the aromatic constituents remains obscure, especially because of doubts concerning the degree of metamorphism to which these rocks have been subjected. The authors found it difficult to explain the amount of aromatic material in terms of biological processes if the metamorphism was as slight as is commonly supposed. The origin of these bodies remains problematical.

Kvenvolden (1972) emphasized certain doubts concerning the biogenic origin of these structures; he reported rod-shaped, bacterium-like microfossils in the Fig Tree Group, and also spherical alga-like microfossils (Pflug, 1966; Barghoorn and Schopf, 1966; Schopf and Barghoorn, 1967). The study of the organic components of rocks from both occurrences has led him to conclude that the evidence supports bacterial and algal origin, biochemical pathways 3000 million years ago being similar to those of the modern biosphere. The major area of doubt is whether the organisms were contemporaneous with the laying down of the Katarchean sediments. Evidence against contemporaneity is, in fact, quite compelling. Systematic studies on many samples may yet resolve this question.

References

Barghoorn, E. S., and Schopf, J. W. (1966). Microorganisms 3 billion years old from the Precambrian of South Africa. *Science, 152,* 758–763.

Cloud, P. E., and Licari, G. R. (1968). Microbiotas of the banded iron formation. *Proc. Nat. Acad. Sci., U.S., 62,* 623–630.

Kvenvolden, K. A. (1972). Organic chemistry of early Precambrian sediments. *Proc. 24th Int. Geol. Congr., 1,* 31–41.

Macgregor, A. M. (1940). A Precambrian algal limestone in Southern Rhodesia. *Trans. Geol. Soc. South Africa, 43,* 9–15.

Nagy, B., and Nagy, L. A. (1969). Early Precambrian microstructures, possibly the oldest fossils on earth. *Nature, 223,* 1226–1227.

Pflug, H. D. (1966). Structures organic remains from the Fig Tree Series of the Barberton Mountain Land. *Econ. Geol. Res. Unit, Univ. Witwatersrand, Inform Circ., 28,* 1–14.

Rutten, M. M. (1962). Geological aspects of the origin of life on earth. Elsevier, Amsterdam.

Schopf, J. W., and Barghoorn, E. S. (1967). Alga-like fossils from the early Precambrian of South Africa. *Science, 156,* 508–511.

_____, Oehler, D. Z., Horodyski, J., and Kvenvolden, K. A. (1971). Biogenicity and the significance of the oldest known stromatolites. *J. Palaeontol., 45,* 477–485.

VI: GENERAL PAPERS ON THE ARCHEAN

Topics of major importance that call for review are the definition of the Archean, cyclicity in the Archean record, the application of plate tectonics to the Archean, and the nature of the primordial crust. The first two are considered in this introduction; brief reviews of the last two topics are presented as Parts VII and VIII.

The Archean is commonly defined as the period prior to the great earth revolution that caused the development of the stable platforms, geosynclines, and median massifs. According to such a definition, the end of the Archean could be delineated by recognizing the oldest stable platform sequences superimposed on it or the first superimposed tensional dike suites or "trappean" volcanics, both of which followed the thickening of the crust to form rigid cratons by the agency of granite invasion. Stable platform sequences and trappean dikes and volcanics should, by virtue of their very nature, be referred to the Proterozoic.

Such a natural-based definition, based on straddling platform sequences and dike development, appears, however, to break down, inasmuch as geochronologic evidence indicates that there may be no globally applicable time plane for the end of the Archean. The stabilization of cratons was, it appears, a rolling event. Characteristic Archean-type greenstone sequences can be recognized from 3600–2500 million years ago, and similar assemblages of later date are known, such as those of the Belomorian in North Russia (2100 million years) and of Arizona (1800 million years), and have been considered to be Archean in type. Granitization leading to protocratonic stabilization commenced, as we have seen, early in the Katarchean and produced more or less circular areas of granite–gneiss terrain of quite limited extent; it reached its peak 1200 million years later, in the late Archean, about 2600 million years ago, forming very extensive tracts. Some Archean granite tracts suffered repeated granitic invasions and were not stabilized until 1700–1800 million years ago, for example, in Guiana.

There is evidence of delay in the poststabilization trappean dike development in cratonic areas that suffered this late stabilization; for example, the dikes of Guiana date at 1700 million years. The earliest fully stabilized cratonic regions were subject to very early invasion by trappean dikes, of which the first known example is the Ameralik dikes of Greenland, which are more than 3600 million years old. Likewise, they are subject to very early superimposition of platform sediments, as is exemplified by the Pongola and Witwatersrand series of South Africa (Paper 35) and the Saamides of North Russia (Paper 22).

Although there remains an area of doubt concerning correlations of radiometric numbers obtained by using different decay methods, the discrepancies are of such an order that there can be little doubt that the first platform sediments were being deposited prior to the volcanicity, sedimentation, metamorphism, and granite invasion that culminated in the cratonization of the Yilgarn Block, Western Australia, in the Kalgoorlie region, of the Superior and Slave Provinces of Canada, and Dharwars–Iron Ore Group in India. The Huronian (Canada) and Fortescue Group (Australia) are the oldest known straddling platform sediment superimpositions in these regions, and yield the expected younger ages (about 2300 million years) than for the entire Witwatersrand, Pongola, and Saamides series. In India, the Singhbhum Series (Sarkar, 1972) appears to be the oldest straddling platform series; it dates from about 2000 million years ago.

Trappean dike swarms invading areas of late stabilization yield ages ranging from 2420 million years (Widgiemooltha, Australia; Turek, 1966), through 1900 million years in Canada (Gates and Hurley, 1973), to as late as 1700 million years in Guiana (McConnell and Williams, 1970). Very ancient major basic–ultramafic complexes of poststabilization age are recognized from Rhodesia (Great Dyke, lower limit 2800 million years; Allsop, 1965), from the Kola Peninsula, USSR (Monchegorsk layered intrusion, 3250 million years; Paper 22), and from Montana, where the Stillwater intrusion invading Archean rocks of the Beartooth Plateau is dated at 2750 million years (Nunes and Tilton, 1971). The Katarchean protocratons were by no means as extensive or thick as those of the late Archean and did not apparently develop such spectacular arrays of transecting dikes, striding purposefully in one direction across immense cratonic expanses. Large dikes do transect the Katarchean of the Pilbara, but appear to be of the same age as the Widgiemooltha Dikes (personal communication, J. R. de Laeter). Earlier trappean invasions of the Katarchean cratonic stabilizations appear to be restricted to a few major intrusive complexes and localized, more minor swarms of dikes.

The fact that the widely accepted definition of the Archean, which is based on the Kenoran orogeny in Canada (2550 million years), is arbitrary emerges from the above discussion. This boundary definition suits Canada, Australia, and India, but not early stabilized regions in South Africa and North Russia, nor later stabilized regions such as Guiana. Nevertheless, such an arbitrary definition remains the only practical one. The life-based definition of the Proterozoic is quite meaningless, for there are life forms found as fossils in both the Swaziland and Bulawayan systems (stromatolites and microbiota). Equally, no worldwide earth revolution terminated the Archean. Thus, only an arbitrary definition can have meaning. In using this arbitrary boundary we must, however, realize that it does not rigidly separate Archean-type greenstone–granite terrain development from Proterozoic-type platform sedimentation and trappean eruptivity. It seems clear that the diachronous end to the Archean protocratonic stabilizations is a reality, and it is certainly quite evident, as Windley remarks (Paper 38), from the intrinsic nature of Archean greenstone sequences that deposition and volcanism were separated by no long span of time from the regional metamorphism that provides the geochronological numbers; thus, we can no longer evade the problem of apparent diachroneity by supposing that the greenstones of the Yilgarn Block, the Superior and Slave provinces, the Dharwars and Iron Ore Group, and Nyanza are Katarchean sequences reworked by a late Archean metamorphism (an assumption that lay behind the abortive attempts by South African geologists to correlate the Kalgoorlie late Archean with the Barberton Katarchean!).

As a basis for methodology, we must accept the following major divisions of geologic time:

Precambrian	4600–3000 million years	Katarchean
	3000–2550 million years	(Late) Archean
	2550–600 million years	Proterozoic
	600 –	Phanerozoic

The shield-forming or "chelogenic" cycles proposed by Sutton (Paper 31) are the longest cycles in terms of duration so far recognized in the geologic record. They are supposed to be related to major shifts in convection patterns. Four such "cycles" were originally proposed, but this was subsequently revised to three (Papers 31a and 38):

1. 3800–2500 to 2000 million years.
2. 2500 to 2000–1000 million years.
3. 1000 million years to present.

Such cycles are supposed to be initiated by discordant transection of stabilized cratonic areas by mobile belts (as opposed to shallow intracontinental basins containing platform sediments). This discordance contrasts with the earlier concordant accretion of belts around cratonic nucleii within the duration of a cycle. Some mobile belts of discordant geometry effected fragmentation of cratons into new, discordantly margined plates, whereas the shallow transecting intracratonic basins only involved incipient fragmentation of the cratons, and some discordant mobile belts only developed to form welts. A new pattern of cratonic blocks is supposed to emerge for each cycle. Each cycle includes a sequence of mobile belt developments and stabilizations.

One may question whether chelogenic cycles are, in fact, cycles? What seems to emerge from a study of the record of the earth, through nearly 4000 million years, is a pattern of three main stages of progressive development rather than three major cycles. The first stage was one of cratonic stabilization superimposed on the products of sedimentation and volcanism within a number of restricted proto-oceanic troughs: the pattern of the troughs and the foci of granite invasion being related to *nuclear* patterns early in the stage, and more continuous linear patterns as the areas of stabilization increased in size, late in the stage. The second stage was characterized by predominant ensialic, platform-type sedimentation, within shallow, to some extent linear, superimposed basins within the cratons; the appearance of continental red bed sequences; the development of trappean volcanics; and the development of linear mobile belts in which granite invasion occurred in much the same manner as in the Archaean stage, but which mostly had the character of linear welts superimposed on the cratonic surface without achieving fragmentation, welts which are analogous to the rift zones of the later geological record. This was the middle, quiet, cratonic phase. The third stage was by the development of vast oceans due to lateral displacements of plates on a scale hitherto not achieved, and by extreme island-arc, geosyncline manifestations. The mobile belts, characterized by granite invasion, are still evident, and are linear, faulted, and upwarped zones parallel to the geosynclinal troughs. The geology of the third stage is the familiar geology of the Phanerozoic, the second stage that of the Proterozoic, and the first that of the Archean. Many processes of geology show repetition in a cyclical fashion through these stages, and some aspects show a progressive directional development through them. The stages themselves do not appear to be cycles, but simply main stages in progressive planetary development.

It is from the concept of chelogenic cycles that Sutton proposes a numerical division of the Precambrian:

- PC1. 3800–2700 million years.
- PC2. 2700–1900 million years.
- PC3. 1900–950 million years.
- PC4. 950–600 million years.

Semenenko (Paper 28) had earlier derived a similar scheme, but fivefold not fourfold. Both schemes are based on the Gastil (1960) radiometric maxima. These maxima have not proved to be a reliable basis for intercontinental correlations; for although they are real maxima of activity, this activity is not of the same kind all over the globe and the maxima do not form a basis for matching similar rock formations. But they, rather than the chelogenic stages, do perhaps provide the best basis for numerical subdivision of the Precambrian, for they represent maxima in time at which rocks all over the globe tended to become closed systems for daughter elements of decay, whether on original formation or subsequent metamorphic modification.

The Gastil maxima require recompilation to incorporate the many age-dating results gathered in since they were first presented in 1960. Such a replot could reveal up to nine maxima in the Precambrian, and could form the basis of a new, realistic ninefold numerical division of the Precambrian. The systems of Semenenko and Sutton were gallant attempts to quantify the Precambrian into stages, but were perhaps presented before their time, in the sense that geochronology of the Katarchean was not then sufficiently advanced to be incorporated in these schemes in a logical manner.

The Gastil cycles reveal a rhythmicity of diastrophism all through the globe, with a variable but not too inconstant periodicity averaging about 400 million years. The possibility that this periodicity is due to cosmic influences rather than internal ones has been suggested by Khain (Papers 30A and 30B); the periodic triggering of long-lived radionucleid decay by cosmic influences has been suggested. Unlike western scientists, Russian authorities tend to discount internal convection as a major process within the earth, being solidly ranged against concepts of continental dispersion. The Gastil cycles are the most definite, well-established cycles in the geologic record, and represent a fundamental rhythmicity possessed by this planet.

Cycles of smaller duration, within the limits of single greenstone sequences, are considered by Anhaeusser (Paper 35); he makes the following definitions:

1. Supercycles: entire greenstone belt assemblages; scale in the order of 10,000 m.
2. Major cycles: scale in the order of 100–1000 m.

3. Minor cycles: scale in the order of 100–1 m.
4. Minicycles: scale in centimeters.

Trendall (1972) recognizes macro-, meso-, and microbanding cycles in banded iron formation; these are cyclothem and varve equivalents, and are analogous to the coarser sedimentation rhythms ubiquitous in some Archean turbidites (Dunbar and McCall, 1971) and to Anhaeusser's minor and minicycles.

Both Anhaeusser and Goodwin (Paper 33) recognize eruptive cyclicity. Goodwin identifies three stages in a major eruptive cycle, which may be repeated more than once in the same greenstone belt:

1. Platform stage: abyssal tholeiite flows predominant, erupted from numerous dispersed vents.
2. Edifice stage: calc-alkaline eruptives dominant, erupted from relatively few localized centers in extensive volcanic piles.
3. Erosional stage: predominantly clastic sedimentation.

The smoothly progressive basic to acid transition of the Abitibi belt seems to be unusual if not unique, and other greenstone belts tend to display a hiatal rhythmicity. In South Africa, andesites do not seem to be very prominent, and in Australia (Kalgoorlie Belt) the bimodality of the rhythms is marked; it is produced by tholeiitic volcanics abruptly followed by acid volcanics or sodic porphyries without any intermediate eruptives intervening. The platform to edifice progression seems to reflect a decrease in the fluidity of the magma, a multitude of orifices giving way to a few central vents and pyroclasis becoming increasingly developed.

References

Allsop, H. L. (1965). Rb-Sr, K-A measurement in the Great Dyke of Rhodesia. *J. Geophys. Res., 70,* 977–984.

Anhaeusser, C. R. (1970). Cyclic volcanicity and sedimentation in the evolutionary development of Archaean greenstone belts and shield areas. *Spec. Publ. 3, Geol. Soc. Aust.,* 57–70.

Dunbar, G. J., and McCall, G. J. H. (1971). Archaean turbidites and banded ironstones of the Mt. Belches area, Western Australia. *Sed. Geol., 5,* 93–133.

Gastil, G. (1960). Continents and mobile belts in the light of mineral dating. *Proc. 21st Int. Geol. Congr., 9,* 162–169.

Gates, T. M., and Hurley, P. M. (1973). Evaluation of Rb/Sr dating methods applied to the Matechewan, Abitibi, McKenzie and Sudbury dyke swarms in Canada. *Can. J. Earth Sci., 10,* 900.

McConnell, R. B., and Williams, E. (1970). Distribution and provisonal correlation of the Precambrian of the Guiana Shield. *Proc. 8th Guiana Geol. Conf., 1969,* 20p.

Nunes, P. D., and Tilton, G. R. (1971). Uranium–lead ages of minerals from the Stillwater igneous complex and associated rocks. *Bull. Geol. Soc. Am., 82,* 2231–2250.

Trendall, A. F. (1972). Revolution in earth history. *J. Geol. Soc. Aust., 19,* 287–311.

Turek, A. (1966). Rubidium–strontium isotopic studies in the Kalgoorlie-Norseman area. Ph.D. thesis, Australian National University, Canberra, 64p.

PAPERS 27 THROUGH 29: COMMENTARY

Papers 27 through 29 illustrate the wide cleavage in opinion among Russian authorities. Salop (Papers 27A and 27B) supports a concept of irreversibility and evolution, overprinted on a pattern of global cycles. He believes in globally universal cyclicity, both in connection with the minor Gastil cycles and with the supposed greatest of all earth revolutions at the end of the Archaean. Salop proposes a totally radical revision of accepted usage for subdividing the Archaean. We mention this for the record, but the proposal appears quite unacceptable to Western geologists.

Pavlovskiy and Markov (Paper 29) also adopt an evolutionary concept, dividing rigidly a first "permobile" stage (equivalent to the Katarchaean and characterized by nuclear terrains) from the linear, "protogeosynclinal" stage of the Late Archaean, which in turn gave way to a "protoplatform" stage of the Proterozoic. Semenenko (Papers 28A and 28B) presents an actualistic concept of cyclicity, believing that there were no unique geologic developments in the Archaean—Archaean geology being explicable in terms of familiar geosynclinal and platform concepts. However, he does seem to accept a rolling end to the Archaean in the form of progressive terminal cratonic stabilization, not a global terminal revolution, synchronous throughout the globe. He believes that the fivefold cyclicity of the Precambrian has been obscured by confusion engendered by use of different radiometric methods of dating, and by lack of consideration of exactly what is being dated in each case and what each number really signifies.

27A: PRE-CAMBRIAN GEOCHRONOLOGY AND SOME FEATURES OF THE EARLY STAGE OF THE GEOLOGICAL HISTORY OF THE EARTH

L. I. Salop

ABSTRACT

Age determinations prove five Pre-Cambrian global tectono-plutonic cycles, aged respectively 2,600–2,800, 1,800–2,100, 1,400–1,500, 1,000–1,100 and 600 million years. These are natural limits, that conclude Archaean, Lower Proterozoic, Middle Proterozoic, Upper Proterozoic and Epiproterozoic eras. The most prominent revolution in the geological history of the earth occurred soon after the termination of the Archaean, which led to the formation of geosyncline system and platforms everywhere, that were similar to those of Neogene.

Monotonous climate and geochemical conditions during Pre-Cambrian (especially during early Pre-Cambrian) gave rise to certain common features in supercrustal complexes of the same age, located in remote regions. Various Pre-Cambrian epochs had their own special sedimentary formations, that were not repeated during geological history. Irreversible evolution of sedimentation conditions is perfectly illustrated by ancient ferriferous, partly dolomitic and high-alumina formations.

In the course of the geological history of the earth duration of anorogene stages became shorter, while tectonic movements increased in intensity and differentiation, speed of sedimentation increased simultaneously.

SUMMARY

Salop's* original paper has been widely cited as a pioneer attempt to correlate the various Precambrian sequences throughout the world, looking outward from Russia where no less than four Archean cratons are recognized, all characterized by a dominance of high-grade metamorphism. These are the Baltic, Ukraine, Aldan, and Anabar shields. Salop recognizes five tectoplutonic cycles in the Precambrian, and he considers that the most prominent revolution in earth history occurred soon after the end of the Archean, leading to the development of geosyncline–platform–median massif patterns that have dominated geotectonics up to the present day.

Paper 27A highlights the need to revise the Gastil data, which revealed definite maxima of activity. Many new radiometric age determinations have been carried out since 1960, and a new diagram is called for, extending right back to the

*Salop writes as follows: My three articles published in this volume were written at different times and over a relatively long interval (1964–1972), which is known for its advances and progress in studies of the Precambrian, and which produced a great quantity of new information and data; consequently, some of my views and ideas were changing in the course of this progress. This concerns especially the dating of geochronologic boundaries of the main Precambrian subdivisions (groups, eras). The main principles of Precambrian subdivision which I am arguing for at the moment, and which are briefly stated in my article of 1972, are given in detail in my monograph entitled *Unified Stratigraphic Scheme of Precambrian,* Leningrad, Nedra, 1973. L. I. Salop.

Reprinted from *Proc. Sec. 10,* 22nd Int. Geol. Congr., India, 1964, p. 131. The summary was prepared by the volume editor.

Katarchean. Geochemical and climatic monotony are assumed by Salop to have characterized the Archean. The nature of the early Archean sedimentation is supposed to have been nonuniformitarian; this is suggested by peculiar, highly aluminous, ferruginous, and in places dolomitic formations. The summary of Archean sedimentation provided in Paper 27A is, perhaps, to some extent influenced by the high-grade character of the metamorphic terrains familiar to Salop. He discounts preservation of clastic texture, of rudites, and of megascopic sedimentary structures in Archean sediments, a conclusion in conflict with recent evidence from South Africa, Australia, and Canada.

The irreversible evolution factor is stressed, and Salop questions whether the peculiar granite dome and brachysyncline patterns can be reconciled with partition of the crust into the familiar geosynclines and platforms of post-Archean systems.

Cyclicity of diminishing cycle length is suggested by Salop, and there appears to have been a belated first appearance in the record of molasse deposits in the Upper Proterozoic. Such deposits were, he believes, developed more and more extensively in later cycles as the vertical component of tectonic movements increased. A progressive increase in the volume of sediments accumulated over a given time span, and secular acceleration in the evolution of geosynclinal systems is also suggested.

27B: A UNIFIED STRATIGRAPHIC SCALE OF THE PRECAMBRIAN

L. I. Salop

ABSTRACT

The records of isotope dating of plutonic and metamorphic rocks are indicative of the existence of five diastrophic cycles of global significance in Precambrian time that took place about 3.5, 2.6–2.8, 1.9–2.0, 1.0–1.1, and 0.65–0.68 $\times$ 10^9 years ago.

Supracrustal formations formed between two diastrophic cycles, together with plutonic rocks of the final diastrophism, form natural complexes that can be conventionally compared with the largest subdivisions of the stratigraphic groups. The Precambrian groups of the same age in various regions are characterized by similar formations, tectonic structures and organic remains.

The greatest difference is between the most ancient group ($>$ 3.5 $\times$ 10^9 years) and younger groups. In accordance with this, two eons (supergroups) are distinguished in the Precambrian: Archean and Protozoic. The Archean eon is composed of one Archean group devoid of definable organic remains; disconformities within it are lacking, although there are evidences of the existence of an earlier gneiss-granite basement that has not been detected anywhere for certain as yet. The Protozoic eon is composed of four groups for which the following names have been proposed: Paleoprotozoic, Mesoprotozoic, Neoprotozoic and Epiprotozoic. Neoprotozoic is divided into three subgroups: the two lower subgroups characterized by global platform magmatism (1.6–1.7 $\times$ 10^9 years), the two upper by characteristic complexes of stromatoliths. The worldwide display of two-phase glaciation is characteristic of Epiprotozoic. Above the Epiprotozoic, Eocambrian deposits occur, as a rule, unconformably; in most regions there is a gradual transition to the Lower Cambrian, which is characterized paleontologically. A short description of the subdivisions is given.

SUMMARY

In Paper 27B Salop has developed a new and revolutionary classification of Precambrian stratigraphy. He recognizes five diastrophic cycles of global significance, with maxima at 3500, 2800–2600, 2000–1900, 1100–1000, and 680–650 million years. In formulating this he appears to have accepted the reality of globally synchronous revolutions (the tradition developed by Suess and refined by Stille for the Phanerozoic, and by Holmes for the Precambrian). Only rocks older than the first date listed above are now referred to the Archean—the Pre-Swaziland rocks of South Africa and the Pre-Pilbara rocks of Australia. The first are only tentatively recognized in the form of vestigial remnants within granitic terrain (Papers 7 and 35) and the second is still hypothetical (Paper 14).

Although Salop's approach is logical, assuming the reality of globally synchronous revolutions, the resultant system savors too much of a *reductio ad absurdum* to be workable. Certainly, there are real maxima of activity throughout the globe, but Salop appears to have accepted that these maxima, recognized by Gastil, reflect an identical development of a revolutionary nature throughout the

Reprinted from *Rept. Sect. 1,* 24th Int. Geol. Congr., Ottawa, 1972, p. 253. The summary was prepared by the volume editor.

globe. This assumption overlooks much evidence in the form of regional geochronologic data which suggests that this was not the case, and that the culminations of the Archean were essentially diachronous.

The proposal of Salop, which takes away virtually the entire population of rocks that geologists have hitherto referred to as Archean, is not accepted by the editor as a basis for this text; the older, admittedly imperfect arbitrary definition of the Archean based on the Kenoran orogeny is preferred and retained.

28A: CORRELATION OF THE HISTORY OF PRE-CAMBRIAN BY THE DATA OF ABSOLUTE GEOCHRONOLOGY

N. P. Semenenko

ABSTRACT

The author establishes the large orogenies of folded belts and the magmatism and metamorphism associated with them.

The Rakhovian (Carpathian foundation) orogeny, occurring during the period of 570 to 700–800 million years corresponds to the last Pre-Cambrian cycle in Europe.

The Hyperborean orogeny took place in the period of 800 to 1100–1500 million years in the Baltic shield and is synchronous with the Grenville orogeny in the Canadian shield and the Chuisky in the Sayano-Baikal region and the Yenisey ridge of Siberia. These orogenies form the folded regions surrounding the East European platform; they form the Baikal folded region, the Katanga belt of Africa and constitute the megacycle, Pre-Cambrian IV.

The Gothian and synchronous Volynian and Ovruch orogenies occurred in the period of 1150 to 1600–1650 million years at the time of the formation of the folded belts constituting the Russian platform, the Stanovoi ridge, the Oka and Daldarmian zones of the Eastern Sayans of Siberia, the Caragve-Ancola and Kibara in Africa.

The Krivoy Rog orogeny in the Ukrainian shield, and the Karelian and Svecofennian in the Baltic shield took place in the period of 1650 to 1900–2000 million years when the formation of these shields was, in the main, accomplished; the Hudson orogeny in the Canadian shield is synchronous with them. These two very large orogenies, occurring when the basic part of the East European platform was formed, comprise the megacycle, Pre-Cambrian III.

The White Sea orogeny in the Baltic Shield, the Bug and Podolian orogenies in the Ukrainian Shield occurred in the period of 1,900/2,000–2,300 million years. The Saamian in the Kola peninsula, the Dnieper in the Ukrainian Shield and the Keneran orogeny in the Canadian Shield occurred in the period of 2,300/2,400–2,650/2,700 million years. These orogenies constitute the megacycle Pre-Cambrian II.

Finally, the megacycle Pre-Cambrian I embraces the interval from 2700 to 3650 million years and older, established in the Ukrainian shield, in the core of the Kola peninsula and in the southern belt of the African shield.

Fifteen age groups are distinguished in the Pre-Cambrian megacycles.

For the oldest parts of the Pre-Cambrian section normal geosynclinal formations have been established in the Konka series of the Ukrainian shield. Fauna with ages of 1900 and 2200 million years have been found in the Pre-Cambrian of the Kola peninsula. Finally, unmetamorphosed platform deposits have been found with an age of 1500 million years and older.

Reprinted from *Proc. Sec. 10,* 22nd Int. Geol. Congr., India, 1964, p. 150. The summary was prepared by the volume editor.

SUMMARY

Semenenko extends the concept of cyclicity in the Precambrian to a numerical system of megacycles (I–V). Cycle I is referred to as the *Katarchean.* Low-grade greenstone metamorphic assemblages, with prominent basic and acid volcanic members and no marked dominance of ultramafic rocks, are present in the oldest Konka–Belozerian zone of the Ukraine. Their character is considered to be consistent with the assumption of atmospheric and lithospheric conditions being similar to those of the present day. Deposition of beds rich in iron (magnetite, hematite) and silica 3550 million years ago indicates, Semenenko believes, that some free oxygen was present in the atmosphere at that very early date.

In outlook, Semenenko contrasts with Salop (Paper 27), for he is actualistic rather than evolutionary. He argues against the concept of a unique early Precambrian "pre-geosynclinal" period (Papers 22 and 29), believing that geosynclinal–platform partition of the crust was a reality already in the Early Archean. He notes the disparity evident in attempted intercontinental correlations of geosynclinal and faulting episodes. Including the Bug–Podolian and Belomorian in Precambrian II, he takes the Archean right up to include the earliest initial stable platform sedimentation elsewhere (e.g., South Africa). In doing so he seems to accept that there were late extensions of the development of Archean-type greenstone–granite terrains, with even later stabilizations of these terrains in certain cratons.

28B: PRECAMBRIAN GEOCHRONOLOGY AND ITS PROBLEMS

N. P. Semenenko

SUMMARY

In this paper, which cannot be reprinted in facsimile because of lack of space, and which is thus only summarized here, Semenenko discusses the significance of radiometric age determinations on zircons. In the Inyul–Ingulets System of the Ukraine, rounded zircons yield ages between 2750 and 2600 million years, whereas monazites give ages of 2000 million years, which is taken to be the true metamorphic age of the injection gneisses that contain them. The zircon age is not the age of the gneiss or even that of the sediments from which it was derived, but the age of the rocks eroded to form those sediments.

Much confusion and ambiguity are introduced by not carefully considering the exact meaning of each radiometric number. In the South Bug River area, dating by U/Th/Pb methods has separated the formation and breakup of the basement blocks, the folding, and the superimposition of new fracturing and mineralization in the Ingulets–Krivoi Rog stage. Semenenko (1968, 1970) believes that much of the ambiguity in geochronology results from different events being dated, and that five megacycles can be distinguished with upper boundaries at 2700, 2000, 1700, 1200, and 550 million years. Gneiss–granulite metamorphism is represented in all five, and high-grade metamorphism has no restriction to the very ancient systems. He doubts whether there is, in reality, a unique Archean geotectonic pattern; he regards the *nuclear* pattern (Papers 27, 29, and 32) of domes and brachyfolds as being due to secondary reorganization of preexisting fold systems, in a way that occurs in some much younger systems (e.g., the Paleozoic of the Erzegebirge in Germany). He cites the very old Konka–Belozerian of the Ukraine, which includes normal linear elements, compressed imbricate elements, and overturned isoclinal folds; some structural elements here are traceable for up to 150 km along their strike.

Semenenko favors the concept of Starik and Sobotovich (1959) of global differentiation by the agency of downward heating from the surficial crust, instead of outward from the core, the radioactive elements that provided the source of this heat having been localized peripherally here during earlier accretion. The geospheres are taken to be of primary origin, not due to secondary differentiation. He believes that the protocrust was differentiated some tens (if not hundreds) of millions of years after the actual formation of the earth. For example, Naydenov (1972) suggests that the earth existed for 200–400 million years prior to the setting of the radiometric clocks. Shales and sands are present in Precambrian I of the Ukraine, Aldan Shield, and in North Africa dated at between 3000 and 3500 million years, and these must, he believes, have been deposited on an even older granite-like substratum, not a basaltic substratum. He rejects the idea that the sialic crust formed from a primordial basic crust of basaltic character or from the femic mantle. Instead, he takes the model of Sobotovich, and invokes differentiation of a protocrust of granitic character, for which he sees much support in isotopic data. During the Precambrian, the continents grew at the expense of oceanic crust, with the platforms so formed suffering subsequent fracturing with the formation of geosynclines within rather than marginal to the cratons. He stresses cyclicity and reversibility.

This summary, from *Int. Geol. Revs.,* **14,** 947–953 (1972), was prepared by the volume editor.

References

Naydenov, B. M. (1972). Calculation of the isochron and the original differentiation of terrestrial substance. *Int. Geol. Rev., 14,* 954–958.

Semenenko, N. P. (1968). Geology of the Ukrainian Precambrian. *Can. J. Earth Sci., 5,* 661–671.

____. (1970). Geochronological aspects of stabilisation of continental Precambrian platforms. *Eclogae Geol. Helv., 63,* 301–310.

Starik, I. Ye., and Sobotovich, E. V. (1959).

29: PECULIARITIES IN THE EARLY TECTONIC EVOLUTION OF THE CONTINENTAL CRUST

Ye. V. Pavlovskiy and M. S. Markov

ABSTRACT

Two stages in the development of the crust are outlined in the early Precambrian. Nuclear (about 3 bln. years)—lava effusions of a basaltic-andesitic composition, accumulation of graywackes and ferruginous-siliceous chemogenic deposits; an absence of intense differentiated movements, the development of isometric structures of brachyform type and an absence of linear structures and big fractures.

The further development proceeded, apparently, in two ways: (1) After a wide granitisation of the nuclear complex the crust became "rigid", was split up by large fractures with which suture structures were associated (about 2.5 bln. years) and was filled by thick clastic and effusive accumulations. (2) On a non-granitised (Aldan shield) or slightly granitised nuclear basement, gigantic troughs (protogeosynclines) originated. In them there has been an accumulation of geosynclinal formations with facies stable throughout extensive areas. An evolution of magmatism similar to geosynclinal is observed.

After the appearance of separate rigid areas with a continental type of the crust (early Proterozoic) protoplatform structures were formed.

The first protoplatform that can be proved geologically appeared about 2 bln. years ago. Locally granito-gneissic domes have been cropping up on them. Sometimes the basement and cover were split by big fractures, along which ultrabasic, basic and even acid magmas penetrated. Protoplatforms did not become extended by the growth of young folded zones along their peripheries. Superimposed geosynclinal troughs are associated with extensive zones of protoplatform faults.

In the process of an irreversible development of the crust, a gradual formation of consolidated areas with a "granitic" layer, are observed, a greater role of fractures, a replacement of isometric and disorderly oriented structures of the early stage by organised linear structures of later stages, and a lesser and linear localisation of granitization processes.

SUMMARY

Pavlovskiy and Markov question whether cycles of the very early Precambrian are the same as those of the Caledonian, Hercynian, and Alpine foldings. A special and peculiar early Precambrian tectonic regime is suggested, one that does not conform to the widely accepted models of geosynclines and platforms developing in subsequent geologic time. Basic effusions made up much of these early sequences and graywackes accumulated in basins separating the volcanic zones. A pattern of more or less isometric, poorly oriented areas of sedimentation, the product of slight differentiation of relief and minor tectonic movement contrasts, was modified by the superimposition of large, isolated domes, which provided the focus for intense migmatization, granitization, and metamorphism of the earlier formed graywackes and volcanics. These domes have a nonlinear grouping and display flat crowns and marginal zones of compression; they are associated with isoclinal, drag, and flowage

Reprinted from *Proc. Sec. 4,* 22nd Int. Geol. Congr., India, 1964, p. 53. The summary was prepared by the volume editor.

folds cut by later faults. This early developmental stage is referred to as the "nuclear stage." The authors believe that it terminated diachronously. In Canada and the Ukraine, termination occurred between 3000 and 2500 million years ago.

Subsequent crustal development is supposed to have taken two forms. First, the rigid continental crust was split by major faults, which initiated the development of narrow trough structures of Saxagen–Timiskaming type, later infilled by clastic washout products from the granitized nuclear complex and basic enclaves. Subsequent folding was accompanied by further granitoid intrusion, and the platform regime became established over extensive continental areas about 2000 million years ago.

Second, where the nuclear crust did not have sufficient rigidity (i.e., where it was composed of basalts and andesites not sufficiently reinforced by granite invasion), extensive depressions or *protogeosynclines* developed, closely resembling younger geosynclines. Characteristically, geosynclinal sedimentary and magmatic (ultrabasic, basic, or acid) sequences accumulated in these depressions, the acid eruptives appearing late in their development. Anatexis, palingenesis, and migmatization combined with high-temperature metamorphism to affect these assemblages. The protogeosynclines were very extensive compared with their more modern analogues.

Typical examples are the Upper Archean protogeosynclines preserved as fragments within the Anabar and Aldan shields of Siberia. These tremendous geosynclinal basins had simple shapes; distinctly expressed intrageosynclinal uplifts are lacking and also there is no facies persistence in the sedimentary units. Folding patterns were quite simple. The first linear structures originated at this time and were persistent in trend across broad regions. Some of these very old fold belts locally lost their mobility after the end of folding, thus supplying the foundation for the first platform structures—the *protoplatforms.*

The basement beneath these can consist of granitized nuclear or protogeosynclinal terrain composed of poorly oriented, isometric or linear structural elements. The first platform sequences (e.g., Witwatersrand or Huronian) were later complicated by the development of the Proterozoic "geosynclines," quite unlike those of the Archean. These early platform sequences were characterized by complication of the central area by superimposed geosynclines (mobile belts?), a variety of magmatic manifestations, local mobilization of the crystalline basement beneath to form isolated granite–gneiss domes, and local contact metamorphism by intrusions and regional metamorphism to greenschist facies assemblages. Thus, they had a more complicated history than modern analogues.

The distinction between *nuclear* and *protogeosynclinal* Archean terrains is perhaps too sharply drawn by these authors. Many geologists now recognize the unusual nonlinearity of terrains older than 3000 million years, but it does not seem to be the case that these patterns were only developed in such ancient systems. There seems to have been a progressive development of linearity as the sedimentary volcanic depositories became more extensive on the larger stabilized cratonic surfaces of the late Archean. The early Archean isometric depositories must be regarded as the earliest protogeosynclines. It does not seem to be unequivocally demonstrated that the small, nonlinear synclines of the early Archean correspond to the actual outlines of the basins of deposition, but in the Slave Province, Canada (Paper 9) and the Shamvaian of Rhodesia (Paper 7) this appears in fact to have been the case, although there is a contrary view (Papers 16 and 35).

There is a critical question: were the early Archean sedimentary volcanic sequences deposited in arcuate basins surrounding the granite domes, or were the patterns secondarily imposed during the ascent of diapirs, which distorted the originally more extensive linear troughs into restricted brachysynclines (Semenenko, Paper 28)? The question deserves further detailed investigation.

The supposition that the later protogeosynclines invariably developed on the older stabilized nuclear surfaces is controversial (see Paper 19: Commentary).

PAPERS 30A AND 30B: COMMENTARY

One must agree with Khain that the extent of early granite terrains in no way precludes a basic primordial crust of oceanic affinities, but such a crust, if it existed at all, must have existed one or more cycles back from the now-well-documented Katarchean of South Africa (Engel, 1968), dated at 3500–3100 million years. It could be incorporated in wisp-like palaeosomes within the earliest known Archean granitic terrains. No basic rocks displayed to us in the Archaean terrains are likely to be unmodified slices of such a primordial crust, and the greenstones are certainly trough eruptives poured out into fault-controlled troughs within an older granite surface: or into troughs opened up between blocks of such a surface.

There appear to have been at least three activity maxima in the Archean, each characterized by greenstone–granitization cycles in some part of the globe. The process of continental stabilization was essentially diachronous, and greenstone associations were developed prior to very early stabilizations in South Africa, the Baltic, the Ukraine and Australia, and in the late Archean in parts of Australia and Canada. They may well have developed even earlier in Greenland and be represented in vestigial remnants within the Amtsoq gneisses of Greenland (3787 ∓ 85) (Papers 20 and 38). Later greenstone trough accumulations had more linearity and continuity than their more ancient analogues, but there seems to be no basis for complete distinction between circular Katarchean Patterns (*nuclear*) and the patterns of the late Archean, such as the Abitibi Belt, which retains a trace of the typically Katarchean nuclear elements showing through a predominantly linear fabric. The linearity seems to have appeared with the development of larger cratonic stabilization elements, which allowed more extensive linear belts to form.

The Ventersdorp lavas are the oldest recorded "trappean" volcanic sequence, but the much older Ameralik dikes of Greenland (Papers 20 and 38) may have once fed an even older "trappean" lava plateau, during tensional fracturing of a small protocraton.

Reference

Engel, A. E. J. (1968). The Barberton Mountain Land–clues to the differentiation of the earth. *Trans. Geol. Soc. South Africa, 71*(annexure), 255–270.

30A: MAIN STAGES AND SOME GENERAL REGULARITIES IN THE DEVELOPMENT OF THE EARTH'S CRUST

V. E. Khain

ABSTRACT

The earliest stages (before 3,500 m.y.) of the evolution of the earth's crust and its primeval composition are not known. The suggestion of a primitive basaltic crust still remains valid, though no less probable is an early appearance of andesites; among the oldest known rocks, the main role belongs to granitogneisses.

The first of the stages of the geological evolution–the Katarchean (3,500–3,000 m.y.), was characterised by the formation of geosynclines amidst oceanic platforms (thalassocratons). During the second, early Archean stage (3,000–2,300 m.y.), nuclei of the future continents were formed, of the type of the larger median masses of younger geosynclines. Toward the end of this stage, primeval Epiarchean platforms (cratons) originated; in late Archean–early Proterozoic (2,500–1,700 m.y.) they were subjected to breaking up leading to the formation of "jaspilitic" geosynclinal troughs and then to reconsolidation creating Epicarelian platforms. One part of the last also turned out to be unstable and was divided by Middle Proterozoic (Gothian, 1,700–1,500 m.y.) geosynclines into separate blocks. Thus, the third stage of the development of the crust (early and Mid-Proterozoic) was characterised already by the existence of large, but unstable platforms, repeatedly subjected to breaking up.

At the beginning of the late Proterozoic (Sinian, Riphean; 1,300–600 m.y.) the structural pattern of the earth's crust took shape, and then steadily developed till the Recent epoch. From this time on, the Pacific Ocean was existing and probably the central parts of the other oceans originated during the previous stage of the distension of the crust. Toward the end of the Pre-Cambrian (600 m.y.) the ancient platforms and mobile belts fringing them definitely took their shape.

The further history of the crust (the fifth stage) consisted of a gradual decline of geosynclinal activity and extension of continental platforms (epeirocratons), but at the beginning of each minor tectonic cycle a partial involving in geosynclinal subsidence of portions of neighbouring platforms was taking place, and at the end of the cycles a spreading of mountain-building in the direction of platforms creating epi-platform orogenic belts. The last was particularly characteristic of the Alpine cycle.

Thus the general trend of the evolution of the earth's crust, consisting of the transformation of oceanic platforms into geosynclinal mobile belts and then into continental platforms, was periodically (at the beginning of major cycles-megacycles and on lesser scale of minor cycles) interrupted by an inverse tendency; this was most probably connected with a replacement of conditions of predominant compression by the conditions of predominant expansion.

Reprinted from *Proc. Sec. 4,* 22nd Int. Geol. Congr., India, 1964, p. 36. The summary was prepared by the volume editor.

SUMMARY

In this paper, Khain suggests that the separation of the Katarchean from the late Archean on lithologic and structural grounds appears quite unrealistic in the light of geochronologic evidence that there is no preference of granite, gneiss and migmatite to the Katarchean, or of basic volcanic–graywacke associations (greenstones) to the late Archean. Terrains of both ages display, the world over, a dominance of granite–gneiss–migmatite terrain, and contain subordinate tracts of greenstone trough assemblages, which are spectacular but always subordinate to the enclosing "sea" of granitic rock.

Khain, like Sutton (paper 31) and Engel and Kelm (1972), recognizes the quiet, cratonic–mobile nature of the middle Proterozoic, which separates the active Archean stage from the reactivation of the late Proterozoic stage. The long period of cratonic stability (2100–950 million years) was characterized by abundant platform sedimentation and dike-forming eruptivity, combined with limited intra-platform mobile belt activity. Contrasting with the earlier and later active periods, it gives the history of the earth a marked trimodality, recognized in Sutton's modified three chelogenic-cycle scheme. There was in this quiet period some fragmentation into separate blocks (e.g., by the Middle Proterozoic Gothian geosynclines of Sweden, 1700–1300 million years), but the present geotectonic pattern of the earth did not really start to take shape until the late Proterozoic.

The general evolutionary trend of the earth's crust, with transformation of oceanic "platforms" (thalassocratons) into geosynclinal mobile belts and then into continential platforms, was periodically reversed, according to Khain, at the beginning of mega and to some extent minor cycles. Conditions of compression were, he believes, replaced by conditions of predominant distension. Periodic compression and distension, volume pulsations, global shape changes, and global rotation rate changes are all attributed to internal, deep-seated processes involving emissions and accumulations of radiogenic heat, coupled with cosmic influences of extraterrestrial origin that mutually interacted with each other and intensified according to the principle of resonance. The greatest rearrangements occurred at the end of the Archean, but none completely destroyed the preexisting structures or completely obscured the earlier record of continental development of the earth. The growth of continental nucleii by accretion is, Khain believes, a major process in the evolution of the earth's crust, but has been overemphasized by some authorites and wrongly discounted by others.

30B: MAIN TRENDS IN THE DEVELOPMENT OF THE EARTH'S CRUST (LITHOSPHERE)

V. E. Khain

ABSTRACT

In the evolution of the lithosphere, four main stages (megastages) become more and more clearly distinguishable. The Early Precambrian (Protogäikum, H. Stille, 4.5–2.5 × 10^9 years) was the time of the most intensive differentiation of the mantle and formation of the crust–oceanic at first (to 4.0–3.5 × 10^9 years) and, after the beginning of granitisation, continental. The stage of primitive ocean was succeeded by one of microcontinents and micro-oceans and then by the formation of a single continental-cratonic mass–Megagea. During the Middle Precambrian (Deuterogäikum, 2.0–1.4 × 10^9 years), this Megagea was probably opposed by the Protopacific ocean. With the beginning of the Late Precambrian the evolution of the lithosphere enters into its third stage–Neogäikum, manifested in the breakup of Megagea and formation of a pattern of meridional (West- and East-Pacific, Atlantic, Ural-Mozambique) and latitudinal (Arctic, Mediterranean, Southern) geosynclinal belts, with cratons in between. The continental crust was broken up in the inner parts of the geosynclinal belts (eugeosynclines) and oceanic crust reappeared instead. Geosynclinal belts then die out gradually, giving place to folded mountain belts with regenerated continental crust. In the southern hemisphere, this process ended with the beginning of the Paleozoic era and the formation of the Gondwana supercontinent; and in the northern hemisphere, with the end of the Paleozoic era and the formation of Laurasia. In the Late Paleozoic–Early Mesozoic, a new stage begins–Epineogäikum–the stage of the disintegration of supercontinents and the formation of new oceans at the site of ancient geosynclinal-orogenic belts, and of renovation of the Pacific crust. This process was initiated by rifting in the axial parts of the belts and was represented by large-scale drawing apart of the lithosphere at the site of future mid-oceanic ridges and its collapse and reworking in the peripheral parts of oceans.

The evolution of the lithosphere is connected with two kinds of related and alternating processes: primary vertical movements originated by the development of deep-seated geospheres and leading to the accretion of the crust; and primary horizontal movements stipulated by the rotation of the earth and causing the break-up of the continental crust. The long existence of the global net of abyssal faults superimpose restrictions on the scale of possible horizontal displacements, but the latter are realised along these faults.

SUMMARY

This paper by Khain includes a suggestion that before the early (Katarchean) geosynclinal stage, the earth's surface may have been dominated by magmatogenic ring structures of endogenetic origin. A thick andesite–basalt volcanogenic protocrust 10–15 km thick is suggested, overlain by ocean some 1.5–2.0 km deep. This entire primitive crust suffered nonuniform granitization, producing the early Arche-

Reprinted from *Rept. Sec. 3,* 24th Int. Geol. Congr., Ottawa, 1972, p. 58. The summary was prepared by the volume editor.

an granite terrains. High water content of the juvenile mantle and high heat flow led to shallow partial melting, which produced the basalt and andesite eruptives. The silica and alkalis that produced the reworking (granitization) also came from the juvenile mantle. Metamorphic differentiation concentrating the low melting fractions high in the crust was widespread over the crust, but is now localized to restricted zones such as island arcs.

The gneiss ovals of Salop (1972), formed in the early and late Archean, are sites of crustal thickening (protocratons) surrounded by regions of subsidence and volcanism (protogeosynclines). Mobile belts accreted on the initial continental nucleii, but were also later developed transverse to them, causing the breakup and destruction of early cratons. Later, intracratonic mobile belts, such as that of the Grenville (Paper 23) were initiated by deep faults. The first consolidated nucleii, of Katarchean age, are supposed by Khain to have had a thin, continental-type crust; its granite may have been formed as long ago as 4000 million years. The first stage of rejuvenation destroyed the greater part of the initial sialic eoplatforms 2800–3000 million years ago, producing the incipient late Archean geosynclines. The late Archean (3000–2300 million years) trough formations consisting of metamorphosed basic lavas and their tuffaceous equivalents and graywackes formed by their erosion are regarded by Khain as analogous with modern geosynclinal systems, although magmatic basites and "hyperbasites" are more profusely developed, and there is a marked preference for an interoceanic situation (compare with Indonesia today). By the end of the Archean granitization and uplifts had replaced this trough magmatism and sedimentation. The original continental nucleii of the Katarchean had developed into continental platforms of considerable areal extent and thickness, and the first intracratonic depressions had already accommodated the Witwatersrand sediments and Ventersdorp trappean lavas.

In this later paper, Khain puts the second stage of earth history as beginning 2400 million years ago and extending to 1400 million years ago, and defines two succeeding stages. The second stage was initiated by an abrupt decrease in radiogenic sources of heat (Paper 3), and was marked by cratogenic predominance. Khain differs from most "Western" authorities (Papers 31A and 31B), in that he does not attribute these major stages in the development of the crust of the planet to changes in internal convection patterns, although his stages realy equate with Sutton's chelogenic cycles, broadly speaking. Cyclicity is accepted as evident in the geological record, certainly in the Gastil cycles if not in the chelogenic cycles or stages; but Khain, although not disputing such cyclicity, attributes it to galactic effects of a gravitational nature productive of periodic tension, the secondary effect of which is increased radiogenic heat output from the interior of the earth. This is also the view of the series editor (Fairbridge, 1972).

References

Engel, A. E. J., and Kelm, D. C. (1972). Pre-Permian global tectonics. *Geol. Soc. Am. Bull., 83,* 2325–2340.

Fairbridge, R. W. (1972). Planetary spin-rate and evolving cores. *Ann. N.Y. Acad. Sci., 187,* 88–107.

Salop, L. (1972). Two types of Precambrian structures: gneiss folded ovals and gneiss domes. *Int. Geol. Rev., 14,* 1209–1228.

PAPERS 31A AND 31B: COMMENTARY

Paper 31A is unfortunately lengthy and not in a format suited to facsimile reproduction and is therefore presented in summary. It is a stimulating text, in which the concept of chelogenic cycles was first advanced. The ideas were developed further in Paper 31B.

31A: THE EXTENSION OF THE GEOLOGICAL RECORD INTO THE PRE-CAMBRIAN

John Sutton

ABSTRACT

Recent developments have made it possible to apply the methods of historical geology to a longer time-span and a wider range of phenomena than could be dealt with by the classic methods of stratigraphy. The paper explores some results of this change, illustrating them by reference to the Pre-Cambrian evolution of the continental masses lying between the Rockies and the Urals. Recent proposals for the division of Pre-Cambrian time are compared and it is seen that their similarities reflect a long-term rhythm in crustal activity related to the chelogenic cycle with a time-span of 800–1000 m.y. The effects of this rhythm on the geological history of the British Isles are examined. The effects of the Katarchaean, Archaean and earlier Proterozoic chelogenic cycles in a wider area of the northern hemisphere are then considered.

It is suggested that orogenic processes of sedimentation, plutonism and deformation were widespread during early stages of each cycle, but subsequently became restricted to parts of the regions initially affected, while stable regions became increasingly extensive. The start of each cycle was marked by the initiation of new networks of mobile belts. Characteristic features of the three cycles are compared and some progressive changes in the behaviour of the crust at equivalent stages of successive cycles are examined.

SUMMARY

As possible bases for subdivision of the Precambrian, Sutton erected four numbered megacycles (in contrast to Semenenko's five). Both authorities (Paper 28) take Precambrian II up to 2000 million years ago, so that the second cycle straddles the Kenoran-based Archean–Proterozoic boundary. Sutton's scheme is time-defined and seems not to carry the connotation that a particular type of crustal development is restricted to a particular megacycle (thus Precambrian I embraces both the Swaziland greenstones and the platform sediments of the Witwatersrand). Both Sutton and Semenenko's systems seem to be related to the Gastil maxima (Paper 27).

The scheme of numbered cycles recalls the French fourfold division of the African Precambrian, but the geochronologic dating represents an advance in Precambrian usage. Nevertheless, a revised scheme based on revision of the geochronologic data following the statistical method of Gastil (1960), and incorporating the new geochronologic evidence concerning the Katarchean, could produce a more realistic numerical system of about nine megacycles covering the Precambrian from 4000 million years ago up.

Sutton further divides the geologic column into four *chelogenic* or shield-forming cycles, the longest term cycles so far recognized in the geologic record. These were later reduced to three (Papers 31B and 38), with the first having a diachronous end. Mobile belts stabilized in most cases within a few hundred million years, but a few remained mobile throughout an entire chelogenic cycle. Stabilization in the last cycle commenced in Devonian times and continued over 400 million years, with Caledonian, Hercynian, Nevadian, and Alpine mountain chains passing successively into the nonorogenic state of stable cratons. The last cycle shows coincidence between onset of stabilization and continental dispersion (as far as the earliest dispersion can be recognized), but such dispersion may or may not have commenced at the onset of stabilization in earlier chelogenic cycles.

Mobile belts develop discordantly on cratonic surfaces composed of stabilized former orogenic belts and are initiated as sedimentary troughs. Such development defines new shield areas, hence the term "chelogenic." This effect counteracts the increase in stable continental area that occurs within a single cycle by outward accretion of mobile belts and their stabilization. In the long term there is a progressive decrease in the amount of crustal surface involved in mountain building in any chelogenic cycle. The concept seems to be one of long term cycles within the duration of which there was no significant fragmentation of the craton, although successive belts were accreted peripherally. Transverse mobile belts do, in fact, seem to have been developed to a marked extent in the Proterozoic, essentially a cratonic-mobile period, but seem to have been largely contained within stable cratonic areas and not to have caused fragmentation and rearrangement of the existing cratonic blocks. Chelogenic cycles are supposed to reflect major shifts in convection patterns within the Earth's mantle, but Russian authorities (e.g., Khain, Papers 30A and 30B) have suggested quite other controls for the earth's megacyclicity, possibly of extraterrestrial, cosmic origin.

Sutton then considers the Precambrian between the Rocky Mountains of North America and the Urals in Russia, an area considered to be the best known for its size and to have been the birthplace of many fundamental concepts. The Archean of the British Isles was placed by him entirely in Precambrian II, no Katarchean being recognized (but compare Paper 20). The Scourian is a high-grade metamorphic complex more than 2600 million years old, definitely Archean. It includes rocks of sedimentary origin, and some, of low grade, at Loch Maree could represent an Archean belt which, like the greenstone belts of other continents, has escaped the full intensity of metamorphism. Katarchean rocks form masses rarely more than a few kilometers across within enveloping migmatitic and granitic gneisses that contain at least some Katarchean material in the Kola Peninsula and Ukraine regions of Russia. They yield ages of 3600–3000 million years, whereas the gneisses give ages of 2700–2000 million years. Katarchean age dates were known only from Ontario and Minnesota, in North America, at the time when Sutton wrote Paper 31, and were restricted to determinations of lead minerals and zircons, both of which could refer to older material recycled. Similar ages have since been obtained from Montana, Wyoming, and Manitoba, and of course from Greenland (Papers 20 and 38).

The likely extent of the Katarchean crust was considered. Sediments of Archean age include clastics that require an older source in sialic crust of continental type (a point stressed by Semenenko, Condie et al. (Paper 17A), Donaldson and Jackson (1965), and McCall et al. (1970), and the Kola Series includes extensive areas of pelitic–psammitic sediments metamorphosed as long ago as 3600–3200 million years. A sialic continental source approaching 4000 million years old is inferred to explain their occurrence.

Sutton believes that there were no extensive shields in existence in Katarchean times, that is, no areas so completely stabilized that they could endure to the present day without subsequent orogenic superimposition and obliteration except for preservation in the form of a few restricted remnants. What source terrains there were for the Katarchean clastic sediments had the familiar character of continental shields, but were of no great degree of stability and underwent reworking that affected the entire crust throughout the globe during late Archean time.

The Archean of Precambrian II (2600–1900 million years) has a great extent compared with the Katarchean in the Northern Hemisphere, the latter appears to have been largely obliterated by reheating and consequent migmatization and metamorphism during the Archean, whereas the late Archean of Precambrian II appears not to have been so destroyed. Reworking may have been more intense in the Archean of Precambrian II than at any time in the period of twice its duration that followed it. Precambrian II was the time of initial stabilization, but evidence from Greenland and the Southern Hemisphere leaves little doubt that some stabilization commenced well back in Precambrian I, although it probably only affected small, localized areas of the crust. In Greenland the remnant of a dike-transected eocraton 3700 million years old reflects this restricted, early stabilization.

The pattern of fold belts in the Northern Hemisphere fans out and may reflect a smaller Archean globe with a radius of about 4400 km. Archean fold belts are not confined between forelands, but can be traced almost indefinitely across their structural trend; they are only interrupted by later transverse belts. The Archean crust seems to have been subjected nearly everywhere to crumpling during Archean times.

The elongated character of the belts of calc-alkaline volcanics, clastic sediments, subordinate iron–jasper and calc-alkaline sediments, and rare limestones, and their separation by hundreds of kilometers of granite–gneiss and granulite terrain are noted. The isograds in the greenstone terrains of characteristic low grade may cut across the greenstone belt structure and thus appear to be related to the granitic rocks. The high-grade granulites of the intervening tracts appear to be the products of metamorphism synchronous with the lower-grade metamorphism within the greenstone belts, but at deeper crustal levels. The granulites were, however, overprinted by migmatization, superimposing amphibolite facies assemblages.

31B: SOME DEVELOPMENTS IN THE CRUST

John Sutton

ABSTRACT

The nature of geological activity has changed during geological time. The major structures have altered, though as at the present time they always appear to have formed parts of a global tectonic arrangement. Two long-term changes are (a) a tendency to increasing rigidity throughout geological time punctuated by (b) the development of shorter cycles lasting about 1,000 million years during each of which a new system of mobile belts formed and decayed. New structural situations in which economic deposits could form became possible during the various stages in the development of the Precambrian crust. Some examples of these are discussed.

SUMMARY

In this paper, Sutton takes his arguments further; he notes that all metamorphic terrains show evidence of folding, but that this does not of necessity indicate compression, as early geologists supposed, but only creep at shallow depth. Two bases are recognized for studying the overall crustal history on a global scale. First are the peaks in radiometric statistics, that is, maxima of determinations of age dates such as recognized by Gastil (1960) and representing times when unusually large amounts of metamorphic terrain cooled finally in different parts of the globe. Second, there is the dating of dike swarms that indicate periods of widespread intracratonic tension just after cratons were stabilized. The Gastil maxima indicate major shifts in metamorphically active regions in a period just before initiation of new mobile belts, whereas the dike swarms define tensional episodes just after stabilization. The first defines globally synchronous events, albeit not all events of the same type throughout the globe; the second stipulates more localized events, confined, possibly, to individual cratonic histories.

Sutton suggests that the patterns of the Archean were components of a global geotectonic pattern not like that of the present day, possibly because the crust was thinner and heat flow was enhanced. His philosophy is evolutionary, like that of Salop (Paper 27). Despite Salop's belief in an increase in sedimentation rates from Archean to present, the early Archean sequences are remarkably thick. In Canada and Scotland, some are immensely thick sequences of sediments, but most are mixed volcanic and sedimentary accumulations.

The origin of the granite–gneiss domes, the tracts that separate the greenstone belts, remains one of the major problems of Archean geology. Sutton accepts that they are broad domical uplifts of a granite basement and represent its reactivation after accumulation of supracrustal rocks of the greenstone belts, which may have suffered migmatization at their basal roots and be sharply contact metamorphosed at their margins. The predominant amphibolite facies assemblages of the rocks within the intervening domes contrast with the greenschist metamorphism of the belts.

Sutton cites the recognition of nappe styles of folding in Rhodesia (Stowe, 1968) as a case of localized instability of greenstone belts under conditions of thin crust and high heat flow. Such styles appear however to be rare in the Archean.

Telescoping of metamorphic zones is evident in Archean terrains and reflects the unusual accession of heat to the surface.

The *protocrust* is discussed—the elusive source of the detritus contained in the earliest sediments. The crust that existed in the earliest Archean times may well have included sediments, but these were reheated to such an extent as to reset the isotopic clocks (see Paper 3). Only in the upper parts of the protocrust, prior to 3400 million years ago, could sediments survive in such a raised geothermal gradient, and such upper levels were rapidly removed by erosion. Even sequences of ages between 3400 and 2700 million years that have survived include no wholly undeformed and unmetamorphosed rocks. The temperatures were still too high for rigid plates with flat, unaltered sedimentary cover to survive, if such plates existed, without removal by erosive agents, and nothing in the nature of a rigid shield has survived this period. The greenstones of the sedimentary troughs of this period are of diverse ages, the locus of deposition and volcanism undergoing shifts during this period.

The preceding statement possibly requires some modification in light of the Archean age now widely accepted for the Pongola and Witwatersrand stable platform series of South Africa and the Saamides of Northern Russia, and also in light of the recognition of virtually unmetamorphosed residuals in Archean terrains of Australia and other regions. However, such modification would not affect the strength of the argument.

High-grade metamorphism characterizes parts of Canada, Greenland, Scotland, and Russia: these high-grade tracts include sediments, amphibolites, and granulite facies assemblages. Anorthosites, also represented in such terrains, seem to be peculiar to the earlier stages of crustal history. In Greenland, pillow lavas in all stages of modification from greenschist to granulite facies assemblages are recognized, which suggests that such high-grade terrains do incorporate greenstone belts. Even so, such relics are rare. There are many other developments of supracrustal relics in such terrains but these are commonly extensively modified by migmatization. More recent work (Commentary, Paper 17A) has, however, suggested that much of the sedimentary material is of a quite different, pelagic facies to that of the greenstone belts, and possibly was laid down in quiet, shallow epicontinental-type seas on cratonic foundations. Likewise, the igneous rocks of many such relics are of quite different type to those of the greenstone belts, including as they do, chromititic anorthosites. The anorthosite sheets of these terrains seem to have exerted a considerable influence in thickening the crust, adding to the effect of granite invasion. The evidence from the high grade terrains favors the superimposition of high-grade metamorphism up to granulite facies on low-grade greenschist or granite basement terrain, at least in some cases.

References

Donaldson, J. A., and Jackson, G. D. (1965). Archaean sedimentary rocks of North Spirit Lake Area, North Western Ontario. *Can. J. Earth Sci., 2,* 622–646.

Gastil, G. (1960). Continents and mobile belts in the light of mineral dating. *Proc. 21st Int. Geol. Congr., 9,* 162–169.

McCall, G. J. H., Braybrooke, J. C., and Middleton, D. D. (1970). The Merougil sub-area. *J. Roy. Soc. Western Aust., 53,* 9–32.

Stowe, C. (1968). Intersecting fold trends in the Rhodesian basement complex. *Trans. Geol. Soc. South Africa, 71*(annexure), 53–78.

PAPER 32: COMMENTARY

This paper is extremely lengthy, so is only summarized here. One can advance some criticisms of this explicit paper: the supposed lack of clastic sandstones and quartzites in the Archean is not borne out by facts. Thick clastic sediment sequences are well known from Australia, South Africa, India, and Canada; some are Katarchean (Papers 5, 7, and 14). The picture of Archean palaeogeography given here applies to chemogenic sequences, of the quiet, pelagic periods intervening between rapid, intermittent greenstone accumulations in troughs; to the shallow seas in other parts of the proto-oceans that no doubt existed elsewhere when these troughs were being filled; and possibly to the earliest shallow seas that in all probability existed prior to the first greenstone developments in troughs (with their coarse clastic sediments). It is not a picture that is universally applicable to Archean sedimentation.

The argument that Proterozoic granite–gneiss terrains are all reworked oldest Archean foundation material is, perhaps, overstated, although it is certain that such terrains do include much very ancient reworked gray granite–gneiss material (Paper 22).

The separation of the Archaean and Proterozoic into *permobile, protogeosynclinal,* and *platform* stages that are sequential in time and abruptly defined is also, perhaps, an oversimplification of what was more in the nature of a complex series of diachronously overlapping stages. Yet the generalization has within it an essential germ of truth.

The supposition that greenstone belts invariably possess a sunken sialic foundation is very difficult to equate with the fact that greenstone sequences may attain thicknesses of up to 35,000 m (Gemuts and Theron, 1973). It seems that some greenstone belts may have been founded direct on oceanic crust within a gulf opened up between separated protocratonic blocks, in the same manner as analogous volcanics (Paper 16) are accruing at the present time on the floor of the opened-up Atlantic Ocean; perhaps all greenstone belts were so founded, except near their foreland margins.

The argument concerning the Stanovoy region appears invalid, for older sial may underlie greenstones locally, yet such a relationship does not establish a universal rule for greenstone belt foundations.

Reference

Gemuts, I., and Theron, A. (1973). Stratigraphy and mineralisation of a portion of the Archaean between Norseman and Coolgardie, Western Australia. *Abstract Vol., Sec. 3, ANZAAS, 45th Congr., Perth,* p. 88.

32: TECTONIC HISTORY AND STRUCTURES OF PLATFORMS AND SHIELDS

L. I. Salop and Yu. M. Scheinmann
Institute of Physics of the Earth, Moscow, U.S.S.R.

ABSTRACT

An attempt is made in this paper to outline the general tectonic development of platforms. We hope that such a review will help to determine the directions which geological-geophysical research should take within the Upper Mantle Project.

We apply the term platform, a synonym of craton, to any large portion of the continental crust bounded by folded belts or sometimes by oceans, and characterized by the following specific features: (a) tectonic stability over a very long period (several hundreds of million years or more); (b) general tectonic homogeneity (despite a complex inner structure and different times of formation of its separate parts); (c) an absence of high gradients of movement.

Shields are elements of platforms which have been characterized by a tendency towards slow uplift over periods of hundreds of million years. They are not sharply delimited.

We have accepted the following ages for the upper boundaries of Precambrian units: Archaean 2,600–2,800 m.y. (million years); Lower Proterozoic (Palaeoproterozoic, according to Salop) 2,100 m.y.; Middle Proterozoic (Mesoproterozoic) 1,850–1,650 m.y.; Upper Proterozoic (Neoproterozoic) 1,100 m.y.; Epi-Proterozoic (Epiproterozoic) 620 m.y.; Eocambrian 570 m.y.

The authors believe, although the material here presented is based mainly on studies of Eurasian structures, that these results can be also applied to all other continents. Reference to data on other areas is always stipulated.

SUMMARY

The authors define platform (craton) as any large portion of continental crust bounded by fold belts or oceans and characterized by tectonic stability over periods exceeding 100 million years, by tectonic homogeneity but internal complexity of structure (different parts being formed at different times), and by the absence of high gradients of movement.

Their definition does not allow for oceanic (thalasso-) cratons (e.g., Pacific). *Shields* are elements of platforms uplifted slowly over hundreds of millions of years without intense deformation.

The Archean of Siberia (2800–2600 million years) displays poor facies zonality and lacks rudites: it consists mainly of metabasites, quartz-carbonate rocks, high alumina rocks, and iron formations. Such an assemblage calls for special conditions, and the authors suggest shallow seas spread over much of the globe with low-lying, frequently flooded islands and unstable shoals. Subaqueous effusion was the predominant igneous manifestation. Platforms and miogeosynclines, characteristic of the Phanerozoic are lacking. Although clastic quartzites and sandstones characterize such Phanerozoic systems, they are only weakly developed in the middle and upper Proterozoic, and the quartzites of the Archean are products of redeposition of material derived from subaerial or subaqueous crusts formed by chemical weathering. Quartzites, high-alumina rocks and iron formations are all believed to have been derived in this way, sillimanite and magnetite being characteristically present in the quartzites. The silica of the crusts was dissolved by heated waters and redeposited elsewhere. Thus, quartzite occurrences of the Archean show no correspondence with outwash areas.

The weathered crusts have been supposed by many authorities (Papers 22 and 30) to only develop over basic volcanics, but the balance of the material that would derive from such sources is not recognized in Archean sedimentary sequences, and this cannot be the entire answer. Indeed, basic rocks were probably not the main source of the siliceous crusts, for rounded clastic zircons are present in very ancient Archean quartzites and could not have such a source. Trace-element distribution is also against such a simple solution.

The authors suggest that the 1000-m-thick lowest quartzite and aluminous slate member of the Aldan sequence in Siberia is consistent with a very slightly dissected relief.

The lack of weakly metamorphosed Archean rocks in Eurasia is discussed. Amphibolite- to granulite-grade rocks predominate and are affected by varying degrees of granitization. This contrasts with the terrains of the southern hemisphere and much of Canada.

The lower Proterozoic–Archean boundary is taken at 2800 million years, an unconventional usage presaging Salop's later revolutionary suggestions concerning geochronologic usage (Paper 27A). The gneisses and granulites of Labrador, the Grenville (Paper 23), Madagascar, Natal, Tanzania, Uganda, and Zambia, which yield Proterozoic ages, are taken by these authors as being reworked equivalents of the Eurasian Archean, modified lowest members of the Archean sequence. This conclusion is, they believe, the only possible one in the light of the small area covered by the greenstone belts and the immense spread of the apparently younger granitoid terrain. Unconformities noted beneath the Keewatin and an older granite foundation in Canada (Steeprock Series: Joliffe, 1966), and between Dhwars and a similar granite–gneiss basement in India, are considered by the authors to support this view, or at least the reality of an ancient gneissic basement to the greenstones. They note that stromatolites occur in the Steeprock Series, but none are known from the Archean of Eurasia.

The gneiss domes of South Africa and Canada are believed to be remobilized areas of older basement, as is shown be relic dating in zircons. The tectonics of the Archean are thought by the authors to be controlled by gregarious granites (MacGregor, 1951), which produce a chaotic interweaving of fold zones, without virgation, and with a preponderance of brachyform and minor domical elements. This pattern is one of high plasticity. The folding is closely related to granitization, deep-seated anatexis leading to diapiric ascent of granite masses. The authors use the phrase "universally *permobile* crustal condition." The terrains are referred to as "nuclear terrains" (Paper 29).

In the "early Proterozoic,"* fragmentation and partial submergence of the Archean gneissic basement occurred, the surface was denuded and peneplaned, and geosynclinal belts coupled with median massifs developed. This phase ended by the Middle to Upper Precambrian, but may have continued in later regenerations.

The first platforms involved the Witwatersrand and Olenek River, Siberia (about

*In conventional Russian usage, the meaning of this time definition corresponds approximately to our late Archean.

2100 million years) sequences. These authors seem to distinguish a Katarchean permobile stage, a 2800–2100 million years protogeosynclinal stage (of late Archean greenstone belts and granite invasions), and a later characteristically Proterozoic platform stage. They suggest that the oldest geosynclines did not have a foundation on primary basalt or directly on the mantle, but were founded on fragmented and foundered blocks of the earliest Archean granite–metamorphic basement, now represented by rejuvenated remnants that give apparent Proterozoic ages.* The fact that greenstones of the Stanovoy System, Siberia, show basement relics beneath them, appears to support their model.

References

Joliffe, A. W. (1966). Stratigraphy of the Steeprock group, Steeprock Lake, Ontario. *Spec. Paper 3, Geol. Assoc. Canada,* p. 75–98.

MacGregor, A. M. (1951). Some milestones in the Precambrian of Southern Rhodesia. *Proc. Geol. Soc. South Africa,* p. xxvii–lxxi.

*Again, remembering conventional Russian terminology, this probably should be read as late Archean to Proterozoic ages.

PAPER 33: COMMENTARY

The Archean of the Superior Province and other Canadian structural provinces is characterized by mafic to felsic sequences of volcanic rocks forming thick volcanic–sedimentary piles. Three stages have been recognized in the standard volcanic cycle:

1. *Platform:* a mafic platform or "basic plate" [R. H. Ridler uses this term, which is preferable because of the other meaning of platform (Paper 8)] formed by widespread submarine effusion, predominantly of tholeiitic basalt (3300–6600 m).
2. *Edifice:* high-rising volcanic piles erected on the platform foundation, in which piles the emissions increase in felsic, predominantly fragmental character (the bottom 1700–3300 m consists of intermediate flows and pyroclasts; the top 700–3300 m is mainly felsic pyroclasts).
3. *Erosional:* denudation produced thick sedimentary blankets and prisms (700–2000 m consisting of sediments and subordinate volcanic rocks).

There may be a single cycle, a single cycle with mafic capping, or, rarely, two or more successive cycles.

The total stratigraphic thickness of volcanic and sedimentary rocks may amount to 12,000 m. The volcanic piles display lateral facies variation. There is widespread evidence of subaqueous accumulation.

Five petrographic classes were erected on the basis of silica content; basalt (52 percent), andesite (52–58 percent), dacite (58–64 percent), rhyodacite (64–71 percent), and rhyolite (71 percent). These appear to be natural groupings for the Archean.

These studies embraced twelve belts, seven in detail and two in preliminary fashion. Similar results were indicated by preliminary study of belts at Yellowknife, Flin Flon, and Saskatoon.

Basalts appear first, followed progressively by the more siliceous rock types in the stratigraphic column of any cycle; the order is of overall decreasing abundance in a cycle and also, overall, through a stack of cycles.

A high degree of similarity was noted right through the province. The volcanics resemble standard tholeiite and calc-alkaline basalt–andesite–rhyolite associations in the literature (continental and island arc). Geochemical peculiarities are high FeO/Fe_2O_3 ratios and CO_2/H_2O contents and low alkali contents, particularly K_2O and TiO_2 contents.

The Archean volcanics are supposed to be differentiates of a tholeiitic, common Archean parent magma; basalt, being predominant, is the most likely parental stem of the five petrographic types. For example, the Birch-Uchi consists of basalt (58 percent), andesite (18 percent), dacite (11 percent), rhyodacite (9 percent), and rhyolite (4 percent) based on spaced sampling. In some areas, acid volcanics were more abundant, but basalts always account for at least 50 percent of the sampling from any single belt.

The sediments include flyschoid turbidites and are of a character suggesting rapid accumulation in tectonically unstable thin crustal environments. They are of two types:

1. *Volcanogenic:* (up to 3300 m) graywacke, conglomerate, breccia, banded ironstone; the clastic component compares closely with the nearby volcanic rocks and a local source is inferred, deposition being in troughs and basins. The banded ironstones are supposed to stem from chemical sedimentation direct from a volcanic source.

2. *Quartzofeldspathic:* thinly bedded impure quartzites, arkoses, argillites, and graywackes, together with banded ironstones, some conglomerates, and breccias form thick successions intercalated with the volcano–sedimentary belts, and equivalent to them in thickness. Volcanic intercalations are minor. Current and graded bedding are widespread. The indications are of turbidity current deposition in deep water. Some shallow-water deposition is indicated by large-scale channeling and current bedding structures in thick, ungraded sandstone sequences.

Compositionally, the average of the sediments of the Superior Province approximates to granodiorite.

The well-bedded nature of the sediments and the character of the conglomerates and breccias indicates a nearby continental source, which included substantial granitic exposures, a cratonic area similar to known exposed Archean granite–gneiss terrain; such a terrain must have been in existence at the time of deposition as a foreland. Potassium-rich granitic phenoclasts are lacking, and this terrain was apparently typically of low K/Na ratio.

The granitic rocks intervening between the volcano–sedimentary belts are weakly to strongly foliate granodiorite gneisses of likely metasedimentary origin, together with schists, migmatites, and igneous plutons. The Archean crustal conditions were apparently conducive to intense pervasive granitic transformation, notably of sedimentary rocks. The following cycle is evident:

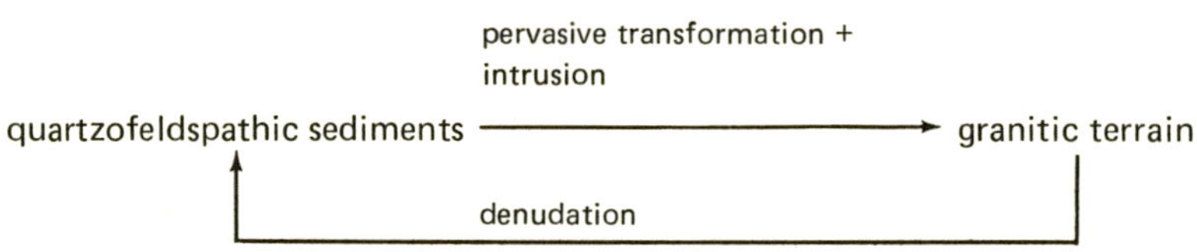

Volcanic-rich assemblages were not greatly affected by this transformation.

Volcanic-rich belts increase in their development from north to south in Ontario; sedimentary belts predominate in the northwest of the province. East–west trends predominate in the Superior and Churchill provinces, but north–south trends in the Slave Province. A pattern of warping around three elliptical protocontinental areas was deduced; the nucleii are situated in northwest Ontario–Manitoba, northwest Hudson's Bay, and Central Slave regions. A fourth, in Ungava, was inferred. The largest Superior protocontinental nucleus may be taken as the type example. It was girdled by volcanic and sedimentary belts, alternating and increasing in width outward. The largest of all, the Abitibi Belt, contained profuse acid volcanics, and also unusual alkali-potassic trachytes, leucitites, and syenites at Kirkland Lake and sodic rhyolites at Noranda. The belts are thought to decrease in age outward from the nucleus, but establishment of detailed geochronology is fraught with problems. The pattern of outward younging of belts seems to be repeated in the Windigo, Amisk, and La Ronge belts in the northwestern part of the same province. Lack of definition is, however, evident in the case of the Hudson and Slave provinces, which seem to display a random, sublinear development of small volcanic and sedimentary belts. Dominance of tholeiites and scarcity of acid volcanics also suggest fundamental differences to the Superior Province.

In the Proterozoic, long linear belts with broad basinal elements replaced the Archean protocontinental pattern.

33: ARCHAEAN PROTOCONTINENTAL GROWTH AND EARLY CRUSTAL HISTORY OF THE CANADIAN SHIELD

A. M. Goodwin

Introduction

Early Precambrian or Archaean volcanic and sedimentary rocks of the Canadian Shield bear the imprint of unstable environments of accumulation characterized by a thin crust which are similar to those of orogenic belts at continental margins, including island arc systems. The character of Archaean lithologic assemblages and the distribution patterns of the assemblages suggest a history of crustal development by growth of several protocontinents. At least four of these are postulated in the Canadian Shield.

Archaean protocontinental growth is thought to have resulted in development of a stable, continental mass characterized by a thicker crust which dominated Late Precambrian or Proterozoic time. Crustal stability in Proterozoic time is evidenced by characteristic stable platformal sedimentary and volcanic assemblages. In addition to these surficial accumulations, the Proterozoic continental mass probably was thickened by addition of crustal material below.

The record of the Canadian Shield, in brief, suggests a crustal evolution in the direction of a larger, thicker, more stable crustal mass. The thesis of protocontinental growth developed in this paper offers a model for consideration in future work.

Archaean protocontinental growth

Volcanic rocks

Archaean volcanic rocks of the Canadian Shield, typically, are distributed in distinctive mafic-to-felsic sequences as described below.

Volcanic sequences: The typical Archaean greenstone assemblage of the Canadian Shield contains thick, complex volcanic-sedimentary piles. The standard Archaean volcanic cycle proceeded in three stages (Fig. 1): (1) the Platform stage — construction of a mafic platform the product of widespread effusion of predominantly tholeiitic basalt; (2) the Edifice stage — marked by increasingly

Reprinted from *Proc. 23rd Int. Geol. Congr., Prague,* 1, 69–89 (1968).

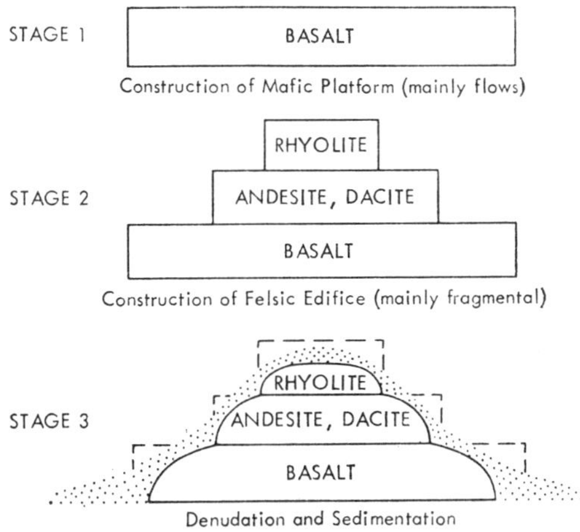

Fig. 1 — Schematic illustration of the three main stages in evolution of an Archaean volcanic pile

felsic, predominantly pyroclastic eruption of material of calc-alkalic chemical affinity leading to erection of numerous high-rising piles resting upon the platform; and (3) the Erosional stage — marked by denudation of the volcanic piles and construction of thick, volcanogenic sedimentary blankets and prisms.

The resulting succession in ascending order common to many Archaean segments may be summarized as follows (see also Goodwin, 1968, Fig. 3): 10,000 to 20,000 feet of basalt flows and associated intrusions; 5,000—10,000 feet of intermediate flows and fragmentals; 2,000—10,000 feet of predominantly felsic pyroclastics; and finally, 2,000—6,000 feet of sediments with subordinate volcanic rocks.

Many Archaean volcanic assemblages contain a single mafic-to-felsic volcanic sequence; some contain a complete mafic-to-felsic sequence together with a younger mafic capping; rarely, two or more superimposed mafic-to-felsic volcanic sequences are present. The total stratigraphic thickness is commonly in the order of 35,000 feet. Lateral facies changes are common in the volcanic piles. Evidence of subaqueous accumulation is widespread. Felsic rocks are commonly in pyroclastic forms.

Despite the destructive influences of erosion, intrusion and deformation, these ancient Archaean volcanic rocks are remarkably well preserved; they are on the whole astonishingly fresh in texture and structure. Their prevailing vertical to steeply inclined attitudes combined with the fresh glaciated erosional surfaces common to the Canadian Shield provide unexcelled stratigraphic sections, particularly of the deep volcanic "plumbing" systems. Also, in some localities gently dipping volcanic rocks provide a partial plan view of the volcanic piles. Thus a substantial threedimesional view of a variety of Archaean volcanic piles is available for study in the Canadian Shield (Fig. 2).

Chemical composition: Because Archaean volcanic rocks are composed almost exclusively of secondary minerals such as chlorite, mica, zoisite, epidote, quartz, carbonate and secondary feldspars, a rock classification based on primary minerals is of little value. Instead, a chemical classification is employed. Until such time as a system based on normative minerals suitable for Archaean rocks has been developed and adequately tested (see Irvine, 1966), a simple classification based on silica content of fresh volcanic rocks is employed. The volcanic divisions are as follows:

basalt	— less than 52 per cent SiO_2,
andesite	— 52—58 per cent SiO_2,
dacite	— 58—64 per cent SiO_2,
rhyodacite	— 64—71 per cent SiO_2,
rhyolite	— greater than 71 per cent SiO_2.

Based on experience to date, these divisions represent natural chemical groupings of Archaean volcanic rocks.

A substantial body of chemical data is available from systematic study and sampling of Archaean volcanic assemblages of the Canadian Shield (see in particular Baragar, 1966, 1967; Goodwin, 1962, 1964, 1965, 1967; Wilson, et al., 1965). Detailed study by the author of twelve volcanic areas (Fig. 3) evenly distributed across the southern part of the Superior province (see Map 4—1965, Geol. Surv. Canada) of the Canadian Shield forms the basis of the following comments on the chemistry of Archaean volcanism. Of the twelve areas, seven (Birch-Uchi, Lake of the Woods, Kakagi Lake, Eagle Lake, Timmins, Kirkland Lake and Noranda) have been studied and sampled in systematic detail, three (Lake St. Joseph, Michipicoten and Swayze) in moderate detail, and two (Pickle Lake, Savant) in preliminary manner only. In each of the twelve areas, the sampling was conducted under stratigraphic control. This has permitted sequential arrangement of the chemical results. Accordingly, in Tables 1—4 the volcanic types within the volcanic piles are arranged sequentially from left to right in generalized ascending stratigraphic order.

The volcanic rock samples, each of which is composed of three to five pounds of fresh rock chips taken at appropriate intervals across a volcanic unit, were analysed in the chemical laboratories of (1) the Geology Department of the University of Manitoba, (2) the Ontario Department of Mines and (3) the Geological Survey of Canada. Details of the chemical procedures and listings of the individual analyses have been published in the references cited or will be published in due course.

Table 1 shows the average chemical compositions and generalized ascending orders of volcanic rocks in the Birch-Uchi Lakes, the Lake St. Joseph and the Pickle Lake areas. All three areas form part of a main east-trending regional volcanic-rich belt in the Superior province (Fig. 3; see also p. 82). The volcanic rock types present and their generalized ascending order are: basalt-andesite-dacite-rhyodacite-rhyolite. The Birch-Uchi Lakes area contains the full comple-

Table 1 — Chemical Composition of Archaean Volcanic Rocks

	Birch-Uchi Lakes					Lake St. Joseph				Pickle Lake		
	Basalt	Andesite	Dacite	Rhyodacite	Rhyolite	Basalt	Andesite	Dacite	Rhyodacite	Basalt	Andesite	Rhyolite
SiO_2	48·6	54·0	60·6	67·9	73·4	46·9	53·6	61·3	66·8	49·4	57·8	71·6
Al_2O_3	15·5	14·5	14·7	13·0	13·1	14·1	13·8	15·4	15·5	15·4	18·4	15·7
Fe_2O_3	2·51	2·81	1·31	1.71	0·75	0·85	0·31	1·15	0·95	1·62	1·74	0·23
FeO	8·84	7·37	4·47	4·27	2·00	10·04	9·23	5·09	2·57	8·55	3·38	1·25
MgO	5·80	4·20	3·09	1·37	0·67	9·37	4·00	2·46	1·80	4·15	1.31	1·31
CaO	8·79	6·19	4·88	2·58	1·17	11·9	5·14	5·01	2·98	8·15	8·75	3·21
Na_2O	2·35	2·83	3·84	3·63	4·30	2·52	2·98	3·29	3·53	3·03	2·59	1·71
K_2O	0·30	1.02	1·38	1·08	1·71	1·22	0·37	0·87	2·48	0·41	0·31	1·16
H_2O }	5·39	5·49	4·35	3·18	1·89	0·89	3·21	1·50	0·74	2·39	1·30	2·01
CO_2 }						0·28	5·35	1.58	0.25	3·44	1·04	1·34
TiO_2	0·89	0·99	0·72	0·54	0·38	0·82	0·71	0·64	0·50	1·10	0·98	0·16
MnO	0·20	0·18	0·11	0·11	0·05	0·18	0·16	0·13	0·16	0·23	0·27	0·02

Table 2 — Chemical Composition of Archaean Volcanic Rocks

	Lake of the Woods					Kakagi-Eagle Lakes					Savant Lake		
	Basalt	Andesite	Dacite	Rhyodacite	Rhyolite	Basalt	Andesite	Dacite	Rhyodacite	Rhyolite	Basalt	Andesite	Dacite
SiO_2	49·2	54·6	60·6	68·1	74·9	48·9	55·4	60·9	67·7	76·1	49·3	52·4	60·9
Al_2O_3	13·6	14·5	15·2	15·2	14·8	14·9	15·4	15·4	15·2	13·3	15·0	14·0	15·4
Fe_2O_3	2·53	2·37	2·11	1·02	0·33	2·92	2·34	1·82	1·15	0·67	1·54	2·95	4·64
FeO	8·37	7·13	4·25	2·09	0·77	8·49	6·59	4·64	1·51	1·59	10·62	9·81	6·37
MgO	7·37	4·89	2·31	1·65	0·78	5·38	4·35	2·76	1·28	0·21	5·62	4·58	1·21
CaO	8·34	5·95	4·92	3·19	0·89	8·71	6·24	5·35	3·31	1.28	6·92	7·60	3·37
Na_2O	2·08	2·75	2·90	2·39	1·85	2·34	2·81	3·47	3·93	3·16	2·17	1·67	1·50
K_2O	0·55	0·51	1·66	1·76	3·29	0·35	0·70	1·21	1·43	1·83	trace	trace	0·67
H_2O	2·57	2·35	1·73	0·91	0·70	5·80	4·83	3·45	3·25	2·13	4·20	4·25	4·15
CO_2	2·62	2·01	3·17	2·22	0·90						2·47	1·88	2·09
TiO_2	0·78	0·83	0·68	0·39	0·25	1·08	0·91	0·65	0·44	0·21	0·87	0·72	0·83
MnO	0·18	0·15	0·10	0·05	0·03	0·22	0·47	0.12	0·08	0·14	0·22	0·24	0·30

Table 3 — Chemical Composition of Archaean Volcanic Rocks

	Michipicoten				Swayze						Timmins-Kirkland Lake-Noranda					
	Basalt	Andesite	Dacite	Rhyolite	Basalt 1	Basalt 2	Andesite	Dacite	Rhyodacite	Rhyolite	Basalt	Andesite	Dacite	Rhyodacite	Rhyolite	Trachyte**
SiO_2	46·4	53·6	59·9	72·3	49·9	48·8	53·1	62·8	66·1	72·3	50·2	55·0	63·7	67·0	76·7	55·5
Al_2O_3	18·5	17·2	16·6	12·1	12·9	16·7	17·1	15·6	15·3	11·6	11·7	12·6	16·8	14·4	10·9	16·5
Fe_2O_3	1·63	1·71	7·28*	5·07*	4·72	1·99	1·32	0·98	1·01	1·00	2·1	2·0	1·3	1·20	0·4	3·5*
FeO	8·23	10·74			10·23	10·02	8·25	5·74	3·56	5·88	7·7	7·4	2·6	4·60	2·5	
MgO	4·39	4·63	2·68	1·66	3·91	5·19	3·53	2·47	1·72	1·25	7·8	6·2	2·1	1·64	0·6	4·5
CaO	4·27	2·56	3·80	1·21	6·18	8·70	6·87	4·75	3·63	2·22	10·8	7·3	3·4	3·03	1·1	5·1
Na_2O	n.d.	1·71	n.d.	1·70	2·02	1·06	2·00	0·66	2·20	0·87	2·4	3·3	4·6	2·49	3·0	4·2
K_2O	n.d.	0·51	n.d.	1·55	0·82	0·78	0·68	1·05	1·79	1·21	0·16	0·38	1·24	0·60	2·77	4·3
H_2O	n.d.	5·48	6·30	2·99	4·14	3·03	2·63	2·48	1·88	1·11	3·6	3·2	1·9	2·16	1·0	n.d.
CO_2	n.d.				2·60	2·39	2·23	2·10	1·55	0·65	1·4	0·99	2·09	0·42	0·6	n.d.
TiO_2	1·43	1·21	0·80	0·40	1·50	0·95	1·09	0·38	0·44	0·26	1·05	1·17	0·47	0·67	0·24	0·76
MnO	n.d.	0·43	0·41	0·33	0·13	n.d.	n.d.	n.d.	0·03	n.d.	0·19	0·22	0·08	n.d.	0·06	0·15

n.d. — not determined
* — all iron determined as FeO
** — present at Kirkland Lake only

ment. However, rhyolite has not been located so far in the Lake St. Joseph area nor have dacite and rhyodacite in the Pickle Lake area.

Table 2 shows the average chemical compositions and generalized ascending orders of volcanic rocks in the Lake of the Woods (the original Keewatin) area, the Kakagi-Eagle Lakes (combined) area[1] and the Savant Lake area. These three areas form part of a second, main east-trending, regional volcanic-rich belt in the Superior province (Fig. 3). Again the volcanic rock types are arranged in generalized ascending order. The Lake of the Woods and Kakagi-Eagle Lakes assemblages contain a diversity of volcanic types. In the Savant Lake area the apparent absence of highly acidic types (rhyolite and rhyodacite) is due to lack of sampling of the upper acidic stratigraphic zones present in the western part of the area.

Similarly Table 3 presents average chemical compositions and generalized ascending orders in the Michipicoten area, the Swayze area, and the Timmins-Kirkland Lake-Noranda (combined) area, all part of a third, main, east-trending, regional volcanic-rich belt (Fig. 3). Again the common sequential order of basalt-andesite-dacite-rhyodacite-rhyolite is repeated. Two types of basalt are present in the Swayze area as described below. At Kirkland Lake a succession of trachytic volcanic rocks which is so far as known unique in the Archaean of the Canadian Shield overlies the normal basalt-andesite-rhyolite assemblage. The lowermost part of the trachytic succession comprises calc-alkaline andesites, trachytes and quartz trachytes; the upper part comprises alkalic trachytes and potassic mafic and leucitic trachytes, tephrites and phonolites (Cooke, 1966).

In Table 4 the average chemical compositions of Archaean volcanic rocks and their generalized ascending order within the volcanic piles of the Superior province based on available information are listed. The averages constitute the arithmetic mean of the average compositions of volcanic rock types in the twelve greenstone areas listed above (Tables 1—3, see p. 72—73). As previously stated, trachytes are restricted to the one area of Kirkland Lake.

Sequential sampling of volcanic rocks in the Yellowknife area of the Northwest Territories in the northern part of the Canadian Shield (Baragar, 1966) has yielded comparable chemical and stratigraphic results. Preliminary sampling of volcanic rocks in the Flin Flon area of Manitoba and Saskatchewan has yielded similar chemical results (unpublished information from E. Froese, Geol. Surv. Canada and L. C. Coleman, University of Saskatchewan). Thus the detailed results obtained in the Superior province are taken as representative of the Canadian Shield-at-large.

The following five characteristics of Archaean volcanic rocks are noted:

[1] The Kakagi Lake and Eagle Lake volcanic components are chemically similar. Therefore, the two areas have been combined for present purposes.

1. The generalized ascending stratigraphic order in a volcanic pile is basalt-andesite-dacite-rhyodacite-rhyolite which is also an order of decreasing abundance.
2. The volcanic rocks in general are chemically similar across the Superior province.
3. Archaean volcanic rocks compare in part with standard tholeiitic types (Turner and Verhoogen, p. 180, Table XVI) and in part with calc-alcalic rocks of the basalt-andesite-rhyolite association (Turner and Verrhoogen, p. 221, Table XXI). Such rocks are characteristic of continental orogenic regions including the island arc systems (Wilson et al, 1965, p. 175).
4. Despite this similarity to recognized world-wide types, Archaean volcanic rocks have the following five distinctive chemical characteristics:
   a) high FeO : Fe_2O_3 ratios, b) low alkalic content particularly K_2O, c) low TiO_2 content, d) high volatile (CO_2 and H_2O) content, and e) limited range (2·8 per cent) in average Al_2O_3 contents (see Table 4).
5. Archaean volcanic rocks appear to have been derived by differentiation of a tholeiitic-type parent magma. This parent magma, which was similar in composition to the common Archaean basalt (Table 4), was probably closely related to the Archaean mantle.

Table 4 — Average Chemical Composition of Archaean Volcanic Rocks in the Superior Province of the Canadian Shield

	Basalt	Andesite	Dacite	Rhyodacite	Rhyolite
SiO_2	48·9	54·7	61·5	67·3	74·3
Al_2O_3	14·5	15·0	15·7	14·8	12·9
Fe_2O_3	2·14	2·00	1·83	1·17	0·74
FeO	9·03	7·64	4·49	3·44	2·22
MgO	6·27	4·50	2·38	1·55	0·85
CaO	8·74	6·39	4·41	3·13	1·48
Na_2O	2·51	2·79	3·15	3·07	2·47
K_2O	0·45	0·55	1·16	1·40	2·10
H_2O	3·34	2·92	2·27	1·56	1·17
CO_2	1·93	1·93	2·18	0·98	0·86
TiO_2	1·06	0·99	0·63	0·51	0·26
MnO	0·21	0·28	0·16	0·08	0·10
P_2O_5*	0·07	0·12	0·12	0·07	0·07
N	162	119	272	177	85

N = number of analyses. Total number is 815

* = P_2O_5 determinations available on 96 samples only

Although generally similar across the Superior province, local chemical variations are present in the twelve greenstone areas. The variations are best seen by comparing the averages of volcanic types in individual areas (Tables 1—3) with the corresponding averages of the province (Table 4). Thus Lake St. Joseph volcanic rocks are comparatively high in ferrous iron; also the basalts are highly

Fig. 2 — Distribution of Archaean rocks of the Canadian Shield

magnesian. In the Kakagi-Eagle Lakes area the acidic volcanic rocks are low in iron particularly ferrous iron. Savant Lake volcanic rocks are high in ferrous iron and water and low in alkalies. Michipicoten volcanic rocks are high in Al_2O_3 (except rhyolite), TiO_2, FeO, MnO, CO_2 and H_2O and low in CaO, Na_2O and SiO_2; these chemical characteristics mainly reflect mild but pervasive chemical alteration involving introduction of Fe, Mn — carbonates and removal of Na_2O

and SiO_2 (see further Goodwin, 1964); on the other hand, high Al_2O_3 and TiO_2 contents in Michipicoten volcanic rocks appear to be primary features. Swayze volcanic rocks are comparatively high in FeO and low in Na_2O. In addition, the Swayze assemblage contains both low-Al_2O_3 basalt (basalt 1, Table 3) and high-Al_2O_3 basalt (basalt 2, Table 3); the former is common in the lower parts of the volcanic assemblage whereas the latter is more commonly associated

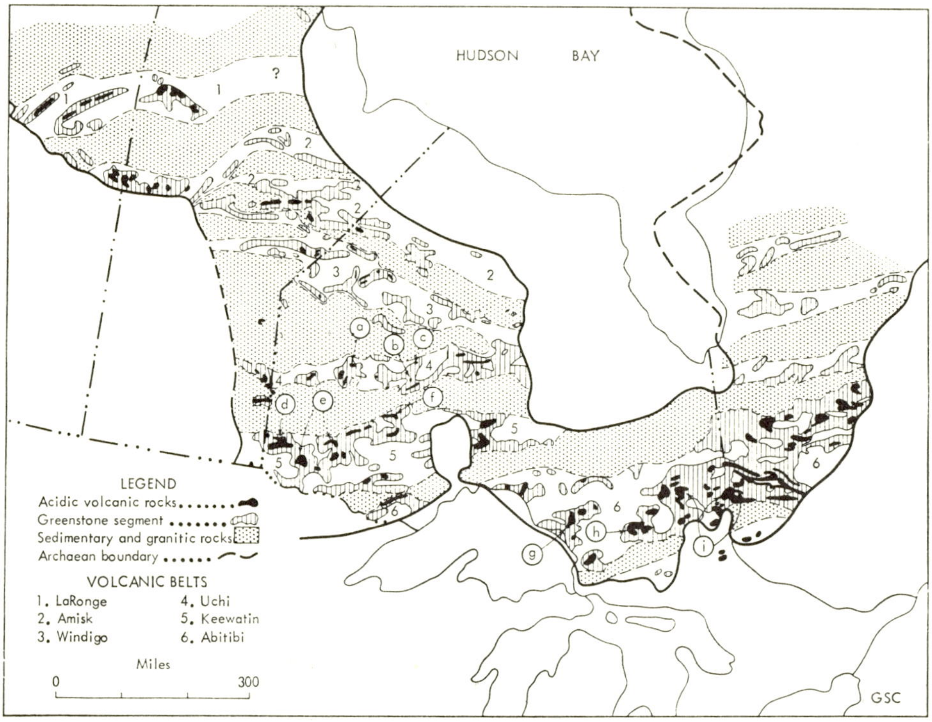

Fig. 3.—Distribution of volcanic rich segments and belts in the southern part of the Canadian Shield. Designated areas are as follows: a — Birch-Uchi, b — Lake St. Joseph, c — Pickle Lake, d — Lake of the Woods, e — Kakagi-Eagle Lakes, f — Savant, g — Michipicoten, h — Swayze, i — Timmins — Kirkland Lake — Noranda.

with the overlying intermediate to acidic components. This relation of low- and high-Al_2O_3 basalts has been noted in other areas of the Superior province (Baragar, 1967, p. 3).

The Timmins-Kirkland Lake-Noranda (combined)[1] volcanic rocks contain high

[1] Detailed volcanic studies in the Timmins-Kirkland Lake-Noranda area have been completed recently by the author. The resulting chemical and stratigraphic data will be published shortly. In the meantime preliminary results have been incorporated in this paper (see further Baragar, 1967, p. 3).

Al_2O_3 (except dacite and rhyodacite) and Na_2O and in the case of basalt, high MgO. Considered alone (not illustrated), Noranda volcanics are notably high in Na_2O, MgO and TiO_2. In addition, the Kirkland Lake trachytes stand apart in their uniquely high K_2O content.

Considered in terms of distribution within the three main, east-trending volcanic-rich belts of the Superior province (Fig. 3), it is seen that the volcanic assemblages of the northern belt (Birch-Uchi, Lake St. Joseph and Pickle Lake areas, Table 1) and central belt (Lake of the Woods, Kakagi-Eagle Lakes and Savant areas, Table 2) are chemically similar. However, the volcanic assemblages of the southern belt (Michipicoten, Swayze and Timmins-Kirkland Lake-Noranda areas, Table 3) differ chemically in the following respects, (1 the presence of both low-Al_2O_3 and high-Al_2O_3 volcanic rocks, and (2 higher TiO_2 and alkalies (both Na_2O and K_2O) particularly in the Timmins-Kirkland Lake-Noranda area. Further work is required to define and assess these regional chemical variations.

The proportions of rock types in the volcanic piles are of the following order as accurately determined in the Birch-Uchi Lakes area of northwestern Ontario (Goodwin, 1967): basalt — 58 per cent; andesite — 18 per cent; dacite — 11 per cent; rhyodacite — 9 per cent; rhyolite — 4 per cent. Although the quantity of acidic volcanic rocks is higher in some other areas, particularly in the south-eastern part of the province (see Fig. 3), it is unlikely that the proportion of basalt in any Archaean volcanic-rich area is less than 50 per cent.

Thus the Archaean volcanic rocks of the Canadian Shield compare closely in nature and composition with those of active continental orogenic belts including the island arc systems. Interpretations of Shield evolution presented below are based upon this fact.

Sedimentary rocks

Archaean sediments, mainly flysch-type deposits with substantial proportions of poured-in turbidites, suggest rapid accumulation in tectonically unstable, thin crustal environments. In broad terms, two main types of Archaean sediments are present — volcanogenic and quartzofeldspathic.

Volcanogenic sediments ranging to 10·000 feet thick, notably greywacke, conglomerate, breccia and banded iron formation, predominate within and near the volcanic assemblages. The clastic components compare closely in composition with the nearby volcanic rocks and have obviously been derived from them mainly by rapid erosion and subaqueous deposition in nearby troughs and basins (see Bass, 1961; Boutcher et al., 1966; Goodwin, 1962; Hewitt, 1963; Moorhouse, 1965). In addition, associated chemical sediments (iron formation) are attributed to a direct volcanic source (Goodwin, 1962, 1964).

Other Archaean sediments are mainly quartzofeldspathic varieties that form major belts intercalated with and equivalent in dimension to the main volcanic-rich belts of the Canadian Shield. These sediments characteristically comprise thin-bedded, fine grained, impure quartzite, arkose, argillite, and greywacke; conglomerate, breccia and banded iron formation are included. Metamorphic equivalents, mainly biotite-oligoclase-quartz schist locally with garnet, cordierite and andalusite, are common. Volcanic intercalations are locally present. Graded bedding is common and crossbedding is present. Sedimentary features characteristic of deposition by turbidity currents in deeper water, "eugeosynclinal-type" environments predominate. However some shallow-water deposition is indicated by largescale channeling and crossbedding in thick, ungraded "sandstones" (Danaldson and Jackson, 1965, p. 641).

Table 5 — Average Chemical Composition of Archaean sediments and granitic rocks

	Archaean Sediments						Archaean Granite Gneiss	
	1	2	3	4	5	6	7	8
SiO_2	65·43	73·58	60·60	67·8	62·5	56·95	64·5	66·3
Al_2O_3	16·1	12·90	18·90	17·4	21·1	16·67	16·1	15·4
Fe_2O_3	1·50	0·24	3·01	0·2	1·4	0·86	1·5	1·3
FeO	4·03	2·72	4·53	1·3	2·6	6·72	2·9	2·9
MgO	2·64	1·16	3·26	0·5	0·5	4·15	2·3	2·1
CaO	1·72	1·38	2·40	2·2	1·9	3·22	3·3	4·0
Na_2O	2·20	2·72	2·96	6·0	2·3	2·94	4·0	3·9
K_2O	2·78	4·10	2·51	2·1	3·4	2·43	2·78	2·3
H_2O	1·40	0·33	0·99	0·8	3·0	4·01	0·8	0·87
CO_2	0·03	nil	0·11	0·2	0·1		0·2	0·15
TiO_2	0·61	0·37	0·59	0·25	0·34	nil	0·48	0·47
P_2O_5	0·11	0·06	0·15	0·09	0·10	nil	0·16	0·13
MnO	0·09	0·04	0·07	0·02	0·06	0·40	0·08	0·07

1. Quartz-mica schist, Lake St. Joseph, Ontario (ref. Goodwin, 1965, Table 3, p. 31, col. 16).
2. Impure arkose Lake St. Joseph, Ontario (ref. Goodwin, 1965, Table 3, p. 31, col. 13).
3. Andalusite-bearing schist, Lake St. Joseph, Ontario (ref. Goodwin, 1965, Table 3, p. 31, col. 14).
4. Arkose, North Spirit Lake, Ontario (ref. Donaldson and Jackson, 1965, Table 3, col. 4).
5. Argillite, North Spirit Lake, Ontario (ref. Donaldson and Jackson, 1965, Table 3, col. 5).
6. Greywacke, SW Rouyn map area, Quebec (ref. Wilson, M. E., 1962, p. 22).
7. Average composition entire New Quebec sample area, 200·000 sq. mi. (Eade, et al., p. 7).
8. Average composition of 40·000 sq. mi., NW Ontario (Eade, et al., p. 7).

Great thicknesses of quartzofeldspathic sediments are present. For example an estimated thickness of 20·000—30·000 feet is present in a main east-trending sedimentary belt lying south of the Birch-Uchi area (Bateman, 1940). Thus the Archaean sedimentary-rich assemblages compare in thickness and extent with the Archaean volcanic-rich assemblages of the Canadian Shield.

The prevailing well-bedded character of the quartzofeldspathic sediments and the presence of conglomerate and breccia suggest subaqueous deposition of detritus derived in part by rapid erosion of nearby source areas of substantial relief. The source areas must have included substantial if not predominant amounts of granitic rock, probably cratonic masses similar in composition to the typical, present day, Archaean granitic terrain. The scarcity of phenoclasts of potash-rich granite in certain sedimentary assemblages of northwestern Ontario emphasized by Donaldson and Jackson (1965, p. 646) is considered to be a valid generalization which accurately reflects prevailing low potash-content of many Archaean granitic rocks.

Chemical compositions of some typical Archaean sediments representative of the broad quartzofeldspathic sedimentary belts in the Superior province of the Canadian Shield are listed in Table 5, columns 1—6. Included are impure quartzite, arkose, greywacke and argillite together with metamorphosed equivalents. Quartz-mica schist (col. 1) is by far the dominant rock type. The others (cols. 2—6) represent common variants including aluminous (cols. 3, 5), quartzitic and potassic (col. 2), and mafic (col. 6) types which collectively compare closely in chemical composition with the dominant quartz-mica schist (col. 1).

The average composition of the common quartzofeldspathic sediment of the Superior province compares closely with that of common granodiorite. As discussed below this relationship points to Archaean provenances of general granodioritec composition.

Granitic rocks

Large areas of the Canadian Shield are underlain by granitic rocks (Fig. 2; see also map 1045A, Geol. Surv. Canada). The rocks consist in large part of weakly to strongly foliated granodiorite gneiss of probable metasedimentary origin together with schists, migmatites, and undoubted igneous plutons (see Eade, Fahrig et al., 1966). The average chemical composition of two large granitic areas of the Canadian Shield, the one in the northeastern (New Quebec) part and the other in the central (northwestern Ontario) part, are listed in Table 5 (cols. 7, 8). On this basis it is seen that the average composition of Archaean granitic rocks approaches that of common granodiorite. The typical Archaean granitic terrain undoubtedly contains igneous products of several superimposed orogenies. Indeed Archaean crustal conditions seem to have been peculiarly conducive to widespread, pervasive granitic transformation (granitization) of pre-existing, notably sedimentary, rocks.

A close chemical similarity exists between the common Archaean quartzofeldspathic sediments (Table 5, cols. 1—6) and the common Archaean granitic rock (Table 5, cols. 7, 8). This chemical similarity may reflect a genetic relationship as follows: the sediments were derived mainly by erosion of Archaean granitic terrains, and the granitic rocks in turn were derived mainly by metamorphic alteration (granitization) of Archaean sediments. Considering the Canadian Shield at large, the process of "granitization" appears to have been selective; it affected the main sedimentary-rich assemblages without greatly affecting the adjoining volcanic-rich assemblages.

Archaean protocontinents

The Canadian Shield contains a large number of Archaean volcanic- and sedimentary-rich belts which are enclosed in granitic rocks (Fig. 2). In the southern part of the Shield these belts are aligned in sub-lineal to lineal patterns of alternating volcanic-rich and sedimentary-rich belts. In the northern part of the Shield the volcanic and sedimentary belts display a comparatively random, discontinuous pattern featuring sub-lineal alignment only.

Volcanic-rich belts are common in the southern part and to a lesser degree in the north-central part of the Canadian Shield. In the northwestern and western parts of the Shield, on the other hand, sedimentary-rich belts are more common (Fig. 2).

For the purpose of constructing the model and in keeping with the fresh, remarkably well-preserved state of the Archaean volcanic rocks, the present trends of the Archaean belts are interpreted as being mainly primary features of an evolving crust rather than due to later deformational and metamorphic activities. On the basis of the distribution patterns of these belts and the nature of the volcanic and sedimentary components a theory of protocontinental development is proposed.

The prevailing trends of Archaean volcanic- and sedimentary-rich belts are easterly in the southern and central parts of the Shield (Superior and Churchill tectonic provinces) and northerly in the northwestern part (Slave tectonic province). Examined in detail, Archaean trends warp around three elliptical areas situated respectively, (1) in northwestern Ontario and the adjoining part of Manitoba; (2) in the northwest part of Hudson Bay; and (3) in the central part of the Slave province. This distribution of Archaean trends about separate nodal areas serves to roughly delineate the nuclei of the three postulated ancient Archaean protocontinents. These are named the Superior, Hudson and Slave protocontinents (Fig. 4). In highly generalized form each protocontinent comprises a nucleus (craton) and marginal orogenic belts (orogens). A fourth Archaean protocontinent, named the Ungava protocontinent, is inferred in the northeastern part of the Superior tectonic province. Of the proposed Archaean protocontinents, the Superior protocontinent is by far the largest and best developed. It serves as the type Archaean protocontinent.

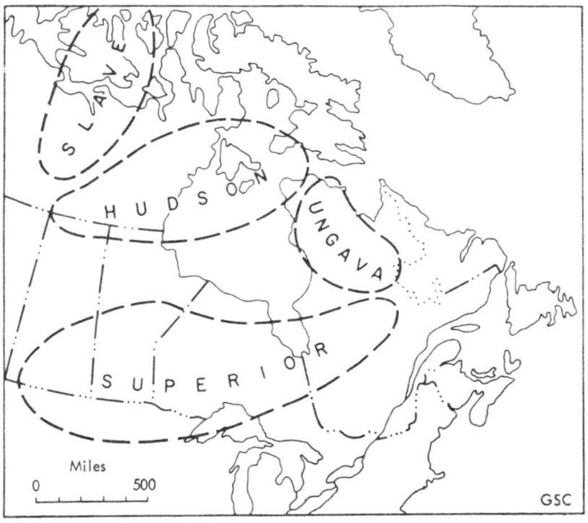

Fig. 4 — Distribution of four Archaean protocontinents of the Canadian Shield

Superior protocontinent: The Superior protocontinent contains a number of alternating, volcanic-rich and sedimentary-rich belts. In the southern part of the protocontinent, three easterly-trending volcanic-rich belts are present (Fig. 3). The three belts are named from north to south, — the Uchi belt, the Keewatin belt, and the Abitibi belt. The three belts are separated by intervening sedimentary-rich belts now composed of metasediments, migmatite, paragneiss and granitic rocks. In the northwestern part of the Superior protocontinent a similar though less well-defined series of easterly to southeasterly-trending volcanic-rich belts is present. These belts are named from south to north — the Windigo belt, the Amisk belt, and the La Ronge belt. These six belts (orogens) are distributed about a nucleus (craton) situated in northwestern Ontario and the adjoining part of Manitoba.

Considering the three main easterly-trending volcanic-rich belts in the southern part of the Superior protocontinent — Uchi, Keewatin and Abitibi — examination of their physical construction and knowledge of their volcanic composition brings out the following trends across the belts considered from north to south, (1) the three volcanic-rich belts increase in width, the widest lying in the south; (2) individual volcanic-rich areas within the belts increase in size, the largest being situated in the east part of the Abitibi belt in the south; and (3) the acidic volcanic content of the volcanic assemblages increases to the south and attains a maximum in the eastern part of the Abitibi belt. In addition, the alkalinity of the acidic volcanic rocks is highest in the Abitibi belt in the south; the increased alkalic content is exemplified by the presence of K_2O-rich trachyte at Kirkland Lake and of Na_2O-rich volcanic rock at Noranda.

It is postulated that these progressive changes across the protocontinent in the direction of increasing size and compositional complexity to the south correspond to progressively younger rocks in that direction. Accordingly the Uchi belt would represent an older Archaean belt centrally located within the protocontinent with respect to the peripherally located Abitibi belt in the southeast. Final proof of such an age sequence awaits absolute dating of the age of deposition of the Archaean volcanic and sedimentary rocks, a major unresolved geochronological problem.

In the northwestern part of the Superior protocontinent the three main volcanic-rich belts — Windigo, Amisk, and La Ronge — show similar though less-striking progressive changes in relative size and acidic volcanic content to the northwest. Again these changes are postulated to be in the direction of younger rock assemblages. It is noted, however, that none of these three northwestern belts correspond in size and complexity to the Abitibi belt in the southern part of the protocontinent.

In brief summary, a skeleton framework of the Superior protocontinent is proposed, a framework comprising a predominantly granitic nucleus or craton with adjoining annular belts or orogens. The core-to periphery increase in size and complexity of volcanic components is postulated to be in the direction of younger rock assemblages. A long and complex orogenic history involving many periods of granitic emplacement is represented.

Hudson and Slave protocontinents: In the northern part of the Canadian Shield Archaean volcanic-sedimentary belts are less well-defined (Fig. 2). Although due partly to limited geological mapping, to metamorphic overprint, and to the presence of Proterozoic cover rocks, the linear pattern of Archaean belts does not appear to have been developed to the same degree as in the Superior protocontinent to the south. Rather, a random to sublinear distribution of relatively small discontinuous volcanic- and sedimentary-rich belts is characteristic. In the Hudson protocontinent easterly trends predominate. In the Slave protocontinent northerly trends predominate.

In the Hudson protocontinent, on the basis of the distribution pattern of sub-linear belts and the presence of granitic masses a craton is postulated to be present in the northwestern part of the Hudson Bay area. In the Slave protocontinent to the northwest, the distinctive northerly, sub-linear trends of the belts may reflect the presence of a cratonic mass now covered by younger rocks to the north.

In both postulated protocontinents, as illustrated in Figure 2, predominantly sedimentary-rich belts alternate along and across the regional trends with relatively narrow and discontinuous volcanic-rich belts. Granitic rocks are comparatively abundant in the Hudson protocontinent.

The volcanic assemblages in the Hudson and Slave protocontinents are compositionally simple and have a high proportion of tholeiitic-type basalt (see Baragar, 1966). The distribution of known acidic volcanic rocks (see Goodwin, in press) suggests that acidic differentiates are comparatively scarce compared with the Superior protocontinent in the south. In short, on the basis of available data a relatively simple, compositionally primitive volcanic association lacking substantial acidic differentiates prevails in both the Hudson and Slave protocontinents.

Ungava protocontinent: Very little is known about Archaean volcanic compositions and orogenic trends in the highly metamorphosed, structurally complex Ungava portion of the Superior tectonic province. However, long linear belts of Proterozoic rocks are known to girdle all but the southwest part of a large area of granitic rocks in central Ungava. This suggests the existence of a northerly trending protocontinent, the Ungava protocontinent.

Other protocontinental units may be present particularly in the northeastern part of the Canadian Shield.

Summary: The typical Archaean protocontinent in generalized form is considered to have comprised a predominantly granitic nucleus (craton) with additions of volcanic-sedimentary belts (orogens). The proposed Superior protocontinent featured a peripheral migration of orogenic belts away from a main craton. Similar patterns of orogenic migration may yet be defined in the other more highly metamorphosed Archaean protocontinents.

In addition to these obvious orogenic activities now marked by the volcanic-sedimentary belts, the Archaean protocontinents probably experienced substantial addition of crustal material below. Sub-crustal deposition (sialic "underplating" of Engel, 1967) may well have contributed substantially to growth and stabilization of the Archaean protocontinents.

Essentially the Archaean protocontinents are viewed as centres of crustal thickening. Whether they constituted independent crustal segments (island continents) or simply thicker (and thickening) parts of a thin continuous crust is not known. In this regard the uniformity of orogenic trends across large parts of the Shield is more in keeping with the latter concept. For this reason a crustal model featuring progressive nodal thickening (the protocontinents) of a thin, continuous crust is preferred. Accordingly, successive orogens would tend to locate at the margins of the growing protocontinents and in the thinner-crustal, intraprotocontinental zones. A complex growth pattern would result, a pattern featuring irregular flowage of orogens about cratonic masses which is as interpreted, the dominant tectonic pattern of the Canadian Shield. With continuous growth the protocontinents would merge one with the other and so lose their individual identities in a larger continental mass.

Proterozoic growth

The distribution of Proterozoic sedimentary and volcanic rocks of the Canadian Shield is shown in Figure 5. In contrast to Archaean rocks, the distribution pattern of Proterozoic rocks is dominated by long linear belts and broad basinal elements. Northeasterly and, in lesser degree, northwesterly trends predominate in contrast to Archaean easterly and northerly trends. Flay-lying to gently dipping strata are common in most Proterozoic assemblages. Proterozoic rocks commonly overlie Archaean basement rocks with a profound structural uncon-

Fig. 5 — Distribution of Proterozoic, sedimentary and volcanic rocks of the Canadian Shield

formity. This is particularly so in the Superior tectonic province. In the Churchill tectonic province the effects of the Hudsonian orogeny (Proterozoic age) have obscured the Archaean-Proterozoic age relationships in many places. As a result an Archaean versus Lower Proterozoic (Aphebian) age of some rock units is in doubt, e.g. northern Saskatchewan and south shore of Baffin Island (see further Stockwell, 1965).

Proterozoic sediments include a high proportion of stable platformal types such as orthoquartzite and sandstone, dolomite and limestone and granular iron formation (Superior type, see Gross, 1965, p. 90). Notable examples are the Aphebian sequences of the Lake Superior and Labrador regions.

The form and habit of Proterozoic rocks is well illustrated by the Labrador "trough", a northwesterly-trending belt at least 800 miles long. It contains a complex assemblage of sediments including iron formation, mafic volcanics and associated intrusions. The shallow-water, shelf facies on the west contains granular cherty oxide iron formation, orthoquartzite and dolomite. The deeper water facies on the east contains fine-grained clastic rocks including pelites, mafic volcanics and associated tuffs.

The Proterozoic assemblages of the Lake Superior region and of the northwestern part of the Shield include substantial proportions of mature, crossbedded and ripple-marked quartzite, clear evidence of stable platformal environments of deposition.

Proterozoic volcanic rocks are represented mainly by thick, non-sequential, tholeiitic basalts. Examples are the Labrador "trough" volcanics, Keweenawan flows of the Lake Superior region, and the Coppermine series of the Northwest Territories. In the cases cited there is no substantial evidence of compositional differentiation as with the Archaean volcanic assemblages. In this respect they resemble the great flood basalts of northwestern United States and of India (Deccan traps). Trachyte are present locally in Proterozoic assemblages of the Canadian Shield as, for example, in the Dubawnt Group of the Northwest Territories (Donaldson, 1965, p. 5).

In summary, a broad and continuous distribution pattern featuring stable paltformal sediments and non-sequential volcanics with preferred northeasterly and northwesterly trends characterizes the Proterozoic rocks of the Canadian Shield. This broad sweep of Proterozoic rocks across the Shield (Fig. 5) clearly indicates the presence of large stable cratonic masses during Proterozoic time.

It is noted that early Proterozoic rocks of the Canadian Shield contain, amongst others, major deposits of uranium (and thorium) and of lead (Fig. 5; see further Goodwin, in press, C.I.M.M.). Further, the late Archaean and early Proterozoic igneous suites contain many alkalic intrusions with substantial K_2O content. For example a large number (estimated at sixty) of alkalic and alkaline ring complexes have been tentatively identified as plugs intruding the Archaean Shield; at least three ages of intrusion spanning late Archaean and Proterozoic time are represented — 2,600, 1,600 and 1,000 million years ago (Wilson, 1966, p. 182). In addition lamprophyre dikes are widespread. These activities in all likelihood reflect a major increase in K_2O content of the Precambrian crust.

The appearance of these major heat-producing elements — uranium, thorium, lead and potassium — in large quantities during late Archaean and early Proterozoic times appears to mark a degree of thermal migration and activation new in the history of the crust. The crustal consequences must have been profound. Thus a major event in crustal evolution distinguishes Archaean and Proterozoic times.

Considered further, the Archaean and Proterozoic of the Canadian Shield stand in such sharp contrast lithologically, orogenically and tectonically as to record some first-order crustal event. In the Canadian Shield, and particularly in the southern part, the Archaean-Proterozoic boundary is marked by the Eparchaean Interval, a ninterval at least 150 million years long (from 2,500 to approximately 2,350 million years ago). Thus the characteristic Archaean crustal conditions terminated with the Kenoran orogeny (2,500 million years ago). Following the Eparchean Interval the characteristic Proterozoic crustal conditions prevailed. The history of this important Interval is obscure. Possibly the Interval was marked by some first-order, mantle-crust differention which had as a main crustal expression the widespread distribution of the major heat-producing elements.

Summary and conclusions

The main stages of Precambrian crustal evolution in the Canadian Shield postulated here are summarized in Figure 6. The suggested trends were from thin-crustal, unstable protocontinents to a thick-crustal, relatively stable continental mass, the respective units representing the pregeosynclinal and early geosynclinal stages of crustal development. The sites of accumulation of the sediments and volcanics progressed from randomly and sub-linearly distributed Archaean basins and trenches to relatively long, continuous Proterozoic troughs, stable platforms and primitive geosynclines. Easterly and northerly trends of

PRECAMBRIAN CRUSTAL EVOLUTION

	ARCHAEAN		PROTEROZOIC
UNITS:	PROTOCONTINENTAL	→	CONTINENTAL
CRUST:	THIN, UNSTABLE	→	THICK, STABLE
STAGE OF DEVELOPMENT:	PRE-GEOSYNCLINAL	→	EARLY GEOSYNCLINAL
SITES OF ACCUMULATION:	BASINS (NUCLEAL TO SUB-LINEAL DISTRIBUTION PATTERNS)	→	STABLE PLATFORMS AND PRIMITIVE GEOSYNCLINES
TRENDS:	EAST; NORTH	→	NORTHEAST; NORTHWEST
LITHOLOGY:			
SEDIMENTS:	IMMATURE QTE, GWKE, BRECCIA	→	MATURE QTE, CARBONATES, GRANULAR IRON FM
VOLCANICS:	SIMPLE, SEQUENTIAL	→	VARIED, NON-SEQUENTIAL

Fig. 6 — Precambrian crustal evolutionary trends

Archaean time gave way to northeasterly and northwesterly trends of Proterozoic time. Immature, flysh-type Archaean sediments were succeeded by mature, shelf-type Proterozoic sediments. Volcanic activity changed from simple and sequential to varied and non-sequential. In addition, a major increase in the crustal content of the main heatproducing elements — uranium, thorium, potassium and lead — is apparent in late Archaean and Proterozoic rocks, an event which in all likelihood influenced all major aspects of crustal history — lithologic, orogenic, tectonic and biologic.

References

Baragar W. R. A. (1966): Geochemistry of the Yellowknife volcanic rocks; Can. Jour. Earth Sciences, vol. 3, No. 1, p. 9—30.

— (1967): Geochemistry of Noranda, Quebec, volcanic rocks; Abstracts of Papers, Geological Association of Canada, International Meeting, Aug 31 — Sept. 2, 1967.

Bass M. N. (1961): Regional tectonics of part of the southern Canadian Shield; Jour. Geol., vol. 69, p. 668—702.

Bateman J. D. (1940): Geology nad gold deposits of the Uchi-Slates Lakes area; Ont. Dept. Mines, vol. XLVIII, pt. 8, p. 1—43.

Boutcher S. M. A., Edhorn, A. S., and Moorhouse, W. W. (1966): Archaean conglomerates and lithic sandstones of Lake Timiskaming, Ontario; Proc. Geol. Assoc. Canada, vol. 17, p. 21—42.

Cooke D. L. (1966): in Report of Volcanological subcommittee; Canadian Geophysical Bulletin, published by the Associate Committee on Geodesy and Geophysics of the National Research Council of Canada, vol. 19, p. 182.

Donaldson J. A. (1965): The Dubawnt Group, Districts of Keewatin and Mackenzie, Geol. Surv. Canada, Paper 64—20, 11 p.

Donaldson J. A., and Jackson G. D. (1965): Archaean sedimentary rocks of North Spirit Lake area, northwestern Ontario; Can. Jour. Earth Sciences, vol. 2, No. 6, p. 622—647.

Eade K. E., Fahrig W. F., and Maxwell, J. A. (1966): Composition of crystalline rocks and fractionating effects of regional metamorphism; Nature, vol. 211, No. 5055, p. 1245—1249.

Engel A. E. J. (1966): The Barberton Mountain Land: Clues to the differentiation of the earth; Information Circular No. 27, Economic Geology Research Unit; University of Witwatersrand.

Goodwin A. M. (1962): Structure, stratigraphy, and origin of iron formation, Michipicoten area, Algoma District, Ontario, Canada; Bull. G. S. A., vol. 73, p. 561—586.

— (1964): Geochemical studies at the Helen iron range; Econ. Geol., vol. 59, No. 4, p. 684—718.

— (1965): Geology of Pashkokogan Lake — Lake St. Joseph area; Ontario Dept. Mines, Geol. Rept. No. 42, 58 pp.

— (1967): Volcanic studies in the Birch-Uchi Lakes area of Ontario; Ont. Dept. Mines, Misc. Paper MP6, 96 pp.

— (1968): Evolution of the Canadian Shield; Proc. Geol. Assoc. Canada, vol. 19, p. 1—13.

— (in press) Growth and Mineralization of the Canadian Shield; Can. Inst. Min. Metallurgy.

Gross G. A. (1965): Geology of iron deposits in Canada, Volume 1: General geology and evaluation of iron deposits; Geol. Surv. Canada, Econ. Geol., Rept. No. 22.

Hewitt D. F. (1963): The Timiskaming Series of the Kirkland Lake area; Can. Mineral., vol. 7, p. 497—523.

Irvine T. N. (1966): Recommendations on the processing and presentation of chemical data on volcanic rocks; Proceedings of the 47th Meeting of the Associate Committee on Geodesy and Geophysics, National Research Council of Canada, Ottawa, p. 1—7 to 1—12.

Moorhouse W. W. (1965): Stratigraphic position of sulphides in the Archaean; Can. Inst. Min. Met. Trans. vol. 68, p. 261—264.

Stockwell C. H. (1965): Tectonic map of the Canadian Shield; Map 4—1965, Geol. Surv. Canada.

Turner F. J. and Verhoogen J. (1951): Igneous and Metamorphic Petrology; McGraw-Hill Book Company, Inc., 602 pp.

Wilson H. D. B. (1966): Alkalic and alkaline ring complexes in the Archaean; Canadian Geophasical Bulletin, vol. 19, p. 182.

Wilson H. D. B., Andrews P., Moxham R. L. and Ramlal K. (1965): Archaean volcanism in the Canadian Shield; Can. Jour. Earth Sciences, vol. 2, No. 3, p. 161—175.

Wilson M. E. (1962): Geology of the Rouyn-Beauchastel map-areas, Quebec; Mem. 315, Geol. Surv. Canada.

PAPER 34: COMMENTARY

Glikson's statement that his first type of assemblage is not recognized in Canada is not now acceptable, because of the discovery of spinifex textured flows of the high magnesia basalt association at Munro Township, Ontario (Pyke et al., 1973). Such igneous rocks are indeed present there, and form Goodwin's platform stage (Paper 33); ultramafic flows and associated high magnesia basalts are less prominent in Canada than in South Africa and Western Australia, whereas calc-alkaline assemblages are much less conspicuous in these regions. (Hallberg, personal communication, has lately recognized a few scattered and small-scale calc-alkaline acid volcanic centers in the Yilgarn Province of Western Australia, but most acid volcanic centers in that region are not calc-alkaline.)

Bimodality has been recognized by Glikson in the suite from Western Australia, and such bimodality seems to be an inherent feature of many alkaline differentiation suites (McCall and Hornung, 1972). The possibility that the sodic-alkaline porphyries of the Kalgoorlie area might be broadly analogous to late-cycle developments of potassic and sodic alkaline volcanics at Kirkland Lake and Noranda (Paper 8) is suggested. The geochronology of the Kalgoorlie belt indicates late Archean metamorphism, and this belt could, in Goodwin's model of protocontinental development, be an outermost girdling belt around a Katarchean protocontinental nucleus situated in the western half of the Yilgarn Block and in the Pilbara Block (McCall, 1972).

Glikson's ultrabasic/basic volcanic association appears to be no different from the universal platform stage (Paper 33) of all Archean greenstone cycles. The *ophiolites* do not appear to have been distinguished from the rocks of this stage elsewhere on petrographic or chemical criteria; they are high magnesia basalts and tholeiites of abyssal type as elsewhere. Similarly, the term "calc-alkaline volcanics or sodic porphyries/rhyolites" would better cover his second stage. The alkaline lavas of Canada and South Africa are prefolding and surely cannot be inserted in the last stage with the potash granites, which are clearly a much later eruptive suite.

The suggestion that high Ni, Co, and Cr content is peculiar to the Archean basalts and indicates incomplete core separation within the globe at this early date is not supported by evidence from the Rift Valley (McCall and Hornung, 1972) of

*Defined by McCall and Leishman (1971): "The name is a field term alluding to the pattern resemblence to grasses colloquially called *spinifex* in Western Australia. In the spinifex rocks the olivines display elongation in two directions to form lamellae measuring up to 10 cm maximum dimension. The lamellae are assorted into domains in which they lie subparallel.

very high trace-element components of these in Quaternary alkalic basalts: some other explanation for these anomalies must be sought. The term "capping" applied to sodic porphyries in this paper covers a number of irregularly developed lenses, high in the cyclic succession, not a continuous, well-developed, terminal capping.

Glikson tends to lean toward Alpine analogies: for example, in the continued use of the term "ophiolite" and in the introduction of the term "molasse." The ophiolite analogy (also earlier used by the editor) now appears strained. And the term "molasse," although used very tentatively by Anhaeusser et al. (Paper 5) in the case of the Moodies Group of Swaziland (they refer to the group as a possible "molasse equivalent") seems out of place. There seems to be little reason other than terminality in the succession to equate conglomerates in the Kalgoorlie succession, which develop progressively in prominance upward through that sucession and are everywhere metamorphosed, with molasse, which is by definition a terminal, postorogenic (postfolding, postmetamorphism, and postuplift) sedimentary accumulation in Alpine sequences.

References

McCall, G. J. H. (1972). The nickel sulphide bearing ultramafic rocks and their environment in the Archaean of Western Australia. *Proc. 24th Int. Geol. Congr., 1,* 354–362.

_____, and Hornung, G. (1972). A geochemical study of Silali volcano, Kenya, with special reference to the origin of intermediate and acid eruptives of the central rift valley. *Tectonophysics, 15,* 97–113.

_____, and Leishman, J. (1971). Clues to the origin of Archaean eugeosynclinal peridotites and the nature of serpentinisation. *Spec. Publ. 3, Geol. Soc. Aust.,* 281–299.

Pyke, D. R., Naldrett, A. J., and Eckstrand, O. R. (1973). Archaean ultramafic flows in Munro Township, Ontario. *Geol. Soc. Am. Bull., 84,* 955–978.

34: PRIMITIVE ARCHAEAN ELEMENT DISTRIBUTION PATTERNS: CHEMICAL EVIDENCE AND GEOTECTONIC SIGNIFICANCE

A. Y. Glikson

Bureau of Mineral Resources, Geology and Geophysics, Canberra, Australia

ABSTRACT

Major and trace element data from low-grade isochemically metamorphosed Archaean igneous suites in Transvaal, Rhodesia, Western Australia and Canada show consistent variations from those of corresponding younger igneous suites. The metabasalts of Archaean basalt ultrabasic-basic assemblages conform to recent oceanic tholeiites with respect to K, Na/K, Sr, Zr and Fe^{3+}/Fe^{2+}, but have considerably lower Al and Ti, and tend to have higher Mn, Ni, Cr, Co, Rb and Fe/Fe+Mg (total iron as FeO) than recent equivalents. Metabasalts of Archaean calc-alkaline volcanic series are more differentiated, and have K, Ba, Sr, Zr and Fe/Fe+Mg values higher than those of metabasalts of ultrabasic-basic assemblages and recent oceanic tholeiites. The evidence suggests a derivation of the early basalts through partial melting of the mantle at depths shallower than 15 km. Possibly, the Archaean mantle had higher levels of siderophile and certain transition elements than the present mantle; this may have been related to a lesser degree of development of the core. The earliest Archaean granites have high Na/K, Ni and Cr levels, and were probably derived from partial melting of the oceanic crust. Granitic cycles showing a progressive decrease of the Na/K and K/Rb ratios reflect an increase in the degree of differentiation and a thickening crust. The pattern and evolution of Archaean volcanic-sedimentary troughs were controlled by early nuclei of sodic granites and granodiorites emplaced in the oceanic crust; these greenstone belts are therefore not analogous to Alpine or island-arc systems. The stabilization of these belts through deformation and metamorphism, resulted in the aggregation of the early granitic nuclei into shields; subsequently high-grade belts developed along shield-ocean transition zones.

SUMMARY

In a review paper (summarized and commented upon here because of restrictions of length) Glikson makes the assumption that igneous and sedimentary rock compositions from low-grade Archean greenstone suites may be taken as original compositions. He cites the lack of composition range in calc-alkaline petrographic groups, the preservation of primary plagioclase, pyroxene and even olivine in unaltered residuals within basic–ultramafic assemblages, and the lack of variation in potassium and calcium content in basalt analyses. Chemistry cannot, he believes, be significantly modified by metamorphism or diagenesis. He attributes the lack of metamorphic modification to the fact that the facies is the greenschist facies of the low andalusite–sillimanite series of Miyashiro, characterized by low volatile partial pressures. This conclusion is supported by lack of chemical variation observed between metamorphosed and nonmetamorphosed specimens of the same igneous rocks sampled within a few meters of one another; furthermore, it is supported by the preservation of sharply defined differentiation trends similar to those of unmetamorphosed suites. Nevertheless, in accepting this as a valid basis for geochemical methodology, as has long been done in Canada and South Africa, account must be taken of possible metamorphically induced complications.

Glikson draws on geochemical studies by Hallberg (Paper 16A) and O'Beirne (1968), dealing with basalts and acid porphyry–rhyolite associations respectively, to revise his previous conclusion (1968) that the Kalgoorlie province displays a progression from basalts through andesites, dacites, and rhyodacites to rhyolites, as in the case of the calc-alkaline suites of Canada. He now divides the Archean greenstones into three types of assemblage:

1. Ultrabasic–basic (tholeiitic).
2. Calc-alkaline.
3. Ophiolite–sodic porphyry.

He now recognizes a lack of continuous progression in the chemistry, and thus introduces the concept of a hiatal or bimodal ophiolite–sodic porphyry type of assemblage in which the ultrabasic–basic platform assemblage (Paper 33) is supposedly capped by extensive lenses of sodic porphyry. The term "ophiolite" introduces an Alpine analogy (see Paper 34: Commentary). The hiatal or bimodal concept is certainly much more realistic than Glikson's original concept of smooth progression through the cycle. Drawing on the results of studies in South Africa (Papers 5 and 6), Glikson considers the characteristic sequence of greenstone–granite associations of the Archean to be:

5. Potassic granites (or late-stage syenite–alkaline lava as in Canada and South Africa).
4. End of folding.
3. Sodic granites.
2. Calc-alkaline volcanics or "ophiolite–sodic porphyry."
1. Ultrabasic–basic volcanics.

Available evidence is summarized showing that Archean tholeiites display lower Al and Ti contents than modern ocean-floor abyssal equivalents: and higher Mn, Ni, Cr, Co, and Rb contents and Fe/Fe + Mg ratios. Calc-alkaline basalts of Archean age show a similar disparity, but, in addition, display higher Ba contents.

References

Glikson, A. Y. (1968). The Archaean geosynclinal succession between Coolgardie and Kurrawang, near Kalgoorlie, Western Australia. Ph.D. thesis, University of Western Australia.

O'Beirne, W. R. (1968). The acid porphyries and porphyroid rocks of the Kalgoorlie area. Ph.D. thesis, University of Western Australia.

PAPER 35: COMMENTARY

This important paper is summarized, for it includes a summarization of much of the material presented in more detail in Papers 5–7. Yet it advances many important new facts about the Archean of southern Africa and also discussions of global significance.

35: THE EVOLUTION OF THE EARLY PRECAMBRIAN CRUST OF SOUTHERN AFRICA

C. R. Anhaeusser

Economic Geology Research Unit, University of the Witwaterstrand, Johannesburg, South Africa

ABSTRACT

The early Precambrian granite–greenstone basement complex of Rhodesia and South Africa constitutes one of the oldest known fragments of the earth's crust. Scattered within the expanse of granitic rocks are numerous greenstone belts some approaching, or even possibly exceeding, ages of about 3400 Ma. These ancient greenstone belt remnants represent the earliest clearly decipherable geological events on the Earth's surface, and any evidence concerning the nature and evolutionary development of the primitive crust must be sought largely from an examination of their lithologies.

The stratigraphy of the early greenstone occurrences in southern Africa is described with particular emphasis being given to the geological record established in the well-preserved Barberton Mountain Land of the Transvaal. The evidence for a primitive sialic versus simatic crust is reviewed and reference is made to the reported pre-greenstone belt basement remnants found in Rhodesia as well as in Swaziland. Comparisons are drawn between the early greenstone belts and modern island arcs, use being made of available geochemical data from the two environments.

It is concluded that the early crust in southern Africa had analogies with the present-day abyssal regions of the oceans. Progressive transformation, involving island arc-like development and protocontinental nucleation, resulted from processes involving partial melting of the primitive lithosphere coupled with the secular mixing of mantle derived, granitic additions. The early thin, unstable, primordial crust underwent progressive thickening processes and, about 3000 Ma ago, stabilized sufficiently to permit the development of the first interior or cratonic-type sedimentary basins. Increases in the areal extent of the developing basins are briefly traced and the more significant events, from the earliest Precambrian to middle Proterozoic times (*ca.* 1700 Ma ago), are contrasted and summarized.

SUMMARY

Anhaeusser considers various alternative hypotheses concerning the nature of the earth's primordial crust. The most popular hypotheses are those stated as alternatives by Engel (1963), the first postulates a thin sialic crust formed very early in the earth's history during a rapid differentiation into core, mantle, and crust; the second postulates a primordial differentiation of a less complete nature, resulting in the development of protocore, protomantle and oceanic basalt crust. Gill (Paper 36) has favored the latter alternative: Engel (1968) has also developed this concept, as has Glikson (Paper 37, and Glikson and Sheraton, 1972).

The important paper by Engel (1968), could have been included here, but this summary must suffice. Engel considered that the greenstone belts of the Swaziland System typify the ancient greenstone belts of the world. The fact that greenstone belts of Canada and Australia show significant differences does not invalidate this generalization. In fact these belts, like those of Rhodesia and the Pilbara, Australia, are typical of the Katarchean greenstone belts, and the late Archean belts are somewhat different in character. The likelihood is that ancient greenstone belts more than 3400 million years old are second or subsequent generations of graywacke–volcanic complexes, their precursors now being no more than mafic or gneissic screens within the oldest granite–gneiss terrains, indistinct paleosomes of which the original nature has been almost obliterated by metamorphism and metasomatism. Continental evolution is envisaged by Engel as occurring, initially, in rift volcanoes or island-arc-like welts within the much earlier, primeval seas. The volcanoes of these linear zones were evolved from a protocrust chemically more primitive than the present ocean floor.

Engel advanced a model involving a major global thermal event in which primary core-mantle differentiation occurred, and potassium, uranium, thorium, and lead migrated outward, such migration being related to the formation of the granite envelopment surrounding the greenstone belts. According to geochemical evidence from the Barberton Mountain Land (part of the Swaziland System), the upper mantle cannot have suffered significant change since 3400 million years ago. The analogy between the Swaziland greenstones and the assemblages of more modern island arcs is emphasized, the following characteristics are shared:

1. Progressive crustal thickening.
2. Progressive decrease in volcanic output.
3. Progressive increase in residuate sedimentary content.
4. Progressive decrease in instability.

The Onverwacht Group of the Swaziland System is envisaged as being initially formed on a crust 12–15 km thick (by analogy with the Aleutians and Kuriles). As the Swaziland System developed, so both greenstone belts and neighboring source area for the detritus thickenened, were invaded by granite, and stabilized: the sediments reflect this progressive change in their mineralogy. The stabilized and thickened block resisted orogenous superimposition for 2500–3000 million years, forming now one of the thickest and most differentiated areas of the crust, characterized by low heat flow as a result of impoverishment of radiogenic sources in the mantle beneath. There is an extreme contrast between the present state of the Kaapvaal craton and the thinnest and least differentiated ocean-floor segments of the present crust. The first eruptions of the Swaziland greenstone sequence were, however, basaltic analogues to the eruptive of the present ocean floors and unusual peridotite flows (respectively partial and virtually entire mantle melting products). Both were derived from a similar potassium, uranium, thorium, and lead impoverished upper mantle.

Anhaeusser remarks that the Kaapvaal and Rhodesian cratons are girdled by younger, curvilinear metamorphic belts (Zambezi, Limpopo, Mozambique, Namaqua–Natal), the Archean protocontinental nucleii being surrounded by Proterozoic mobile belts.

In Rhodesia, three successive volcanic sedimentary sequences are recognized—Sebakwian, Bulawayan, and Shamvaian—all Archean. The lowermost Sebakwian consists of mafic and ultramafic rocks with sedimentary interlayers and ultramafic intrusions, all invaded by granites and locally granitized. The overlying Bulawayan includes pillowed and unpillowed basalts, andesite breccias and interbedded sediments, whereas the Shamvaian forms a sedimentary overlay of quartzites, conglomerates, graywackes, and shales. "Pre-Sebakwian" rocks recognized by Stowe near Selukwe are regarded by Anhaeusser as controversial; they have the form of scattered remnants of mafic and ultramafic material and banded ironstones within gneiss–migmatite terrain. Similar rocks in South Africa, devoid of sedimentary material, occur as scattered relics in the Kaapvaal craton (Johannesburg–Pretoria dome) and as definite greenstone belts (Murchison, Pietersburg). The latter show a very poor development of sedimentary relics.

The Ancient Gneiss Complex of Swaziland is regarded as older than the Swaziland System by Hunter, but neither this terrain of kinzigites, quartzites, tachonites, and amphibolites, siliceous diopsidic granulites, ultrabasics set within extensive tonalitic gneisses and migmatites, nor any other of the possible older terrains in southern Africa have yielded geochronologic ages supporting inferences of prior age to the Swaziland System or Sebakwian. The conclusion of Hunter was based on a lack of diapiric doming and abrupt metamorphic contrast to the Barberton sequence. Complex developments of granite to the north of the Barberton greenstone belt–including gneisses, migmatites, homogeneous and nebulitic granodiorites, adamellites, and pegmatites, have yielded ages from 3160 to 2990 million years, suggesting that they, like the Ancient Gneiss Complex, are not older basement or at least not older basement incompletely reworked. These ages are consistently younger than the diapiric granites that invade the Barberton!

The Ancient Gneiss Complex is interpreted by Viljoen and Viljoen as a complementary response to vast outpourings of largely primitive mafic and ultramafic lavas in the greenstone belt nearby. All these lavas are notably deficient in alkali and other lithophile element content. Like Anhaeusser, these authors do not accept the conclusion of Hunter, but consider the Ancient Gneiss Complex to be part of the main granite emplacement that postdated the Barberton sequence accumulation.

Anhaeusser asks the questions: "Were granitic rocks, always a dominant part of Archean cratons, present in some form or another from the very earliest beginnings? Or were they the result of a later invasion of granodioritic material locally into certain regions of a primary oceanic crust?" The very early existence of continental crust is now beyond question (to about 4000 million years ago), but he concludes that this question remains unresolved. There is no direct evidence that the Ancient Gneiss Complex and its analogues are formed by reworking older basement, but either this is the case, and the basic relics are derived from a system still older than the Onverwacht Group, or they are simply unassimilated fragments derived from that group itself.

Anhaeusser considers that an unconformity recognized in Rhodesia beneath the Bulawayan sediments could be of an intraformational nature within the threefold succession, and thus might not provide evidence for a much older granite basement. He considers the pre-Sebakwian could, likewise, prove to be no more than digested Sebakwian relics within an invading granite. Abrupt changes in metamorphic grade cited both in that case and in the case of the Ancient Gneiss Complex are a feature of the Archean the world over and can be explained in terms of telescoped metamorphic aureoles. There is no need to invoke two ages of metamorphism simply because of abrupt juxtapositions of contrasting grades. The best evidence for an oceanic primordial crust lies, Anhaeusser believes, in geochemical evidence of lack of crustal contamination in some very old sequences such as those of the Slave Province and the Morton and Montevideo gneisses in Minnesota.

Anhaeusser still accepts the analogy with island arcs suggested by Engel (see above). He compares the Onverwacht Group with the Aleutians and Kuriles, island arcs with crustal thicknesses of 12–18 km. He compares the entire Swaziland System with the Kamchatka, New Caledonia, and some Philippine arcs (15–25 km), and mentions the Yellowknife-Fossa Magna (Japan) analogy (Paper 11). The Fossa Magna is regarded as the type example of the birthplace of continental crust at the expense of oceanic crust. The mafic–felsic cycles of Barberton have calc-alkaline affinities, the only possible analogues being abyssal and island arc calc-alkaline suites. In South Africa both types seem to be represented, as in Canada, but the sequence is marked by an unusual development of primitive, high magnesia basalts (komatites). Initial strontium ratios of 0.700 and very low potash contents suggest that abyssal basalts are certainly represented. The initial strontium values are consistent with patterns recognized in modern island-arc suites.

The geosynclines must have been very deep, as is shown by the immense thicknesses. The troughs are comparable with modern trenches bordering island arcs. The greenstone sequences are only preserved now in downfolded synclines, and their possible wider extension is discussed. They could have covered wide tracts of oceanic crust at this early date: the proliferation of such belts or scraps of belts supports this, but insuperable geochronologic difficulties have so far prevented testing this hypothesis. All that can be said is that all such greenstone inclusions seem to be older than the enclosing granites, superimposition relationships are unknown.

Anhaeusser favors a basic primordial crust with later sialic underplating occurring at least 4000 million years ago. Such secular granite additions are supposed to have occurred away from the downsagging island-arc environments, and were accompanied by fragmentation, metamorphism and migmatization, changes that commenced with trondjhemitic and tonalitic liquids, and involved K, U, and Th streaming from the mantle.

Anhaeusser illustrates the later superimposition of platform sequences on the Archean of southern Africa, and emphasizes the progressive increase in basin size. The transition was from initial oceanic crust (5–7 km), through island arc crust (15–25 km), to a final cratonic thickness of 35–37 km in the final phases of progressive stabilization. The superimposed Proterozoic-type platform basins of deposition (which commenced with the Pongola and Witwatersrand in a time period that is by definition late Archean) were formed on this fully thickened, cratonic crust. The broad stable platform basins were filled with flat-lying to gently dipping strata (orthoquartzite, conglomerate, sandstone, limestone, dolomite, chert and granular iron formations). Thick and extensive nonsequential volcanics within these sedimentary successions include flood basalts of continental tholeiite chemistry with subordinate andesites, trachytes, and rhyolites.

References

Engel, A. E. J. (1963). Geological evolution of North America. *Science, 140,* 143–152.

_____. (1968). The Barberton Mountain Land—clues to the differentiation of the earth. *Trans. Geol. Soc. South Africa, 71*(annexure), 255–270.

Glikson, A. Y., and Sheraton, J. W. (1972). Early Precambrian Trondjhemitic suites in Western Australia and North West Scotland, and the geochemical evolution of shields. *Earth Planetary Sci. Letters, 17,* 227–242.

PAPER 36: COMMENTARY

In this brief, but important paper, Gill attacks one of the most pervasive concepts in geology, that of a continuous, primeval granite crust or "basement." The origins of the concept of a granitic protocrust are traced to blast furnace slags and observations of the ancient gneiss–granite nucleii in every continent. That such basements are not everywhere so very ancient has been demonstrated, for example, in East Africa, where the characteristic "basement" (known as the Mozambiquian) yielded ages transitional between Precambrian and Paleozoic. The concept is thus only tenable today if it is accepted that such terrains consist largely of reworked, little modified primordial crust (Papers 35A and 35B); they certainly do include ancient material reworked.

Gill does not favor this concept of a primordial granite crust. He considers two models for the formation of the earth, either cooling of a liquid or liquid surfaced globe or formation from solid particles without surface liquifaction. Neither model is found to be compatible with the formation of a siliceous protocrust. The continents could conceivably have formed by plication of a thin primordial crust, but Gill regards this as unlikely. He believes that the mechanism for formation of granitic and gneissic terrains was by means of repeated geosynclinal volcanic and sediment accumulation, with subsequent deformation, metamorphism, and injection by predominantly siliceous invasive material, and, further, by the transformation of sediments to gneiss with differential melting to form granitic permeations and intrusions.

Gill contends that the average composition of the continental crust is more nearly basic than acidic, and that the first sedimentary sequences, although siliceous, could have been derived from an original basaltic skin, together with contributions from basalts, andesites, gabbros, and peridotites of the first geosynclines (here, one could perhaps add primary rhyolites). Gneissification and granitization processes are regarded by Gill as repeated cyclical events without primeval significance.

Gill raises the question of the nature of the nucleii of the first land masses, and suggests volcanic eminences, in the case of his first, the liquid surface model, and volcanic eminences together with initial surface irregularities in his second, the solid surface model.

36: GRANITE CRUST

J. E. Gill

ABSTRACT

The expression "granite crust" is misleading because available evidence indicates that such a composition for any complete section of the earth's crust as it is now is highly improbable.

The continental platforms are certainly not simple remnants of a disrupted original crust.

No real support has been found for the idea that the earth once had a felsic skin. A composition near basalt appears more probable.

Derivation of the crust from material with an average composition of basalt through transformation in the exogenic cycle and in orogenic belts appears to be in accord with available facts. Contribution of granite or granitizing fluids directly from the mantle may not be necessary.

Thirty years ago it was commonplace for geologists to refer to the original granite crust of the earth and to identify it in their minds with the abundant material in the complex basal material found in all the continents and referred to as "basement" or "basement complex". True, there were some skeptics, especially amongst workers in the Precambrian, but too little was known about the continental nuclei at that time to provide a base for either denial or confirmation of the existence of a continuous felsic crust in the early stages of earth history.

The idea evidently arose from consideration of the results of cooling and solidification of a liquid earth or at least a liquid outer envelope. Observations of felsic fractions of layered mafic intrusives, separations effected in blast furnaces and information provided by experimental work on basalts seemed to leave little doubt that granite, or some similar felsic product of low density, is the principal late-stage solid product of gradual cooling and crystallization of intermediate to mafic magmas such as might have formed a liquid outer envelope of the infant earth. It appeared reasonable, therefore, to conclude that the first solid outer layer of the crust was formed of granite or something similar in composition.

With the passage of time much has been learnt about the structure and composition of the continental platforms and, more recently, about the ocean basins. It is now well established that the oceanic crust has little, if any, felsic material in it. Its physical properties, so far as they have

Reprinted from *Proc. Sec. 10,* 22nd Int. Geol. Congr., India, 1964, pp. 355–363.

been measured, suggest a few kilometres of basalt or serpentine (Hess, 1962) underlain by peridotite.

The continents all show complex structure. The so-called basement complex has been mapped in detail in many places and everywhere the oldest rocks are folded and faulted sediments and volcanics, more or less metamorphosed, resting on younger intrusives. Some of the intrusives are granite but many other mafic and felsic types occur. The average bulk chemical composition is near that of granodiorite. Seismic evidence indicates no persistent horizontal layering going downward, but rather a persistence of the structural complexity and a gradation to more mafic material with depth. If the crust is considered to extend down to the Mohorovičic discontinuity, at an average depth of 33 km, the average composition may well be nearer to basalt than granite. In any case use of the expression 'granite crust' is somewhat misleading, for there is no real evidence that a continuous layer of this composition ever existed.

In recent years more and more geologists have concluded that there is little to support the view that the earth once had a complete felsic crust, but this idea is far from dead. Such widely read authors as Umbgrove, Carey and Beloussov have supported it in recent years, so it appears worthwhile to examine the pros and cons at this time.

Umbgrove (1947, p. 253) visualized a thin original felsic crust which, during Precambrian time was folded and thickened, by the action of convection currents at its base, to form the continental nuclei.

Our knowledge is too limited for one to deny the possibility of working over an original thin granite crust to form part of the continents as they now exist. The 3 to 1 shortening required could have been accomplished by repeated mountain building. However, complete removal by such processes from two-thirds of the earth's surface appears so improbable that the absence of sizeable remnants of sial in the ocean basins indicates strongly that there never was a complete felsic skin.

Carey (1950, p. 322) advocates the view that the earth was originally much smaller—so small that the continental plates would have completely covered it. Expansion of the interior to some 20 times its initial size split the skin apart and caused separation into the continents.

Beloussov (1962, p. 682) thinks that huge segments of an original continuous felsic crust have been removed by a process called "basification' or 'basaltization' to produce the ocean basins. The hypothesis requires sudden and radical changes in normal processes and trends, the causes of which are not explained.

Both Carey and Beloussov would require an original world-girdling crust similar in make-up to the existing continental crust. Radiometric dates now prove that continents are made up of large strips and segments that have remained relatively stable over periods of time ranging from less than a billion years to 3.0 billion years or more, and that different parts became stable at different times. Clearly this crust must have developed gradually through a long succession of events, not as an initial skin.

Other features of the Carey and Beloussov hypotheses are designed to account for the present distribution of felsic crust and the origin of the ocean basins. These problems are interesting and important, but have no direct bearing on the topic under discussion. They will not, therefore, be discussed further here.

Ideas about the nature of the original crust depend on hypotheses of earth's origin. So far as our problem is concerned we need only deal with two situations—

1. formation by cooling of a liquid earth, or at least one that had a complete liquid outer envelope at one stage in its history, and
2. formation on an earth built up of solid particles and never completely liquefied in its outer part.

Considering first the formation of a solid crust from an outer liquid envelope, we may assume that since it must have contained materials that eventually went into the present continental and oceanic crusts and a substantial part of the mantle, it would have been more mafic than basalt. Strong tides and convection currents would have kept it thoroughly mixed, and the first crusts formed at the surface would have been of average composition. They must, therefore, have been denser than the liquid, and would have sunk into it only to be remelted at depth. These processes should eventually have brought the whole mass to a situation where the temperature was near the melting point at each level. Solidification would then have taken place relatively rapidly with less opportunity for crystal settling because of this and the increasing viscosity.

At some late stage a surface crust would have become thick enough to maintain continuity by virtue of its strength, and the rapid healing of fractures by penetration and chilling of liquid already near the melting point. This crust would have had the composition of the liquid at that stage. Removal of mafic silicates by accumulation at the base would have left a magma less mafic than the original, but separation of a granite layer under the conditions described appears highly improbable. Accumulation of local clots of light minerals above descending currents of convection cells is conceivable. Local exudations of felsic material formed by differentiation of pockets of magma and forced out through a mafic skin also seems possible (see Buddington, 1943). For the rest, a composition near basalt appears more probable.

Rittmann (1962, pp. 272-273) has suggested that a continuous shell of "pegmatite" consisting of quartz, feldspar, muscovite and rare elements condensed from an atmosphere heavily charged with water-vapour and other volatiles. This felsic shell, he deduces, would have been deposited on a solid crust of olivine amphibolite. Later, he imagines that large parts of it were swept into primitive seas condensed from the atmosphere by further cooling and this, through orogenesis became incorporated into the continents.

If condensation of felsic materials did occur as supposed by Rittmann, it appears unlikely that they contributed more than a small fraction to the building of the continents. More probably the bulk of such condensation would have occurred before a permanent solid crust developed and these materials would have become mixed with the rest of the silicate liquid-solid fluid silicate mass which eventually formed the original crust.

If the earth were built up from solid particles and never completely fused in the outer part, the composition of the outer part would probably have been non-uniform and mafic, considering what is known of the average composition of the earth and of meteorites. A felsic crust could form only by heating to produce mobility and permit the rise of light fractions toward the surface. This would require temperatures near the melting point, presumably developed locally and irregularly. Formation of a continuous felsic skin would be highly improbable with this earth model.

From the considerations outlined above it is concluded that very little, if any, felsic material could have existed when the first oceans formed. A few clots of limited extent are conceivable. If they were present, they would have formed islands and could have been nuclei from which continents grew. Continents could, however, have started equally well from volcanic piles similar to those seen in the modern ocean basins. Early oceans would have been shallow. On the liquid earth model the first land areas protruding through the seas would have been due to volcanic eruption. Fragmentation and vesiculation produce volume increase and permit building up to thousands of feet above the normal level with maintenance of isostatic adjustment in a regional sense.

On the solid earth model initial topographic irregularities may well have provided land areas. Volcanoes would eventually have risen to augment these.

Since the continental platforms cannot reasonably be taken to be remnants of an original felsic crust, and because the upper parts of them now contain substantial amounts of felsic material, we must conclude that this felsic part has developed from more mafic materials. To accomplish this, it is obvious that the original solid crust must have been remobilized by having its temperature raised to near or above the melting point. Such mobility and the production of felsic rocks on a grand scale have occurred characteristically in the mountain systems of the world.

As pointed out earlier, radiometric dates and the internal structure of continental nuclei indicate that continents have grown through much of geologic time. Precambrian mountain systems can be identified in all the continents by their internal structures. Prolonged erosion has reduced their topographic relief to inconspicuous levels, but their make-up and histories are similar in a broad way to those of younger orogens. The principal difference is the higher proportion of high grade metamorphic rocks and felsic intrusives. Students of the Precambrian have almost all come to regard these merely as evidence of deep truncation to levels where temperatures approached or reached the melting points of parts of the geosynclinal accumulation (Sederholm, 1925; Collins, 1936; Wahl, 1936 and James, 1960). Wyllie and Tuttle (1960) estimate this to occur starting at around 600°C at a depth of approximately 20 km. The sources of the heat and of the silica and alkalis, in which these rocks are enriched, are, of course, subjects for continuing debate.

From what has been learnt about the histories of areas underlain by Precambrian rocks, it appears that earth processes 2.5 billion years ago differed but little from those operating today. Thick accumulation of sedimentary and volcanic rocks formed in geosynclines and were later deformed, metamorphosed and injected by intrusives of a wide range of compositions, felsic batholiths being prominent (Goldich *et al.*, 1961). Considering the vast stretches of time represented, the limited quantities of identifiable sediments and the large volumes of gneiss and "granite", the conclusion appears inescapable that large quantities of sediments have been converted into gneiss and "granite" during orogenesis, and that this occurred time after time as the continents grew.

The numerous occurrences and varied relations of granite and the many complex problems presented have been reviewed at length by Read (1956) and Raguin (1957). After considering all the suggestions so far made, the author is most impressed with the combination of transformations in the exogenic cycle and reorganization at elevated temperature during orogenesis as a mechanism for producing not only the granite masses, but the wide variety of metamorphic rocks associated with them.

The transformation of sediments to gneiss can be seen to have occurred in many places. Derivation of granite by differential melting of sediments is harder to prove by observation in the field, though some highly suggestive transitions occur. Experimental work on silicate systems and experiments on melting of sedimentary and igneous rocks have shown conclusively that, with rising temperatures, and in the presence of water, silica and the alkalis go into solution first (Tuttle and Bowen, 1958; Kranck and Oja, 1960; Winkler and von Platen, 1962). It is obvious that upward migration of these solutions through any escape routes provided during the prolonged and complex adjustments characteristic of mountain building would be a natural consequence. Reaction of such solutions with rocks at higher levels must result in enrichment of these in feldspar and quartz, thus contributing toward the formation of a felsic layer. Liquefaction of larger masses of varying composition is a reasonable expectation with continuing rise of temperature. Such magmas could at certain times and places be injected to higher levels after which they may crystallize under conditions of relative stability. However, during crystallization, squeezing at intervals, collapse of part of the roof of the magma chamber or increasing internal gas pressure could cause injection of fractions into the surrounding rocks, forming a suite ranging, in sequence of age, from mafic to felsic.

Earlier in this paper it was pointed out that the bulk composition of the local crust down to the *M* discontinuity might be near that of basalt. This appears reasonable since the volcanic piles in the Pacific Ocean, derived from the upper mantle by differential melting, are of this composition. Also it should be noted that basaltic and andesitic flows, gabbroic sheets and peridotite masses are common in geosynclinal accumulations.

As a first approximation let us assume that the complete continental crust has been formed from basalt through breakdown and redistribution of materials in the exogenic cycle, followed by folding, faulting and reorganization chemically and mineralogically during repeated episodes of mountain building. On this assumption all the material seen at the surface of continents, and that indicated to exist at depth to the Mohorovičic discontinuity, with those dissolved in the oceans and deposited on their floors, should have the same overall mass and bulk composition as olivine basalt. A calculation has been made using available data and certain assumptions set out below to get a rough idea of what the composition of the lower inaccessible part of the continental crust would be if this were true—

Assumptions: 1. The average composition of the continental crust is that of olivine basalt;

2. The upper part of the crust shows a linear gradation from the average composition of granodiorite at the

surface to a more mafic composition X at 20 km. depth, beyond which the average composition is X to 33 km. depth.

Data used: Figures for average compositions of rocks given by Poldervaart (1955). Water and other materials that are liquid or volatile at the surface were ignored. These undoubtedly played important parts in the transformations, but our principal concern here is with materials that are solid at the surface.

Area of stable continental platforms	105×10^6 km^2
Area of younger folded mountain belts	42×10^6 km^2
Area of shelves and slopes	93×10^6 km^2
Specific gravity of granodiorite	2.7
Specific gravity of basalt	2.9

The details of this computation have been published elsewhere (Gill, 1961). Composition X, representing the average for the lower 13 km. of the continental crust was found to be:

SiO_2 38.7%, TiO_2 4.3%, Al_2O_3 15.3%, Fe_2O_3 4.3%, FeO 11.3%, MnO 0.2%, MgO 11.0%, CaO 12.0%, Mo_2O 2.4%, K_2O 0.1% and P_2O_5 0.3%.

This material would be a residuum from the differential melting and degassing that goes on during the reorganization of geosynclines during mountain building. Wide variations in local composition would undoubtedly occur and the structure would, no doubt, be complex. It would be devoid of fluxes, so the melting point should be well above that of fresh basaltic magma from the mantle. This may account for the marked stability of the continental platforms.

A search through Washington's compilation of analyses revealed only two, classed as basalt, resembling the composition given above. This also is in accord with the hypothesis, since remobilization of such inert material would be expected to occur only under exceptional circumstances.

From the considerations outlined it is concluded that the continental crust can have formed from an original basaltic skin plus contributions to geosynclines of basalts, andesites, gabbros and peridotites without any direct contribution of granitic magmas or "granitizing" fluids or gases from the mantle.

Summary

1. The use of "granite crust" to refer to the continental crust is

misleading. A complex layer of 5 or 10 km. may have an average composition near that of granodiorite. The average composition down to the Mohorovičic discontinuity should be nearer to gabbro.

2. The continental platforms are certainly not simple remnants of a disrupted original felsic crust.

3. No real evidence has been found supporting the idea that the earth's original crust was felsic. An average composition near that of basalt appears probable if the earth ever had a liquid outer layer. If it formed from solid particles without subsequent liquefaction the composition could have varied widely from place to place but the average would probably have been more basic than granodiorite.

4. Calculations using available quantitative data indicate that the whole outer crust in its present form and composition could have been derived from material with the average composition of olivine basalt. Weathering and erosion, redistribution of the products, partly into geosynclines, contribution of basic volcanics and intrusives to geosynclines from the mantle, and the reorganization of the geosynclinal accumulations during mountain building are the principal processes involved. This complex milling-smelting process could, it would appear, produce the existing continents without any direct contribution of granite magma or "granitizing" fluids from the mantle.

REFERENCES

Beloussov, V. V. (1962) : Basic Problems in Geotectonics; McGraw-Hill Book Co.

Buddington, A. F. (1943) : Some Petrological Concepts and the Interior of the Earth; *Amer. Mineralogist,* 28, pp. 119-140.

Carey, S. W. (1950) : Continental Drift, A Symposium; Geol. Dept., University of Tasmania.

Collins, W. H. (1936) : Derivation of Granitic Rocks. *Int. Geol. Congr., 16th Session,* I, pp. 271-282.

Gill, J. E. (1961) : The Origin of Continents. *Royal Soc. Canada Proc.,* Sec. 3, Vol. 55, pp. 103-113.

Goldich, S. S., Nier, A. O., Baadsgaard, H., Hoffman, J. H. and Krueger, H. W. (1961) : The Precambrian Geology and Geochronology of Minnesota; *Minn. Geol. Survey, Bull.* 41.

Hess, H. H. (1962) : History of Ocean Basins; in Petrologic Studies. *Geol. Soc. Amer., Buddington Volume,* pp. 599-620.

James, H. L. (1960) : Problems of Stratigraphy and Correlation of Precambrian Rocks with Particular Reference to the Lake Superior Region. *Amer. J. Sci.*, 258A, pp. 104-114.

Kranck, E. H. and Oja, R. V. (1960) : Experimental Studies of Anatexis. *Int. Geol. Congr., 21st Session,* pp. 16-29.

Poldervaart, Arie (1955) : Chemistry of the Earth's Crust. *Geol. Soc. Amer., Spec. Paper* 62, pp. 119-144.

Raguin, E. (1957) : Géologie du Granite, Deuxième Édition, Masson et Cie, Éditeurs, Paris.

Read, H. H. (1956) : The Granite Controversy. Thos. Murby & Co., London.

Rittmann, A. (1962) : Volcanoes and their Activity. John Wiley and Sons, New York, London.

Sederholm, J. J. (1925) : The Average Composition of the Earth's Crust in Finland. *Comm. Géol. Finlande Bull.* 12, 70.

Tuttle, O. F. and Bowen, N. L. (1958) : Origin of Granite in the light of Experimental Studies in the System $NaAlSi_3O_8$-$KAlSi_3O_8$-SiO_2-H_2O. *Geol. Soc. Amer. Mem.*, 74.

Umbgrove, J. H. F. (1947) : The Pulse of the Earth. Martinus Nidjhoff. The Hague.

Wahl, W. A. (1936) : The Granites of the Finnish Part of the Sveco-Fennian Archean Mountain Chain, *Comm. Géol. Finlande Bull.* 115, pp. 489-505.

Winkler, H. G. F. and Von Platen, H. (1961) : Experimentelle Gesteinmetamorphose—V, Experimentelle anatektische Schmelzen und ihre petrogenetisch Bedeutung, *Geochim. et Cosmochim. Acta,* Vol. 24, pp. 250-259.

Wyllie, P. J. and Tuttle, O. F. (1960) : Melting in the Earth's Crust, *Report of 21st Session, Int. Geol. Congr.,* Part XVIII. pp. 227-235.

(*Manuscript received on December 27, 1963*)

[*Editor's Note:* The Comments section has been omitted owing to limitations of space.]

PAPER 37: COMMENTARY

Glikson equated his "ultrabasic–basic" association (Goodwin's "platform" association, Paper 33) with primordial oceanic crust. He attempted to fix maximum depths for the formation of igneous rocks of the greenstone belts by partial melting, but admitted to uncertainties due to enhanced heat flow, geothermal gradient and possible variations in gravity. All the evidence suggests that the crust was thin, but exact calculation of thickness is impossible. The granite nucleii were formed in the earliest events following the formation of the original (primordial) crust, he believed, and their emergence controlled the configuration of the intervening subsiding zones (greenstone belts), which were loci of calc-alkaline volcanism, sedimentation, folding and metamorphism.

The analogy with Alpine geosynclines was favored (ophiolites; also oceanic-crust-derived basal units, flysch, molasse), but Glikson did not regard the greenstone troughs as having been situated in continental margin positions, but rather to have been marginally related to small granite nucleii. He considered the possibility that sodic granites were derived as a result of fractional melting of amphibolitic material, a process demonstrated as experimentally possible by other workers. Partial melting below subsiding segments of oceanic crust and rapid rise of the magma generated to eliminate differentiation was suggested, and the high Ni, Cr, and Co content of these granites was considered to support such a hypothesis.

The ultrabasic–basic associations were attributed to partial melting of the suboceanic mantle, the calc-alkaline associations to such melting involving a considerable oceanic crust component beneath the "intrabatholithic" troughs. Variations between Archean eruptives of different provinces—for example the Superior and Slave provinces of Canada—were attributed to variations in the amount of oceanic crust component.

37: EARLY PRECAMBRIAN EVIDENCE OF A PRIMITIVE OCEAN CRUST AND ISLAND NUCLEI OF SODIC GRANITE

A. Y. Glikson

Commonwealth of Australia, Bureau of Mineral Resources, Geology and Geophysics, Canberra, Australia

ABSTRACT

The primordial crust consisted of a stratiform oceanic-type ultrabasic-basic assemblage, relicts of which are retained at the base of Archaean sequences of eastern Transvaal, Rhodesia, and Western Australia. Commonly isochemical metamorphism renders the chemistry of well-retained segments of the meta-igneous rocks significant to their original compositions. Archaean tholeiitic metabasalts comply with the principal criteria for oceanic tholeiites, but tend to have higher Fe/Mg and Mn, and lower Al, Ti, K/Rb, and Fe^{+3}/Fe^{+2} than average recent oceanic tholeiites. These geochemical characteristics can be interpreted in terms of either shallow-level differentiation and (or) a mantle relatively depleted in certain siderophile and transition elements, which possibly indicates a lesser degree of core segregation in the Archean. Metabasalts occurring within basic to acid calc-alkaline volcanic cycles at higher stratigraphic levels than the ultrabasic-basic oceanic assemblages are chemically distinct from metabasalts incorporated in the latter; these rocks have higher K, Sr, Zr, and Y values than recent oceanic tholeiites, and may have originated from high degrees of melting of an underlying oceanic substratum. A model is suggested whereby the evolution from an oceanic crust to greenstone belts proceeded through the rippling of the oceanic crust and the subsidence of linear zones; partial melting below the troughs resulted in the emergence of sodic granites and in cyclic calc-alkaline volcanism. High Ni/Mg and Cr/Mg ratios in the earliest sodic granites support their derivation from the oceanic crust. The granites display cyclic decrease in Na/K, K/Rb, and Sr/Rb with time, trends which reflect crustal thickening. The plutonism resulted in the superposition of sialic nuclei on early linear structural trends, and in a consequent formation of spatially discrete volcanic-sedimentary depositories, or greenstone belts, in which sequences show an Alpine-type trend of evolution from ophiolites into turbidites and late-stage conglomerates. Notwithstanding this similarity, important differences between Archaean greenstone belts and Alpine or island-arc belts are indicated. Significant similarities exist between this model and the evolutionary pattern of the Fijian archipelago.

SUMMARY

In this paper Glikson discusses the alternative hypotheses for the primordial crust. The *sialic model* depends on a slag analogy (although there is also a novel hypothesis of extraterrestrial provenance for the sial). Oceans are taken in this model to be sunken areas of continent initiated by fractures, according to Glikson, but this is not a complete picture, for the sialic hypothesis can accommodate opening up of such fracture controlled zones to form tracts floored by new oceanic crust–as has manifestly happened in the case of the Atlantic ocean. This seems indeed to be the only way in which new oceanic crust is forming today.

Glikson favors the second, *oceanic primordial crust* hypothesis and raises the following objections to the hypothesis of a sialic primordial crust:

1. Lowermost Archean ultramafic assemblages are never seen to rest unconformably on granite or sial-derived sediments that include granite-derived clasts.
2. Granite greenstone contacts are always intrusive and marked by aureoles.
3. Geochemistry of Archean granite–gneiss terrain seems to require a source of greenstone terrain and subsequent granitization.
4. Low initial strontium ratios in such terrains favor incorporation of mantle derived greenstone assemblages and not an origin in primordial sial.
5. Lack of meteorite bombardment effects, which might be expected if it were a primordial crust, is noted in Archean granite–gneiss terrain.
6. The coarse texture of the granites of the earliest Archean terrains contrasts with the fine texture to be expected of a primordial sialic crust.
7. Lithophile elements are low in greenstone assemblages, which suggests that they have migrated into the granitic terrains.
8. There is a lack of sialic contamination in the greenstone volcanics such as would be expected if they were erupted through a granitic substratum.
9. Experimental evidence suggests that a primordial sialic crust could not stem from direct melting of mantle material, but would require a two stage process.
10. Planetary mantles appear to be little differentiated for major elements, and this does not favor large-scale separation of a sialic crust from the earth's mantle.

The possibility that a thin granophyric crust might have been formed is accepted, but such a crust has not been recognized. It could conceivably have been obliterated by effects of erosion, impacts or subduction, but may never have existed.

Glikson does not favor the arguments of Green (Paper 4), who attribute the early greenstone depositories to mare-type earth impacts and abrupt adiabatic melting.

Glikson endeavors to draw an analogy between Precambrian greenstone-belt assemblages and the Troodos Mountains of Cyprus, which are widely accepted as a thrust-elevated segment of Mesozoic oceanic crust.

Reviewing the possible origin of granites and calc-alkaline Archean magmas, Glikson considers the process of the streaming of lithophile elements causing large-scale mantle differentiation, as suggested by Engel (1968). He notes the possible objection that the granite seems to have formed in successive events; also the sodic–potassic progression (Paper 5) does not support this concept. Instead, he favors two-stage melting from the mantle, citing rare-earth data published by Condie and Lo (1970) that support this origin for granodiorites in Wyoming, the probable derivation being from eclogite. Amphibolite, favored by Green and Baadsgaard as a parent to calc-alkaline melts at Yellowknife, is also considered. Glikson believes that porphyries could be derived from partial melting of basic rocks.

Nowhere have sodic granites been found to intrude the higher sediments of Archean greenstone sequences, and Glikson suggests that the granites formed the first island nucleii, superimposed in greenstone-belt terrain that marked megaripples in the global oceanic crust. These megaripples were viewed as subsiding troughs above subduction zones. The granite batholiths invaded this crust over a long period. He regards the *nuclear terrains* (Papers 27, 29, and 32) as a purely secondary modification of original *linear terrain.* Some of the granite ascent may have accentuated rather than obliterated the linearity.

Glikson suggests three stages in Archean terrain development:

1. Initial subsidence on linear zones.
2. Deformation consequent to pluton emplacement (sodic granites).
3. Open folding and low-grade metamorphism consequent on subsidence of greenstones to warmer geothermal depths.

Alpine sequential and environmental analogies are considered to be valid, but special features of the Archean include the facts that (1) granite nucleii control the geometry, (2) the deformation patterns are unique, and (3) metamorphism possesses a low-grade, low-pressure uniformity in the greenstone belts.

Glikson does not believe one can apply plate-tectonic models and convection current concepts to Archean terrains. This would require separate subduction zones beneath each trough, an unlikely pattern. However, there may well have been convection-initiated divergence beneath greenstone–granite systems.

Enumerating the evidence for the oceanic nature of the primordial crust, Glikson leaves little doubt that a greenstone sequence founded either on sial or on oceanic crust–mantle cannot be reconciled with the concept of primary sial, of a sialic primordial crust. In rejecting this concept he enumerates a number of arguments, yet the case must still be considered unproved. The analogy drawn with the Troodos Mountains appears strained: this complex displays a multitude of sheeted dikes in its basal zone. The analogy can validly be drawn between the Troodos pillow lavas and those of the Archean greenstone belts, but all we learn from this analogy is that both include basic eruptives of the ocean floor. The lack of sheeted dikes and the presence of interbedded sediments in Archean greenstone suites suggests that we seldom, if ever, see the lower part of the oceanic crust transitional to the upper mantle in such suites, as we seem to be seeing in Cyprus. This may perhaps be taken as evidence favoring greenstone superimposition on foundered sialic blocks rather than oceanic crust! But it does not rule out greenstone superimposition direct on a hidden oceanic crust overlying the suboceanic mantle. Even so, there is no evidence that such a crust, if it existed, was primordial.

References

Condie, K. C., and Lo, H. H. (1971). Trace element geochemistry of the Louis Lake Batholith of early Precambrian age. *Geochim. Cosmochim. Acta, 35,* 1099–1120.

Engel, A. E. J. (1968). The Barberton Mountain Land—clues to the differentiation of the earth. *Trans. Geol. Soc. South Africa, 71*(annexure), 255–270.

PAPER 38: COMMENTARY

In Paper 38, Windley stresses the need for more emphasis on high-grade terrains in Archean studies, putting the case for a silicic primordial crust in its most up to date form. Granite–gneiss terrains, he notes, in general yield higher radiometric numbers than greenstone terrains, a fact that he advances in favor of a sialic primordial crust. Despite doubts concerning the validity of this argument, higher latitude factors for granitic terrains, and the incompleteness of Windley's data for greenstone terrains, no geologist would dispute the 3787 ± 85 age determined for the Amtsoq gneiss near Godthaab in Greenland; these are gray granite gneisses (Paper 20) that have been locally subjected to a later, more intense granulite metamorphism, but have preserved their very high radiometric numbers in the type area. That such granite–gneiss terrains existed at all at such an early date would seem to be a significant fact, although 800 million years *after* the starting point.

None of the enclosed predominantly amphibolitic tracts are of the character of the classic later Katarchean and late Archean greenstone sequences. They lack the thickness and progressive changes—they are rarely more than 1 km or so wide—and the charcteristic synclinal form. There has been much argument concerning their origin and true nature.

The lack of penetration of some of these amphibolite tracts by dikes that cut the enclosing gneisses has suggested infolding of superimposed cover. This interpretation seems to be the accepted one for the Malene supracrustals, which include large, layered chromitic anorthosite bodies (synformal in at least one case, at Fisknaesset) and less metamorphically modified pillowed amphibolites. Yet Bridgwater et al. (Paper 20) conclude that some paleosomes are represented that are of material older than the gneiss host. Windley has shown the picture to be even more complicated in a later report. He suggests that the amphibolitic tracts may be slices of oceanic crust, analogous in their components to ophiolites of similar origin, mixed up by interthrusting with continental gneiss areas in such a way that there is no systematic time or stratigraphic interrelationship.

Despite these uncertainties, the Godthaab gneiss complex seems to be a very ancient, primitive protocontinental nucleus, and there seems to be no certainty that primordial crust, whether sialic or oceanic, has been recognized in it. The problem has been taken back one step in time; the primordial crust remains elusive.

Granulite formation on a very large scale, said by Windley to be restricted to the Archean, is considered to be due *either* to upward movement of volatile elements leaving the granulite zone depleted in the low crust, *or* to isochemical metamorphism of preexisting low water, potassium, uranium, and thorium content tonalites.

The less metamorphosed terrains of the Archean provide evidence that granite invasion is not axial to greenstone belts but parallel and lateral to them in the hinge zone between positive basement and negative greenstone basins. This fact, accepted by Windley, tends to obscure the greenstone, granite–gneiss contact; the capricious variation of radiometric numbers across the Yilgarn Block (Paper 24) and the similar pattern recognized by Bridgwater et al. (Paper 20) in the North Atlantic craton are compatible with repeated granitic invasion and metamorphism, each invasion enlarging the respective Archean cratons. The fact that the North Atlantic Craton has been dispersed by plate movements must be borne in mind; in its original form it was probably not very different in size and nature from the western part of the Yilgarn Block, Western Australia. The latter area calls for very detailed study, because it can be studied as a continuous unfragmented terrain.

The Katarchean *nuclear patterns* are attributed by Windley to movements caused by the role of gravity in inversed density systems, rather than plate movements. Dense volcanics set on a sialic foundation will produce such overturn effects, leading to mass transfer and diapiric uprise of light material. However, there seems to be a cause and effect problem here; the first process in any greenstone system is the formation of the trough. The model favored by Windley allots secondary status to the diapiric uprise of the granite domes. What initiates the trough subsidence in the first place? Could not the whole system of domes be due to a system of intersecting crustal fractures with volatiles moving from the periphery toward the intersection points (the sites of the domes) and initiating the up and down swell pattern? The gravity role would then be essentially augmentive, causing continuance of the process and eventual diapirism.

The idealized greenstone sequence derived from the evidence in the Barberton Mountain Land, with a thick series of ultramafic volcanics at the base, is discussed by Windley. Some doubt has now been engendered in Western Australia by the tentative recognition of thick, older sequences beneath the sequence that carried the main development of high-magnesia basalts (komatiites) and nickel mineralization. This Barberton model seems to have limited applicability. Two models for greenstone belt evolution appear tenable to Windley:

1. Downsagging of basins on thin sialic crust forming intracontinental minigeosynclines, predecessors of the early Proterozoic protooceans; the gneiss terrains occur below or between the greenstone belts.
2. Greenstone belts forming as oceanic crust directly upon the mantle without a sialic foundation, and separating sialic plates.

The question posed by the first alternative is what is meant by intracontinental; were there any large continental areas at this early date? The second alternative is given, in place of the original statement, in which the Katarchean greenstone belts were supposed to be oceanic and precede any sialic crust (Paper 6); this is not favored in the light of the Godthaab discovery. The likelihood would seem to be that the greenstone belts could either be founded on fragmented and foundered older sial or on oceanic crust and mantle in the manner of present-day abyssal ocean-floor suites, which in their first platform sequences they so closely resemble.

The least acceptable proposal of Windley is the suggested analogy between greenstone belts and oceanic ridges. This ignores the fact that vast volcanic and sediment accumulations, all submarine, *could only accrue in a subsiding environment.* Rifts and oceanic ridges are, in the editor's opinion, features of positive environments, subsidence being secondary. The distinction between *primary* upwarps and subsiding geotectonic environments is crucial to an understanding of both Archean and modern geotectonic environments. Tensional activity can be associated with a positive arch and a negative trough, and in either case magmatic ascent may be triggered. Greenstone volcanics and probably modern oceanic volcanics are likely to be related to tensional feeders on the marginal zones of primary descending troughs. Later, edifice-stage volcanics may be superimposed in the positive zones. The eruptives of the intervening granitic area are the ones that show analogies to rift zones and oceanic ridges. The analogy drawn between the Keeweenawan lavas and the greenstone sequences seems quite out of place. The Keeweenawan is an intracontinental trappean sequence and is related in petrography and geotectonic associations more to the late tensional dikes than to the greenstones.

Windley notes the varying ages of the first stable platform basin and tensional dike superimpositions, and distinguishes two types of mobile belt in the Proterozoic—those linear belts filled with low-grade, deformed geosynclinal sediments and the completely mobile zones. He regards mio- and eugeosynclinal distinction as becoming apparent for the first time in the Coronation Geosyncline and Labrador Trough (2000 million years ago), although it must be noted that tentative recognition of such a distinction has been made by some geologists in Archean terrain (e.g., Dunbar and McCall, 1970).

Turning to plate tectonic analogies, Windley suggests that the Great Dyke and great regional tensional dike suites, such as the Widgiemooltha suite of Western Australia, are related to upswell in the mantle and are analogues of primitive plate boundaries. These structures are supposed to be closely related to rift valleys and mid-ocean ridges where the initial fracturing begins the separation of the plates. The analogy between the Grenville mobile belt and rift valleys is mentioned by Windley, and this is of critical importance. The elements of the Grenville are clearly those of the Archean granitic terrains bordering the greenstone belts, and this suggests that both granitic invasion and later dike development are features of swell geotectonics and analogous to rifts and oceanic rises. Windley does not recognize any possible analogues of the Grenville in later systems, but such analogues can surely be seen in the Peruvian batholith (Pitcher, 1972) and in the granites of the Himalayas (Ashgirei, 1972).

The evolutionary trends of Earth history are emphasized by Windley. Calc-anorthosites, widespread granulites, nonlinear nuclear patterns, and greenstone-belt development characterize the Archean and fade away in the Proterozoic; banded ironstones appear in the Archean, have a maximum development in the early Proterozoic, and disappear; massive anorthosites and rapakivi granites are peculiar to the Proterozoic. It might be added that alkali basalt and alkalic intrusive complexes come in strongly in the Phanerozoic.

As to the three chelogenic cycles of Sutton's modified statement (Paper 31B)—ending at 2500–2000 million years ago—Windley questions whether the decay of rift zones leading to immense lateral dispersion systems of plates was ever fully achieved in the two earlier chelogenic cycles.

References

Ashgirei, G. D. (1972). The problem of the origin of the continental earth crust. *Proc. 24th Int. Geol. Congr., 3,* 3–9.

Dunbar, G. J., and McCall, G. J. H. (1971). Archaean turbidites and banded ironstones of the Mount Belches area, Western Australia. *Sed. Geol., 5,* 93–133.

Pitcher, W. S. (1972). The coastal batholith of Peru—some structural aspects. *Proc. 24th Int. Geol. Congr., 2,* 156–163.

38: CRUSTAL DEVELOPMENT IN THE PRECAMBRIAN

B. F. Windley
Department of Geology, University of Leicester, Leicester, England

CONTENTS

The oldest Archaean rocks in most shield regions are largely granulites and gneisses, and in west Greenland there is evidence of 1000 Ma of crustal history before the final high-grade metamorphism. Archaean greenstone belts are mostly younger than the high-grade terrains although in some areas, such as southern Africa, this has not yet been proved reliably. The greenstone belts may have developed as oceanic crust in connexion with plate movements, the earlier continents being represented by the more deeply eroded high-grade regions.

Stabilization of the Archaean cratons is signalled by continental-scale intrusion of dolerite dyke swarms. Proterozoic mobile belts are exposed at two structural levels. Some early linear basins have mio- and eu-geosynclinal parts and may have been located along Proterozoic suture lines. More deeply eroded mobile belts are often floored by extensive, partly reworked, crystalline basement and probably developed along linear rifted zones which acted as loci for high heat flow and igneous activity; they lack ophiolites and are difficult to interpret as collision-type mountain belts. Most probably there were intra-continental plate movements in the Proterozoic.

INTRODUCTION

With the recent rapid advances in understanding of tectonic relationships, and in radiometric dating, it is appropriate at this time to take stock of what is known about Precambrian evolution. We are at the point when plate tectonic models are beginning to be applied to Precambran belts and it is now possible to follow the chronological evolution of the continents, albeit in a crude form, from about 3800 to 600 Ma ago – about seven-eighths of geological time. The aim of this paper is to provide an introduction or preview for this volume in the form of a framework for the Precambrian within which individual aspects may be seen in context.

The main time boundary used here between older Archaean and Proterozoic is 2500 to

2100 Ma. This was a period of stabilization in several continents, representing a fundamental break in Earth evolution, far more significant than that between the Precambrian and the Phanerozoic.

The earliest Precambrian crust

There are broadly two types of Archaean terrains exposed today (Windley & Bridgwater 1971): those dominated by high-grade gneisses and migmatites, and those by low-grade greenstone belts and granites.

Table 1. Selected age determinations from Archaean granulite–gneiss terrains

	country	rock unit	age (Ma)	method	reference
1	Swaziland	hornblende gneiss ancient gneiss complex	3340	Rb/Sr	Allsopp *et al.* (1969)
	Swaziland	granodioritic gneiss	3440 ± 300	Rb/Sr	Allsopp *et al.* (1962)
2	Minnesota, U.S.A.	Montivideo gneiss	3550	Rb/Sr	Goldich *et al.* (1970)
3	W. Greenland	Amitsoq gneiss Godthaab	3980 ± 170	Rb/Sr	O.I.G.L. & McGregor (1971)
			3620 ± 100	Pb/Pb	
		tentative age of Fiskenaesset complex	3600	Rb/Sr	Evensen *et al.* (in prep.)
		granulite-grade metamorphism of Fiskenaesset complex	2900	Rb/Sr	
		granulite facies metamorphism of Fiskenaesset complex and hypersthene gneisses	2850 ± 100	Pb/Pb	O.I.G.L. and Windley (in preparation)
4	Scotland	Scourian gneisses	2900 ± 100	Pb/Pb	Moorbath *et al.* (1969)
5	N. Norway, Lofoten islands	granulite facies gneisses	2800 ± 85	Rb/Sr	Heier & Compston (1969)
6	Aldan shield, U.S.S.R.	granulite facies gneisses	3300 ± 200	Pb/Pb	Rudnik *et al.* (1969)
7	Ukrainian shield	Yamburg migmatites	3500–3600		Semenenko (1967)
8	Guyana, S. America	Imataca complex gneisses and granulites	3000		Hurley *et al.* (1968*a*)
9	Sierra Leone	pyroxene granulites	3000		Hurley *et al.* (1968*b*)
10	S. India	oldest gneisses, granulite facies	3065 ± 75	Rb/Sr	Crawford (1969)
11	W. Australia	Yilgarn block gneisses	2800–3100	Rb/Sr	Arriens (1971)

For many years popular opinion has been that the oldest continental rocks are volcanics and sediments belonging to the greenstone belts (Goodwin 1968; Anderson 1969; Anhaeusser, Mason, Viljoen & Viljoen 1969; Glikson 1971; Green & Baadsgaard 1971; White, Jakes & Christie 1971; Cloud 1971). However, the time is ripe for a major reorientation in the assessment of early crustal material because recent reliable age determinations show that the oldest rocks in most continents are in a high-grade metamorphic state and that most greenstone belts are comparatively young.

Sutton (1967) pointed out that the oldest ages are found in migmatitic granitic gneisses, such as in the northern Eurasian shields, which have been through a long history of tectonic reworking; therefore they might contain even older material. Windley & Bridgwater (1971) proposed that more emphasis be placed on these high-grade terrains, which are often extensive and present on all continents, in order to redress the balance in appraising Archaean evolution. Over-emphasis in terms of volume of detailed studies has been placed on the greenstone belts.

The following points are relevant to this discussion:

(1) Shackleton (1970) concluded that greenstone belts in eastern Africa accumulated on older gneissic crust of continental type, while Bliss (1969) and Wilson (this volume, p. 389) suggested that relics of pre-Sebakwian gneissic basement occur in the Transvaal craton of Rhodesia, and Hunter (1970) suggested that the Swaziland gneisses pre-date the Barberton Mountain Land greenstone belt.

Table 2. Selected age determinations from Archaean greenstone belt–granite terrains

	country	rock unit	age/Ma	method	reference
1	Barberton Mountain Land, South Africa	Onverwacht group Swaziland system	3375 ± 20	Rb/Sr	Hurley *et al.* (1972)
		Hooggenoeg formation Swaziland system	3360 ± 100	U/Pb	Van Niekerk & Burger (1969)
		Fig Tree shales Swaziland system	2980 ± 20	Rb/Sr	Allsopp *et al.* (1968)
2	Rhodesia	pegmatite in Bulawayan granite	3370 ± 70	K/Ar	Vail & Dodson (1969)
3	Canada	Yellowknife group Great Slave Province	2650	U/Pb	Green & Baadsgaard (1971)
		Rainy Lake Superior Province	2750 ± 30	U/Pb	Hart & Davies (1969)
4	S. India	Dharwar lavas	2345 ± 60	Rb/Sr	Crawford (1969)
5	W. Australia	Kalgoorlie greenstone belt	2600–2700	Rb/Sr	Turek & Compston (1971)

(2) Catanzaro (1963) and Stern (1964) obtained zircon dates from the Morton gneiss in the Minnesota River Valley. When plotted on a concordia diagram their data define a straight line with an upper intercept at about 3600 Ma (Bickford & Wetherill 1965). More recently, Goldich, Hedge & Stern (1970) reported a 3500 Ma Rb/Sr isochron age for the Montivideo gneisses in the Minnesota Valley. The information from this area is particularly important as it shows that there was high-grade sialic crust in existence 1000 Ma before the nearby greenstone belts of the Superior Province were formed.

(3) Oxford Isotope Geology Laboratory and McGregor (1971) found that amphibolite facies gneisses in West Greenland had a Rb/Sr isochron age of 3980 Ma (a more accurate, revised date is 3787 ± 85 Ma, Moorbath, personal communication) demonstrating that 'windows' of extremely old crustal material may be found in Archaean high-grade terrains.

(4) Several unconformities are known between greenstone belts and older high-grade gneisses and granites: in the Superior province of Canada at Steeprock Lake (Jolliffe 1966) and at Cross Lake (Rousell 1965); in India where the Dharwar greenschists overlie a granulite-gneiss basement (Nautiyal 1965). At Que Que, in Rhodesia, the Bulawayan basal conglomerate lies against granitic gneisses; and at Point Lake in the Slave province of northern Canada a basal conglomerate of the Yellowknife Supergroup overlies granite bordering the greenstone belt (Stockwell 1933); McGlynn & Henderson (1970) review other data suggestive of a basement to the Yellowknife Supergroup.

Although K/Ar dates have been widely used in assessing the age of Precambrian rocks, their value has been questioned (Moorbath 1967) as they are considered to represent, not the peak of orogenic activity, but the termination of an ^{40}Ar diffusion event related to the uplift and stabilization of continental blocks (Green & Baadsgaard 1971). Continuous diffusion of radiogenic argon may take place at depths of as little as 3000 m in orogenic belts (Hurley, Hughes, Pinson & Fairbairn 1962). Therefore Windley & Bridgwater (1971) predicted that the oldest rocks would be found with more sophisticated dating techniques in the high-grade Archaean terrains, because they would not become closed systems to radiogenic ^{40}Ar diffusion until their uplift following that of the higher level greenstone belts. For these reasons it is important to rely on more accurate radiometric age determinations made by the mineral and total rock Rb/Sr, Pb/Pb and zircon U/Pb methods.

In tables 1 and 2 age determinations are listed from all continents demonstrating that the higher grade parts of most Archaean complexes tend to be older than the lower grade greenstone belts. Several points of interest emerge from these tables:

(1) Whereas the radiometric determinations on the greenstone belts give an approximate idea of the age of their formation, those on the gneissic terrains are mostly dating the age of the last high-grade period of metamorphism; the recent evidence from west Greenland shows that the 2850 to 2900 Ma granulite-grade metamorphism was the culminating high-grade event of 1000 Ma of earlier crustal history. Therefore, while the high-grade rocks in many shield regions have isotopic ages older than those of nearby greenstone belts, if their evolution has been in any way similar to that in West Greenland, they may be expected to contain evidence of very early crustal history.

(2) The isotopic data suggest that most continents experienced a period of high-grade metamorphism in the general period 2800 to 3200 Ma. Most probably there were local diachronous hot spots developed as a result of early thin crust, steep geothermal gradients and fundamental fracture zones.

(3) Greenstone belts in Canada, Australia and India formed in the period 2345 to 2750 Ma. These are the oldest dates on the earliest volcanics; associated granites have younger ages. Since the Precambrian is now known to extend back as far as about 3800 Ma, these greenstone belts must be seen as belonging to the latest Archaean, or even the early part of the mid-Precambrian.

(4) The ages of events in southern Africa are significantly older than those in other continents, but it is emphasized that the geological relationships (i.e. the types of rock belts) are not different. The greenstone belts apparently developed diachronously, those in southern Africa forming before those in other shield regions. Although it is considered by Anhaeusser *et al.* (1969) and Viljoen & Viljoen (1969) that the Swaziland System of the Barberton Mountain Land greenstone belt contains the oldest rocks in South Africa, there are three groups of nearby high-grade gneisses that may well be older: the Swaziland gneisses, as suggested by Hunter (1970): the Nelspruit gneisses of the Barberton Mountain Land; and the gneisses of the Limpopo mobile belt, which form the basement to the Messina Formation (Mason, this volume p. 463). Note from tables 1 and 2 that the Swaziland gneisses are as old as the Swaziland System.

The most reasonable conclusion to be drawn from the available evidence is that the continental crust as a whole evolved diachronously. The development in what is now southern Africa appears to have been consistently ahead of that in other continents: some of the earliest gneisses are to be found there (but they have so far received very little detailed attention); the earliest greenstone belts formed there; and the early stabilization of the Archaean craton, consisting of the greenstone belts/granites and the gneisses, allowed platform deposits such as the Dominion Reef System to be laid down by 2800 Ma (Van Niekerk & Burger 1969), i.e. at a time when greenstone belts were still forming on several other continents.

The high-grade gneissic terrains did not go through continuous mobility or plutonism as they contain relic layers of meta-supracrustal rocks, evidence of high-level crustal stability before the high-grade metamorphism and deformation. In West Greenland the Malene supracrustals in the Godthaab area probably formed not long after 3800 Ma, and in the Fiskenaesset area the preliminary age of the formation of the major anorthosite complex is 3600 Ma (Evenson, Rama Murthy & Windley, in prep.), in which case the supracrustal rocks in which it formed (now metavolcanic amphibolites with rare marbles and schists) are older. The Swaziland gneisses contain relic layers of supracrustal rocks which, according to Hunter (1970), pre-date the Swaziland System. The high-grade gneiss terrains also contain abundant agmatitic basic fragments, largely of amphibolitic composition; Windley & Bridgwater (1971) regard these as the broken-up remains of very early supracrustal rocks which, if they had been preserved in basin-shaped form, would be termed 'greenstone belts'. In other words, the earliest supracrustal basins or greenstone belts were fragmented as their underlying crystalline basements were still mobile at such early stages of Earth history. What we interpret as 'greenstone belts' today are merely those that have survived.

Archaean granulite/gneiss complexes

Although some regional granulite facies metamorphism occurred in the Proterozoic, most occurred in the early Archaean, in particular in the period 2800 to 3200 Ma. Some high-grade Archaean complexes are extensive and well preserved, others are small and relic. For many years Russian geologists (e.g. Salop & Scheinmann 1969) have emphasized that the northern Eurasian Archaean complexes are in a high-grade metamorphic state, as greenstone belts are rare or absent there. The North Atlantic craton (Bridgwater, Watson & Windley, this volume, p. 493), extending from Labrador through West and East Greenland to the Scourian of Scotland, is one of the more extensive and best exposed terrains of this type from which many of the following details are drawn. Four important features of these high-grade complexes are:

(a) *The reworking of earlier crystalline material*

It is commonly implied that regional granulites formed from the metamorphism of an already crystalline basement; Winkler (1967), for instance, suggested that the dry nature of the rocks may be due to their secondary origin by remetamorphism of a partially dried out suite of crystalline rocks. However, in West Greenland the results of O.I.G.L. & McGregor (1971) show that the granulite-grade metamorphism was late in a long sequence of events and thus it was superimposed on already gneissic/granitic rocks. Bridgwater, Watson & Windley (this volume, p. 493) emphasize that high-grade gneisses were derived more by the remetamorphism of older granites and gneisses than by isochemical recrystallization of supracrustal rocks.

There may be a common reason why much of the early Precambrian crust underwent granulite-grade metamorphism. As indicated by Ray (1970), geothermal gradients were no doubt steeper than at present (figure 1) – a response to a probable early thin crust and a high rate of radiogenic heat production, which Dickinson & Luth (1971) show was several times greater 3000 Ma ago than today (figure 2).

Archaean granulites have low contents of H_2O, K, U and Th compared with equivalent amphibolite facies rocks (Lambert & Heier 1967; Eade & Fahrig 1971: Heier, this volume, p. 429). The relative basification of the granulites may be due to at least two processes:

(i) Upward loss of the more volatile elements due to chemical fractionation of the crust during high-grade regional metamorphism: this is the more accepted view. A corollary of this mechanism is that granites would be produced in the catazone and would migrate to higher crustal levels. If the model of Fyfe (1970) were applied to the Archaean, the granulites would be regarded as a the low crustal residue from which granites were derived: these migrated upwards to invade the borders of the overlying greenstone belts.

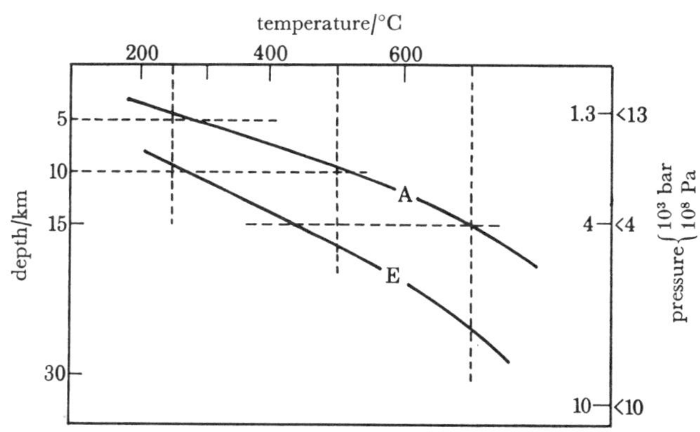

FIGURE 1. Geothermal gradients for (A) the Archaean and (E) the present (after Ray 1970).

(ii) O.I.G.L. & McGregor (1971) conclude that the greater part of the Godthaab area consists of tonalites which were emplaced as igneous complexes about 3100 Ma ago, before the granulite-grade metamorphism about 2850 to 2900 Ma ago. This means that the low potash content of many granulite facies gneisses may be due to isochemical metamorphism of pre-existing tonalites. Bridgwater, Watson & Windley (this volume) consider this to be a plausible mechanism for the derivation of many granulite facies gneisses, which lack potash feldspar and are poor in biotite, in the North Atlantic craton.

(b) Layered igneous complexes with prominent anorthosites

Calcic anorthosites (with bytownite to anorthite where best preserved) form a significant component of many high-grade Archaean terrains (Windley, in press). They are particularly common in west Greenland (at least 500 km strike length) and in the Limpopo mobile belt of southern Africa (at least 100 to 150 km strike length) and also occur in the Scourian of Scotland, India, Madagascar, West Africa and probably East Africa and Australia. The anorthosites occur with other rock types in conformable layers with a maximum thickness of 8 km in high-grade gneisses. The Sittampundi complex of southern India contains layers of chromite and was regarded by Subramaniam (1956) as a metamorphosed layered igneous body. In west Greenland the Fiskenaesset intrusion, which is at least 200 km long (strike length of the folded layers), is shown by Windley, Herd & Bowden (in press) to have an internal stratigraphy passing upwards from lower ultramafics (dunites, peridoties and pyroxenites) through mafic gabbros, cumulate gabbros, anorthosites, chromitites to garnet anorthosites, and a bulk igneous differentiation trend which is well preserved despite subsequent granulite and amphibolite facies metamorphism. Cryptic variation trends of individual minerals, such as plagioclase, olivine, orthopyroxene and hornblende, are also intact (Windley & Smith, in press). The complex formed from a magma with the composition of a hydrous high-alumina basalt and the tentative age of its formation is 3600 Ma (Evenson *et al.* in prep.). As these are the first major basic intrusions in West Greenland, it is tempting to think that the early mantle was degassing extensively at this time. Just as there is a marked uniformity in the character of Archaean greenstone belts in several continents (Anhaeusser 1971), so too is there a remarkable similarity in these anorthositic complexes – most likely there was a uniform, diachronous development of these rocks in different parts of the early continental crust.

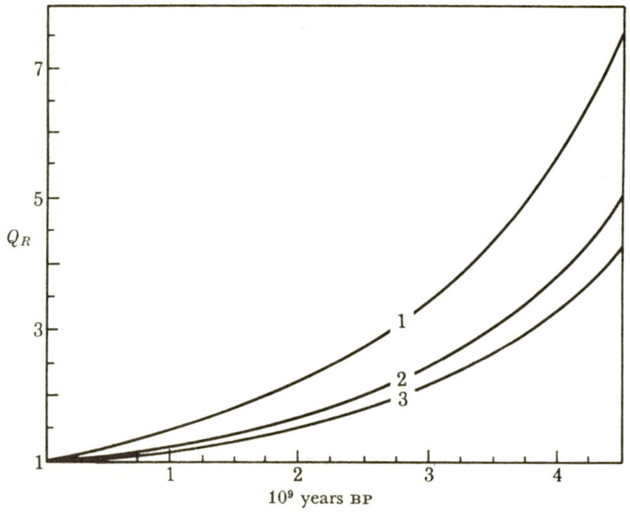

FIGURE 2. Relative radiogenic heat production in the past (after Dickinson & Luth 1971). Q_R is the ratio of heat production at any time in the past to that currently observed. BP is time before the present. Curves indicate values for Q_R calculated for following Earth models: (1) the chrondrite model, (2) the carbonaceous chrondrite or Orgueil model, and (3) the Wasserburg model.

(*c*) *Infolding of cover sequences into underlying basement gneisses*

Wilson (1958) suggested that infolded cover rocks would have the form of synclinal keels at the present levels of exposure. In high-grade gneisses these infolded rocks are most commonly represented by layers of amphibolite, mica schist and quartzite, but it is usually impossible to demonstrate that these sequences have a synclinal form, largely because lack of stratigraphy in the rocks prevents their structure being determined. However Windley *et al.* (in press) have demonstrated that parts of the Fiskenaesset intrusion, which is bordered by meta-supracrustal rocks, has a duplicated stratigraphy repeated in reverse order with internal synclinal closures and thus that the intrusion, and the cover rocks into which it had been emplaced, were infolded synclinally into a gneissic basement.

Another way of demonstrating this age relationship has been found by McGregor (this volume, p. 343) who, by observing that amphibolite dykes are absent in meta-volcanic amphibolite layers but are present in adjacent gneisses, concluded that the cover rocks must have been tectonically emplaced into their underlying gneissic basement.

(*d*) *Regional structural style*

The granulite/gneiss terrains are characterized by complex fold interference patterns that lack a linear regional trend. The constituent folds may be domes, basins, luniform-shaped or refolded isoclines, and they may be several kilometres across (Windley & Bridgwater 1971). This type of structural style occurs, for example, throughout the Archaean block of West Greenland, the Limpopo belt of southern Africa and the Aldan shield of Russia. Salop & Scheinmann (1969) point out that the folding process was closely associated with the formation of granitic masses by partial melting with consequent diapiric rise of granites in the cores of domal folds. Somewhat similar relationships exist in West Greenland where the cores of some domes consist of more homogeneous gneiss than is found in surrounding gneiss layers, demonstrating structural control of selective homogenization of the gneisses. This tectonic style and close association with partial melting of granitic gneisses is indicative of a high degree of plasticity during the regional deformation and metamorphism.

The formation of these non-linear structures over enormous areas is less likely to have been caused by lateral deformation associated with continental drift than by vertical movements caused by the role of gravity in a system with inverted density stratification as modelled by Ramberg (1967). It is particularly significant that the cover rocks are dominated by dense basic volcanics that were infolded into less dense granitic gneiss basement. The domes tend to reach a maximum size of about 15 km and, according to Ramberg's theory, the diameter of the domes and the spacing between them should be proportional to the thickness of the source layers. This might provide a means of assessing the thickness of the early Precambrian crustal layers.

Archaean greenstone belt/granite complexes

Much has been written on these low-grade belts and only brief mention will be made here of some main problems concerned with their relationship with the high-grade complexes just described and their mode of origin.

Where their original form is best preserved, as in western Australia, the greenstone belts form parallel linear belts and they ideally are basin-shaped. Anhaeusser *et al.* (1969) have demonstrated that their primary shape has often been modified by the emplacement of bordering late granitic plutons so that their boundaries are arcuate. An eventual stage may be reached where star-shaped basins are located between three granitic bodies.

The greenstone belts have an idealized stratigraphy ranging from lower ultramafics through largely basic volcanics to an upper, larger clastic, sedimentary group capped with chemically precipitated banded iron formations and cherts. Anhaeusser *et al.* (1969) and Anhaeusser (1971) have emphasized that the stratigraphy and the structures of the greenstone belts are sufficiently similar on many continents to indicate a uniform stage of development of the crust.

One of the major problems concerning the mode of occurrence and origin of the belts lies in the surrounding 'granitic' rocks which have received relatively little attention compared with the supracrustal sequences. Although they were earlier thought to be homogeneous granitic plutons, a major advance was made by the Geological Survey of Rhodesia in distinguishing different types of granitic/gneissic rocks with variable age relationships (Wilson, this volume, p. 389). Some granitic areas contain so much relic gneiss material that they are considered to be partially reactivated basement gneisses but other, more massive or homogeneous, granites have intrusive relationships. McGlynn & Henderson (1970) suggest that some of the granitic rocks bordering the Yellowknife greenstone belt of northern Canada represent a possible basement to the belt. They point out that a positive terrain of sialic composition must have existed nearby to provide by erosion the tremendous quantities of silicic sediments infilling the basins. They also suggest that diapiric granite bodies were intruded preferentially between the positive (basement) and negative (greenstone basins) regions of the crust – an idea very similar to that of Windley & Bridgwater (1971). The actual contacts of the greenstone belts with the high-grade gneiss complexes are thus obscured by a bordering screen of partly to completely mobilized granitic plutons. This relationship can be demonstrated in Sierra Leone, where Wood (1972) confirms Wilson's (1965) conclusion that the Mano Moa granulites constitute the basement to the Kambui Schist belts. The latter have the general form of an isoclinal syncline bordered conformably by reactivated basement. The actual boundary between the Schist belts and the basement is occupied by homogeneous granite in clear contrast to the more distant granulite-migmatite basement. Future ideas on the mode of formation of the greenstone belts are as likely to come from studies on the marginal 'granites' as they are from the supracrustals themselves.

Pettijohn (1970) makes the interesting observation that the late Palaeozoic or Mesozoic greenstones, volcaniclastic sediments and granitic plutons of the Sierras of California are reminiscent of those of the Archaean terrains. He suggests that this pattern of granite and volcano-sedimentary basins will develop anywhere and any time a thick sequence of denser lava rests on lighter sialic material. A kind of convective readjustment takes place with rising diapiric granitic stocks and downsagging of greenstone basins.

Windley & Bridgwater (1971) have interpreted the greenstone belt/granite terrains as having formed a high crustal level, whereas the granulite/gneiss terrains formed at relatively deep levels. Windley *et al.* (in press) estimate a pressure of at least 700 MPa (7 kbar) for the formation and metamorphism of the Fiskenaesset complex, West Greenland. It is important to distinguish between different erosional levels that are at the surface of the crust today and thus it is conceivable that some late Archaean belts of low-grade basic volcanic sequences in the high-grade terrains are the infolded root zones of greenstone belts. It is also conceivable that granulites occur under the high level greenstone belt/granite regions, such as the Superior Province of Canada, the Yilgarn and Pilbara blocks of Australia and the Transvaal and Kaapvaal cratons of southern Africa.

Models of greenstone belt evolution

So far two main models have been proposed to explain the mechanism of formation of Archaean greenstone belts.

(1) The greenstone belts are postulated by Anhaeusser *et al.* (1969), Martin (1969), and McGlynn & Henderson (1970) to have formed in downsagging basins on a thin crust. According to this model the greenstone belts are intra-continental mini-geosynclines (Clifford, in press), which are predecessors of early Proterozoic proto-oceans and the high-grade gneissic terrains occur below and/or between the greenstone belts.

(2) The greenstone belt sequences are considered by Glikson (1971), Anhaeusser (this volume, p. 359), and Talbot (this volume, p. 413) to have formed as oceanic crust between separating continental fragments. Although these authors only deal with the oceanic setting,

this plate tectonic model implies the prior existence of continents which accordingly would consist of the high-grade gneissic terrains described above. This model views the drifting sialic masses as mini-continents (Talbot, this volume).

Although these two models satisfy many requirements when tested against the data shown by the rocks, neither is totally satisfactory. The first takes no regard of current plate tectonic models (there were probably at least proto-plates in the Archaean), whilst the second requires

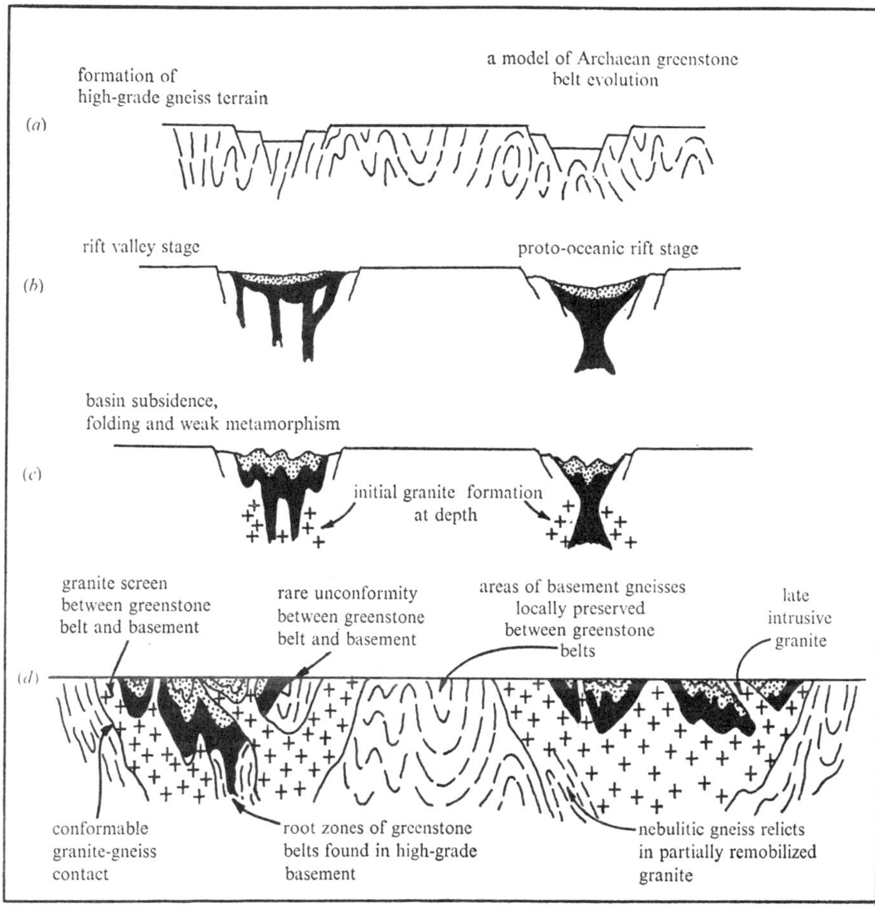

FIGURE 3. Proposed stages in evolution of Archaean greenstone belts.

extensive continental drift and ocean floor spreading, for which in fact there is no evidence (a serious drawback of Talbot's model is that the basins are symmetrical and are infilled with coarse, clastic sediments which implies the existence of nearby eroding source areas on both sides and deposition in subsiding basins rather than over widespread oceans). Neither takes into account the earlier high-grade gneissic terrains nor uses the types of ore deposits in the belts as petrogenetic indicators. The two models approach extremes of interpretation, the one implying no, and the other extensive, continental drift. Most probably the answer lies somewhere between these extremes and thus a model is proposed here which satisfies the essentials of both (figure 3).

There has been much discussion as to whether the greenstone belts are analogous to recent island arc or oceanic environments. A major advance was made by Glikson (1970, 1971) who, by detailed geochemical comparisons, was able to show that the basaltic assemblages conform to recent oceanic tholeiites with respect to K, Na/K, Sr, Zr and Fe^{3+}/Fe^{2+}. A geochemical aspect bearing on this problem that has been largely neglected concerns the well-known ore deposits in the greenstone belts – in particular Au–Ag, the tellurides and Cu–Zn. These are accumulations of elements which have just as an important bearing on the mode of origin of the associated volcanic rocks as the potassium and sodium, etc. Where comparisons are made with Recent environments similar deposits occur, for example, in the Basin and Range province, a high heat flow rift zone which is the northward continental extension of the East Pacific Ridge, and in the Red Sea, a proto-oceanic basin. In other words they are characteristic of early extensional plate boundaries.

A further important tectonic feature to consider is that the volcanic complexes in the greenstone belts have a linear distribution and probably formed as fissure eruptions parallel to the margins of the belts (Goodwin & Shklanka 1967; Goodwin & Ridler 1970).

If an attempt is now made to relate the greenstone belts to the modern plate tectonic model, when all the above points are taken into consideration they best fit one type of tectonic environment – a proto-oceanic ridge zone (figure 3), the nearest present-day equivalent of which is probably the Red Sea basin. The fact that the basalts have chemical affinities with recent oceanic tholeiites need not mean there was extensive continental drift and ocean floor spreading, but merely the formation of a proto-oceanic rift system involving only small amounts of crustal separation. In terms of the general concept of orogenesis according to lithospheric plate tectonics (Bird 1970) the incipient stage in the separation of crustal plates is represented by linear graben-like rifted zones which are the site of intense volcanic activity and which are later infilled with rapidly deposited molasse or turbidity sequences from the nearby continental areas – the ideal environment for greenstone belts. Subsidence of these high density accumulations would give rise to downsagging basins with vertical stretching and upright folds, typical features of the belts, while the contact zones between the negative basins and the positive areas would be the site of remobilization of the adjacent gneissic crust to form granitic plutons.

It is suggested that this model of a proto-oceanic ridge system best explains the featues of the greenstone belts. Although one type of mechanism is envisaged, there was probably more crustal separation and more oceanic crust generated in some belts than in others. There may in fact have been all gradations from ensialic graben-like rift zones to narrow, linear oceanic basis caused by relatively small amounts of 'drift'. Broad belts such as the Abitibi in southern Canada probably involved separation of crustal segments of a few hundred kilometres.

Proterozoic mobile belts

During the Proterozoic, taken here to extend from about 2500 to 600 Ma, there were many complex events too numerous to document in detail here. Particular rock groups or types of mobile belt are often broadly similar on several continents and formed at definite stages in the development of the crust (Salop & Scheinmann 1969).

The Archaean shield areas underwent considerable erosion and uplift in the initial stages of stabilization, which generally took place in the period 2500 to 2000 Ma. However, because the southern African cratons themselves stabilized slightly earlier than those in other continents, they were subjected to intrusion and sedimentation at an earlier time: the Dominion Reef System was deposited 2800 Ma ago (Van Niekerk & Burger 1969) and the Great Dyke of Rhodesia was intruded about 2550 Ma ago (Allsopp 1965). These were followed by the deposition of the Witwatersrand, Ventersdorp (2300 Ma, Van Niekerk & Burger 1964) and Transvaal Systems, the emplacement of the Bushveld Igneous Complex 1950 Ma ago (Nicolaysen, de Villiers, Burger & Strelow 1958), and the intrusion of the Umkondo and Mashonaland dolerites in the period 1600 to 1850 Ma (Vail & Dodson 1969; Wilson, this volume, p. 389).

In other continents there was intrusion of extensive dyke swarms, most with tholeiitic but some with ultrabasic compositions: in western Australia there is the Widgiemooltha Dyke Suite including the Jimberlana Dyke (2420 Ma, McCall 1971), and the Binneringie Dyke, at least 320 km long (McCall & Peers 1971); there is a prominent northwest trending dyke swarm extending across the Canadian Shield from Ungava to the Slave Province (2000 Ma, Clifford 1968). In the North Atlantic craton there is the Scourie Dyke swarm of the Lewisian of Scotland, more than 250 km wide (2400 to 2200 Ma, Bridgwater, Watson & Windley, this volume, p. 493), the Kangamuit Dyke Swarm of West Greenland, and the dyke swarm in East Greenland (Bridgwater & Gormsen 1969); these dykes were intruded in fracture zones that later became the site of the Archaean–Proterozoic tectonic boundaries and thus were subjected to varying degrees of amphibolitization and deformation (Bridgwater, Escher & Watterson, this volume, p. 513).

The overall change in the character of the crust from Archaean to Proterozoic time was progressive stabilization or cratonization, well documented in Africa by Clifford (1970). Stable areas became larger enabling extensive platform deposits to form and mobile zones became more confined and linear.

There are broadly two types of Proterozoic mobile belts: linear belts, consisting largely of low-grade deformed geosynclincal sediments and volanics (e.g. the Labrador trough and the Coronation geosyncline); and mobile zones, consisting mostly of medium- to high-grade reactivated basement, maybe with some infolded supracrustal rocks (e.g. the Grenville of Canada and the Nagssuqtoqidian of Greenland). With respect to the orogenic belts in Africa – the Eburnian and Huabian (1850 + 250 Ma), the Kibaran (1100 + 200 Ma) and the Damaran-Katangan or Pan-African (550 + 100 Ma) – Clifford (in press) has concluded that the mobile zones are floored by older crystalline basement and that no geosynclinal sequences were deposited on extensive areas of oceanic crust: he suggested that these two types represent different structural levels through an idealized mobile belt. Shackleton (this volume, p. 317) makes the important point that there are no offsets of earlier structures across these belts.

For the first time in the history of the crust miogeosynclinal and eugeosynclinal sequences can be distinguished (Salop & Scheinmann 1969). Hoffman (this volume, p. 547) has suggested that the stratigraphic and structural evolution of the Coronation geosyncline (2000 Ma) in northern Canada are sufficiently similar to those of late Phanerozoic geosynclines to warrant the conclusion that it is located along a Proterozoic plate boundary. Likewise the Labrador Trough (2000 Ma) has recognizable mio- and eu-geosynclinal facets (Dimroth 1970) thought to be located on a Precambrian suture line by Gibb & Walcott (1971).

There were appreciable changes in the kind of sediments and volcanics deposited in Proterozoic compared with Archaean basins. Chemically precipitated banded iron formations and cherts form minor components of greenstone belts but they reach a peak of development in the 2000 Ma old basins (e.g. Hamersley, western Australia, Mauritania, and the Labrador trough).

With progressive photosynthesis and increased oxidizing conditions the former gave way to red beds in mid- to late-Proterozoic shallow basins (Cloud 1968). Stromatolites, rare in greenstone belts, become abundant in early Proterozoic basins (Hoffman, this volume, p. 547), and in Russia are even used as a mid-Proterozoic zone fossil. Due to the high carbon dioxide content of the Proterozoic atmosphere and hydrosphere limestones and in particular dolomites were formed in abundance.

Archaean volcanism was characterized by low-potash basalts chemically comparable with present-day abyssal tholeiites (Glikson 1971), more primitive ultramafic types, and repeated cyclicity with late pyroclastic phases. In contrast, Proterozoic volcanics consisted typically of potash-enriched, continental type tholeiitic basalts, trachytes and rhyolites (Anhaeusser, this volume, p. 359). There was a comparable increase with time in the potash (and TiO_2 and Na_2O) contents of Canadian dolerite dykes (Eade & Fahrig 1971).

Gibb (1971) has suggested that the Slave Province of Canada drifted northwards from the Superior Province, that early Proterozoic oceanic crust was generated near the axis of rotation and that the Churchill Province formed by accretion of new material. This concept is difficult to reconcile with the fact that the axis area in northwest Greenland consists entirely of early Proterozoic high-grade gneisses and granites (see Tectonic/Geological Map of Greenland by the Geological Survey of Greenland, 1970), while the Churchill Province apparently consists of reworked Archaean greenstone belts and granites.

Following the early Proterozoic geosynclinal, tectonic and metamorphic activity in the northern hemisphere in the period 2000 to 1800 Ma, there was a major period of post-orogenic magmatic activity between 1700 and 1000 Ma in a belt extending from the Ukraine to the western United States characterized by massive labradorite to andesine anorthosites, rapakivi granites and related intermediate to acid plutons, acid volcanics, tholeiitic basalts and dykes, and finally alkaline complexes (Bridgwater & Windley, in press). A similar anorthosite belt in the southern hemisphere (Herz 1969) is confined to the Proterozoic mobile belts extending from South America, through Africa, Madagascar, India to Australia. The emplacement of the early intrusions and extrusions (1700 to 1400 Ma) in the 'North Atlantic' belt was controlled by east–west trending major tectonic lineaments which acted as guidelines for the formation of late graben structures which, in turn, controlled the deposition of continental sandstones and basic to trachytic lavas (*ca.* 1300 Ma) and the intrusion of late alkaline complexes (1300 to 1000 Ma, Allaart, Bridgwater & Henriksen 1969). Bridgwater, Escher & Watterson (this volume, p. 513) show how these rocks in south Greenland are arranged in east–west zones which are parallel to and just north of the Grenville Front on a pre-continental map. In fact they lie in the marginal foreland of the Grenville mobile belt, which was clearly sited on a crystalline sialic basement as it contains abundant remains of partly reworked Archaean and early Proterozoic rocks (Wynne Edwards 1969). For example, the banded iron formations of the Labrador trough (2000 Ma) can be traced as distinctive markers well into the Grenville belt and the massive anorthosite bodies (1400 Ma) are well preserved within it.

Plate tectonics in the Proterozoic

The recent plate tectonic concept provides a wonderful opportunity to understand the mode of formation of some Precambrian mobile belts. That is, if their character can be found to be sufficiently similar to warrant application of the model. However, before embarking on the exercise it is worth noting that, since many types of Precambrian sediments, igneous intrusions, metamorphism, structures and mobile belts were significantly different from their late Phanerozoic counterparts, it might be expected that the types of plates and plate movements were different in the past. In other words, the recent sea floor spreading models may have to be flexibly modified to take account of the special conditions operating earlier in crustal evolution. There were most probably plates of some kind or another; the problem is to establish their character and assess their application.

The simplest type of Precambrian stable 'plate boundary' is represented by the Great Dyke of Rhodesia, which Bichan (1970) proposes developed during the stabilization stage of the Rhodesian craton in a major rift zone sited above a convective mantle updraft. Major basic dyke swarms are common in most shields, and Clifford (1968) suggests that they indicate tensional zones of continental proportions and that they are best explained by subcrustal flow in the form of mantle cell movements with the tensional axis of the dyke swarms (and associated plateau basalt provinces) localized over the position of the convective upwell. This is the most reasonable interpretation of the Karroo dykes and basalts emplaced in Trias times when early rifting was taking place along the axis of the convective updraft in East Africa. But most of the Precambrian dyke zones did not develop into a rift valley stage.

McConnell (1969, 1970) has emphasized that recent rift valley fault zones have been controlled by earlier Precambrian fracture systems and thus that the rift valleys overlie fundamental mantle lineaments. In a similar way some basic dyke swarms were localized along fundamental zones of weakness in the crust. For example, the Kangamiut swarm in West Greenland was intruded into a zone which later became the site of the Archaean-Proterozoic tectonic front (Bridgwater, Escher & Watterson, this volume, p. 513).

Late Phanerozoic plate movements have passed through two basic stages of development. First, basic dyke intrusion, formation of rift systems with a high heat flow, and finally (in the case of the East Africa rift valley) alkaline complexes and carbonatites; and secondly, lateral migration of continental crust, formation of new oceanic crust, drift of the continental plates and formation of mobile belts along continental margins. Rift valleys are sited over arrested convective upwellings and thus have undergone very little tensional separation (Harris 1970). The Rhine graben, for example, has opened only 4.8 km since middle Eocene time (Illies 1967).

The Keweenawan trough of central North America is a late Precambrian rift valley (Pettijohn 1970). It is almost 160 km wide and 1600 km long, is filled with continental red clastics and 15000 m of basic amygdaloidal lavas, and is expressed by the prominent mid-continental gravity anomaly. As suggested by Sumner (1969) it has the appearance of a late Proterozoic oceanic ridge crest, a rift zone, filled with high density lava flows, which has subsided. The offsets in the anomaly pattern might even be transform faults.

There is marked similarity in the evolutionary pattern of some Proterozoic high-grade mobile belts, such as the Grenville, and recent rift systems. The Grenville belt went through stages of rifting, basic dyking, formation of steep geothermal gradients giving rise to granulite-grade metamorphism in a narrow linear zone, and formation of late alkaline complexes and carbonatites (Doig 1970), distributed within, or near, the central high heat flow zone (figure 4). Shackleton (1969) concluded that heat has migrated upwards along this type of mobile belt as it should overlie rising and divergent convection currents. The Grenville belt does not resemble any of the present-day continental margin mountain belts and the Grenville front has no large linear belts of basic rocks such as characterize divergent or convergent junctures

(Irving, Park & Roy 1972). In fact the mobile belt has all the major ingredients of an arrested rift system and it clearly did not reach the continental drift-ocean floor spreading stage, as it is floored by reactivated sialic basement.

Ringwood (1970) pointed out that generation of new crust from the mantle has proceeded via two distinct types of differentiation processes: first, vertical mass transport involving fractional melting of the mantle and rise of magmas into the crust giving rise to high heat flow and intrusions and extrusions; and secondly, lateral mass transport associated with recent sea floor spreading, oceanic basalts being transported down subduction zones where they are fractionally

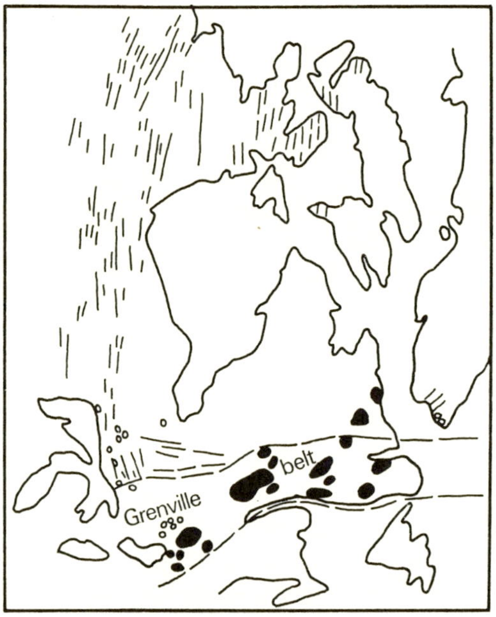

Figure 4. Northeast Canada and south Greenland on a pre-continental drift map showing the major 1300 to 1100 Ma old dolerite dyke swarms and the Grenville mobile belt formed 1000 Ma ago. Anorthosites in black. Circles, alkaline complexes 1100 to 850 Ma old (after Doig 1970). There are many more alkaline complexes 450 to 600 Ma old along the border of and within the Grenville belt (see Doig 1970).

melted to give rise to the calc-alkaline igneous suite that was intruded into the crust. Ringwood suggested that it is probable that the relative importance of these processes has varied throughout geological time and that the first type was more common in the Precambrian.

Many Proterozoic mobile belts have been uplifted and so deeply eroded that essentially their root zones are exposed. At the present deep structural levels it can be seen that they clearly contain abundant relic masses of earlier, partly reactivated, crystalline basement, e.g. the Eburnian and Huabian of Africa, the Nagssuqtoqidian and Ketilidian of Greenland, the Grenville of Canada. It might be suggested that these are collision-type mountain belts which should have sialic roots as a result of underthrusting of continental crust upon closing of the ocean (Bird 1970). This type of belt should contain ophiolites, remnants of the oceanic crust, in flysch-mélange suture zones. However, these Proterozoic belts lack ophiolites and in several it can be shown that there was no period of volcanicity or sedimentation before the 'orogenesis'; no oceanic crust preceded the plutonism. The most reasonable interpretation of these belts is that they are narrow, linear, rifted zones which acted as loci for the rise of geothermal gradients with consequent high heat flow and 'regional' metamorphism and abundant igneous activity. The zones are often associated with early fracture systems and underwent sufficient, although limited, intra-continental block (plate) movements to cause the deformation of basic dyke swarms and infolded sediments. Ringwood's first type of differentiation process – vertical mass transport along intra-continental linear rifted zones – with or without basin deposits best satisfies the formation of this type of mobile belt.

Figure 5 shows movement directions of three ensialic blocks (plates) in the North Atlantic region during the Proterozoic. The Nagssugtoqidian belt moved southwards overriding the Achaean North Atlantic craton (Bridgwater, Escher & Watterson, this volume), the Labrador Trough was thrust westwards (Dimroth 1970), and the Grenville belt underwent a northerly movement (Stockwell *et al.* 1970).

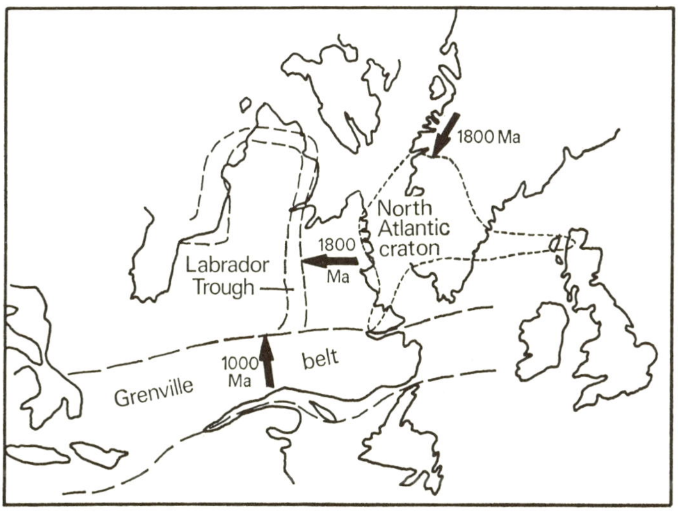

FIGURE 5. The North Atlantic region on a pre-continental drift map, showing the movement directions of three intra-continental blocks in the Proterozoic.

DISCUSSION AND CONCLUSIONS

The evolution of the Archaean has to be seen in two parts, belonging to a single type of crustal development which was not repeated in the same way either in the Proterozoic or Phanerozoic. Many Archaean rock groups and structures (e.g. calcic anorthosites, widespread granulites, nonlinear structures on an 'orogenic' scale, greenstone belts) were seldom, if ever, developed in the Proterozoic while many Proterozoic rock groups, such as banded iron formations, massive Adirondac-type anorthosites and rapakivi granites, were seldom, if ever, developed in the Phanerozoic. Major alkaline complexes and carbonatites are most common in Sutton's latest chelogenic cycle. Much evidence suggests that several distinctive rock groups and structures developed at particular periods in the history of the continents; that is, the crust has evolved with time going through definite stages when certain rock associations formed in response to particular environments.

Many writers have demonstrated that the continents have been through progressive stabilization, the cratonic areas becoming larger with time. In particular, platform deposits became more extensive in the Proterozoic while mobile belts became correspondingly more confined and linear. Conversely, the early crust appears to have undergone a high degree of mobility, although the earliest Precambrian areas, characterized by granulites and gneisses, contain evidence of very old sediments and volcanics that represent a period of stability before the high-grade metamorphism. A major point of controversy at present is 'Where were the old granulite-gneiss complexes when the greenstone belts were formed?' Were they underneath and between them, or were they in the drifting continents at the time of formation of the greenstone belt oceanic crust? Remarkably little detail is known of the actual contact relations between these two types of terrains.

The oldest rocks known at present are 'granites' in West Greenland, found by McGregor in a low deformation and metamorphic area within the high-grade North Atlantic craton. One of the most interesting points to emerge from this work is that this high-grade terrain had been through 1000 Ma of crustal history before the granulite/high amphibolite facies metamorphism. It can be predicted that other high-grade terrains in the world will in future be found to contain very old material – much older than the greenstone belts. Ideas on the mode of formation of the high-grade terrains have been slow to evolve in comparison with those stemming from studies on greenstone belts because of the great difficulty in breaking through the 'conformability barrier' in these areas. Ramberg (1952) realized that one of the major problems of the catazonal regions is that all main rock types are mutually conformable. It is now known that this is a result of deformation, that all stratigraphy is tectonic, and that the conformable layers of meta-supracrustal rocks, i.e. amphibolites and schists, in gneisses commonly hide obliterated unconformities. Thus, as McGregor has done, it is possible to break through the late high-grade metamorphism and deformation to older stable environments.

Sutton (1963) proposed a concept of chelogenic cycles which are long-term shield-forming periods. With modifications due to recent advances the boundaries of the cycles are:

Archaean	*ca.* 3800 to 2500/2000 Ma with a diachronous end.
Proterozoic	*ca.* 2500/2000 to 1000 Ma.
Eocambrian–Phanerozoic	*ca.* 1000 Ma to present day.

In essence the latest cycle involved the establishment of a mobile belt system and as it decayed a rift system developed and continental drift began (Sutton 1971). The question arises 'Did the earlier cycles go through this complete sequence?'

The Archaean is now seen to have been through an early mobile phase, giving rise to high-grade terrains, followed by formation of greenstone belts, and finally basic dyking. The model proposed here is that the greenstone belts represent a stage of continental scale mega-fracturing with formation of rift zones, incipient oceanic ridges and a relatively small amount of oceanic crust. Tensional basic dyke swarms finalized this first shield-forming cycle.

The Proterozoic started with the formation of more linear mobile belts, some low-, some high-grade, it went through a major period of post-orogenic calc-alkaline igneous activity and culminated in a series of arrested rift systems with little evidence of extensive ocean floor spreading. The Keweenawan trough, for example, is very reminiscent of the Archaean greenstone belts and likewise may well have been a proto-oceanic ridge crest: it developed in the final stages of the second shield-forming cycle which was terminated by trans-continental basic dyke swarms.

Finally, a word for the Precambrian. Since so may features of Precambrian rocks are different from their late Phanerozoic counterparts, plate tectonic models should be applied with caution. In recent mobile belts the types of sedimentary, igneous, metamorphic and structural régimes are related to the associated plate boundaries and their movements. In the Precambrian geothermal gradients, shapes of mobile belts and their internal structures, sedimentary, igneous and metamorphic activity resulted from distinctive conditions at particular times in earth history which were often different from those operating in post-Mesozoic times. Therefore the types of plate tectonics should also have been different

In this preview paper it has been necessary to quote many publications, and if I have made any serious omissions, it has been entirely unintentional. I am grateful for many stimulating discussions over several years with my colleagues from the Geological Survey of Greenland, and in particular with D. Bridgwater. R. K. Herd kindly suggested improvements to the manuscript.

References (Windley)

Allaart, J. H., Bridgwater, D. & Henriksen, N. 1969 Pre-Quaternary geology of southwest Greenland and its bearing on North Atlantic correlation problems. In North Atlantic–Geology and continental drift. *Mem. Am. Ass. Petrol. Geol.* **12**, 859–882.

Allsopp, H. L. 1965 Rb–Sr and K–Ar age measurements on the Great Dyke of southern Rhodesia. *J. geophys. Res.* **70**, 977–984.

Allsopp, H. L., Davies, R. D., de Gasparis, A. A. A. & Nicolaysen, L. O. 1969 Review of Rb–Sr age measurements from the Early Precambrian terrain in the south-eastern Transvaal and Swaziland. *Spec. Publ. geol. Soc. S. Afr.* **2**, 433–443.

Allsopp, H. L., Roberts, H. R., Schreiner, G. D. L. & Hunter, D. R. 1962 Rb–Sr age measurements on various Swaziland granites. *J. geophys. Res.* **67**, 5307–5313.

Allsopp, H. L., Ulrych, J. J. & Nicolaysen, L. O. 1968 Dating some significant events in the history of the Swaziland System by the Rb–Sr isochron method. *Can. J. Earth Sci.* **5**, 605–619.

Anderson, D. L. 1969 The evolution of terrestial-type planets. *Appl. Optics* **8**, 1271–1277.

Anhaeusser, C. R. 1971 Cyclic volcanicity and sedimentation in the evolutionary development of Archaean greenstone belts of shield areas. *Spec. Publ. geol. Soc. Aust.* **3**, 57–70.

Anhaeusser, C. R. 1973 The evolution of the early Precambrian crust of southern Africa. *Phil. Trans. R. Soc. Lond.* A. **273**, 359–388 (this volume).

Anhaeusser, C. R., Mason, R., Viljoen, M. J. & Viljoen, R. P. 1969 A reappraisal of some aspects of Precambrian shield geology. *Bull. geol. Soc. Am.* **80**, 2175–2200.

Arriens, P. A. 1971 The Archaean geochronology of Australia. *Spec. Publ. geol. Soc. Aust.* **3**, 11–24.

Bichan, R. 1970 The evolution and structural setting of the Great Dyke, Rhodesia. In *African magmatism and tectonics* (ed. T. N. Clifford and I. G. Gass), pp. 51–72. London: Oliver and Boyd.

Bickford, M. E. & Wetherill, G. W. 1965 Compilation of Precambrian geochronological data for North America. In Geochronology of North America. *Nat. Acad. Sci. Nat. Res. Counc. Nucl. Sci. Ser. Rep.* **41**, 21–27.

Bird, J. M. 1970 General concepts of orogenesis in terms of lithospheric plate tectonics. *Abstr. geol. Soc. Am.* for 1970, pt 7, 733–734.

Bliss, N. W. 1969 Thermal convection in the Archaean crust? *Nature, Lond.* **222**, 972–973.

Bridgwater, D., Escher, A. & Watterson, J. 1973 *Phil. Trans. R. Soc. Lond.* A. **273**, 513–533 (this volume).

Bridgwater, D. & Gormsen, K. 1969 Geological reconnaissance of the Precambrian rocks of south-east Greenland. *Rapp Grønlands Geol. Unders.* **19**, 43–50.

Bridgwater, D., Watson, J. & Windley, B. F. 1973 The Archaean craton of the North Atlantic region. *Phil. Trans. R. Soc. Lond.* A **273**, 493–512 (this volume).

Bridgwater, D. & Windley, B. F. in press. Massive anorthosites, post-orogenic granites, acid volcanic rocks and crustal development in the North Atlantic during the mid-Proterozoic. 'Granite 71' Symposium, Salisbury, Rhodesia, 1971.

Catanzaro, E. J. 1963 Zircon ages in south-western Minnesota. *J. geophys. Res.* **68**, 2046–2048.

Clifford, P. M. 1968 Flood basalts, dike swarms and sub-crustal flow. *Can. J. Earth Sci.* **5**, 93–96.

Clifford, T. N. 1970 The structural framework of Africa. In *African magmatism and tectonics* (ed T. N. Clifford and I. G. Gass), pp. 1–26. London: Oliver and Boyd.
Clifford, T. N. in press. The evolution of the crust of Africa. Conference on the Precambrian, Morocco, 1970.
Cloud, P. 1968 Atmosphere and hydrospheric evolution of the primitive earth. *Science, N.Y.* **160**, 729–736.
Cloud, P. 1971 The Precambrian. *Science, N.Y.* **173**, 851–854.
Crawford, A. R. 1969 Reconnaissance Rb–Sr dating of the Precambrian rocks of southern peninsular India. *J. geol. Soc. India* **10**, 117–166.
Dickinson, W. R. & Luth, W. C. 1971 A model for plate tectonic evolution of mantle layers. *Science, N.Y.* **174**, 400–404.
Dimroth, E. 1970 Evolution of the Labrador geosyncline. *Bull. geol. Soc. Am.* **81**, 2717–2742.
Doig, R. 1970 An alkaline rock province linking Europe and North America. *Can. J. Earth Sci.* **7**, 22–28.
Eade, K. E. & Fahrig, W. F. 1971 Geochemical evolutionary trends of continental plates – a preliminary study of the Canadian shield. *Bull. geol. Surv. Can.* **179**, 1–51.
Evensen, N. M., Rama Murthy, V. & Windley, B. F. in prep. Rb–Sr dating of the Fiskenaesset complex, West Greenland. *Geol. Jl* Sp. Issue no. 2.
Fyfe, W. S. 1970 Some thoughts on granitic magmas. In *Mechanism of igneous intrusion* (ed. G. Newall and N. Rast), pp. 201–216. Liverpool: Gallery Press.
Gibb, R. A. 1971 Origin of the Great Arc of eastern Hudson Bay: a Precambrian continental drift reconstruction. *Earth Pl. Sci. Lett.* **10**, 365–371.
Gibb, R. A. & Walcott, R. I. 1971 A Precambrian suture in the Canadian shield. *Earth Pl. Sci. Lett.* **10**, 417–422.
Glikson, A. Y. 1970 Geosynclinal evolution and geochemical affinities of early Precambrian Systems. *Tectonophysics* **9**, 397–433.
Glikson, A. Y. 1971 Primitive Archaean element distribution patterns; chemical evidence and geotectonic significance. *Earth Pl. Sci. Lett.* **12**, 309–320.
Goldich, S. S., Hedge, C. E. & Stern, T. W. 1970 Age of the Morton and Montevideo gneisses and related rocks, southwestern Minnesota. *Bull. geol. Soc. Am.* **81**, 3671–3696.
Goodwin, A. M. 1968 Archaean protocontinental growth and early crustal history of the Canadian shield. *23rd int. geol. Congr., Prague* **1**, 69–89.
Goodwin, A. M. & Shklanka, R. 1967 Archaean volcano-tectonic basins: form and pattern. *Can. J. Earth Sci.* **4**, 777–795.
Goodwin, A. M. & Ridler, R. H. 1970 The Abitibi orogenic belt. In *Symp. on Basins and geosynclines of the Canadian shield* (ed. A. J. Baer). *Geol. Surv. Can. Pap.* **70–40**, 1–30.
Green, D. C. & Baadsgaard, H. 1971 Temporal evolution and petrogenesis of an Archaean crustal segment at Yellowknife, N.W.T., Canada. *J. Petrol.* **12**, 177–217.
Harris, P. G. 1970 Convection and magmatism with reference to the African continent. In *African magmatism and tectonics* (ed. T. R. Clifford and I. G. Gass), pp. 419–438. London: Oliver and Boyd.
Hart, S. R. & Davis, G. L. 1969 Zircon U2–Pb and whole-rock Rb–Sr ages and early crustal development near Rainy Lake, Ontario. *Bull. geol. Soc. Am.* **80**, 595–616.
Heier, K. S. 1973 Geochemistry of granulite facies rocks and problems of their origin. *Phil. Trans. R. Soc. Lond.* A **273**, 429–442 (this volume).
Heier, K. S. & Compston, W. 1969 Interpretation of Rb–Sr age patterns in high-grade metamorphic rocks, north Norway. *Norsk geol. Tidsskr.* **49**, 257–283.
Herz, N. 1969 Anorthosite belts, continental drift and the anorthosite event. *Science, N.Y.* **164**, 944–947.
Hoffman, P. 1973 The Coronation geosyncline: lower Proterozoic analogue of the Cordilleran geosyncline in the northwestern Canadian shield. *Phil. Trans. R. Soc. Lond.* A **273**, 547–581 (this volume).
Hunter, D. R. 1970 The ancient gneiss complex in Swaziland. *Trans. geol. Soc. S. Afr.* **73**, 107–150.
Hurley, P. M., Nagy, B. & Pinson Jr, W. H. 1972 Ancient age of the Middle marker horizon, Onverwacht Group, Swaziland Sequence, South Africa. *Science, N.Y.*
Hurley, P. M., Hughes, H., Pinson, W. H. & Fairbairn, H. W. 1962 Radiogenic argon and strontium diffusion parameters in biotite at low temperatures from Alpine fault uplift in New Zealand. *Bull. geol. Surv. Can.* **26**, 67–80.
Hurley, P. M., Kalliokoski, J., Fairbairn, H. W. & Pinson, Jr, W. H. 1968*a* An age of 3000 (±) m.y. for the Imataca Series, Venezuela. *16th Ann. Rep.* (1968) *Dep. Geol. Geophys., Mass. Inst. Tech.* pp. 67–69.
Hurley, P. M., Laing, E. M., Fairbairn, H. W. & Pinson Jr, W. H. 1968*b* Age determinations in the older basement rocks of Sierra Leone. *16th Ann. Rep.* (1968) *Dep. Geol. Geophys., Mass. Inst. Tech.* pp. 83–84.
Illies, H. 1967 Gestalt und Fundament des Oberrheingrabens im Modell – ein Beitrag zur Taphrogenese *Oberrhein. geol. Abh.* **16**, 1–10.
Irving, E., Park, J. K. & Roy, J. L. 1972 Palaeomagnetism and the origin of the Grenville Front. *Nature, Lond.* **236**, 344–346.
Jolliffe, A. W. 1966 Stratigraphy of the Steeprock group, Steeprock Lake, Ontario. *Precambrian Symp. geol. Ass. Can. Spec. Pap.* **3**, 75–98.
Lambert, I. B. & Heier, K. S. 1967 The vertical distribution of uranium, thorium and potassium in the continental crust. *Geochim. cosmochim. Acta* **31**, 377–390.

Martin, H. 1969 Problems of age relations and structure in some metamorphic belts of southern Africa. *Spec. Pap. geol. Soc. Can.* **5**, 17–98.

Mason, R. 1973 The Limpopo mobile belt – southern Africa. *Phil. Trans. R. Soc. Lond.* A **273**, 463–485 (this volume).

McCall, G. J. H. 1971 Some ultrabasic and basic igneous rock occurrences in the Archaean of western Australia. *Spec. Publ. geol. Soc. Aust.* **3**, 429–442.

McCall, G. J. H. & Peers, R. 1971 Geology of the Binneringie dyke, western Australia. *Geol. Rdsch.* **60**, 1174–1263.

McConnell, R. B. 1969 Fundamental fault zones in the Guiana and West African shields in relation to presumed axes of Atlantic spreading. *Bull. geol. Soc. Am.* **80**, 1775–1782.

McConnell, R. B. 1970 The evolution of the Rift system of eastern Africa in the light of Wegmann's concept of tectonic levels. In *Graben problems* (ed. J. H. Illies and S. Mueller), pp. 285–290. Stuttgart: Schweitzerbartsche Verlags.

McGlynn, J. C. & Henderson, J. B. 1970 Archaean volcanism and sedimentation in the Slave structual Province. *Geol. Surv. Can. Pap.* **70–40**, 31–44.

McGregor, V. R. 1973 The early Precambrian gneisses of the Godthaab district, West Greenland. *Phil. Trans. R. Soc. Lond.* A **273**, 343–358 (this volume).

Moorbath, S. 1967 Recent advances in the application and interpretation of radiometric data. *Earth-Sci. Rev.* **3**, 111–133.

Moorbath, S., Welke, H. & Gale, N. H. 1969 The significance of lead isotope studies in ancient, high-grade metamorphic basement complexes, as exemplified by the Lewisian rocks of northwest Scotland. *Earth Planet. Sci. Lett.* **6**, 245–256.

Nautiyal, S. P. 1965 Precambrian of Mysore plateau. *53rd Indian Sci. Congr. Calcutta, Sect. Geol. Geograph.* 1–14.

Nicolaysen, L. O., de Villiers, J. W. L., Burger, A. J. & Strelow, F. W. E. 1958 New measurements relating to the age of the Transvaal System and of the Bushveld Igneous Complex. *Trans. geol. Soc. A. Afr.* **61**, 137–163.

Oxford Isotope Geology Laboratory & McGregor, V. R. 1971 Isotopic dating of very early Precambrian amphibolite facies gnesses from the Godthaab district, West Greenland. *Earth Pl. Sci. Lett.* **12**, 245–259.

Pettijohn, F. J. 1970 The Canadian shield; a status report 1970. *Geol. Surv. Can. Pap.* **70–40**, 239–265.

Ramberg, H. 1952 *The origin of metamorphic and metasomatic rocks.* Chicago: University Press.

Ramberg, H. 1967 *Gravity, deformation and the Earth's crust.* London: Academic Press.

Ray, S. 1970 The plutonic concept, *J. geol. Soc. India.* **11**, 54–60.

Ringwood, A. E. 1970 Composition of the crust and upper mantle. *Cal. Tech. Rep.* 1967–9, pp. 164–165.

Rousell, D. H. 1965 Geology of the Cross Lake area. *Man. Dep. Mines natn. Res. Publ.* pp. 62–64.

Rudnik, V. A., Sobotovich, E. B. & Terent'yev, V. M. 1969 Archaean age of the oldest rocks of the Aldan complex. *Dokl. Acad. Sci. U.S.S.R.* **188**, 93–96.

Salop, L. I. & Scheinmann, Yu. M. 1969 Tectonic history and structures of platforms and shields. *Tectonophysics* **7**, 565–597.

Semenenko, N. P. 1967 Absolute geochronology and history of formation of folded Precambrian zones in the east European platform. In *Chemistry of the Earth's crust* (ed. A. P. Vinogradov). Israel Progr. Sci. Trans. Jerusalem.

Shackleton, R. M. 1969 Displacement within continents. In Time and place in orogeny. *Spec. Publ. geol. Soc. Lond.* **3**, 1–8.

Shackleton, R. M. 1970 On the origin of some African granites. *Proc. Geol. Ass.* **81**, 549–559.

Shackleton, R. M. 1973 *Phil. Trans. R. Soc. Lond.* A. **273**, 317–320 (this volume).

Stern, T. W. 1964 Isotopic ages of zircon and allanite from the Minnesota River Valley and La Sol mountains, Utah. *Abstr. Trans. Am. Geophys. Un.* **45**, 116.

Stockwell, C. H. 1933 Great Slave Lake-Coppermine River area, Northwest Territories. *Geol. Surv. Can. Sum. Rep.* 1932, pt C, 37–63.

Stockwell, C. H., McGlynn, J. C., Emslie, R. F., Samford, B. V., Norris, A. W., Donaldson, J. A., Fahrig, W. F. & Currie, K. L. 1970 Geology of the Canadian Shield, in Geology and economic minerals of Canada. Douglas, R. J. W. (ed) *Geol. Surv. Can. Econ. Geol. Rep.* **1**, 43–150.

Subramaniam, A. P. 1956 Mineralogy and petrology of the Sittampundi complex, Salem district, Madras State, India. *Bull. geol. Soc. Am.* **67**, 317–390.

Sumner, J. S. 1969 *Geophysics, geologic structures and tectonics.* Iowa: W. C. Brown and Co.

Sutton, J. 1963 Long-term cycles in the evolution of the continents. *Nature, Lond.* **198**, 731–735.

Sutton, J. 1967 The extension of the geological record into the Precambrian. *Proc. Geol. Ass.* **78**, 493–534.

Sutton, J. 1971 Migration of high temperature zones in the crust. In *Palaeogeophysics* (ed. S. K. Runcorn), pp. 365–375. London: Academic Press.

Talbot, C. J. 1973 A plate tectonic model for the Archaean crust. *Phil. Trans. R. Soc. Lond.* A. **273**, 413–427 (this volume).

Tectonic/Geological map of Greenland. 1970 Compiled by A. Escher, Geological Survey of Greenland.

Turek, A. & Compston, W. 1971 Rubidium-strontium geochronology in the Kalgoorlie region. *Spec. Publ. geol. Soc. Aust.* **3**, 72 (abstr.).

Vail, J. R. & Dodson, M. H. 1969 Geochronology of Rhodesia. *Trans. geol. Soc. S. Afr.* **72**, 79–113.

Van Niekerk, C. B. & Burger, A. J. 1964 The age of the Ventersdorp System. *Annal. geol. Surv. S. Afr.* **3**, 75–86.

Van Niekerk, C. B. & Burger, A. J. 1969 A note on the minimum age of the acid lava of the Onverwacht Series of the Swaziland System. *Trans. geol. Soc. S. Afr.* **72**, 9–21.

Viljoen, M. J. & Viljoen, R. P. 1969 An introduction to the geology of the Barberton granite-greenstone terrain. *Spec. Publ. geol. Soc. S. Afr.* **2**, 9–27.

White, A. J. R., Jakes, P. & Christie, D. M. 1971 Composition of greenstones and the hypothesis of sea-floor spreading in the Archaean. *Spec. Publ. geol. Soc. Aust.* **3**, 47–56.

Wilson, A. F. 1958 Advances in the knowledge of the structure and petrology of the Precambrian rocks of south-western Australia. *J. Proc. R. Soc. West Aust.* **41**, 57–83.

Wilson, J. F. 1973 The Rhodesian craton – an essay in cratonic genesis. *Phil. Trans. R. Soc. Lond.* A **273**, 389–411 (this volume).

Wilson, N. W. 1965 Geology and mineral resources of part of the Gola Forests, south eastern Sierra Leone. *Bull. geol. Surv. Sierra Leone* **4**, 1–102.

Windley, B. F. & Smith, J. V. in press. The Fiskenaesset complex, West Greenland. Pt 2: General mineral chemistry. *Bull. Grønlands geol. Unders.*

Windley, B. F. in press. Archaean anorthosites: a review with the Fiskenaesset complex, West Greenland as a model for interpretation. 'Granite 71' Symp. Salisbury, Rhodesia, 1971.

Windley, B. F. & Bridgwater, D. 1971 The evolution of Archaean low- and high-grade terrains. *Spec. Publ. geol. Soc. Aust.* **3**, 33–46.

Windley, B. F., Herd, R. K. & Bowden, A. A. in press. The Fiskenaesset complex, West Greenland. Pt 1: A preliminary study of the stratigraphy, petrology and whole-rock chemistry. *Bull. Grønlands geol. Unders.*

Winkler, H. G. F. 1967 *Petrogenesis of metamorphic rocks.* (2nd Edit.) Berlin: Springer-Verlag.

Wood, R. 1972 Early Precambrian Kambui schist belt, southern Sierra Leone, and the surrounding basement. *Nature, Phys. Sci.* **236**, 14.

Wynne Edwards, H. 1969 Tectonic overprinting in the Grenville Province, south-western Quebec. *Spec. Pap. Geol. Ass. Can.* **5**, 163–182.

VII: PLATE TECTONICS AND THE ARCHEAN

A recent review (Armstrong, 1973) of a volume produced from papers read at a symposium on the Archean regrets the restriction of plate tectonics to a single mention. That volume (Glover, 1971) included no papers written later than 1970, and the critic overlooked the rapidity of development of modern ideas in this field; ideas on plate tectonics were then barely past their initial statement, and no attempt had yet been made to extrapolate them to the Precambrian. Such attempts have since been made by several authors.

Continents are essentially conservative, and once formed tend to be preserved through potentially destructive events, whereas oceanic plates are nonconservative in a regime of plate tectonics, and newly formed oceanic crust moves from its origin in mid-oceanic ridges to suffer eventual subduction and recycling at oceanic margins (Abelson et al., 1972). Unless oceanic rocks become incorporated within continents, they cannot be indefinitely preserved under such a regime.

The anomalous preservation of oceanic volcanics of abyssal character (Paper 16) in Archean greenstone associations suggests that we should perhaps be looking more to an Atlantic–Indian Ocean analogy than a Pacific analogy, to which the conclusion of Abelson et al. refers. The Atlantic and Indian oceans are separated by the continent of Africa without extensive marginal subduction zones. They form essentially distensive systems—giant "rhombochasms," to borrow Carey's term.

Another anomaly is the preservation of numerous fold belts of island-arc character in the Archean (Papers 8, 10, and 11), despite the lack of tectonic patterns compatible with Pacific-style, ocean-margin subduction zones and other overwhelming evidence against the existence of such zones in the Archean.

Engel and Kelm (1972) have made a special study of plate tectonics with reference to possible extrapolation to the early Precambrian. The nature of the Proterozoic mobile belts seems to preclude such movements, for they were ensialic in nature. Both their component character and their interrelationships with Archean terrains show that they evolved mainly on and between closely spaced Archean protocontinental clusters, which were deformed and refractionated, with thickening as the orogens evolved, but suffered no widespread fragmentation, no dispersion and collision of continental blocks, such as characterizes Phanerozoic plate tectonics.

Concerning the earliest, Archean phase of movements, Engel and Kelm demonstrate that they must have been very restricted in lateral displacement and of another type to the dominant type of post-Permian plate movement. The very old mobile belts cannot conceivably be analogous to the Himalayan belt (as has been suggested by Abelson et al., 1972), which is a thick continental collision suture unroofed by erosion. If "unrolling" to eliminate the effects of post-Permian plate tectonics is attempted, the pre-Mesozoic configuration of the cratons shows surprisingly little evidence of prior relative disturbance; lineaments line up neatly, craton to craton, in an orderly fashion. Both rotation and lateral migration relative to one another seem to have been on a second order of magnitude.

Although analogies have been drawn between Archean calc-alkaline volcanic belts and modern island arcs such as that of Japan (Paper 11), and a similar polarity has been recognized (Abelson et al., 1972), the similarity of igneous rock associations does not in itself warrant acceptance of the geotectonic analogy with subduction zones at margins of deep and wide oceans, where overriding of plates occurs in modern systems. The tectonic regime in the Archean was quite different, as we can see from study of the structures themselves. The nuclear terrains of the early Archean have been mentioned (Papers 22, 29, and 32), and the greenstone belts of both early and late Archean age display peculiar tectonic styles of closely associated open and closed folds, with steep axial surfaces, recumbent styles, and overthrusting; in such a setting the possibility of major lateral plate displacements and plate overriding is relatively remote (see Editor's Comments on Paper 7).

The predominant tectonic style of the greenstone belts is only compatible with limited lateral movement of plates relative to one another, with, possibly, limited opening and closing of the protooceanic troughs. As the Archean terrains are characterized by successive troughs, commonly contemporary, and more or less regularly spaced across large expanses of granite and gneiss, movements of opening and closing in individual troughs cannot have been of great magnitude, in aggregate amounting only to a large number of second-order adjustments (Engel and Kelm, 1972). Major lateral displacements of cratonic plates relative to one another or overriding of plates seem to have been rare exceptions at the most in the Archean.

The freeboard or relief argument (Abelson et al., 1972) also supports this conclusion. The thickness of continental plates must regulate, isostatically, the depth of the oceans. The early sialic plates were thin (Paper 3), of small areal extent, and numerous, and such a condition requires small and shallow complementary oceanic basins.

The model illustrated by Wyllie (1972) of such a cluster of protocratons like a basket of eggs surrounded by large areas of deep ocean can have no application to the Archean; the protocratons and complementary greenstone depositories must have covered the entire global surface. The very earliest seas may have covered most of the globe with only islands and shoals emergent (Paper 32), but they were almost certainly shallow epicontinental seas thinly covering protocontinental foundations. Freeboard contrast will have been minimal in the earliest Archean when such a paleogeography applied, and it seems doubtful if the contrast could have increased to anything like its present magnitude when the first narrow protooceanic troughs, the greenstone depositories, were formed no later than 3400 million years ago, either on subsided zones of protocontinental sialic crust or opened up troughs separating blocks of sialic crust. Without freeboard contrast, major plate overriding seems inconceivable.

We see the Archean, then, as a period of incipient, second-order plate movements. There appears to have been numerous small protocontinental plates and complementary protooceanic troughs, once the first greenstone depositories developed 3400 million years ago or more. This regime lasted until the end of the Archean, for more than 1000 million years, although movements intensified intermittently at the Gastil maxima (Paper 27). Diastrophism, as in the later history of the earth, was an episodic, intermittent event. The greenstone troughs probably formed initially under a regime of divergence, whereas the cycles terminated with some degree of closing, coupled with the development of tension within the intervening cratons. We can envisage a slight opening up of the existing, thin sialic crust and subsequent jostling of the separated blocks. The full development of plate tectonics came only in the post-Permian, for the Proterozoic was a quiet period of intracratonic basins and welt-like mobile belts complementary to them, analogous to rift zones; these were the sites of spreading initiation in the post-Permian, but never significantly opened up in these early developments. In Sutton's system (Papers 31A and 31B) only the last part of the third chelogenic cycle (or better, perhaps, *stage*) was characterized by extensive continental dispersion, large deep oceans, and extensive freeboard contrast.

This seems to be as good a rationale as we can derive at the present stage of knowledge for Archean plate tectonics; for although there is paleomagnetic evidence for rapid relative movement of Archean plates (Abelson et al., 1972), extension of paleomagnetic methodology to the Archean involves much uncertainty due to the nature of the basic assumptions. At present, the geologic evidence for limited movements seems preferable.

Plate tectonics theory is today coupled with a concept of a globe maintaining relatively constant dimensions, but geologic evidence of the nature of Archean plate movements is in line with evolutionary concepts, which could include global expansion (Carey, quoted in Abelsen et al., 1972). This would obviate the anomaly that continental accretion through geologic time should have largely engulfed the oceans (despite any counteraction by transecting fragmentation and oceanization); yet only in the last few hundred million years of geologic history, the post-Permian, has the globe apparently undergone the development of vast, deep oceans. There is much to be said for the postulated 4400-km radius of the Archean globe (Dearnley, 1966), and for seeing plate tectonics as the result of asymmetrical expansive effects. And there would be complementary localized contractive effects (subduction) of a lower order of magnitude. Only in the post-Permian did these effects attain their present magnitude.

If we envisage an expansive regime operating during the Archean, repeated Atlantic-type opening and the more limited closing of the troughs is understandable. Newly formed oceanic crust may have suffered downward displacement (subduction), and calc-alkaline suites of island-arc affinities may have been developed within the troughs (Paper 8), but in the manner of the Indonesian arcs without the complications of continental margin overriding. The lesson to be learned from the comparison of modern plate tectonics with the Archean may be that the subduction that produces island-arc suites by recycling oceanic suites is in no way dependent on continental margin overriding. All it requires is a downcurrent in a convective system. The Archean was apparently characterized by numerous small cells, possibly originated by the initial fracturing of the thin primordial crust (see Part VIII).

In a recent version of the plate tectonics model, Ringwood (1972) recognizes two distinct types of crustal generation from the mantle: vertical differentiation and lateral mass transport. The first involved fractional melting of the protomantle and diapiric rise, due to buoyancy into the crust above, and characterized the "permobile" stage (Paper 32). The process was related to heat dissipation of ^{40}K and ^{238}U (Paper 3) and is supposed to have extended to 3500 million years ago. Isometric to slightly elongated gneissic domes characterized the permobile phase with high plasticity and the integration of folding and granite invasion with diapiric uprise. It was essentially magmatogenic.

Lateral differentiation is supposed to have characterized the second stage in the Earth's tectonic history from 2100 million years onward. This cratonic–mobile belt stage has persisted to the present, and is marked primarily by cratonic accretion. Lateral mass transport is considered to be the dominant process; lithosphere is formed by partial melting in the mantle in a tensional regime at oceanic ridges. This process would be followed by a sliding away of plates from the ridges on an inclined, thermally induced boundary between lithosphere and asthenosphere (Egyed's concept). Subduction of lithospheric plates is supposed to have generated calc-alkaline volcanics from oceanic basalt.

These new ideas of Ringwood are summarized here at length because they are helpful in understanding the relationships between modern plate tectonics and the Archean. If one accepts the permobile–cratonic-mobile bimodality of earth history, then plate tectonics prior to 2100 million years ago is not to be expected, for lateral mass transport is said to be the causal mechanism of plate movements. However, the model seems to be incompletely stated, in the sense that it seems to be concerned with an established oceanic conveyor belt system of the Pacific type, with marginal subduction zones. The critical question must surely be the manner of initiation of the vast conveyor belt.

The Atlantic Ocean provides clues; it must surely have started out as a rift zone, a narrow semimobile belt traversing a craton that included parts of Africa and America. The halves of the old craton must have diverged from this welt in the same manner as Africa and Arabia are doing even now along the Red Sea rift. New magma was progressively generated in the oceanic ridge zone; it is important to note that generation in this zone is essentially a vertical process and analogous to that which predominated over the globe in the permobile stage. The lateral dispersion process is the separation of the newly formed lithosphere; it is not generative. The divergent sea-floor spreading allowed the new lithosphere to move away from the ridge in the wake of the cratonic halves.

It may be suggested that the original semimobile belt formed, in the case of the Pacific, as an upwarp with complementary lateral downwarps. The welt or upwarp is analogous to the nonlinear zones of upwarping in Archean nuclear terrains (the intervening granitic areas), to the linear mobile belts of the Proterozoic (Grenville, Kapuskasing, Fraser Range, etc.), and to the East African Rift zone, an abortive ocean initiator (McCall and Hornung, 1972). The postulated lateral troughs would be analogous to the greenstone depositories of the Archean and later geosynclines. A geosyncline or protogeosyncline (Archean greenstone belt) is a primary zone of subduction. In the editor's opinion, the complementary pattern of welt and trough must be related to *convection cells.* In the Archean these were small and numerous, and never developed beyond this. The nature of the immense post-Permian conveyor belt systems of the modern oceans can only be reconciled with linear convection cells of small lateral dimension that have progressively developed to their present immensity of lateral dimension. The geologic facts seem to require this, if thermal convection is involved at all.

This progressive development model seems to be the only one that can reconcile Archean and post-Permian geology. It makes understandable the fact that carbonatite complexes occupy both rift zones and Proterozoic mobile belts. It makes understandable the similarity of Archean abyssal and arc suites to those of the post-Permian, despite the evidence of limited Archean lateral displacements and lack of plate collisions and overridings. It makes understandable the real analogy between the greenstone associations of the protogeosynclines and the associations of the post-Permian geosynclines, and also their points of disparity.

The permobile phase is the early evolutionary stage of the earth in which linearity was subdued. Throughout the entire history of the globe the essential processes have been similar, but there have been progressive evolutionary changes marked by the dominance of one or another type of structure with variations in the details of the processes themselves. With evolutionary development, the geotectonic patterns have changed radically, and only the patterns of the last few hundred million years have allowed the spectacular developments that form the subject matter of plate tectonic studies.

References

Abelson, P. H., et al. (1972). Plate tectonics and the evolution of continents. *Geotimes, 17*(8), 18–19.

Armstrong, R. L. (1973). Symposium on Archaean rocks, J. Glover, ed. Review: *Econ. Geol., 68,* 718–719.

Dearnley, R. (1966). Orogenic fold belts and a hypothesis of earth evolution. *Phys. Chem. Earth, 7,* 1–114.

Engel, A. E. J., and Kelm, D. C. (1972). Pre-Permian global tectonics. *Geol. Soc. Am. Bull., 83,* 2325–2340.

Glover, J. E. (ed.). (1971). Symposium on Archaean rocks. *Spec. Publ. 3, Geol. Soc. Aust.,* 469 pp.

McCall, G. J. H., and Hornung, G. (1972). A geochemical study of Silali Volcano, Kenya: with special reference to the origin of the intermediate-acid eruptives of the Central Rift Valley. *Tectonophysics, 15,* 97–113.

Wyllie, P. J. (1972). *The dynamic earth.* Wiley, New York, 416p.

VIII: THE NATURE OF THE PRIMORDIAL CRUST

The possible nature of the primordial crust has long formed the subject of continuing controversy, which hinges on the real nature of the granite–gneiss tracts of the Archean and their included greenstone belts. The alternatives have been presented by Engel (1968); arguments for an oceanic primordial crust have been put forward by Gill (Paper 36), Viljoen and Viljoen (Paper 6), and Folinsbee et al. (Paper 11), and have lately been summarized by Glikson (Paper 37).

The suggestion that the komatiites of the Onverwacht Group of the Swaziland System (about 3400 million years old) could be almost direct melting derivatives of the oceanic mantle and the entire group could form part of a primordial oceanic crust has lost support. Similar rocks are recognized from late Archean and even from the earliest Proterozoic (Pyke et al., 1972; McCall, Paper 16C). Such unusual eruptives were apparently components of greenstone volcanic sequences erupted at widely spaced intervals in time through the Archean, and not products of a unique, primordial event. The discovery of a more ancient gneiss terrain in Greenland (Papers 20 and 38) showed that even the most ancient known development of komatiites could not have antedated the first protocratons.

Engel (1968) seems to have come close to the truth in relating the granite envelopment of greenstone belts to potassium, uranium, thorium, and lead migration outward from the first major thermal event, and in accepting that even the oldest known greenstone belts are second- or third-order generations of volcanic–graywacke complexes, their precursors being no more than indistinct paleosome vestiges in the oldest granite–gneiss terrains with their original character almost obliterated by metamorphism. The Onverwacht Group and its analogues are products of diastrophism, of rapid eruption of magma, and of sedimentation of pelagic and clastic type in fault-controlled, dominantly volcanic filled trough depositories of subsidence, analogous to later geosynclines (see Paper 29: Commentary). The ultramafic flows, of unique character, although probably products of almost direct mantle melting, cannot be part of that mantle, separated as they are by sedimentary screens and lacking a sheeted dike association. They can only be part of a primitive, but not primordial, oceanic crust.

Should proof of superimposition of the greenstones direct on the mantle be obtained, it would not mean that the greenstones were founded on a primordial oceanic crust, for such a relationship is quite consistent with new opening along fractures transecting a protocontinental crust. The opening so produced would be floored by oceanic crust analogous to that flooring the modern Atlantic Ocean. We have not recognized oceanic crust of primordial character; if it ever existed, we must seek it well back in the geologic record, possibly beyond 4000 million years.

The suggestion has been advanced, too, that the very ancient gneisses of Greenland (about 3800 million years; see Papers 20 and 38) are samples of primordial, acid crust. Such a crust has been long favored by geologists, mainly because of the furnace-slag analogy; it is part and parcel of the "Basement Complex" philosophy (Paper 36). However, the gneisses of Greenland are no more likely to provide a sample of primordial crust than the basic rocks of the Onverwacht group. These gneisses may well include reworked primordial crust, but there is no evidence that this is so. The dominantly basic paleosomes in this gneiss terrain could equally incorporate primordial oceanic crust, but there is no evidence that this is so, either.

Studies by Wynne-Edwards of the Grenville mobile belt (Paper 23) have shown that similar granite–gneiss terrains of Proterozoic mobile belts do include much reworked granite and gneiss material from the Archean; gray gneisses are virtually indestructable and are incorporated in later and later metamorphic terrains without significant modification. Much of this reworked gneiss stems from reworked, originally igneous granitic material, not transformed sediments, and there is no reason why the starting point for such repeated reworking should not have been an igneous, primordial granite crust to the Earth.

Yet this remains no more than one possibility. The only certainty is that granites and quite ordinary gneisses existed together with basic igneous rocks and possibly primitive sediments in a sialic, continental crust nearly 4000 million years ago.

It has been argued (Paper 38) that the fact that these terrains tend to yield the highest radiometric numbers is an argument for a sialic protocrust. This seems fallacious, for in any Archean terrain we only see well preserved the products of the last cycle of greenstone and granite development; the very old cycles are largely obscured or obliterated, and yield the radiometric numbers of the oldest granitic rocks or of later metamorphism superimposed on them. The paleosomes in such very ancient terrains are mere narrow wisps and could not conceivably have retained their even higher radiometric numbers. And no one can deny that they may represent an oceanic primordial crust, quite deprived of its original radiometric numbers.

Again, we have the proposal that recognition of a sialic intervention between greenstones and mantle would establish the reality of the postulated acid primordial crust. Yet such a foundation to the greenstone belts could simply be the product of fracture-controlled downwarping without opening up of a protocratonic surface older then the greenstones; it could reflect the incipient development of an Atlantic-type oceanic trough. There is no reason to assign a primordial status to such a sialic foundation, should it be locally demonstrated to exist.

The immense volume of the Archean granites presents a problem that must be reconciled in any acceptable theory. As yet the very old granite–gneiss terrains are inadequately studied in detail, and detailed studies of all such terrains are required to determine how much mantle-derived new material, how much regenerated material from even older protocratons, and how much migrated, mobile, alkali- and silica-rich material from the protooceans contributed to the formation of such terrains.

Geologists appear to be faced with two alternatives, each of which has more than one variant:

1. The primordial crust was akin to the modern oceanic crust. This concept draws strength from many arguments, not the least being the analogy with the entirely basaltic surface of the Moon, essentially oceanic rather than sialic in nature. The fact that granite invasions within the ancient crust invariably seem to follow basic eruptivity of fundamental type also suggests that granite invasion is an essentially secondary phenomenon. Engel (1968) has suggested that continental evolution commenced with rifts and island arcs amid primeval seas on a protocrust more primitive than the modern ocean floor.

2. The contrary view is that the Earth had a primitive acid skin enveloping the globe overall. Cosmic origin by accretion for such a skin or outer shell has been favored by Donn et al. (1965) and Starik and Sobotovich (see Paper 28a); all other versions of this concept of a sialic primordial crust require it to be an endogenous segregation from mantle and core, at the end of the first major heating event. Certainly, such a crust is unlikely to have possessed the coarse texture of granite, as Glikson has remarked (Paper 37); it can at the most have been finely granophyric or rhyolitic at the surface. But this is a minor difficulty.

What seems to have been overlooked is the possibility of a middle-of-the-road solution to the problem. Geologists seem to have been looking for primordial crust too late in the geologic record. There is a "Hadean" (Paper 2) span of 1200 million years prior to the emplacement of the Onverwacht Group, a span that has as yet yielded very little material suitable for direct geologic study (the Greenland evidence notwithstanding).

We seem to be missing from the record all but the merest glimmerings of the story of the development of the first small granitic nucleii and their complementary, quite shallow and narrow, primitive protooceans. Both were probably quite small entities scattered over the surface of the globe, the granitic nodes being controlled by an intimate proliferation of intersecting shrinkage fractures. Immediately after the first major thermal event, shrinkage must surely have been a dominant geotectonic process (possibly rapidly superseded by global expansion related to outgassing).

The products of this early stage of Earth development are modified, obscured, and virtually indecipherable at present. Even so, it seems certain that we must go back to the end of the first major thermal event for the beginnings of the protocontinent–protoocean dichotomy. The sial of the continents may never have formed an enveloping slag mantle to the globe, but its segregation from the mantle may have been nodal and controlled by shrinkage fracture intersections. Thus, it seems likely that long before 4000 million years ago incipient separation of acid and basic crust may have commenced, and both may be regarded as coexistent primordial entities. The truth may be that the nuclear terrains of the Katarchean reflect a permobile regime, that is, the same regime that produced the original continental–oceanic crust dichotomy. The Katarchean nuclear terrains may be later echoes of a heterogeneous, nodal, primordial crust pattern.

In the later history of the earth, such nodal development continued. In the Katarchean it was nuclear; in the late Archean, more linear controls by more extensive fracture systems resulted in an irregular blanket pattern of granite invasion and stronger, more linear troughs. In the Proterozoic, the vast cratons formed in the late Archean underwent granite invasion in mainly internal, linear mobile belts.

In the last major stage of Earth development, granite invasion has occurred mainly in geanticlinal zones of orogenic belts such as those of Peru. The processes remain essentially the same, but the geotectonic controls are vastly different, and there is a progressive change in the nature of the eruptives with time. The fundamental question remains: how much of the granite accession to these varied geotectonic environments was primary and mantle derived, and how much was due to reworking? It seems certain that the first protocontinental nodes were of mantle-derived material, and much additional primary material must have been added in the late Archean to form those vast granitic expanses. It is likely that mantle-derived material predominated in the formation of the vast batholiths of Peru (Pitcher, 1972).

Thus, it seems that each granitic accession involved some new primary material, together with much reworking of older material. It seems quite impossible to equate the entire record of granite accession to form continental crust with subduction and reworking of oceanic crust in island-arc situations, and it must be concluded that the production of the immense Archean granite tracts and the production of acid volcanics in island arcs from oceanic crust are quite separate processes.

References

Donn, W. L., Donn, B. D., and Wilbur, G. V. (1965). On the early history of the earth. *Geol. Soc. Am. Bull., 76,* 287–306.

Engel, A. E. J. (1968). The Barberton Mountain Land—clues to the differentiation of the earth. *Trans. Geol. Soc. South Africa, 71* (annexure), 255–270.

Pitcher, W. S. (1972). The coastal batholith of Peru—some structural aspects. *Proc. 24th Int. Geol. Congr., 2,* 156–163.

AUTHOR CITATION INDEX

SUBJECT INDEX

ABOUT THE EDITOR

GERALD JOSEPH HOME McCALL is currently Senior Consultant Geologist for Paragon-Contech, Teheran, where he is engaged in a major regional mapping program for the Geological Survey of Iran. In addition to having served as an independent consultant to various other organizations, he was Director of Exploration at Research and Exploration Management Pty Ltd. in Melbourne from 1970 to 1972 and spent ten years prior to this as Reader in Geology at the University of West Australia and ten as geologist in the Kenya Geological Survey.

Dr. McCall earned his Bachelor of Science degree in geology from London University in 1949 and also holds Ph.D. (1951) and D.Sc. (1968) degrees in geology from that institution. The author of over 90 scientific papers and of *Meteorites and Their Origins* (David & Charles Ltd.), he is active in a number of scientific societies and has just completed a term as Vice President of the International Association of Planetology.